Advanced Optical Technologies for Communications, Perception, and Chips

Advanced Optical Technologies for Communications, Perception, and Chips

Guest Editors

Tianxu Xu

Kaiheng Zou

Cong Liu

Yang Yue

MDPI

Basel • Beijing • Wuhan • Barcelona • Belgrade • Novi Sad • Cluj • Manchester

Guest Editors

Tianxu Xu
School of Electrical and Information Engineering
Zhengzhou University
Zhengzhou
China

Kaiheng Zou
Department of Electrical Engineering
University of Southern California
Los Angeles
USA

Cong Liu
Department of Electrical Engineering
University of Southern California
Los Angeles
USA

Yang Yue
School of Information and Communications Engineering
Xi'an Jiaotong University
Xi'an
China

Editorial Office
MDPI AG
Grosspeteranlage 5
4052 Basel, Switzerland

This is a reprint of the Special Issue, published open access by the journal *Sensors* (ISSN 1424-8220), freely accessible at: https://www.mdpi.com/journal/sensors/special_issues/9B77B07HKF.

For citation purposes, cite each article independently as indicated on the article page online and as indicated below:

Lastname, A.A.; Lastname, B.B. Article Title. *Journal Name* **Year**, *Volume Number*, Page Range.

ISBN 978-3-7258-5175-1 (Hbk)
ISBN 978-3-7258-5176-8 (PDF)
https://doi.org/10.3390/books978-3-7258-5176-8

Contents

About the Editors

Tianxu Xu

Tianxu Xu received B.S. and M.S. degrees in electronic science and technology and optical engineering from Zhengzhou University, Zhengzhou, China, in 2015 and 2018, respectively, and a Ph.D. degree in optical engineering from Nankai University, Tianjin, China, in 2021. She is currently a Lecturer at the School of Electrical and Information Engineering, Zhengzhou University. She has published more than 10 articles and conference papers. Her current research interests include image processing and computer vision.

Kaiheng Zou

Kaiheng Zou received B.S. and M. Eng. degrees from Peking University, Beijing, China, in 2014 and 2017, respectively. He also received his Ph. D. degree from the University of Southern California, Los Angeles, CA, in 2023. His research interests include coherent and direct detected optical communication systems, optical signal processing systems, free-space optics, and applications of optical frequency combs.

Cong Liu

Cong Liu received her B.S. degree in physics from the University of Science and Technology of China in 2014 and her Ph.D. in physics from the University of Southern California in 2021. She is currently a Hardware Engineer at Nokia Bell Labs in Murray Hill, USA. Her research interests span optical communications and networking, free-space and fiber optics, quantum optics, and nonlinear optics. Dr. Liu has authored approximately 40 peer-reviewed journal papers—including publications in Nature Photonics and Nature Communications—and around 50 conference proceedings, with a total citation count of ~2900 and an h-index of 28. She has contributed two book chapters, holds two issued patents, and is an active member of SPIE, Optica, and the IEEE Photonics Society. Dr. Liu also serves as a reviewer for over 30 leading journals and has acted as a Guest Editor for two Special Issues.

Yang Yue

Yang Yue received B.S. and M.S. degrees in electrical engineering and optics from Nankai University, China, in 2004 and 2007, respectively. He received his Ph.D. degree in electrical engineering from the University of Southern California, USA, in 2012. He is currently a Professor at the School of Information and Communications Engineering, Xi'an Jiaotong University, China. He is the Founder and current PI of Intelligent Photonic Application Technology Laboratory (iPatLab). Dr. Yue's current research interest is intelligent photonics, including optical communications, optical perception, and optical chips. He has published over 300 papers in numerous journals (including Science), conference proceedings with over 13,000 citations, two books, eight edited books, two book chapters, over 70 issued or pending patents, and more than 200 invited presentations (including 1 tutorial, over 30 plenary and over 80 keynote talks). Dr. Yue is a Fellow of SPIE, and a Senior Member of IEEE and Optica. He also ranks among the top 2% Scientists List Worldwide by Stanford University. He is an Associate Editor for IEEE Access and Frontiers in Physics, Editor Board Member for four other scientific journals, and Guest Editor for more than 10 journal Special Issues. He also served as Chair or Committee Member for over 100 international conferences and a Reviewer for over 80 prestigious journals.

Preface

In the intelligent era, diverse scenarios are driving explosive development and growth in optical technology and industry. This Reprint, entitled "Advanced Optical Technologies for Communications, Perception, and Chips", captures cutting-edge research and developments that highlight the interdisciplinary convergence of photonics, electronics, and artificial intelligence.

The scope of this Reprint spans theoretical, experimental, and applied studies in high-speed optical communication, novel sensing and perception systems, and integrated photonic and optoelectronic chips. Particular emphasis is placed on innovations that enhance performance, scalability, and integration across these domains.

The motivation behind this Reprint is derived from the growing need for high-bandwidth, low-latency, and energy-efficient solutions across applications such as next-generation networks, autonomous systems, and intelligent computing. By compiling this Reprint, we hope to provide an overview of the emerging trends and challenges in advanced optical technologies, aiming to bridge the gap between fundamental research and practical applications, and foster further collaboration and innovation among researchers and engineers in academia and industry.

This Reprint is intended for researchers, graduate students, and professionals working in photonics, information science, integrated circuits, sensing technologies, and related fields. We believe it will serve as a valuable reference and inspiration for those engaged in pushing the frontiers of optical science and engineering.

We extend our sincere gratitude to all the authors and reviewers who contributed their expertise and time to ensure the high quality of this Reprint.

Tianxu Xu, Kaiheng Zou, Cong Liu, and Yang Yue
Guest Editors

sensors

MDPI

Editorial

Special Issue on Advanced Optical Technologies for Communications, Perception, and Chips

Tianxu Xu [1,*], Kaiheng Zou [2], Cong Liu [2] and Yang Yue [3]

1 School of Electrical and Information Engineering, Zhengzhou University, Zhengzhou 450001, China
2 Department of Electrical Engineering, University of Southern California, Los Angeles, CA 90089, USA; kai-hengz@usc.edu (K.Z.); liucong@usc.edu (C.L.)
3 School of Information and Communications Engineering, Xi'an Jiaotong University, Xi'an 710049, China; yueyang@xjtu.edu.cn
* Correspondence: xutxu@zzu.edu.cn

1. Introduction

With the iterative upgrade and popular application of new information technologies such as 5G, cloud computing, big data, and artificial intelligence (AI), the global data traffic and the demand for computing power has ushered in explosive growth. Traditional data transmission speed, information capacity, and chip computing performance can no longer meet the processing needs of big data [1]. Recent progress in optical technologies has highlighted a feasible development route for these contradictions. In particular, optical communication, optical perception, and optical chips are considered to be the most promising research directions to explore. At present, optical communication focuses on core technical issues, including larger bandwidth, lower latency, and smaller packet loss rate, to ensure that high-quality networks meet and support emerging applications (e.g., 4K/8K broadcasting, VR/AR, and free-view video). Optical perception technology has the advantages of convenient acquisition, low cost, and a large amount of information. With the continuous innovation of basic theory and analysis technology, researchers are devoted to developing perception systems, constructing multi-dimensional image knowledge systems, and promoting the development of image theory and application to a deeper level [2,3]. As a design that closely matches the optical computing architecture and the AI algorithm, optical chips offer the advantages of high-speed parallelism and low power consumption and can solve many problems in the application fields such as long processing time and high power consumption [4]. The above-mentioned research covers the core technologies in key fields to provide high speed, wide bandwidth, large volume, and low power consumption.

This Special Issue focuses on the state-of-the-art advancements in optical technologies for communication, perception, and chips, with a specific focus on digital, electrical, and optical signal processing theories, AI, integrated chips, devices, and subsystems/systems. The included papers successfully achieve these goals by contributing cutting-edge design, analysis, computation, and experiments with remarkable results. Two review papers, one communication, and nine research articles comprise the Special Issue, and these curated contributions highlight remarkable innovations while revealing critical knowledge gaps and future research imperatives. Throughout this editorial, numbers in parentheses (e.g., (1), (2), etc.) refer to the contributions in this collection, details of which can be found in the "List of Contributions" section. For clarity, we have grouped the works by topic and summarized the latest results in the field. This editorial is only a brief overview of the outstanding works included, and we encourage readers to explore this Special Issue in full.

2. Optical Communication: Speed, Security, and Novel Architectures

In optical communications, the pursuit of ultra-low latency and ultra-high bandwidth continues to drive innovation in fiber and free-space systems. Cao et al. designed a graded-index ring-core fiber (contribution 1), which includes a GeO_2-doped silica ring core and a SiO_2 cladding, to achieve near-vacuum optical propagation. Ullah et al. proposed a novel central office (CO) for the optical metropolitan access network (OMAN) with an affordable and unique switching system (contribution 2), thereby facilitating a 1.2 Tbps downlink-capable metropolitan access network. Currently, broader breakthroughs include ultrafast optical time-domain transformation (UO-TDT) techniques achieving MHz-rate imaging through spectral-to-temporal mapping, crucial for real-time monitoring in 5G/6G networks [5]. In pursuit of high-speed optical communication, Zhu et al. adopted a new flexible organic photodetector (OPD) and demonstrated a high data transmission rate of 80 MHz with a bit error rate of 3.5×10^{-4}, meaning it has great potential in next-generation high-speed flexible optical communication systems [6].

With security being critical to a system, quantum augmentation solutions are moving from theory to practice: Quantum key distribution (QKD) networks have now implemented quantum-secured optical layer encryption (contribution 3). To enhance the confidentiality of visible-light communication (VLC) systems, Shi et al. constructed a distributed LED VLC system with multiple users and an eavesdropper, proposing a tabu search-based LED selection algorithm using perfect channel state information (CSI) to avoid local optima (contribution 4). In addition, a loss-tolerant quasi-quantum secure direct communication (QSDC) protocol was demonstrated by using single photons that simultaneously transmitted information and exchanged keys, achieving a world-record 2.38 kbps secure rate over 104.8 km fiber [7].

3. Optical Perception: Sensing, Intelligence, and Applications

In optical perception, computational imaging and AI-driven sensing are overcoming traditional limitations. This Special Issue highlights the first ultrafast time-stretch optical coherence tomography (TS-OCT) system to employ reservoir computing (RC) for direct temporal signal analysis—eliminating the need for Fourier transform techniques. By leveraging the temporal characteristics of interference signals while ignoring frequency chirp, this approach offers an efficient solution to nonlinear wavelength sweeping challenges (contribution 5). Related studies also reflect that computational imaging and sensing are redefining resolution and functional analysis. For example: tip-based ultrafast nano-spectroscopy enables higher levels of spatial resolution with femtosecond temporal precision [8]. Chen et al. proposed a partially coherent optical neural network (PCONN) transmission model using mutual intensity modulation, enabling natural-light-based computational reasoning via simple filtering to achieve end-to-end optical processing from signal acquisition to inference [9]. Li et al. prototyped and integrated an all-optical mechanical sensor that mimics mechano-sensitive hair-like sensilla (MSHS) based on thin-walled glass microbubbles as flexible whispering gallery mode resonators. This mechanical sensor can be used as a real-time, directional mechanical sensing whisker in a quadruped cat-like robot, showing its potential for innovative mechanical transduction, artificial perception, and robotic applications [10]. Meanwhile, spatiotemporal OCT (STOC-T) permits hemo-dynamic monitoring at 112 volumetric frames/sec directly from structural data, by-passing complex flow-processing algorithms [11]. For depth information application, Wang et al. proposed a new non-contact mask recognition method based on time-of-flight (ToF) camera, and can further classify mask types, such as medical masks and N95 masks (contribution 6). Shan et al. developed a hemispherical photoelectric memristor array based on Ag-TiO_2 nanoclusters/sodium alginate film, which realized depth perception and motion detection based on binocular parallax [12]. To further improve the accuracy of feature extraction, the processing of optical information tends to explore the complementarity of cross-modal features. Through deep learning algorithms, it enhances the robustness of sensing systems to achieve adaptive fusion and intelligent processing of multi-dimensional information [13,14]. In addition, an optical microfiber array skin (OMAS) was reported to mimic the interlaced structure of tactile vesicles for tactile visualization and object reconstruction perception, which was characterized by high sensitivity (-0.83 N/V) and fast response time (38 ms) [15]. These advances mark a shift to computational perception, where AI extracts potential information from optical signals.

4. Optical Chips and Computing: Devices and Integration

In the photonic device, Yang et al. compared the effect of output waveguide configurations on the performance of array waveguide gratings (AWGs), and found the AWG with an output waveguide converging on the grating circle had larger crosstalk and lower nonuniformity (contribution 7). Based on the above research, a 1×8 arc-shaped AWG on a SOI platform with a central operation wavelength around 1550 nm was designed and fabricated. Experimental results indicated that the fabricated AWGs outperform in terms of insertion loss and nonuniformity. For fiber Bragg gratings, Cięszczyk et al. reduced the effect of noise on the results by cumulative pretreatment. This method can be used directly or in combination with other algorithms and is much simpler than the spectrum downgrade method (contribution 8). Krause et al. proposed a pioneering optimal double titanium nitride heater model for thermo-optic phase shifters at 1550 nm using standard silicon-on-insulator technology (contribution 9), which reduces system energy cost by achieving a π-phase shift with only 19.1 mW power, low thermal crosstalk (0.404), and minimal optical losses over 1 mm, addressing MZM imbalance issues for data centers and long-range optical communication transmitters. In addition, Liu et al. designed a graded exponential ring fiber featuring a GeO_2-doped silica ring core and SiO_2 coating (contribution 10). The structure suppresses spin–orbit coupling effects, thereby reducing power transfer between modes and enhancing orbital angular momentum (OAM) mode purity. Specifically, the purity of the $OAM_{1,1}$ mode increases from 86.48% to 94.43%.

In the field of optical chips, Li et al. proposed and demonstrated an on-chip directional multiplexing strategy for high-capacity 3D multi-planar projection in free space [16]. By exploring and exploiting the directional dependence of guided wave driven detour phase manipulation, they achieved 16-channel on-chip holography, where different images are encoded independently on different z-planes in four illumination directions. Lin et al. experimentally demonstrated an on-chip device capable of generating optical skyrmions with tailored topological invariants [17], which is a breakthrough in chip technology and brings new light to optics and materials science. Gagino et al. demonstrated an integrated optical phased array (OPA) based on an InP photonic integration platform, embedding built-in optical amplifiers and phase modulators to provide beamforming, gain, and beam steering functions in the 1465–1590 nm wavelength range [18]. It has the highest on-chip gain and output power level among active OPAs (i.e., OPAs with amplification functions). These research works show that optical chips are evolving from a single transmission medium to an integrated platform for intelligent computing and communications. The current research focus has shifted to room-temperature operation mechanisms, high-integration systems, and high-dimensional quantum state regulation, breaking through the limitations of traditional low-temperature environments and free-space optics. It promotes the integrated design of quantum light sources, detectors, and chips, laying a foundation for the scalable application of quantum communication and computing [19,20]. In addition to photonic devices and integration, research continues on developing electro-optical nonlinear materials to achieve low-power, high-speed dynamic optical regulation, and the application areas include ultrafast lasers, topological photonic devices, and integrated optical computing [21,22].

Moreover, Bravo et al. first presented a comprehensive state-of-the-art study on hybrid networks based on radio frequency (RF) and VLC under ideal conditions (contribution 11). The characteristics and implementation of these hybrid networks in outdoor scenarios with adverse conditions are then analyzed. Finally, the main challenges and future directions of these hybrid networks are summarized. Furthermore, Isiaka A. et al. provided a comprehensive overview of the enabling technologies, challenges, trends, and future prospects of FSO communications in the context of 5G and beyond (contribution 12). The atmospheric turbulence (AT), weather-induced signal attenuation, and alignment issues faced by current FSO systems are analyzed, and key enabling technologies such as adaptive optics, modulation schemes, and error correction codes are explored. In addition, the advantages of integrating FSO with emerging technologies are discussed. Finally, the prospects of FSO communications in the evolving 5G and future network landscape are emphasized, and key challenges and future research directions for fully realizing the potential of FSO in next-generation communication systems are identified.

5. Conclusions

With the rapid development of the digital economy, the application of advanced optical technology in the fields of communication, perception, and chips has become increasingly important. In terms of optical communications, the scale of the global optical communications market is constantly expanding and maintaining steady growth, and more high-performance networks and devices will be developed. In the field of optical perception, the rapid development of deep learning and artificial intelligence has promoted the continuous breakthrough and iteration of core perception technologies, and has also accelerated the leap from 2D imaging to 3D perception, making optical perception technology a key common technology for intelligent upgrading in each related industry. Optical chips have broad application prospects in emerging fields such as artificial intelligence, high-performance computing, and quantum computing. They can not only improve data transmission speed and reduce energy consumption and costs, but also promote technological innovation and industrial upgrading in relevant fields.

This Special Issue crystallizes critical advancements across optical communication, perception, and chip technologies—each domain propelled by distinct yet interconnected innovations. The collected works not only address immediate technical bottlenecks but also illuminate pathways toward optics-dominated computing and sensing infrastructures. Future breakthroughs will emerge in the areas of device innovation, computational frameworks, and system-level co-engineering. We envision a new era where optical technologies transcend traditional roles, becoming the backbone of intelligent, secure, and sustainable global infrastructures.

Author Contributions: Conceptualization, T.X., K.Z., C.L. and Y.Y.; methodology, T.X. and C.L.; formal analysis, K.Z. and Y.Y.; investigation, T.X. and Y.Y.; writing—original draft preparation, T.X.; writing—review and editing, Y.Y., K.Z. and C.L.; supervision, C.L. and Y.Y. All authors have read and agreed to the published version of the manuscript.

Funding: This research received no external funding.

Conflicts of Interest: The authors declare no conflicts of interest.

List of Contributions:

1. Cao, X.; Luo, M.; Liu, J.; Ma, J.; Hao, Y.; Liu, Y. Speed of Light in Hollow-Core Photonic Bandgap Fiber Approaching That in Vacuum. *Sensors* **2024**, *24*, 6954. https://doi.org/10.3390/s24216954.
2. Ullah, R.; Ullah, S.; Almadhor, A.; Alwageed, H.S.; Al-Atawi, A.A.; Ren, J.; Chen, S. A high-capacity optical metro access network: Efficiently recovering fiber failures with robust switching and centralized optical line terminal. *Sensors* **2024**, *24*, 1074. https://doi.org/10.3390/s24041074.
3. Velasco, L.; Ahmadian, M.; Ortiz, L.; Brito, J.P.; Pastor, A.; Rivas, J.M.; Barzegar, S.; Comellas, J.; Martin, V.; Ruiz, M. Scenarios for Optical Encryption Using Quantum Keys. *Sensors* **2024**, *24*, 6631. https://doi.org/10.3390/s24206631.
4. Shi, G.; Cheng, W.; Gao, X.; Wei, F.; Zhang, H.; Wang, Q. Enhancing Security in Visible Light Communication: A Tabu-Search-Based Method for Transmitter Selection. *Sensors* **2024**, *24*, 1906. https://doi.org/10.3390/s24061906.
5. Liao, W.; Luan, T.; Yue, Y.; Wang, C. Ultrafast Time-Stretch Optical Coherence Tomography Using Reservoir Computing for Fourier-Free Signal Processing. *Sensors* **2025**, *25*, 3738. https://doi.org/10.3390/s25123738.
6. Wang, X.; Xu, T.; An, D.; Sun, L.; Wang, Q.; Pan, Z.; Yue, Y. Face Mask Identification Using Spatial and Frequency Features in Depth Image from Time-of-Flight Camera. *Sensors* **2023**, *23*, 1596. https://doi.org/10.3390/s23031596.
7. Yang, C.; Zhou, Z.; Gao, X.; Xu, Z.; Han, S.; Chong, Y.; Min, R.; Yue, Y.; Duan, Z. Compact silicon-arrayed waveguide gratings with low nonuniformity. *Sensors* **2024**, *24*, 5303. https://doi.org/10.3390/s24165303.
8. Cięszczyk, S.; Kida, M.; Panas, P. Demodulation of Fibre Bragg Grating Sensors by Using Cumulative Sum as a Preprocessing Method. *Sensors* **2025**, *25*, 634. https://doi.org/10.3390/s25030634.
9. Krause, E.E.; Malka, D. Optimizations of double titanium nitride thermo-optic phase-shifter heaters using SOI technology. *Sensors* **2023**, *23*, 8587. https://doi.org/10.3390/s23208587.
10. Liu, Y.; Wang, Y.; Geng, W.; Zhao, W.; Zhang, H.; Zhang, W.; Pan, Z.; Yue, Y. Parabolic-index ring-core fiber supporting high-purity orbital angular momentum modes. *Sensors* **2023**, *23*, 3641. https://doi.org/10.3390/s23073641.
11. Alvarez, L.B.; Montejo-Sánchez, S.; Rodríguez-López, L.; Azurdia-Meza, C.; Saavedra, G. A review of hybrid vlc/rf networks: Features, applications, and future directions. *Sensors* **2023**, *23*, 7545. https://doi.org/10.3390/s23177545.
12. Alimi, I.A.; Monteiro, P.P. Revolutionizing Free-Space Optics: A Survey of Enabling Technologies, Challenges, Trends, and Prospects of Beyond 5G Free-Space Optical (FSO) Communication Systems. *Sensors* **2024**, *24*, 8036. https://doi.org/10.3390/s24248036.

References

1. Al-Jumaili, A.H.A.; Muniyandi, R.C.; Hasan, M.K.; Paw, J.K.S.; Singh, M.J. Big Data Analytics Using Cloud Computing Based Frameworks for Power Management Systems: Status, Constraints, and Future Recommendations. *Sensors* **2023**, *23*, 2952. [CrossRef]
2. Jeon, H.B.; Kim, S.M.; Moon, H.J.; Kwon, D.H.; Lee, J.W.; Chung, J.M.; Han, S.K.; Chae, C.B.; Alouini, M.S. Free-space optical communications for 6G wireless networks: Challenges, opportunities, and prototype validation. *IEEE Commun. Mag.* **2023**, *61*, 116–121. [CrossRef]
3. Szymanowicz, S.; Rupprecht, C.; Vedaldi, A. Splatter image: Ultra-fast single-view 3D reconstruction. In Proceedings of the IEEE/CVF Conference on Computer Vision and Pattern Recognition, Seattle, DC, USA, 17–21 June 2024; pp. 10208–10217.
4. Belgaonkar, V.V.; Triveni, C.L.; Sundaraguru, R. Free space optical communication using OQPSK in the presence of strong atmospheric turbulence and losses. *Opt. Quantum Electron.* **2024**, *56*, 1669. [CrossRef]
5. Zhang, Y.; Tao, C.; Luo, S.; Lau, K.Y.; Zheng, J.; Huang, L.; Zhang, A.; Sheng, L.; Ling, Q.; Guan, Z.; et al. Ultra-fast optical time-domain transformation techniques. *Nat. Rev. Methods Prim.* **2025**, *5*, 11. [CrossRef]
6. Zhu, Y.; Chen, H.; Han, R.; Qin, H.; Yao, Z.; Liu, H.; Ma, Y.; Wan, X.; Li, G.; Chen, Y. High-speed flexible near-infrared organic photodiode for optical communication. *Natl. Sci. Rev.* **2024**, *11*, nwad311. [CrossRef]
7. Pan, D.; Liu, Y.C.; Niu, P.; Zhang, H.; Zhang, F.; Wang, M.; Song, X.T.; Chen, X.; Zheng, C.; Long, G.L. Simultaneous transmission of information and key ex-change using the same photonic quantum states. *Sci. Adv.* **2025**, *11*, eadt4627. [CrossRef]

8. Zhao, Z.; Kravtsov, V.; Wang, Z.; Zhou, Z.; Dou, L.; Huang, D.; Wang, Z.; Cheng, X.; Raschke, M.B.; Jiang, T. Applications of ultrafast nano-spectroscopy and nano-imaging with tip-based microscopy. *eLight* **2025**, *5*, 1. [CrossRef]
9. Chen, R.; Ma, Y.; Zhang, C.; Xu, W.; Wang, Z.; Sun, S. All-optical perception based on partially coherent optical neural networks. *Opt. Express* **2025**, *33*, 1609–1624. [CrossRef]
10. Li, Y.; Guo, Z.; Zhao, X.; Liu, S.; Chen, Z.; Dong, W.-F.; Wang, S.; Sun, Y.L.; Wu, X. An all-optical multidirectional mechano-sensor inspired by biologically mechano-sensitive hair sensilla. *Nat. Commun.* **2024**, *15*, 2906. [CrossRef] [PubMed]
11. Liżewski, K.; Tomczewski, S.; Borycki, D.; Węgrzyn, P.; Wojtkowski, M. Imaging the retinal and choroidal vasculature using spatio-temporal optical co-herence tomography (STOC-T). *Biocybern. Biomed. Eng.* **2024**, *44*, 95–104. [CrossRef]
12. Shan, X.; Wang, Z.; Xie, J.; Han, J.; Tao, Y.; Lin, Y.; Zhao, X.; Ielmini, D.; Liu, Y.; Xu, H. Hemispherical retina emulated by plasmonic optoelec-tronic memristors with all-optical modulation for neuromorphic stereo vision. *Adv. Sci.* **2024**, *11*, 2405160. [CrossRef]
13. Li, Y.; Ji, Q.; Jiao, S. Prompt learning and multi-scale attention for infrared and visible image fusion. *Infrared Phys. Technol.* **2025**, *145*, 105671. [CrossRef]
14. Ji, T.; Fang, H.; Zhang, R.; Yang, J.; Wang, Z.; Wang, X. Plastic waste identification based on multimodal feature selection and cross-modal Swin Trans-former. *Waste Manag.* **2025**, *192*, 58–68. [CrossRef] [PubMed]
15. Zhou, H.; Zhang, C.; Nong, H.; Weng, J.; Wang, D.; Yu, Y.; Zhang, J.; Zhang, C.; Yu, J.; Zhang, Z.; et al. Multi-photon neuron embedded bionic skin for high-precision complex texture and object reconstruction perception research. *Opto-Electron. Adv.* **2025**, *8*, 240152. [CrossRef]
16. Li, Z.; Liu, Y.; Zhang, C.; Qiao, Y.; Deng, R.; Shi, Y.; Li, Z. On-Chip Direction-Multiplexed Meta-Optics for High-Capacity 3D Holography. *Adv. Funct. Mater.* **2024**, *34*, 2312705. [CrossRef]
17. Lin, W.; Ota, Y.; Arakawa, Y.; Iwamoto, S. On-chip optical skyrmionic beam generators. *Optica* **2024**, *11*, 1588–1594. [CrossRef]
18. Gagino, M.; Millan-Mejia, A.; Augustin, L.; Williams, K.; Bente, E.; Dolores-Calzadilla, V. Integrated optical phased array with on-chip amplification enabling programmable beam shaping. *Sci. Rep.* **2024**, *14*, 9590. [CrossRef] [PubMed]
19. Huang, J.; Mao, J.; Li, X.; Yuan, J.; Zheng, Y.; Zhai, C.; Dai, T.; Fu, Z.; Bao, J.; Yang, Y.; et al. Integrated optical entangled quantum vortex emitters. *Nat. Photon.* **2025**, *19*, 471–478. [CrossRef]
20. Chen, B.; Zhou, Y.; Liu, Y.; Ye, C.; Cao, Q.; Huang, P.; Kim, C.; Zheng, Y.; Oxenløwe, L.K.; Yvind, K.; et al. Integrated optical vortex microcomb. *Nat. Photon.* **2024**, *18*, 625–631. [CrossRef]
21. Lv, C.; Meng, F.; Cui, L.; Jiao, Y.; Jia, Z.; Qin, W.; Qin, G. Voltage-controlled nonlinear optical properties in gold nanofilms via electrothermal effect. *Nat. Commun.* **2024**, *15*, 6372. [CrossRef]
22. Chen, R.; Yan, W.; Liu, W.; Cheng, W.; Lu, Q.; Tan, Y.; Chen, F. Nonlinear tuning of multiple topological edge states in photovoltaic photonic lattices. *Sci. Bull.* **2025**, *70*, 1605–1610. [CrossRef] [PubMed]

sensors

MDPI

Article

Ultrafast Time-Stretch Optical Coherence Tomography Using Reservoir Computing for Fourier-Free Signal Processing

Weiqing Liao [1,2], Tianxiang Luan [2], Yuanli Yue [2] and Chao Wang [2,*]

1 Photonics Information Innovation Center and Hebei Provincial Center for Optical Sensing Innovations, College of Physics Science & Technology, Hebei University, Baoding 071002, China; wl258@kent.ac.uk
2 School of Engineering, University of Kent, Canterbury CT2 7NT, UK; tl436@kent.ac.uk (T.L.); yy282@kent.ac.uk (Y.Y.)
* Correspondence: c.wang@kent.ac.uk

Abstract: Swept-source optical coherence tomography (SS-OCT) is a widely used imaging technique, particularly in medical diagnostics, due to its ability to provide high-resolution cross-sectional images. However, one of the main challenges in SS-OCT systems is the nonlinearity in wavelength sweeping, which leads to degraded depth resolution after Fourier transform. Correcting for this nonlinearity typically requires complex re-sampling and chirp compensation methods. In this paper, we introduce the first ultrafast time-stretch optical coherence tomography (TS-OCT) system that utilizes reservoir computing (RC) to perform direct temporal signal analysis without relying on Fourier transform techniques. By focusing solely on the temporal characteristics of the interference signal, regardless of frequency chirp, we demonstrate a more efficient solution to address the nonlinear wavelength sweeping issue. By leveraging the dynamic temporal processing capabilities of RC, the proposed system effectively bypasses the challenges faced by Fourier analysis, maintaining high-resolution depth measurement without being affected by chirp-introduced spectral broadening. The system operates by categorizing the interference signals generated by variations in sample position. This classification-based approach simplifies the data processing pipeline. We developed an RC-based model to interpret the temporal patterns in the interferometric signals, achieving high classification accuracy. A proof-of-the-concept experiment demonstrated that this method allows for precise depth resolution, independent of system chirp. With an A-scan rate of 50 MHz, the classification model yielded 100% accuracy with a root mean square error (RMSE) of 0.2416. This approach offers a robust alternative to Fourier-based analysis, particularly in systems prone to nonlinearities during signal acquisition.

Keywords: optical coherence tomography; reservoir computing; photonic time stretch

1. Introduction

Since its introduction by D. Huang et al. in 1991 [1], optical coherence tomography (OCT) has garnered significant attention due to its non-invasive nature and high-resolution imaging capabilities, making it widely applicable in fields such as biology and industry [2–4]. Building on this foundation, a research team from MIT described Swept Source OCT (SS-OCT) in 1994 [5,6], followed by a team from the University of Vienna in 1995 [7,8], which demonstrated the use of tunable lasers and spectrometer-based interferometry to measure intraocular distances. SS-OCT seeks to combine some of the advantages of standard Time Domain (TD) and Spectral Domain (SD) OCT. In SS-OCT, the spectral components are encoded not by spatial separation but rather by temporal encoding, where the

 https://doi.org/10.3390/s25123738

spectrum is either filtered or generated in single continuous frequency steps, subsequently reconstructed prior to Fourier transformation. By utilizing a frequency-scanning light source (i.e., a frequency-swept laser), the optical setup becomes simpler compared with SD-OCT, but the challenge of scanning is effectively shifted from the TD-OCT reference arm to the SS-OCT light source. Swept laser sources can achieve very narrow instantaneous bandwidths (linewidth) at extremely high frequencies [9].

The sweep rate of traditional SS-OCT systems is constrained by the laser repetition rate, leading to the development of various laser technologies aimed at improving the A-scan rate [10–12]. Conventional swept lasers incorporate a tunable filter placed in the external cavity of the laser, which controls the wavelength of light allowed to oscillate within the cavity. By altering the properties of the filter, the laser can gradually change its output wavelength, enabling wavelength sweeping. However, the scan rate is limited to the kilohertz range, as the laser must regenerate from spontaneous emission with each round trip. In 2011, to further enhance SS-OCT scanning speed, researchers integrated widely tunable micro-electromechanical systems (MEMSs) vertical-cavity surface-emitting lasers (VCSELs) with SS-OCT, achieving a significant increase in sweep rates from the kilohertz to the megahertz range [13]. The MEMS-VCSEL-based SS-OCT leverages the high tuning speed and precision of MEMS technology and VCSELs to enable rapid, stable, and high-resolution three-dimensional imaging, particularly suitable for dynamic and in vivo biological imaging. However, mechanical tunable filters used in swept lasers are limited by inertia, resulting in tuning speeds typically in the kilohertz range, which can be alleviated by adopting non-mechanical or passive scanning methods.

In recent years, Time-Stretch Dispersive Fourier Transform (TS-DFT) has been proven to be a promising method for increasing OCT scanning speed [14]. In TS-DFT, a dispersive medium (e.g., a dispersive optical fiber) is used to temporally stretch the incoming short-pulse light. Due to chromatic dispersion, different wavelength components within the light pulse propagate at different speeds, causing the pulse to stretch during transmission. This means that the time distribution of the wavelength components within the pulse is expanded, making them temporally separated. The resulting chirped pulse sequence functions as a passive swept source, enabling an axial scanning rate in SS-OCT equivalent to the laser's pulse repetition rate [15–17]. However, the wavelength-to-time mapping in a time-stretch system is typically nonlinear due to the inherent high-order dispersion in optical fibers [18]. Frequency chirp is introduced to the interference waveform due to nonlinear mapping, leading to degradation in OCT depth resolution and potential image distortion or blurring after Fourier transformation. Alternative digital post-processing techniques have been proposed, including resampling algorithms such as linear interpolation and polynomial fitting [19,20], though these methods exhibit limited accuracy when addressing complex nonlinearities.

Existing methods primarily focus on correcting nonlinear sweeping and mitigating spectrum broadening after Fast Fourier Transform (FFT) [21,22]. However, these approaches suffer from high computing costs and increased processing time. This is particularly challenging in TS-OCT systems as the normal sampling rate is extreme (tens of GSa/s). Compressive sensing has been successfully applied in TS-OCT to reduce the data sampling rate and to estimate the frequency information of the interference waveform without Fourier transform analysis [23], but it still suffers from chirp-induced spectrum broadening.

On the other hand, direct time-domain signal analysis without relying on computationally intensive Fourier transform offers a promising alternative demodulation solution for a time-varying system [24]. The temporal waveform of the OCT signal changes inevitably with varying axial sample distances. Therefore, based on temporal signal analysis, precise distance measurement can be achieved by distinguishing waveforms corresponding to different distances through sequence data classification and time series prediction methods. Current approaches often rely on Recurrent Neural Networks (RNNs) [25] and Long Short-Term Memory (LSTM) networks [26] for modeling temporal dependencies in sequential data; however, they suffer from the vanishing gradient problem, which limits their ability to capture long-term dependencies. LSTMs were developed to overcome this issue by introducing gating mechanisms that enable them to retain information over longer sequences. While effective, LSTM networks require extensive training through backpropagation, which is computationally intensive and slow, especially for tasks involving complex, long-term dependencies. Reservoir computing (RC) has been developed as a special variant of RNN [27,28]. RC has gained recognition as a highly efficient solution for complex time-series analysis and classification tasks [29–31]. RC stands out for its remarkable capability in handling nonlinear temporal dependencies. Unlike conventional RNN models, RC only requires training of the output weights, thus significantly reducing the time and computational resources needed for model optimization, while still maintaining competitive accuracy. RC is particularly well-suited for tasks involving time-series data, as the reservoir's recurrent structure allows it to capture both short-term and long-term temporal dependencies. For example, in applications such as optical coherence tomography (OCT) or frequency-modulated continuous-wave (FMCW) LiDAR systems [32], where the input signals exhibit complex nonlinear dynamics, RC can quickly classify and process the time-series signals to produce accurate results.

Moreover, RC's simplicity in structure and computation enables it to outperform other methods in terms of speed and efficiency. For instance, while traditional RNN models suffer from backpropagation through time (BPTT) with a complexity of $O(N^2)$, RC reduces the overall computational complexity to $O(N)$, making it a more scalable choice for real-time applications. This advantage is especially prominent in high-frequency temporal signal analysis, where the reduced computational load of RC allows for faster processing and real-time implementation without compromising performance. As shown in Table 1, the RC model demonstrates significant advantages over RNN and LSTM networks, particularly in terms of computational efficiency, processing speed, and scalability.

Table 1. Comparative performances of RC, RNNs, and LSTMs in TS-OCT.

	Classification Accuracy (%)	Computational Complexity	Training Time (s)
RC	100%	$O(N)$	9
RNN	94%	$O(N^2)$	116
LSTM	96%	$O(N^2)$	175

In this study, we propose and validate an innovative Fourier-free TS-OCT system that employs an RC model for highly efficient time-domain signal processing without relying on Fourier transformation. By fully utilizing the temporal characteristics of the interference signal with the presence of nonlinear sweeping, we achieve precise depth measurements in OCT via the classification of temporal waveforms [33]. Unlike traditional Fourier-based TS-OCT systems, our novel system avoids complex signal linearization steps and spectrum broadening issues. In addition, RC is much less computationally intensive than fast Fourier transform (FFT) [29]. For instance, while the computational complexity of FFT for a dataset with a sample size N is $N \log_2^N$, the complexity of RC is merely N. Therefore, through the introduction of RC technology, the system can quickly and accurately extract depth information from time-domain signals while reducing the computational burden of processing.

2. Principles

2.1. Fourier-Free TS-OCT System

The experiment was set up based on the TS-OCT system design shown in Figure 1a. This proof-of-concept system uses a passively mode-locked fiber laser (MLL, Calmar Mendocino FP laser) as the light source, producing a train of ultrashort pulses with a duration of 800 femtoseconds (fs) and a spectral bandwidth of 12 nm. The laser operates at a repetition frequency of 50 MHz, generating high-frequency pulse trains for rapid imaging. The experiment was set up based on the TS-OCT system design shown in Figure 1a. This proof-of-concept system uses a passively mode-locked fiber laser (MLL, Calmar Mendocino FP laser) as the light source, producing a train of ultrashort pulses with a duration of 800 femtoseconds (fs) and a spectral bandwidth of 12 nm. The laser operates at a repetition frequency of 50 MHz, generating high-frequency pulse trains for rapid imaging. Initially, the ultrashort pulses propagate through a dispersion-compensating fiber (DCF) exhibiting a total chromatic dispersion amount of -1.04 ns/nm, which stretches each 800 fs pulse to 8 ns. This time stretch process effectively creates a broadband, passive wavelength-swept optical source. The resulting wavelength-sweeping effect is essential for high-throughput imaging in OCT.

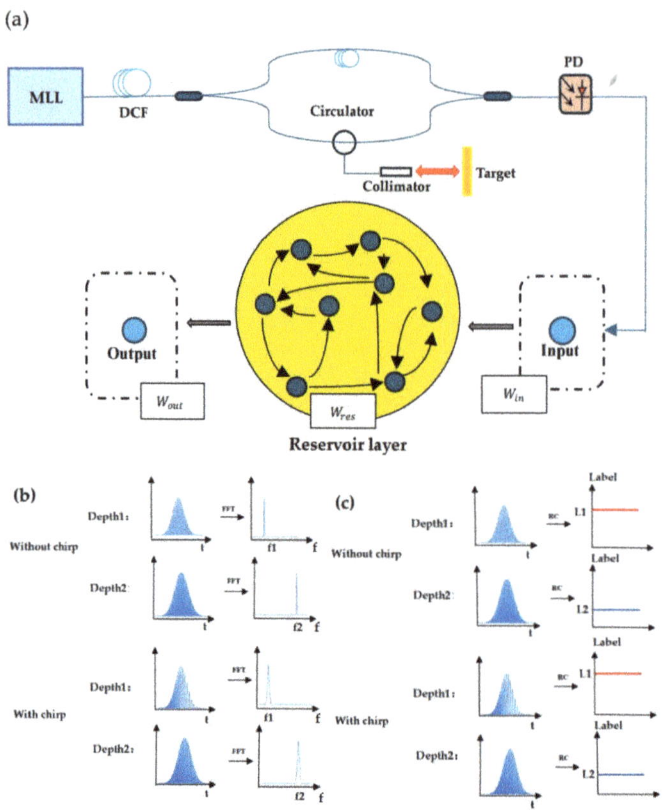

Figure 1. (**a**) Schematic diagram of the proposed Fourier-free TS-OCT system employing an RC module for temporal signal analysis. (**b**) Results of traditional Fourier analysis: nonlinear wavelength sweeping introduces frequency chirp and broadens the FFT peaks, hence poor depth resolution. (**c**) RC-based temporal signal analysis directly determines the depth information without being affected by the frequency chirp.

Following this, the wavelength-swept light is split by a 50:50 coupler and directed into a Mach–Zehnder Interferometer (MZI) structure, which forms the core of the OCT system. The MZI has two arms: the reference arm and the measurement arm. In the reference arm, a 100-m segment of optical fiber is added as an extra dispersion element to enhance nonlinear effects in the interference waveforms [33], which improves the RC model's accuracy in depth measurements by creating a more complex temporal pattern in the reference signal. In addition, this introduced additional dispersion creates a frequency-chirped interference signal that enables testing and validating the RC model's chirp-tolerant capabilities. Meanwhile, in the measurement arm, the light is routed through an optical circulator and directed to a collimator. The collimated beam is then reflected by an axially movable mirror, enabling depth variation within the measurement path. The backscattered light from the mirror is recoupled through the circulator and directed back toward the beam splitter.

The light from the reference and measurement arms interferes with the beam splitter, producing time-domain interference fringes. This interference pattern reflects the depth-resolved information required for OCT imaging. The interference signal is then detected by a high-speed photodetector (PD), capable of capturing the high-frequency fringes produced by the TS-OCT system. A real-time oscilloscope (Tektronix DSA73304D) records the signal for further analysis. The oscilloscope enables high-resolution temporal recording, ensuring that the fine details of the interference signal are preserved. The signal acquired by the oscilloscope is subsequently processed using the RC model. The fundamental principle of RC is to project the input data into a high-dimensional state space via a fixed, dynamic reservoir. Acting as a nonlinear dynamical system, the reservoir maps sequential input data into a set of intricate internal states that effectively encapsulate temporal dependencies. A comprehensive explanation of the RC methodology will be provided in Section 2.2.

In a chirp-free system, where the dispersion-compensating fiber (DCF) does not introduce higher-order dispersion, a uniform interference waveform with a single carrier frequency is generated. This results in a clean and stable signal, enabling accurate and high-resolution depth measurements through Fourier transformation, as illustrated in Figure 1b. However, when nonlinear wavelength sweeping occurs due to higher-order dispersion effects in long DCFs, the interference waveform becomes frequency-chirped. This chirping introduces broadened peaks in the FFT output, which significantly degrades the axial resolution of OCT imaging and reduces depth measurement accuracy.

To address this issue, our proposed system bypasses FFT analysis altogether, instead employing an RC model for real-time distance measurement by directly analyzing the time-domain signals. This Fourier-free approach allows the system to maintain high accuracy without needing to compensate for frequency chirp. By classifying time-domain patterns, the RC model provides output labels that correspond to specific depths, effectively mapping the temporal signal features directly to depth information, as shown in Figure 1c. In Figure 1c, the output labels from the RC model represent the axial depths of the sample, where the blue and red lines indicate different depths, with the red line corresponding to shallower depths and the blue line representing deeper regions. This method eliminates the need for re-sampling or chirp compensation, making the system highly efficient and robust against the distortions typically introduced by nonlinear sweeping. As a result, this RC-assisted TS-OCT system offers a streamlined solution for high-resolution OCT imaging, delivering reliable depth measurements even in the presence of chirp.

2.2. RC Approach for Temporal Signal Processing

Traditional optical coherence tomography systems typically require fast FFT to obtain axial depth information. However, the proposed method in this study, based on direct time-domain signal analysis using an RC model, enables a direct classification of interference waveforms from different depths. This approach allows for precise depth detection, with the target depth information determined from the classification results, regardless of the presence of frequency chirp, as shown in Figure 1c. Unlike FFT-based methods, which face challenges due to frequency linearization, increasing the system's complexity, the proposed RC-based method avoids such issues, simplifying the depth detection process.

To process the acquired signal, the RC model utilizes a structured approach to map and analyze sequential data. The architecture of RC is designed to capture the temporal dynamics of the input efficiently while minimizing computational overhead. This is achieved through the interaction of its key components: the input layer, reservoir layer, and output layer.

The input layer projects the input sequence $u(t)$ into the reservoir, using a randomly initialized input weight matrix W_{in}. The reservoir layer consists of interconnected units whose connections are defined by the reservoir weight matrix W_{res}, which is also randomly initialized but fixed after initialization. The input and reservoir weights are not trained during the learning phase, making RC highly efficient in terms of computational complexity.

At each time step t, the state of the reservoir is updated based on the input signal and the previous state of the reservoir. This is expressed mathematically by the following update equation:

$$s(t+1) = f(W_{in}u(t) + W_{res}s(t)) \tag{1}$$

where $s(t)$ represents the state of the reservoir at time t, W_{in} is the input weight matrix, W_{res} is the internal reservoir connection matrix, and $f(\cdot)$ is a nonlinear activation function, typically chosen as the hyperbolic tangent $f(x) = \tanh(x)$.

The reservoir layer's behavior is governed by its echo state property (ESP), which ensures that its internal states asymptotically depend on the input sequence, rather than on initial conditions. To maintain stability, the spectral radius $\rho(W_{res})$ of the reservoir matrix, defined as the largest absolute value among its eigenvalues, is constrained to be less than 1. For example, if the eigenvalues of W_{res} are {0.5, −0.8, 1.2}, then the spectral radius $\rho(W_{res})$ = 1.2. By ensuring $\rho(W_{res}) < 1$, the reservoir achieves stability, allowing its states to converge over time. This guarantees that the influence of past inputs decays over time, allowing the reservoir to focus on the most recent inputs. Additionally, the past inputs, referring to signals from earlier time steps such as $u(t-5)$, $u(t-4)$, decay over time due to the reservoir's dynamics. This decay ensures that the influence of these earlier signals diminishes, allowing the reservoir to focus on more recent inputs, such as $u(t)$, $u(t-1)$. By constraining the spectral radius $\rho(W_{res})$ to be less than 1, the reservoir achieves this balance, ensuring stability and relevance to the current input sequence.

Unlike traditional recurrent neural networks, only the output layer weights in RC are trained, which significantly reduces computational complexity. The output of the reservoir system is computed as a linear combination of the reservoir states, as follows:

$$\hat{y}(t) = W_{out}s(t) \tag{2}$$

where W_{out} is the output weight matrix and $\hat{y}(t)$ is the predicted output at time t.

The training process involves adjusting the output weights W_{out} by minimizing the error between the predicted outputs and the target outputs. This is typically done using ridge regression with a regularization term to prevent overfitting. The output weight matrix is computed as

$$W_{out} = \left(S^T S + \lambda I\right)^{-1} S^T Y \tag{3}$$

where S is the matrix of reservoir states, S^T is its transpose, Y is the matrix of target outputs, I is the identity matrix, and λ is the regularization parameter. The parameter λ controls the complexity of the model by adding a penalty term to the magnitude of the weights, helping prevent overfitting.

Once the output weights W_{out} have been trained, the RC model can be used for prediction. For any new input sequence, the reservoir states are updated, and the output is calculated using the trained output weights. This simple linear combination of the reservoir states makes RC highly efficient for real-time processing of time-series data, such as in dynamic systems with complex temporal dependencies.

The experiments were conducted on a system equipped with a 13th Gen Intel(R) Core (TM) i7-13650HX CPU (Intel Corporation, Santa Clara, CA, USA) operating at 2.60 GHz, 16 GB of RAM, and an NVIDIA GeForce RTX 4060 Laptop GPU (NVIDIA Corporation, Santa Clara, CA, USA), using Python 3.12 (Python Software Foundation, Beaverton, OR, USA).

3. Results

3.1. TS-OCT Signal Analysis

Using the Fourier-free TS-OCT system mentioned earlier, 50 consecutive cycles of interference signals were captured by the oscilloscope within a 1 μs duration. These signals correspond to optical path lengths of 0.59 mm and 2.46 mm, as illustrated in Figure 2a,b, respectively. These time-domain interference signals represent different imaging depths and highlight the system's capability to capture depth information at a refresh rate of 50 MHz in real time. A closer look at the interference waveforms within a single pulse period is provided in Figure 2c,d, which clearly demonstrate significant frequency chirping effects within each interference signal. This chirping is indicative of pronounced wavelength dispersion caused by the DCF and the extra optical fiber in the reference arm in the TS-OCT system.

To further analyze the impact of this chirp, a Fourier transform was applied to the interference patterns in Figure 2c,d, with the resulting frequency spectra shown in Figure 2e,f. These spectra reveal substantial broadening, which confirms the presence of high-order dispersion in the system [34]. This spectral broadening directly affects the system's axial resolution, as the frequency components become less distinct, leading to degraded depth precision. These observations emphasize the challenges posed by nonlinear sweeping in achieving high-resolution OCT imaging and underline the necessity of a chirp-tolerant approach, such as the proposed reservoir computing (RC) model, which enables direct depth measurement without relying on Fourier analysis.

The collected interference signals were divided into two datasets: a training set and a test set. The training set was used to teach the reservoir computing (RC) model to recognize specific temporal patterns within the time-domain signals, while the test set was reserved for evaluating the model's classification accuracy and generalization capabilities. By processing these datasets, the RC model learns to identify depth information from temporal features directly as a classification task, without requiring Fourier transform-based analysis. In the following sections, we detail how the waveform recognition-based RC model processes these datasets and assess the model's performance in accurate, real-time signal classification, highlighting its potential in high-throughput TS-OCT imaging applications.

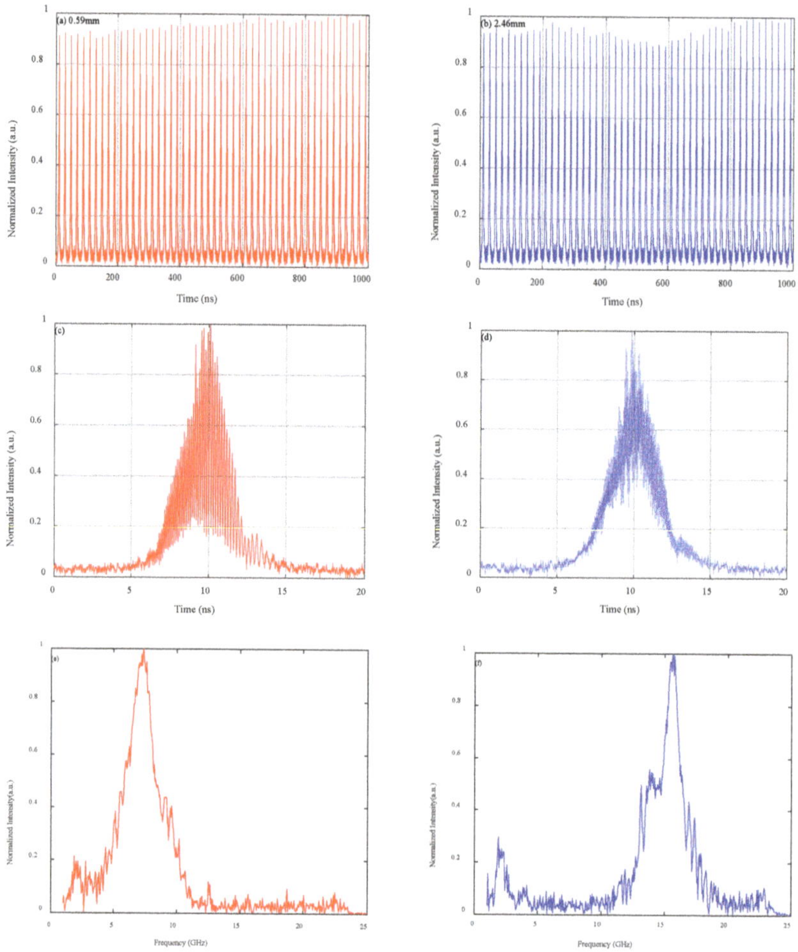

Figure 2. The measured 50 consecutive interference signal waveforms corresponding to a target depth of (**a**) 0.59 mm and (**b**) 2.46 mm. (**c**,**d**) are zoomed-in signals illustrating different temporal features within one single scanning period. To further investigate the impact of the chirp in the system, Fourier transforms were applied to the zoomed-in signals in (**c**,**d**), and the resulting frequency spectra are shown in (**e**,**f**).

3.2. Overcoming Challenges of Fourier Analysis in Nonlinear TS-OCT Using RC

Using the RC model for time-domain signal classification, the proposed system achieves depth measurements directly from the interference signal, bypassing the need for Fourier analysis. This novel approach validates the effectiveness of Fourier-free signal processing in TS-OCT, demonstrating that the RC model can accurately interpret depth information from the temporal characteristics of the interference waveform alone. The experimental setup combines dispersion-compensating fiber (DCF), an MZI structure, and high-speed detection, providing a robust framework for depth imaging even in the presence of significant dispersion and frequency chirping. This configuration highlights the RC model's potential in enhancing OCT imaging data processing efficiency and resolution without the typical computational burden of Fourier-based analysis.

In particular, the RC model derives distance information by classifying the intermediate frequency patterns associated with varying depths in the MZI structure. Specifically, we implemented the RC model using an Echo State Network (ESN) architecture [35], which offers a highly efficient method for processing sequential data. The reservoir size was set to 1000 units to balance complexity with computational efficiency, and the reservoir weight matrix was designed with a spectral radius of 0.8 to ensure stability while retaining temporal memory. An input scaling factor of 0.2 adjusts the sensitivity of the reservoir to incoming signals, while a connection sparsity of 0.3 maintains a manageable level of connectivity among reservoir units, reducing computational costs. To improve generalization, a regularization parameter of 500 was applied during training.

To construct the training dataset, interference waveforms were randomly sampled across a 5 mm axial range, with each individual distance assigned a unique output label. This labeled dataset enables the RC model to learn the association between waveform features and specific depths. Figure 3 presents interference signal waveforms at 14 different distances, clearly demonstrating the strong dependence of temporal features on the axial distances. Due to the presence of nonlinear wavelength sweeping, frequency chirps have been evidenced in all the interference waveforms. These variations typically require complex nonlinear corrections when using Fourier-based analysis.

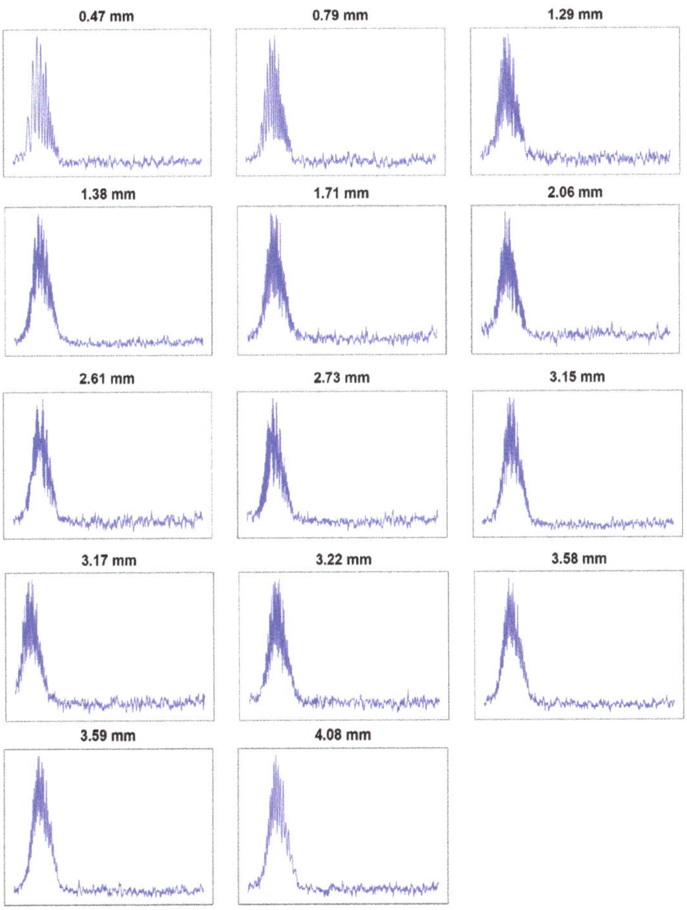

Figure 3. The measured interference signals corresponding to different axial distance differences.

In this work, the RC model directly analyzes the temporal features in each waveform, capturing depth information efficiently without additional processing. By training on this diverse set of chirped signals, the RC model learns to classify depth-related patterns, making it a highly effective and computationally efficient solution for OCT depth extraction, even under challenging nonlinear wavelength sweeping conditions. This Fourier-free approach thus offers a simplified, powerful alternative for obtaining accurate depth information, making it well suited for real-time OCT applications.

Under such circumstances, using RC for temporal signal analysis proves to be an effective approach, as it eliminates the need for complex nonlinear frequency sweep correction. This method enables accurate extraction of depth information directly from the temporal interference signals, facilitating efficient signal processing and analysis in the context of varying fringe densities.

Following the training phase, we created a test dataset comprising 14 waveforms, each corresponding to a unique depth, as depicted in Figure 3. To ensure comprehensive evaluation, the test dataset includes four periods for each depth, where each period is represented by a data vector of length 500×1. Consequently, the total training set is made up of 56 samples, ensuring a robust model training process.

For testing, periods were randomly selected from the 14 distinct depths, ensuring that none of these specific periods overlapped with those in the training dataset. Instead, alternate periods matching the required length were used to construct the test dataset. By randomizing the sequence of periods, we aim to simulate real-world conditions and assess the RC model's ability to generalize accurately to new data. This randomized order provides an additional layer of verification for the model's depth classification performance, reinforcing its potential to maintain reliability even in variable conditions. Through this approach, the RC model's robustness in interpreting time-domain OCT signals is validated, showcasing its effectiveness in reliably distinguishing depth without prior Fourier-based preprocessing, making it a strong candidate for real-time depth measurement applications in optical coherence tomography.

Figure 4 presents the testing results from the RC classification task. In the upper half of the figure, we see the model's output represented by the probabilities of each output label. The lower half of the figure shows the true output labels indicating each sample's actual depth. The RC model makes classification decisions based on the highest probability value within the output vector, where the sum of all output values equals 1, meaning that each individual value is relatively small but collectively adds up to a probability distribution. As shown in Figure 4, the RC model's output labels align precisely with the actual labels across all 14 depth classes, achieving a classification accuracy of 100%. This high accuracy can be attributed to the controlled conditions under which the waveforms were recorded; all waveforms used for both training and testing were captured within a brief timeframe (20 ns per waveform). Consequently, environmental variations or system instability, which could otherwise introduce waveform distortion, was minimized. This stable setup allowed the RC model to classify each depth accurately, highlighting its effectiveness in a well-controlled environment. The results confirm the RC model's potential for high-precision depth measurement in OCT systems, particularly when external factors are stable.

This probabilistic classification approach allows the RC model to perform accurate depth detection by assigning each signal to its most likely depth class. The model's ability to produce distinct class probabilities enables a reliable depth classification, even in cases where the interference signal may have noise or minor distortions. Such a capability demonstrates the RC model's robustness, validating its effectiveness in distinguishing between depths in OCT applications without requiring traditional Fourier-based analysis for signal processing.

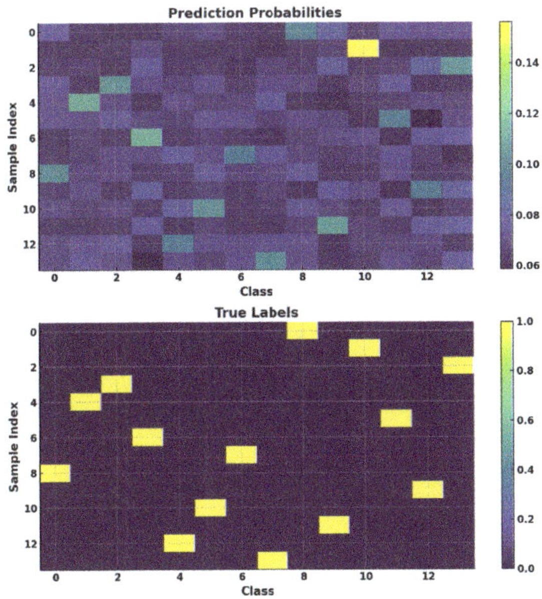

Figure 4. RC classification results of interference waveforms.

To expand the dataset, we conducted 50 measurement cycles for each of the 13 different distance values, resulting in a total of 650 signal samples. Each of the 13 distance values was assigned a unique label, ranging from 1 to 13, to facilitate the classification task. For model training and testing, 80% of the dataset was allocated to training, while the remaining 20% was reserved for evaluating the classification accuracy of the RC system. The final test results, shown in Figure 5 via a confusion matrix, illustrate the system's classification performance across various distances. The results reveal that the RC system effectively classified TS-OCT interference waveforms, achieving a 100% accuracy. This high level of performance demonstrates the robustness and precision of the RC model in depth measurement tasks, even under conditions involving multiple depth classifications. Such accuracy underscores the potential of RC-based approaches for real-time, high-resolution imaging in TS-OCT applications.

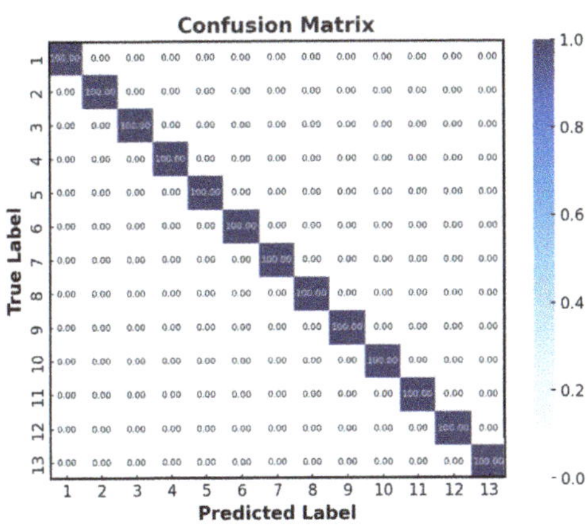

Figure 5. Identification results from RC.

3.3. Waveform Classification Based on RC Model

To further evaluate the classification performance of the RC model on the captured interference signals, we conducted three sets of experiments. In each set, signals corresponding to 13 different depths were collected, and 4 periods of signals were selected within a 1 μs time window, resulting in a total of 156 samples used for training. To ensure robustness, the test set consisted of 52 periods from the 13 depths in a random order. These 52 periods were not selected from the training set but were instead other periods of the required length, randomly arranged to validate the reliability of the algorithm.

We conducted three experiments, each using different training and testing sets. We selected one of the classification results as shown in Figure 6. In each figure, the left half displays the test results, while the right half represents real labels at different distances, with different colors indicating different depths.

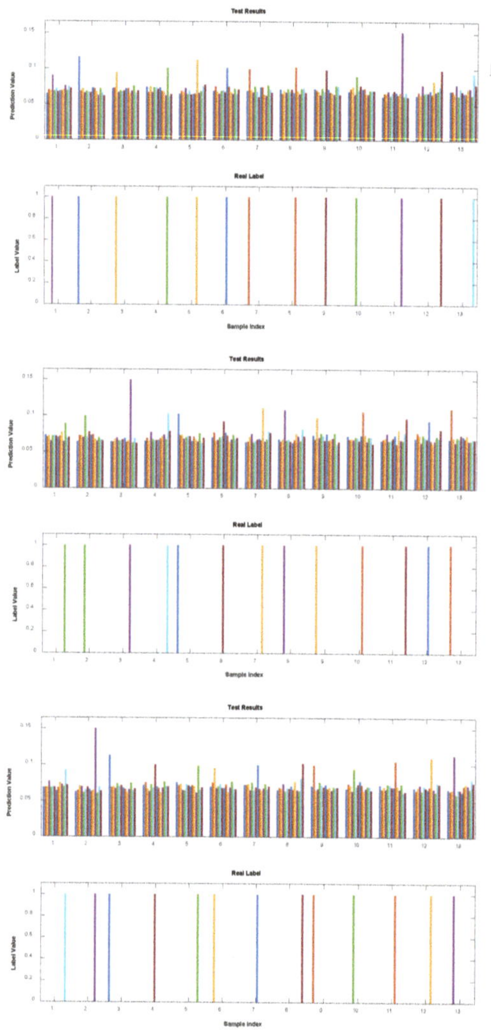

Figure 6. Classification results of the RC model for interference waveforms corresponding to 13 different depths from three sets of experiments. Different colors represent waveforms corresponding to different depths.

The experimental results demonstrated that the proposed RC model effectively identifies the interference waveforms corresponding to different target depths, achieving an overall accuracy of 100%. Furthermore, we assessed the classification performance of the RC model by calculating the root mean square error (RMSE) of the experimental results and comparing it with the RMSE of each training set. The results are summarized in Table 2 below. The average RMSE of the testing set is 0.2416, which indicates the strong classification capability of the RC model in accurately distinguishing interference signals from different depths.

Table 2. The RMSE of classification results.

	Experiment 1	Experiment 2	Experiment 3
RMSE for Testing Sets	0.241178	0.242061	0.241625
RMSE for Training Sets	0.239507	0.239518	0.239424

3.4. Classify RC Using Different Datasets

In the context of temporal signal waveform classification using reservoir computing (RC), we adopt a methodology that involves organizing labels in both sequential and randomized configurations, as shown in Figure 7a,b.

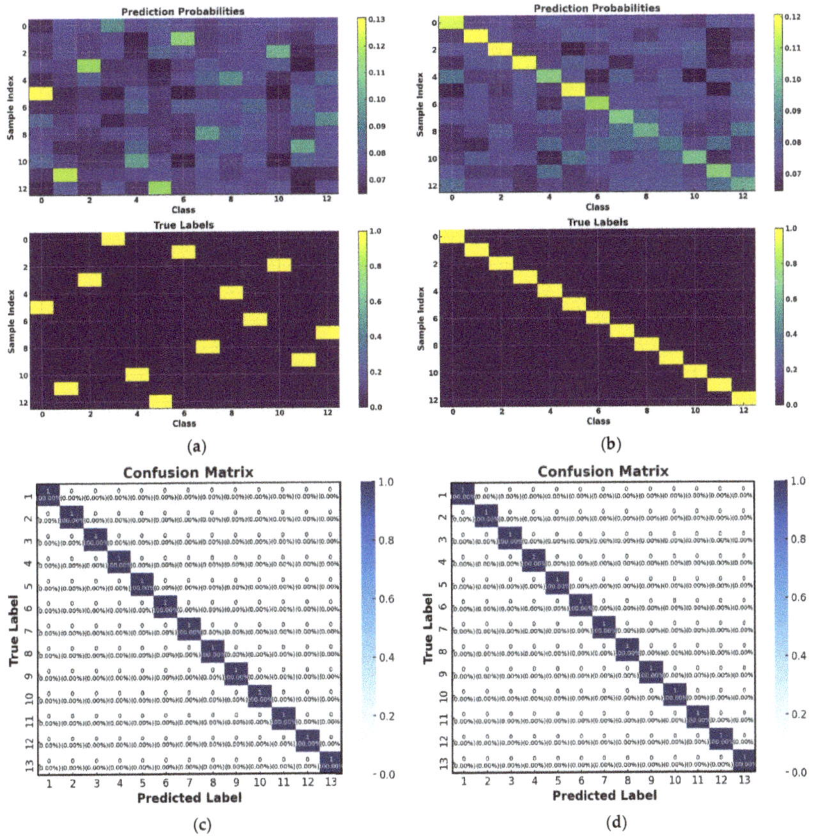

Figure 7. (**a**,**b**) are RC classification results of interference waveforms. (**c**,**d**) are identification results from RC via confusion matrix.

This approach allows for a more granular evaluation of the model's sensitivity to temporal dependencies. A substantial performance improvement on sequentially ordered data relative to randomized data would suggest a strong dependence on temporal structures. In contrast, if the model maintains comparable performance on randomized data, it indicates a higher degree of independence in capturing global or local features beyond temporal sequencing.

Figure 7c,d illustrate two distinct representations of the confusion matrix for the same set of experimental results. In Figure 7c, the class labels are arranged sequentially, while in Figure 7d, the label order is randomized. Under the sequential arrangement of labels in Figure 7c, the deep blue regions along the diagonal of the confusion matrix signify that the model correctly classifies all samples, indicating perfect alignment between predicted and true values for each class, thereby achieving 100% classification accuracy. In contrast, when the label order is randomized, as shown in Figure 7d, the overall structure of the confusion matrix change s, but the model's predictive performance remains consistent, demonstrating its robustness to variations in label ordering.

Furthermore, incorporating both sequential and randomized data enables the model to explore a broader feature space, thereby mitigating the risk of the model being overly reliant on a singular pattern, such as strict temporal dependencies. This diversification fosters a more robust generalization capability for the model when applied to unseen data.

Consequently, in real-world scenarios where the input time-domain waveforms may contain noise, exhibit missing segments, or experience perturbations in sequence, the model can still maintain a high level of classification accuracy.

Table 3 presents the RMSE of the model on both the training and test sets under different data arrangements: sequential and randomized. The table is divided into two columns, representing the RMSE results for sequential (Sequence) and randomized (Random) data arrangements. Although the RMSE for the sequential arrangement is slightly lower than that for the randomized arrangement, the difference is minimal, indicating that the data arrangement order has an insignificant impact on the model's performance on the test set.

Table 3. The RMSE of sequential and randomized classification results.

	Sequence	**Random**
RMSE for Testing Sets	0.257623	0.258119
RMSE for Training Sets	0.249171	0.249102

The RC model exhibits strong robustness and memory capacity when processing temporal data, resulting in low sensitivity to the arrangement order of the input data. The RMSE values for both sequential and randomized arrangements remain at comparable levels, demonstrating that the RC model can effectively capture the features of the input data without being substantially influenced by the order of data presentation.

According to the results in the table, the RC model maintains stable performance when handling both sequentially ordered and randomly ordered data. This stability arises from the unique structure of the reservoir network, which projects the dynamic characteristics of the input data into a high-dimensional space. As a result, the model can learn the intrinsic patterns of the data under varying input sequences, ensuring robust performance regardless of the order of the data arrangement.

3.5. *Effect of Noise on the Robustness of the RC System*

In TS-OCT systems, noise significantly impacts the robustness of classification models. Common noise sources include shot noise, detector noise, excess photon noise, and relative intensity noise (RIN). Shot noise arises from the discrete nature of photon or electron arrival events, which follow Poisson distribution. However, at higher light intensities, the Poisson distribution can be approximated by a normal distribution, making shot noise resemble Gaussian noise under such conditions. The root mean square (RMS) value of shot noise can be expressed as $i_{shot} = \sqrt{2qI\Delta f}$, where q represents the electron charge, I is the photocurrent, and Δf is the measurement bandwidth. By introducing shot noise with varying intensities into the input signals, the model's performance under low-light conditions can be systematically evaluated.

Detector noise includes thermal noise, dark current noise, and readout noise. Thermal noise, caused by the thermal motion of electrons within the detector, has an RMS value given by $V_{thermal} = \sqrt{4k_BTR\Delta f}$, where k_B is the Boltzmann constant, T is the temperature, and R is the resistance. Dark current noise arises from the intrinsic current fluctuations present in the detector even in the absence of light, while readout noise originates from random fluctuations within the electronic readout circuitry. These noise sources are particularly impactful under low-light conditions, where their relative contributions are more pronounced. Simulating these different types of detector noise allows for a detailed assessment of their effects on the overall system noise and classification accuracy.

Excess photon noise predominantly occurs in coherent light sources, originating from mode competition or gain fluctuations within the laser cavity. This noise manifests as irregular intensity fluctuations, which can significantly interfere with the feature extraction process of RC models, especially when analyzing weakly reflective samples or deep tissue layers. By generating intensity signals with random amplitude and phase variations, the characteristics of excess photon noise can be effectively modeled, facilitating the evaluation of its impact on classification performance.

RIN results from random intensity fluctuations in the laser output, typically described by the power spectral density formula $RIN = \frac{\Delta I^2}{\langle I \rangle^2 \Delta f}$, where ΔI denotes intensity fluctuations and $\langle I \rangle$ is the mean intensity. By adding RIN of varying intensities and frequency distributions to the signals, the model's performance under dynamic intensity variations can be assessed. While the classification performance remains stable at low noise levels, it progressively deteriorates as noise intensity increases. At high noise levels, accuracy may approach that of random guessing, indicating the limits of the model's robustness.

Overall, these noise sources can be approximated as Gaussian noise under certain conditions. This approximation is based on the Central Limit Theorem (CLT) in statistical theory, which states that when a random variable is the sum of many independent and identically distributed random variables, its distribution approaches a normal distribution, regardless of the original distribution of each variable. By systematically introducing and evaluating Gaussian noise, the robustness limits of RC models can be thoroughly analyzed, providing valuable insights and data support for improving the noise resilience and performance optimization of the models.

To evaluate the performance of the RC model under different noise levels, Gaussian noise of varying amplitudes was introduced into the test dataset. The addition of Gaussian noise simulates real-world stochastic interference, allowing an assessment of the RC model's stability and accuracy in processing uncertain and noisy signals.

As illustrated in Figure 8, the classification accuracy of the RC model was tested under various Gaussian noise amplitudes, quantified as the signal-to-noise ratio (SNR) in decibels (dB). The x-axis represents the SNR, ranging from 20 dB to 4 dB, while the y-axis indicates the classification accuracy in percentage. The results demonstrate a progressive decline in accuracy as the SNR decreases (i.e., as noise amplitude increases), underscoring the non-negligible impact of noise on signal quality and its substantial influence on model performance under lower SNR conditions.

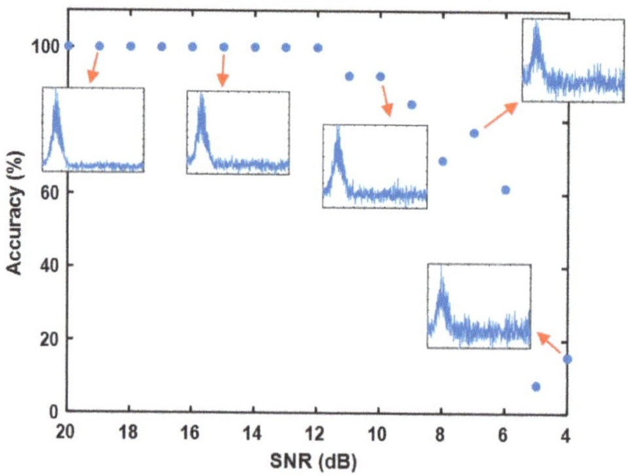

Figure 8. Classification accuracy as shown in blue dots vs. SNR of temporal signals with spectral insights at different noise levels.

The inset plots in Figure 8 highlight the spectral characteristics of the signals under different SNR levels, providing an intuitive understanding of the impact of noise. At high SNR values (e.g., 19 dB and 15 dB), the signal peaks are clearly discernible in the spectrum, with minimal noise interference. Under these conditions, the RC model effectively captures the signal features, achieving classification accuracies close to 100%. This observation indicates that the dynamic architecture of the RC model enables robust feature extraction from time-series signals under low-noise conditions, maintaining high classification performance.

However, as the SNR decreases to 11 dB and 7 dB, the spectral noise components become increasingly pronounced, making the signal features less distinguishable. At an SNR of 8 dB, the interference from noise on the primary signal peaks becomes apparent, leading to a noticeable drop in classification accuracy. This suggests that under moderate noise levels, the robustness of the RC model is challenged, as it struggles to extract sufficient distinguishing information from signals submerged in noise.

When the SNR is further reduced to 4 dB, the spectral noise nearly obscures the signal features entirely, resulting in a sharp decline in classification accuracy. At an SNR of 4 dB, the accuracy approaches zero, indicating that severe noise disrupts the signal structure to such an extent that the RC model fails to effectively recognize and classify the signals. This finding highlights the limitations of the RC model in handling high-noise environments, where its dynamic architecture becomes ineffective in preserving classification capabilities under severe signal distortion. Reducing the SNR below 4 dB is unnecessary, as the results already show that the RC model's classification accuracy drops to near zero at this level. At 4 dB, noise completely dominates the signal, making further reductions redundant. If the system operates at even lower SNRs, pre-processing techniques such as using a balanced detector and applying filtering should be employed first to enhance the signal quality before classification.

In summary, the RC model exhibits a marked disparity in robustness when operating under varying noise levels. At high SNR values (greater than 15 dB), the model's classification performance is virtually unaffected by noise, maintaining near-perfect accuracy due to its ability to efficiently extract relevant features from time-series data. Conversely, when the SNR falls below 11 dB, noise significantly impacts the signal feature extraction process. This effect becomes particularly severe at SNR values below 7 dB, where the model's performance deteriorates sharply, reflecting its inherent limitations under high-noise conditions. These findings provide valuable insights into the RC model's noise tolerance and highlight the need for potential enhancements to improve its resilience in noisy environments.

3.6. Maximum Resolution and Detection Depth of the RC-Based Fourier-Free System

The maximum resolution of the Fourier-free TS-OCT system is determined by the classification capability of the RC model when processing interference signals with minimal axial distance difference. The core principle of the RC model lies in projecting time-domain signals into a high-dimensional dynamic state space and mapping the signal features using fixed nonlinear activation functions. During this process, the RC state space amplifies differences in the input signals, enabling effective processing and classification of complex nonlinear signals.

However, when the depth axial distance difference is minimal, the interference signal waveforms in the time domain become very similar, leading to highly overlapping mapping results in the RC state space. This feature space overlap reduces the distinguishability between state characteristics, thus limiting the RC model's classification capability for these signals.

Specifically, the classification capability of the RC model relies on the dynamic response features of the input signals in the state space. When the differences between input signals are insufficient to induce significant state changes, the RC output weight matrix cannot effectively classify the signals, resulting in classification errors. Therefore, the minimum resolution of the Fourier-transform-free TS-OCT system can be defined as the smallest depth difference that the RC model can reliably distinguish, i.e., the minimum input signal variation that can still form discernible features in the state space.

Numerical simulations were conducted to investigate the effect of interference signals at different depths on fringe density. Specifically, we performed a simulated analysis of interference signals for four different axial distance differences: 1 mm, 0.1 mm, 0.01 mm, and 1 μm, with the results shown in Figure 9. The simulation results indicate that as the axial distance difference decreases, the variation in fringe density of the interference signals becomes progressively weaker, imposing higher demands on the classification capability of the RC model.

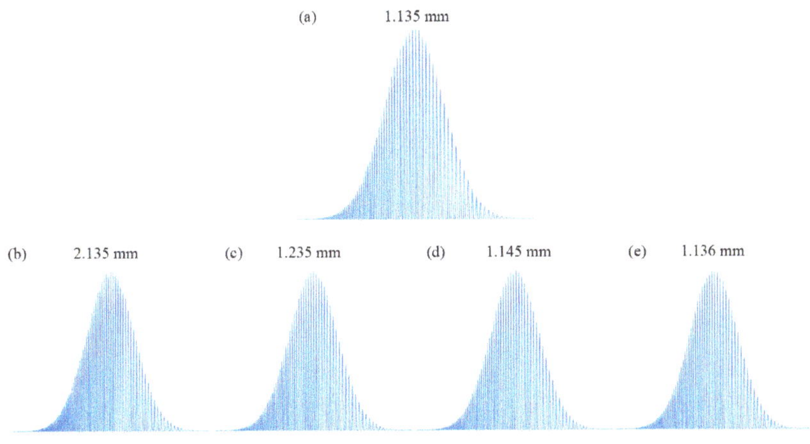

Figure 9. Simulated interference signal patterns corresponding to different axial depths: (**a**) 1.135 mm, showing a standard interference fringe pattern; (**b**) 2.135 mm, with a distinct change in fringe density; (**c**) 1.235 mm, demonstrating slightly denser fringes; (**d**) 1.145 mm, with minimal variation from the standard; and (**e**) 1.136 mm, highlighting the challenge of resolving closely spaced depths.

To evaluate the RC model's classification performance at different scales, we used simulated signals as inputs and conducted classification experiments. At scales of 1 mm, 0.1 mm, and 0.01 mm, the RC model successfully identified signals from different depths with 100% classification accuracy. However, at the ultra-small scale of 1 μm, the classification accuracy of the RC model dropped significantly to only 5.71%. This demonstrates that when the distance difference approaches the system's resolution limit, the fringe density variation of the interference signals becomes insufficient to form separable features in the high-dimensional dynamic state space of the RC model, resulting in a substantial decline in classification performance.

To ensure the statistical reliability of the experimental results, we expanded the training and testing datasets by simulating 70 different distances at each scale. These signals were used for RC model training and testing, and the classification accuracy results are summarized in Table 4.

Table 4. Classification accuracy of the RC model at different axial distance scales.

Axial distance difference	1 mm	0.1 mm	0.01 mm	1 μm
Classification accuracy (%)	100.00	100.00	100.00	5.71

In OCT systems, the Fourier transform is widely used to convert frequency-domain data into depth-domain images. Therefore, the maximum detection depth of the system is fundamentally determined by the bandwidth of the PD and the oscilloscope. For this system, the oscilloscope's bandwidth is 23 GHz, making it the primary limiting factor for the maximum detection depth.

Due to the presence of nonlinear dispersion effects in the system, the relationship between frequency and depth cannot be directly calculated. Instead, the Time Warp Theory is employed to accurately measure the connection between time and frequency, enabling the determination of the actual detection depth.

The relationship between the pulse signal's central wavelength and its time position is shown in Figure 10. By fitting the $t(\lambda)$ curve using a parabolic model, the following expression is obtained:

$$t(\lambda) = t(\lambda_0) + LD(\lambda_0)(\lambda - \lambda_0) + LD'(\lambda - \lambda_0)^2 + LD''(\lambda - \lambda_0)^3 \quad (4)$$

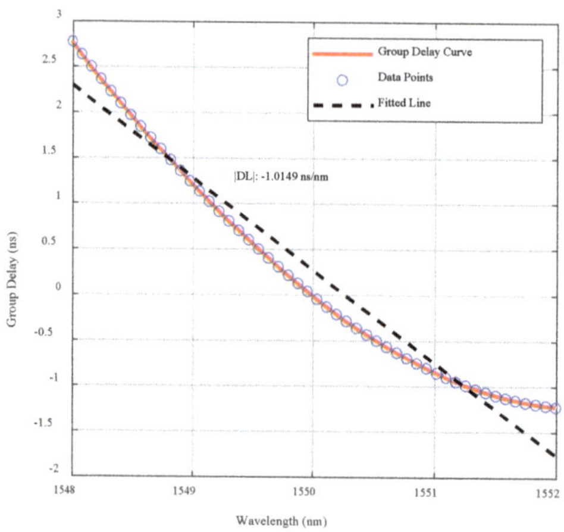

Figure 10. Group delay curve and fitted line for dispersion measurement.

The formula for $t(\lambda)$ is a Taylor expansion around the reference wavelength λ_0, where $t(\lambda)$ represents a physical parameter (e.g., time delay) as a function of wavelength. The first term, $LD(\lambda_0)(\lambda - \lambda_0)$, indicates the value of this parameter at the reference wavelength λ_0. The second term, $LD'(\lambda - \lambda_0)^2$, describes the linear variation when λ deviates from λ_0, while the third term, $LD''(\lambda - \lambda_0)^3$, reflects the nonlinear curvature near λ_0.

The relationship between the frequency shift Δf of the Fourier transform peak and the actual depth is given by

$$\Delta f = 1/\Delta T \tag{5}$$

where ΔT is the pulse duration, determined by the spectral range $\Delta\lambda$, system dispersion $D(\lambda)$, and fiber length L, as follows:

$$\Delta T = D(\lambda)L\Delta\lambda \tag{6}$$

This formula indicates that pulse broadening depends not only on the intrinsic properties of the dispersive medium but also on the spectral range and the medium's length. By experimentally measuring ΔT and $\Delta\lambda$, the system's Free Spectral Range (FSR) can be calculated, and from the formula

$$\Delta\lambda = \lambda/2nl, \tag{7}$$

the Optical Path Difference (OPD), or the system's maximum imaging depth l, is determined. Here, λ is the center wavelength, n is the refractive index, and l is the OPD. A quadratic fit of the dispersion curve provides the linear relation between wavelength and time, with $|DL| = -1.015$ ns/nm. Using the oscilloscope's maximum bandwidth and the above formulas, the system's maximum detection depth is calculated as l = 27.14 mm.

Experimental results show that within a depth difference range of 1 mm to 0.01 mm, the RC model exhibits excellent classification capability, achieving 100% accuracy in distinguishing interference signals at different depths. However, when the depth difference is reduced to the 1 μm level, the RC model's classification performance significantly declines, with accuracy dropping below 10%. This indicates that under ultra-small depth differences, the feature differences between interference signals are too subtle to form separable features in the high-dimensional dynamic state space of the RC model, leading to reduced classification accuracy.

These findings validate the resolution limit of the RC model in the Fourier-transform-free TS-OCT system, defined as the minimum depth difference the RC can reliably distinguish. Furthermore, based on the system's dispersion properties and Fourier transform relationships, the maximum detection depth of this Fourier-transform-free TS-OCT system is determined to be 27.14 mm, governed by the spectral range, dispersive medium properties, and oscilloscope bandwidth.

4. Conclusions

In this paper, we propose and demonstrate an innovative Fourier-free TS-OCT system that leverages RC for temporal signal processing. This system effectively addresses the primary challenges faced by traditional Fourier-based OCT methods, including signal distortion and reduced depth resolution caused by nonlinear frequency sweeping. By directly analyzing time-domain signals and utilizing the dynamic temporal processing capability of RC, the system achieves high-resolution depth measurement without relying on Fourier transform techniques.

Experimental results show that the RC-based system can accurately classify interference waveforms corresponding to different sample depths, achieving a classification accuracy of 100%. The system is capable of distinguishing depth variations at a scale as small as 0.01 mm, with a low RMSE. Compared with traditional Fourier-based OCT systems, the RC-based approach simplifies the data processing pipeline, reduces computational complexity, and eliminates the need for complex linearization and chirp compensation processes. This not only improves system efficiency but also enhances real-time signal processing capabilities, making it particularly suitable for high-speed imaging applications.

The system also demonstrates strong robustness and adaptability to variations in input data, ensuring stable performance across different datasets. Further robustness analysis was conducted by introducing varying levels of Gaussian noise to the test data. The results demonstrated that the RC model maintains high classification accuracy at higher SNR levels, with gradual performance degradation observed only at extremely low SNRs. This highlights the system's ability to operate reliably in noisy environments, further extending its practical applicability. This capability makes the system well suited for applications in complex and dynamic environments where traditional methods may struggle. Furthermore, the relatively low computational requirements of the RC model enable effective handling of large datasets, positioning it as a promising alternative for real-time imaging and signal classification in biomedical and industrial environments. Moreover, experimental results show that in the range of axial depth differences from 1 mm to 0.01 mm, the RC model exhibits excellent classification capability, achieving 100% accuracy in distinguishing interference signals at these scales. However, as the depth difference reduces to the scale of 1 μm, the RC model's classification performance drops significantly, with accuracy falling below 10%. This limitation is due to the insufficient distinction between interference signal patterns at extremely small depth differences, reflecting the resolution limit of the RC model. At the same time, the maximum detectable depth of the system, calculated using dispersion characteristics and the oscilloscope bandwidth, is 27.14 mm. This result highlights the robustness of the RC model over a wide depth range and its constraints in resolving ultra-small depth differences.

Future work can explore optimizing the RC model architecture and training procedures to further enhance performance. Additionally, integrating this approach with other photonic signal processing techniques could lead to the development of more advanced optical imaging systems. Potential applications include non-invasive medical diagnostics, high-speed industrial inspection, and environmental monitoring, where precise imaging is critical.

Author Contributions: Conceptualization, C.W. and W.L.; methodology, W.L. and C.W.; software, W.L. and T.L.; validation, W.L., T.L., and Y.Y.; writing—original draft preparation, W.L.; writing—review and editing, C.W. and W.L.; supervision, C.W. All authors have read and agreed to the published version of the manuscript.

Funding: This research was partially funded by EPSRC under Grant EP/S005625/1.

Institutional Review Board Statement: Not applicable.

Informed Consent Statement: Not applicable.

Data Availability Statement: Data are contained within the article.

Conflicts of Interest: The authors declare no conflicts of interest. The funders had no role in the design of the study; in the collection, analyses, or interpretation of data; in the writing of the manuscript; or in the decision to publish the results.

References

1. Huang, D.; Swanson, E.A.; Lin, C.; Schuman, J.S.; Stinson, W.G.; Chang, W.; Hee, M.R.; Flotte, T.; Gregory, K.; Puliafito, C.A.; et al. Optical coherence tomography. *Science* **1991**, *254*, 1178–1181. [CrossRef]
2. Drexler, W.; Fujimoto, J.G. *Optical Coherence Tomography: Technology and Applications*; Springer Science & Business Media: Berlin/Heidelberg, Germany, 2008.

3. Fercher, A.F.; Drexler, W.; Hitzenberger, C.K.; Lasser, T. Optical coherence tomography—Principles and applications. *Rep. Prog. Phys.* **2003**, *66*, 239–303. [CrossRef]
4. Fujimoto, J.G.; Swanson, E.A. The development, commercialization, and impact of optical coherence tomography. *Investig. Ophthalmol. Vis. Sci.* **2016**, *57*, OCT1–OCT13. [CrossRef] [PubMed]
5. Swanson, E.A.; Huang, D.; Fujimoto, J.G.; Puliafito, C.A. Method and Apparatus for Optical Imaging with Means for Controlling the Longitudinal Range of the Sample. U.S. Patent 5321501, 14 June 1994.
6. SChinn, R.; Swanson, E.A.; Fujimoto, J.G. Optical coherence tomography using a frequency-tunable optical source. *Opt. Lett.* **1997**, *22*, 340–342.
7. Fercher, A.F.; Hitzenberger, C.K.; Kamp, G.; El-Zaiat, S.Y. Measurement of intraocular distances by backscattering spectral interferometry. *Opt. Commun.* **1995**, *117*, 43–48. [CrossRef]
8. Lexer, F.; Hitzenberger, C.K.; Fercher, A.F.; Kulhavy, M. Wavelength-tuning interferometry of intraocular distances. *Appl. Opt.* **1997**, *36*, 6548–6553. [CrossRef]
9. Potsaid, B.; Huang, Y.; Choi, J.S.; Jayaraman, A.; Fujimoto, J.G.; Fermann, M.E. Ultrahigh speed 1050 nm swept source/Fourier domain OCT retinal and anterior segment imaging at 100,000 to 400,000 axial scans per second. *Opt. Express* **2010**, *18*, 20029–20048. [CrossRef] [PubMed]
10. Yun, S.H.; Tearney, G.J.; de Boer, J.F.; Iftimia, N.; Bouma, B.E. High-speed optical frequency-domain imaging using frequency-swept lasers. *Opt. Express* **2003**, *11*, 2953–2963. [CrossRef]
11. Fechtig, D.J.; Grajciar, B.; Schmoll, T.; Blatter, C.; Werkmeister, R.M.; Drexler, W.; Leitgeb, R.A. Line-field parallel swept source MHz OCT for structural and functional retinal imaging. *Biomed. Opt. Express* **2015**, *6*, 716–735. [CrossRef]
12. Wieser, W.; Eigenwillig, C.M.; Klein, T.; Leitgeb, R.A. Megahertz OCT for ultrawide-field retinal imaging with a 1050 nm Fourier domain mode-locked laser. *Opt. Express* **2010**, *18*, 14685–14704. [CrossRef]
13. Potsaid, B.; Huang, Y.; Choi, J.S.; Jayaraman, M.; Fujimoto, J.G.; Cable, A.E. MEMS-VCSELs for swept source optical coherence tomography at 1060 nm. *Opt. Express* **2011**, *21*, 695–700.
14. Mahjoubfar, A.; Churkin, D.; Barland, D.S.; Broderick, N.; Turitsyn, S.K.; Jalali, B. Time stretch and its applications. *Nat. Photonics* **2017**, *11*, 341–351. [CrossRef]
15. Goda, K.; Solli, D.R.; Tsia, K.K.; Jalali, B. Theory of amplified dispersive Fourier transformation. *Phys. Rev. A* **2009**, *80*, 043821. [CrossRef]
16. Grelet, S.; Jiménez, A.M.; Engelsholm, R.D.; Montague, P.B.; Podoleanu, A. 40 MHz swept-source optical coherence tomography at 1060 nm using a time-stretch and supercontinuum spectral broadening dynamics. *IEEE Photonics J.* **2022**, *14*, 1–6. [CrossRef]
17. Vergnole, S.; Lévesque, D.; Lamouche, G. Experimental validation of an optimized signal processing method to handle non-linearity in swept-source optical coherence tomography. *Opt. Express* **2010**, *18*, 10446–10461. [CrossRef]
18. Wang, C.; Yao, J. Photonic generation of chirped millimeter-wave pulses based on nonlinear frequency-to-time mapping in a nonlinearly chirped fiber Bragg grating. *IEEE Trans. Microw. Theory Tech.* **2008**, *56*, 542–553. [CrossRef]
19. Iftimia, N.; Brugge, W.; Upton, R.; Bouma, B. Spectroscopic optical coherence tomography for real-time tracking of microbubbles. *Biomed. Opt. Express* **2011**, *2*, 2252–2262.
20. Jiang, S.; Liu, B.; Wang, S. A dispersion compensation method based on resampling of modulated signal for FMCW Lidar. *Sensors* **2021**, *21*, 249. [CrossRef]
21. Barman, C.R.; Todorovic, F.; Lee, S.H.C.; Quigley, H.; Gooding, A.E. Automatic segmentation of optical coherence tomography retinal layers for localization of retinal diseases. *IEEE Trans. Med. Imaging* **2008**, *27*, 1164–1170.
22. Wieser, K.; Ralston, T.S. Real-time time-stretch optical coherence tomography. *Opt. Express* **2007**, *12*, 1–3.
23. Mididoddi, C.K.; Bai, F.; Wang, G.; Liu, J.; Gibson, S.; Wang, C. High-throughput photonic time-stretch optical coherence tomography with data compression. *IEEE Photonics J.* **2017**, *9*, 1–15. [CrossRef]
24. Jaeger, H. The 'echo state' approach to analysing and training recurrent neural networks. *GMD-Forschungszentrum Informationstechnik* **2001**, *148*, 13.
25. Hewamalage, H.; Bergmeir, C.; Bandara, K. Recurrent Neural Networks for Time Series Forecasting: Current status and future directions. *Int. J. Forecast.* **2021**, *37*, 388–427. [CrossRef]
26. Hua, Y.; Zhao, Z.; Li, R.; Chen, X.; Liu, Z.; Zhang, H. Deep Learning with Long Short-Term Memory for Time Series Prediction. *IEEE Commun. Mag.* **2019**, *57*, 114–119. [CrossRef]
27. Inubushi, M.; Yoshimura, K. Reservoir Computing Approaches for Representation and Classification of Multivariate Time Series. *arXiv* **2018**, arXiv:1803.07870.
28. Tanaka, S.; Yamane, T.; Nakajima, K. A Survey on Reservoir Computing and its Interdisciplinary Applications. *IEEE Access* **2023**, *11*, 12345–12358.
29. Luan, T.; Zhang, R.; Yue, Y.; Liu, S.; Zhang, A.; Wang, C. A frequency modulated continuous wave LiDAR system based on Reservoir Computing. In Proceedings of the 2023 International Topical Meeting on Microwave Photonics (MWP), Nanjing, China, 15–18 October 2023.

30. Brigham, E.O. *The Fast Fourier Transform*; Prentice Hall: Englewood Cliffs, NJ, USA, 1974.
31. Daniel, B.; Soriano, M.C.; Van der Sande, G. Photonic reservoir computing. *Gruyter* **2019**, *8*, 19.
32. Vandoorne, K.; Mechet, P.; Van Vaerenbergh, T.; Fiers, M.; Morthier, G.; Verstraeten, D.; Schrauwen, B.; Dambre, J. Experimental demonstration of reservoir computing on a silicon photonics chip. *Nat. Commun.* **2014**, *5*, 3541. [CrossRef]
33. Yue, Y.; Mididoddi, C.K.; Jing, N.; Wang, C. Reservoir computing assisted ultrafast user localization in beam steering optical wireless system. *IEEE Photonics Technol. Lett.* **2021**, *33*, 1030–1033. [CrossRef]
34. Chi, H.; Wang, C.; Yao, J. Photonic generation of wideband chirped microwave waveforms. *IEEE J. Microw.* **2021**, *1*, 787–803. [CrossRef]
35. Lukoševičius, M. A practical guide to applying echo state networks. In *Neural Networks: Tricks of the Trade*; Springer: Berlin/Heidelberg, Germany, 2012; pp. 659–686.

sensors

MDPI

Article

Demodulation of Fibre Bragg Grating Sensors by Using Cumulative Sum as a Preprocessing Method

Sławomir Cięszczyk *, Marek Kida and Patryk Panas

Institute of Electronics and Information Technology, Lublin University of Technology, Nadbystrzycka 38A, 20-618 Lublin, Poland; m.kida@pollub.pl (M.K.); p.panas@pollub.pl (P.P.)
* Correspondence: s.cieszczyk@pollub.pl; Tel.: +48-538-4309

Abstract: Fibre Bragg gratings are one of the most popular sensors with a huge number of applications. Their most important advantage is signal modulation consisting in shifting the spectrum in the wavelength domain. Determining the wavelength shift is the most important issue in precise measurements of various quantities. New demodulation methods are constantly being developed. Many of them have good properties, but they do not gain much polarity. This is partly due to their high complexity and partly to a small improvement in the accuracy of determining the wavelength shift in relation to classical methods. Cumulative preprocessing is a very simple method of spectrum processing with the property of reducing the influence of noise on the result. The method can be used directly or with additional algorithms. In this article, we demonstrate the advantages of this method and the possibilities of combining it with other signal processing methods. We show that this method is much simpler than the spectrum denoising methods and additionally simplifies the next stage of the algorithm, i.e., determining the wavelength shift itself.

Keywords: FBG sensors; demodulation algorithm; wavelength shift

1. Introduction

Fibre Bragg gratings are periodic structures with a significant number of sensor applications. Such gratings have all the positive properties of fibre optics, such as small size or resistance to electromagnetic radiation and harsh environments. The main advantage of periodic structures is the encoding of information in wavelengths. A physical quantity such as temperature or strain affecting the sensor element causes a shift in the optical spectrum without distorting it. The most popular is a grating with a Gaussian shape of the optical spectrum. Depending on the internal structure, the shape of the spectrum will change. To determine this shift, appropriate interrogation systems or general-purpose spectrum analysers are used. In the case of spectrum measurements, the next step is to use algorithmic procedures for the precise determination of the wavelength shift. The basic problem in achieving a good result is the level of the signal-to-noise ratio. In the case of dedicated interrogators, it may also be important to perform spectral measurements with high resolution. Newly proposed FBG demodulation methods use increasingly advanced algorithms, including those that use artificial intelligence [1–3]. However, the problem here is still developing simple and efficient algorithms that would gain recognition among the scientific and engineering community and enter into widespread use.

Research on demodulation algorithms can be divided into three main parts. The first group is searching for better methods of spectrum denoising. The second group is

https://doi.org/10.3390/s25030634

identifying the simplest methods, in particular with low computational complexity. Finally, the third group is identifying the methods with the best metrological parameters regardless of computational complexity. Among the methods that minimise the impact of noise on demodulation, algorithms such as empirical mode decomposition [4], improved wavelet transform [5], wavelet packet decomposition [6] or Dynamic Statistical Threshold Detection algorithm [7,8] are used. In some FBG applications, in addition to noise, spectral distortions may appear, which may additionally affect the quality of determining the wavelength shift depending on the algorithm used [9]. Recently, scientific publications have also discussed issues related to the simultaneous demodulation of several FBGs [10,11]. Among them, there are also two sensors with mutually overlapping spectra [12,13]. For the simultaneous interrogation of many sensors, simple algorithms with low numerical complexity are preferred. Such solutions include special methods of implementing the second-order polynomial [14] with its additional simplification in order to reduce the computation time [15]. Simple and effective algorithms also include the three-point estimator [16].

The theoretical background of the proposed cumulative spectrum processing method will be presented in the following part of the article. Next, a comparison of several versions of the new algorithm will be presented and their evaluation with the commonly used cross-correlation method and Hilbert transform for simulation data. In the next step, the application of the algorithms for experimental data will be compared. Development and specification of the algorithm properties consisted of the following steps:

- generation of data for testing the algorithms,
- development of the basic cumulative sum algorithm with logistic and linear approximation,
- development of reference algorithms: Hilbert transformation and cross-correlation with Hilbert transformation, testing and comparing the developed algorithms with the reference algorithms,
- modification of the algorithms: wavelet filtration, calculation of the square value of the basic spectrum, difference in cumulative spectra calculated from left to right and from right to left,
- testing of the modified algorithms,
- analysis of experimental data.

2. Analysis of Selected Demodulation Algorithms

The Bragg wavelength of gratings depends on the effective refractive index of the core n_{eff} and grating period Λ:

$$\lambda_B = 2 \cdot n_{eff} \cdot \Lambda. \tag{1}$$

The simplest model of the spectrum is the Gaussian shape, where the reflection spectrum can be written as follows:

$$R(\lambda) = R_0 exp\left(-4ln2\left(\frac{\lambda - \lambda_B}{\Delta\lambda_B}\right)^2\right), \tag{2}$$

where R_0 is the peak amplitude and $\Delta\lambda_B$ is its full width at half maximum. The Gaussian wavelength is achieved by using appropriate apodisation techniques, which are designed to reduce the intensity of the side lobes.

The cumulative curve method can be classified as a method that transforms the spectrum before calculating the shift. Such methods include the Hilbert transformation and the differential filter. The Hilbert transformation can be written as follows:

$$\hat{R}(\lambda) = \frac{1}{\pi}\text{p.v.}\int_{-\infty}^{+\infty} R(\lambda)\frac{1}{\lambda - \tau}d\tau. \tag{3}$$

The basic observed change in the spectrum is the transformation of the resonance peak into a curve passing through the zero value. The Hilbert transformation can be preceded by calculating the cross-correlation function between the reference spectrum R_0 and the analysed spectrum R_1:

$$C_j = \sum_{i=0}^{N-1} R_0(\lambda_i) R_1(\lambda_{i+j}), \tag{4}$$

for $j = 0, 1, 2, \ldots, (2N - 1)$. This method is referred to as the zero-crossing algorithm [17]. An additional process can be the use of low-pass FIR filters [18].

Another proposed transformation of the optical spectrum is the determination of the first derivative after the Hilbert transformation. Such an algorithm is used to distinguish several peaks of gratings recorded on one optical fibre [19]. The most popular methods using signal transformation include the fast phase correlation method [20].

The cumulative sum method is difficult to classify directly into one selected group of methods [21,22]. The spectrum transformation itself can be classified as a broadly understood filtration method. This group includes all preliminary denoising methods, the Hilbert transformation and the linear phase operator [23]. However, this operator is a differentiating filter with a smoothing component. Such methods operate in the wavelength domain of the measured spectrum without using parameters after transformation to another (Fourier) domain. In the second step of the algorithm, it is necessary to match the appropriate function to the modified spectrum. After the cumulative filtration, approximation with a linear curve or a logistic function can be used.

The cumulative sum of points of measured spectra can be written as:

$$A(\lambda_i) = \sum_{i=0}^{i} R_0(\lambda_j). \tag{5}$$

From the perspective of digital signal processing, such an operation can be performed using an integrator, whose difference equation can be written as follows:

$$y[n] = \sum_{k=0}^{n} x[k], \tag{6}$$

where $x[k]$ is input signal and $y[n]$ is output signals. Alternatively, integrator with a feedback form, i.e., as a recursive system:

$$y[n] = x[n] + y[n-1]. \tag{7}$$

The impulse response of such a filter is a unit step, and in the frequency domain it is a low-pass filter (Figure 1). This also means that such a filter significantly attenuates higher frequencies and thus, attenuates noise. It should also be noted that this is linear phase filter meaning that it maintains a constant delay for all frequencies.

The cumulative spectrum method with additional preliminary smoothing filtering is used for spectral analysis in UV spectroscopy [24]. However, it is used is to improve the analytical signal and not to determine the spectral shift. After cumulative processing of the spectrum, the resulting curve can be approximated by a logistic curve. This curve is described by three parameters:

$$N(t) = \frac{M}{1 + ce^{-at}}, \tag{8}$$

or in an equivalent form:

$$N(t) = \frac{M}{1 + e^{-r(t-t_0)}}, \tag{9}$$

if $c = \exp(rt_0)$, M—maximum value of the function, r—rate of growth, t_0 is the point with the highest rate of growth of the curve and is located at the half of the maximum height of the function. The graphical representation of the logistic function is shown in Figure 2.

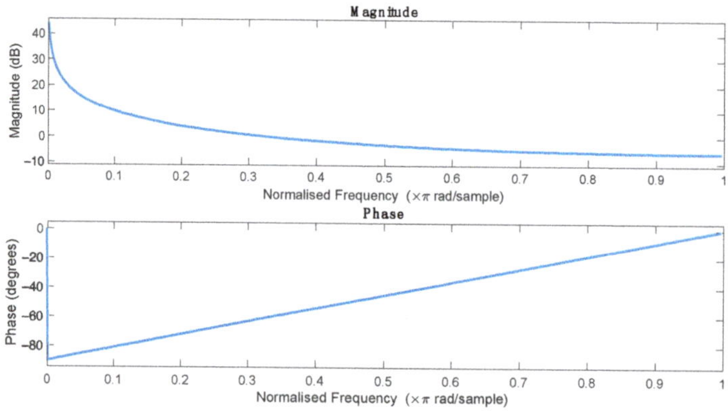

Figure 1. Magnitude and phase response of discrete-time integrators.

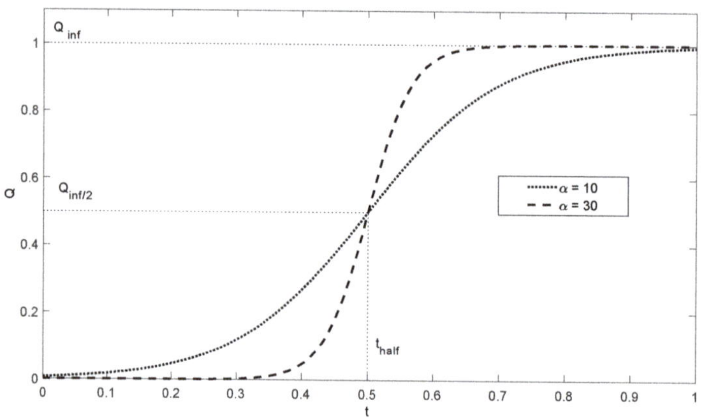

Figure 2. Graphic illustration of model logistic curves.

In a typical application of the logistic curve, we are interested in two of its parameters, namely the growth rate and the t_0 point. In the case of FBG spectra analysis, only the inflexion point is important because the growth rate of the function does not change. Additionally, since the cumulative FBG spectrum is a symmetric and odd function with respect to the inflexion point, the shift in this curve can only be approximated in the range in which its significant growth occurs. Such approximation can also be performed with a linear function. Figure 3 shows five Gaussian shape spectra of a Bragg grating with full width at half maximum of 0.5 nm. Subsequent spectra differ from each other by a shift of 0.24 nm. A spectral resolution of 0.01 nm was assumed. Such spectral parameters allow for explaining the principle of operation of the algorithm consisting in summing subsequent spectrum samples.

The shape of the spectrum after the integrating filter is shown in Figure 4. At the same time, the cumulative curves were approximated using a logistic function. The fit of the cumulative spectra and their logistic approximation is very good in the entire range. In the range of the greatest increase in the value of the cumulative functions, only their shift occurs for individual spectra. In this range of the cumulative spectrum, instead of the logistic function, approximation using a linear function can be used. The determined linear function will shift similarly to the logistic function (Figure 5).

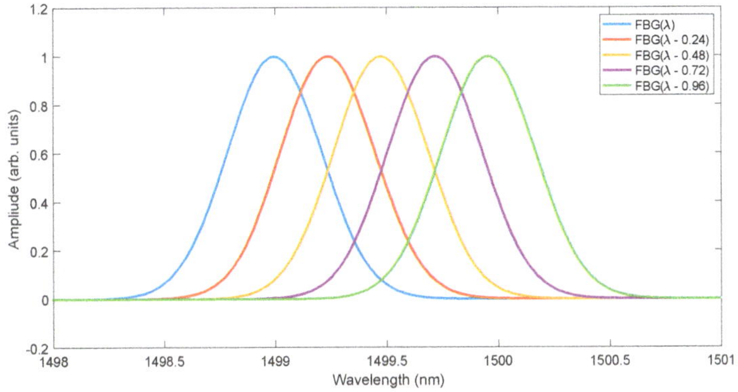

Figure 3. Example FBG spectra with a wavelength shift.

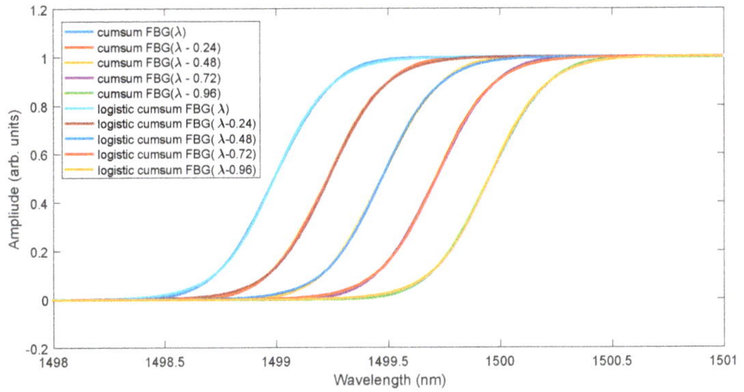

Figure 4. Cumulative sum of FBG spectra calculated from the spectra of Figure 3 together with their approximations using logistic curves.

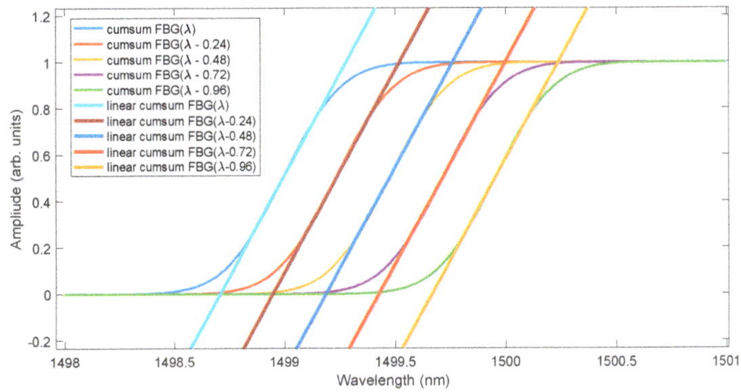

Figure 5. Approximation of slopes of cumulative curves using a linear function.

Figure 6 shows the quality of the fit of the logistic function and the linear function to the cumulative spectrum. For better visualisation, a single FBG spectrum and its cumulative version normalised to unity are shown. The differences between the cumulative spectrum and its approximations are magnified tenfold. The errors for both types of approximations pass through zero at the Bragg wavelength. The maximum approximation error using the logistic function does not exceed 1% in the entire wavelength range. The approximation error using the linear function grows very quickly outside the full width at half maximum of the Gaussian curve.

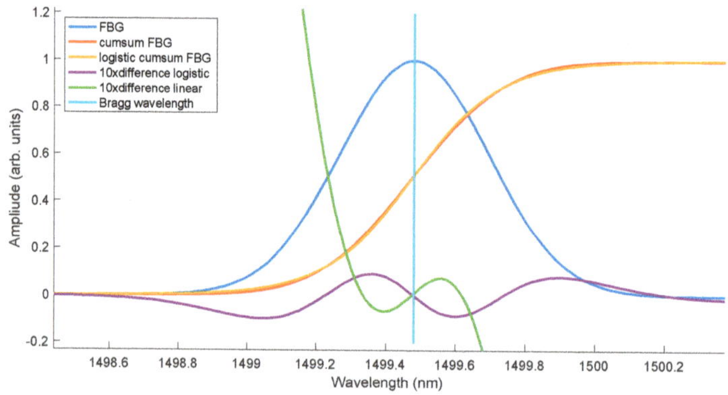

Figure 6. Difference between the cumulative curve and its approximation using a logistic curve and a linear function.

In processing the spectra of Bragg gratings, the Hilbert transform can be used. It can be applied directly to the measured FBG spectrum or, with better results, to the calculated cross-correlation function. A comparison of the Hilbert transform of the modelled FBG spectrum with the scaled values of the cumulative spectrum is shown in Figure 7. The Hilbert transform is a global transform, i.e., each individual value on the wavelength axis is determined using all values of the spectrum subjected to the transformation. In the case of the cumulative sum, the values of all previous points of the processed spectrum are necessary to calculate the value of each point on the wavelength axis. The method of calculating the cumulative curve as an integral filtering has strong properties of attenuating higher frequencies. The effect of signal smoothing can be observed for the FBG spectrum model for 30 dB SNR shown in Figure 8. The noise is visible both for the basic FBG spectrum and for its Hilbert transform. The scaled cumulative curve normalised to the Hilbert transform level is, however, smooth.

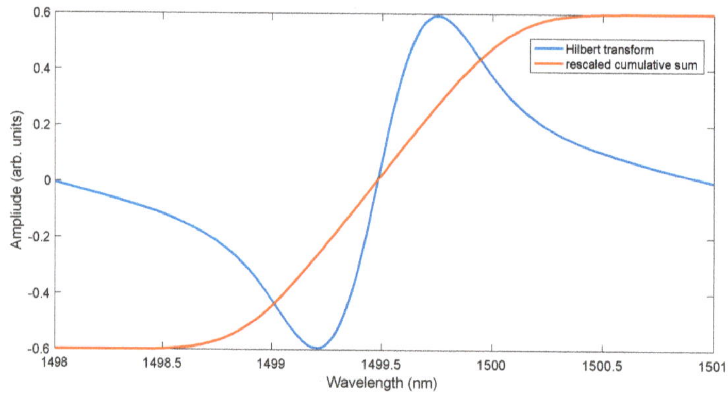

Figure 7. Comparison of the rescaled cumulative curve with the Hilbert transformation of the FBG spectrum.

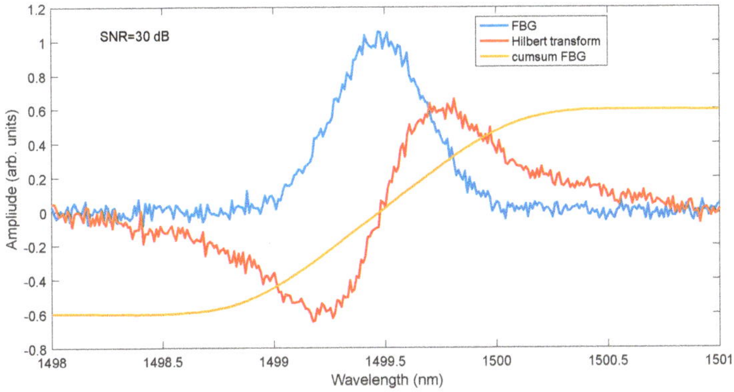

Figure 8. FBG spectrum and comparison of the rescaled cumulative curve with the Hilbert transformation of the FBG spectrum.

3. Properties of the Cumulative Sum Algorithm and Its Modifications

To assess the properties of the proposed method for determining the Bragg wavelength shift, simulations of spectra were performed for a 0.5 nm wide grating model for 1001 Bragg wavelength shift values in the range from 0 to 1 nm. A spectral resolution of 0.01 nm was assumed. Simulations were performed for seven noise levels from 30 to 60 dB. The entire simulation set therefore included 7007 spectra. As a metric for comparing the accuracy of determining the spectral shift for individual algorithms, we used the root mean square error (RMSE) from M simulation spectra:

$$RMSE = \sqrt{\frac{\sum_{m=1}^{M}(\Delta\lambda_t - \Delta\lambda_e)^2}{M}}, \quad (10)$$

where $\Delta\lambda_t$ is the actual (given) wavelength shift and $\Delta\lambda_e$ is the wavelength shift estimated by the specified algorithm.

Figure 9 presents a comparison of the two proposed approximation methods (linear and logistic function) and the Hilbert transformation zero crossing determination method. Figure 10 shows the mean square error value for determining the wavelength shift on a logarithmic scale. The figure shows the two proposed methods, and a three-step method commonly used for shift estimation, which can be considered a reference method. This method consists of performing a Hilbert transformation from the cross-correlation function between the reference spectrum and the analysed spectrum. Then, the zero position is determined from the Hilbert transform using a linear approximation between its minimum and maximum value. The error in determining the shift is smaller for the reference method for smaller SNR values in the range of 30–45 dB. For larger SNR values, the logistic function approximation method is better.

Before the actual analysis of the Bragg wave shift, additional denoising can be applied. One of the better methods in this field is the wavelet transform. Figure 11 shows the noisy spectrum and the denoised spectrum using the wavelet transform. The analysis method using the Symlet wavelet of the sixth order was used. The denoised spectrum contains the sum of the fifth and sixth level of decomposition.

Figure 12 shows that denoising has a small effect on the error of the linear function approximation method and does not change the logistic function approximation error at all. This is due to the good denoising properties of the cumulative sum method itself and the additional averaging, which is the fit of both the logistic and linear models to the cumulative curve. However, a reduction in the mean square error can be achieved by squaring the individual points of spectrum before the cumulative sum calculation.

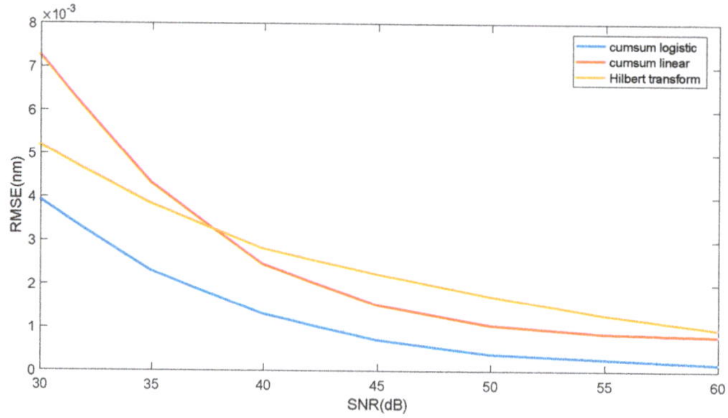

Figure 9. Comparison of the root mean square error for three algorithms: cumulative sum—linear approximation, cumulative sum—logistic approximation, Hilbert transform—zero crossing.

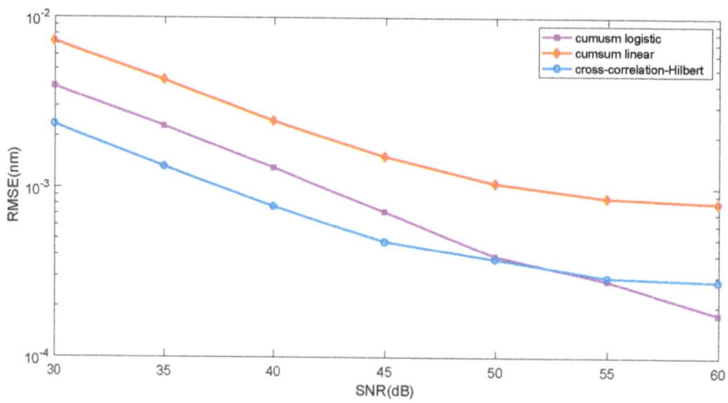

Figure 10. Comparison of root mean squared error for three algorithms: cumulative function—linear approximation, cumulative function—logistic approximation, cross-correlation—Hilbert transform—zero crossing.

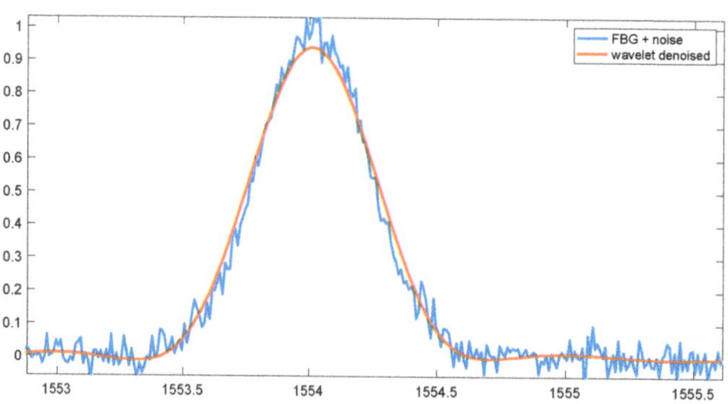

Figure 11. FBG spectrum with 30 dB noise and wavelet-denoised spectrum.

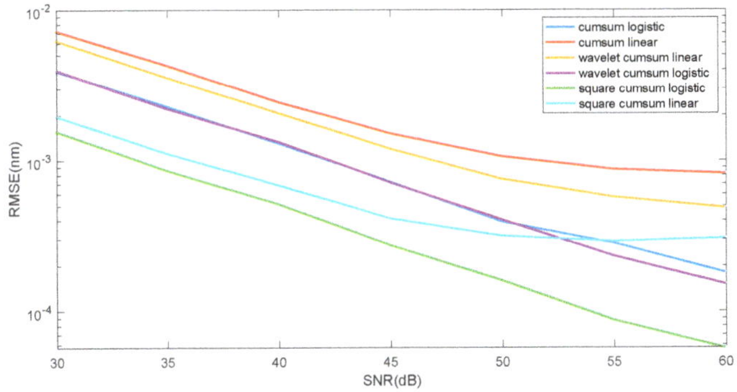

Figure 12. Comparison of the properties of the Bragg wavelength shift determination algorithms in the case of using wavelet filtering and squaring the cumulative signal amplitude.

The cumulative sum method can also be modified in a way other than wavelet filtering. The summation of spectrum samples can also be performed from the side of larger wavelength values. In the next step, subtraction can be performed between the cumulative sum calculated from the left and right side. As a result, the curve has odd symmetry with respect to the point whose coordinates have the value of the Bragg wavelength on the wavelength axis and zero on the amplitude axis (Figure 13). The most important advantage of such a modification of the algorithm is obtaining the mean square error of the linear approximation of the difference curve at the level of the error of the approximation with the logistic function of a single cumulative curve (Figure 14). However, for the difference in cumulative sum, the use of linear and logistic approximation does not change the level of the obtained root mean square error.

The higher resolution of the spectrum measurement means more points with information about the shape of the Bragg grating spectrum and its position. Reducing the resolution causes an increase in the root mean square error of the Bragg wavelength estimation. Figure 15 shows the dependence of the mean square error on the signal-to-noise ratio for five resolution values. The comparison concerns the linear approximation method for the left-side and right-side difference in the cumulative sum.

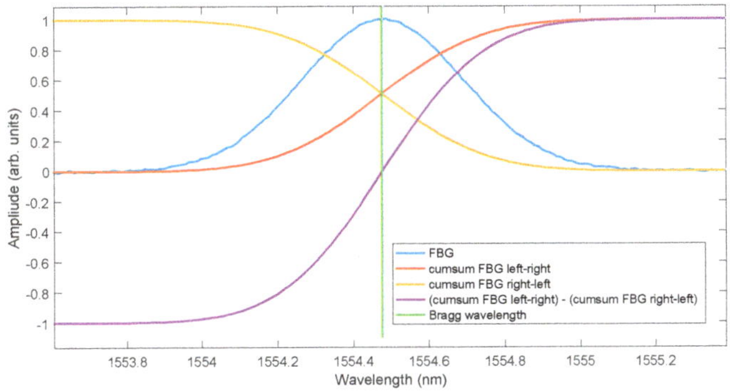

Figure 13. Normalised cumulative spectra calculated from left to right and from right to left.

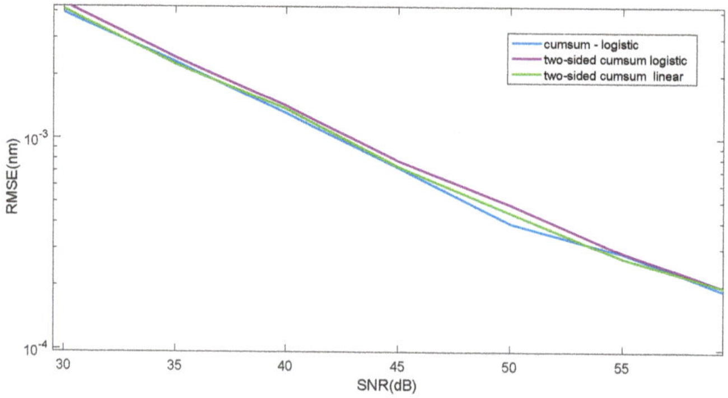

Figure 14. Comparison of the properties of the Bragg wavelength shift determination algorithms using right- and left-sided cumulative sum.

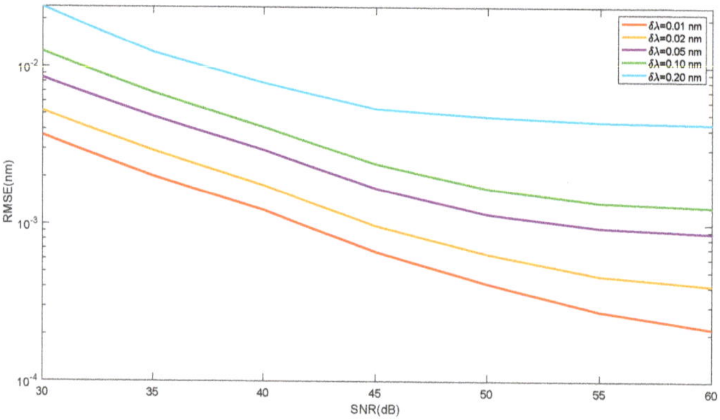

Figure 15. Root mean square error of Bragg wavelength shift determination for different values of spectrum resolution for the left- and right-side cumulative sum subtraction method.

As a result of the FBG manufacturing process, deviations of their spectra from Gaussian-like symmetric curves may occur. Distortions also appear during measurements due to stress inhomogeneity along the grating length. In order to verify the properties of the cumulative sum algorithm for asymmetric spectra, a modification of the generated data were performed. The spectrum model used for the analysis of distorted FBG signals [9] is based on the combination of two Gaussian shapes with different half-widths [25]:

$$R(\lambda) = \begin{cases} R_0 exp\left(-4ln2\left(\frac{\lambda-\lambda_B}{\Delta\lambda_B}\right)^2\right), \lambda < \lambda_B \\ R_0 exp\left(-4ln2\left(\frac{\lambda-\lambda_B}{\chi\cdot\Delta\lambda_B}\right)^2\right), \lambda \geq \lambda_B \end{cases} \tag{11}$$

where χ is responsible for the degree of asymmetry. The shape used to model asymmetric spectra and their shift under the influence of the measured quantity is shown in Figure 16. The figure contains an asymmetric spectrum and two symmetric spectra with FWHM of 0.33 nm and 0.66 nm, respectively. The left half of the asymmetric spectrum has a symmetric spectrum shape with a FWHM of 0.66 nm while the right half has a symmetric spectrum shape with a FWHM of 0.33 nm. The resulting FWHM value of the asymmetric spectrum is 0.5 nm.

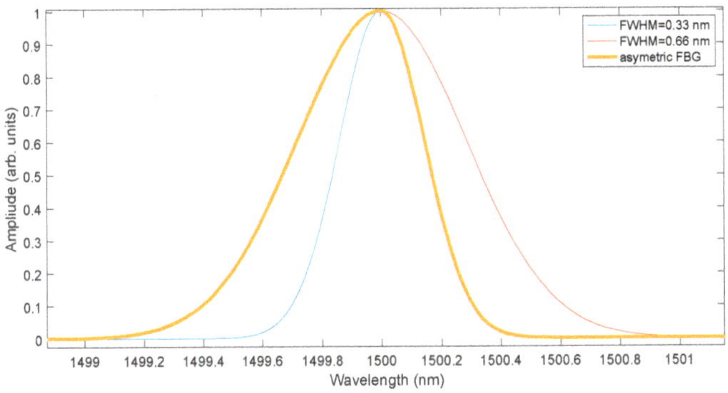

Figure 16. Asymmetric FBG spectrum as a composite of two symmetric spectra, one with twice the width of the other.

It follows from Figure 17 that the distortion from the symmetry of the spectrum does not affect the deterioration of the accuracy of the demodulation of its shift. The smallest mean square error is characterised by the logistic approximation algorithm of the cumulative sum calculated from the square value of the spectrum intensity. This method is better than the algorithm for calculating the Hilbert transform from cross-correlation.

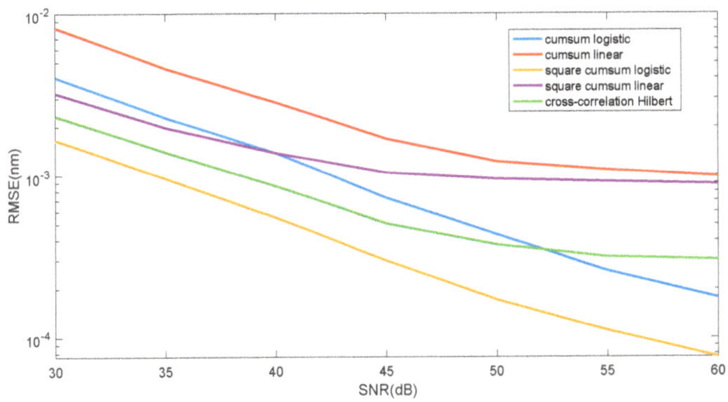

Figure 17. Mean square error in determining the asymmetric shift in the FBG spectrum for five algorithms: cumulative function—linear approximation, cumulative function—logistic approximation, spectrum square—cumulative function—linear approximation, spectrum square—cumulative function—logistic approximation, cross-correlation—Hilbert transform—zero crossing.

In principle, this comparison of methods can only be based on the comparison of RMSE plots for different SNR values. Therefore, the classification of the developed methods together with the reference methods can be presented in the following order:

(a) Hilbert transform—linear approximation,
(b) cumulative sum—linear approximation,
(c) wavelet—cumulative sum—linear approximation,
(d) cumulative sum—logistic approximation,
(e) difference in two-sided cumulative sum—linear approximation,
(f) wavelet—cumulative sum—logistic approximation,
(g) cross-correlation—Hilbert transform—linear approximation,
(h) spectrum squared—cumulative sum—linear approximation,
(i) spectrum squared—cumulative sum—logistic approximation.

4. Experimental Data Analysis

The experiment was conducted using standard laboratory equipment. Measurements were performed in reflective mode with the following components: An Optical Spectrum Analyzer (OSA) AQ6370D (Yokogawa Electric Corporation, Kanagawa, Japan), a Superluminescent Diode S5FC1550P-A2 (SLD) (Thorlabs Inc., Newton, NJ, USA), an Optical Circulator 6015-3-APC (Thorlabs Inc., Newton, NJ, USA), and a Fibre Bragg Grating (FBG) used as a temperature sensor inscribed in the hydrogen loaded single-mode optical fibre using an excimer laser (Coherent Inc., Santa Clara, CA, USA). Light emitted from the SLD source was transmitted through an optical fibre to the first port of the optical circulator and subsequently directed to the second port, which was connected to the FBG sensor placed inside a temperature chamber. The reflection spectra were measured using the OSA, with a resolution of 0.04 nm, connected to the third port of the circulator.

The spectra of the experimental grating for several measured temperatures are shown in Figure 18. The grating has a characteristic spectrum with a significant side lobe amplitude for a shorter wavelength than the main lobe. Figure 19 shows the normalised cumulative spectrum and its approximation by means of a logistic function.

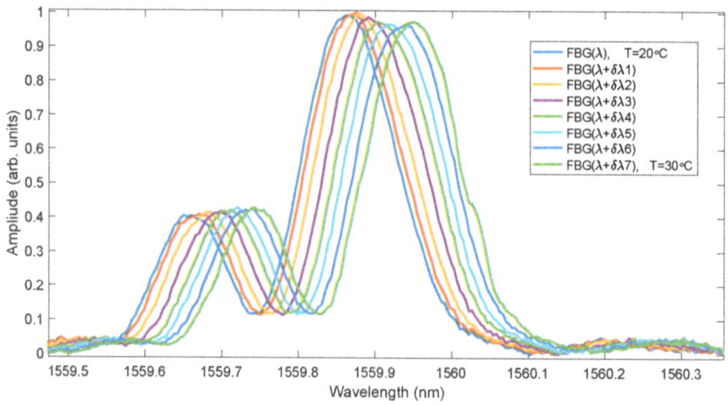

Figure 18. Experimental spectra of the FBG measured during heating in a thermal chamber.

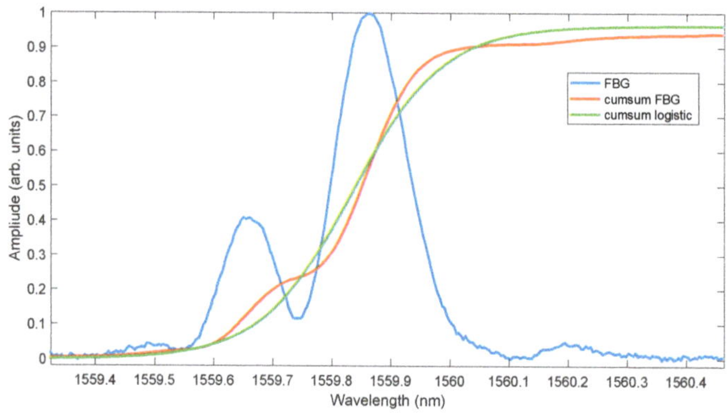

Figure 19. Approximation of the cumulative sum of the experimental spectrum using a logistic curve.

Figure 20 shows the shape of the autocorrelation function of the experimental grating spectrum and the Hilbert transformation of this function. Interestingly, the autocorrelation function is an even function, despite the lack of symmetry of the FBG spectrum itself. The Hilbert transformation is an odd function for which it is possible to determine the zero position. The method combining the cross-correlation functions and the Hilbert transform can be used to compare the Bragg wave shift values with the values calculated using the cumulative sum.

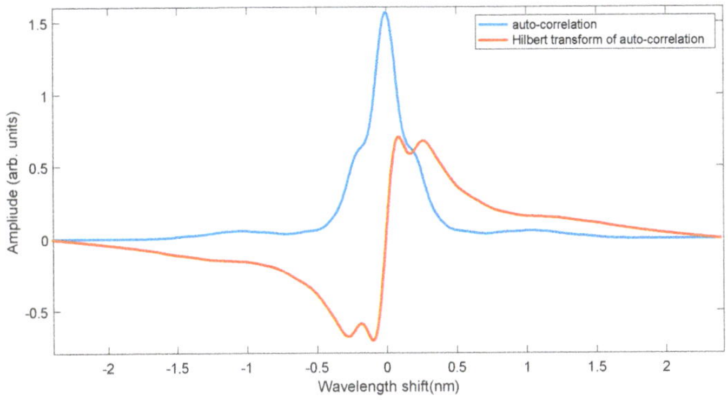

Figure 20. Autocorrelation of the first measured spectrum and its Hilbert transformation.

During the experiment, the grating was placed in a heating chamber at a temperature of 20 °C and then the temperature was set to 30 °C. Figure 21 shows the course of the Bragg wavelength shift from the final moment of heating through the temperature stabilisation phase to the moment of setting the temperature in the heating chamber to 40 °C. The heating chamber maintains the temperature with an accuracy of 1 °C. The spectral resolution of the measurement was 0.04 nm, which can be seen by analysing the course of changes in the Bragg wavelength determined by the maximum value of the spectrum method. Individual changes in the shift value are much smaller for the other two methods. Analysing the courses of the correlation method and the cumulative sum method, it can be seen that there are some impulsive deviations from the smooth trend curve of the temperature change for both methods.

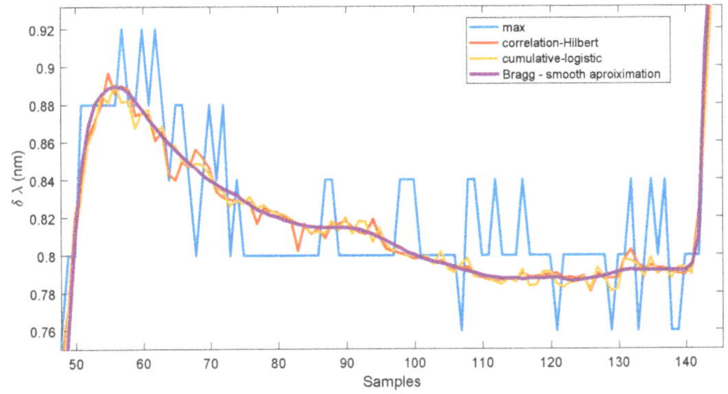

Figure 21. Bragg wavelength shift estimation results for experimental measurements.

5. Conclusions

Using a simple method of summing successive spectral points, a new approach to determining the shift in the Bragg grating spectrum was proposed. The new spectrum transformation together with the linear function approximation algorithms and the logistic function demonstrate good properties for determining the wavelength shift. The results of the proposed method were compared with popular methods such as the cross-correlation algorithm and the Hilbert transform. Additional denoising performed using the wavelet transform does not change the properties of the proposed method. An additional reduction in the root mean square error can be obtained by squaring spectrum before calculating the cumulative sum and then performing the approximation with a logistic or linear curve. By calculating the difference between the cumulative sum calculated from the left and right side, the shape of the odd function with zero crossing can be obtained. This allows achieving the root mean square error for linear approximation at the same level as for the logistic function approximation. The numerical analyses carried out show that the proposed method can be used due to its simplicity and properties comparable to more complex methods.

Author Contributions: S.C. contributed to the idea of this work. S.C., M.K. and P.P. performed data analysis, M.K. and P.P. perform experiments. The content of the article was written by S.C. All authors have read and agreed to the published version of the manuscript.

Funding: This work was supported by the Lublin University of Technology (grant number: FD-20/EE-2/301).

Institutional Review Board Statement: Not applicable.

Informed Consent Statement: Not applicable.

Data Availability Statement: Data are contained within the article.

Conflicts of Interest: The authors declare no conflicts of interest.

References

1. Agliullin, T.; Anfinogentov, V.; Misbakhov, R.; Morozov, O.; Nasybullin, A.; Sakhabutdinov, A.; Valeev, B. Application of Neural Network Algorithms for Central Wavelength Determination of Fiber Optic Sensors. *Appl. Sci.* **2023**, *13*, 5338. [CrossRef]
2. Cao, Z.; Zhang, S.; Xia, T.; Liu, Z.; Li, Z. Spectral demodulation of fiber Bragg grating sensor based on deep convolutional neural networks. *J. Light. Technol.* **2022**, *40*, 4429–4435. [CrossRef]
3. Li, B.; Tan, Z.W.; Shum, P.P.; Wang, C.; Zheng, Y.; Wong, L.J. Dilated convolutional neural networks for fiber Bragg grating signal demodulation. *Opt. Express* **2021**, *29*, 7110–7123. [CrossRef] [PubMed]
4. Shang, Q.; Fan, S.; Qin, W.; Yao, G. Pink noise removal and spectral distortion correction based fiber Bragg grating demodulation algorithm. *Opt. Express* **2022**, *30*, 1066–1080. [CrossRef] [PubMed]
5. Chen, Y.; Chen, L.J.; Liu, H.L.; Wang, K. Research on FBG sensor signal wavelength demodulation based on improved wavelet transform. *Optik* **2013**, *124*, 4802–4804. [CrossRef]
6. Lv, Z.; Wu, Y.; Zhuang, W.; Zhang, X.; Zhu, L. A multi-peak detection algorithm for FBG based on WPD-HT. *Opt. Fiber Technol.* **2022**, *68*, 102805. [CrossRef]
7. Cibira, G.; Glesk, I.; Dubovan, J. SNR-based denoising dynamic statistical threshold detection of FBG spectral peaks. *J. Light. Technol.* **2022**, *41*, 2526–2539. [CrossRef]
8. Cibira, G.; Glesk, I.; Dubovan, J.; Benedikovič, D. Impact of Reducing Statistically Small Population Sampling on Threshold Detection in FBG Optical Sensing. *Sensors* **2024**, *24*, 2285. [CrossRef]
9. Zhang, W.; Lei, X.; Yu, Z.; Shao, B.; Chen, W. Evaluating the effect on demodulation with a comprehensive model of distortions of fiber Bragg grating sensing signals. *J. Light. Technol.* **2018**, *36*, 5823–5832. [CrossRef]
10. Guo, Y.; Yu, C.; Ni, Y.; Wu, H. Accurate demodulation algorithm for multi-peak FBG sensor based on invariant moments retrieval. *Opt. Fiber Technol.* **2020**, *54*, 102129. [CrossRef]
11. Xu, O.; Liu, J.; Tong, X.; Zhang, C.; Deng, C.; Mao, Y.; Fang, D. A multi-peak detection algorithm for Fiber Bragg Grating sensing systems. *Opt. Fiber Technol.* **2020**, *58*, 102311. [CrossRef]
12. Liu, F.; Tong, X.; Zhang, C.; Deng, C.; Xiong, Q.; Zheng, Z.; Wang, P. Multi-peak detection algorithm based on the Hilbert transform for optical FBG sensing. *Opt. Fiber Technol.* **2018**, *45*, 47–52. [CrossRef]
13. Kumar, S.; Sengupta, S. Detection of peak wavelength of multi-FBG using higher-order derivative of wavelets multiresolution analysis and maximum likelihood estimation. *Opt. Commun.* **2023**, *544*, 129621. [CrossRef]
14. Huang, T.; Wang, J.; Duan, F.; Jiang, J.; Fu, X.; Xu, X. A simple algorithm for the implementation of second-order-polynomial based peak-tracking methods. *Opt. Fiber Technol.* **2019**, *47*, 192–196. [CrossRef]
15. Wang, J.; Huang, T.; Duan, F.; Cheng, Q.; Zhang, F.; Qu, X. Fast peak-tracking method for FBG reflection spectrum and nonlinear error compensation. *Opt. Lett.* **2020**, *45*, 451–454. [CrossRef]

16. Sung, J.Y.; Chen, J.K.; Liaw, S.K.; Kishikawa, H. Accurate Peak Detection for Optical Sensing with Reduced Sampling Rate and Calculation Complexity. *Sensors* **2021**, *21*, 2306. [CrossRef] [PubMed]
17. Theodosiou, A.; Komodromos, M.; Kalli, K. Accurate and fast demodulation algorithm for multipeak FBG reflection spectra using a combination of cross correlation and Hilbert transformation. *J. Light. Technol.* **2017**, *35*, 3956–3962. [CrossRef]
18. Theodosiou, A.; Komodromos, M.; Kalli, K. Improvements on the cross-correlation algorithm used for tracking fractional Bragg grating wavelength shifts in multimode fibres. *Opt. Fiber Technol.* **2018**, *46*, 36–42. [CrossRef]
19. Liu, W.; Zhou, W.; Li, H. Demodulation of the overlapping reflection spectrum of serial FBGs based on a weighted differential evolution algorithm. *Opt. Express* **2023**, *31*, 27696–27710. [CrossRef] [PubMed]
20. Lamberti, A.; Vanlanduit, S.; De Pauw, B.; Berghmans, F. A novel fast phase correlation algorithm for peak wavelength detection of fiber Bragg grating sensors. *Opt. Express* **2014**, *22*, 7099–7112. [CrossRef] [PubMed]
21. Tosi, D. Improved KLT Algorithm for High-Precision Wavelength Tracking of Optical Fiber Bragg Grating Sensors. *J. Sens.* **2017**, *2017*, 5412825. [CrossRef]
22. Tosi, D. Review and analysis of peak tracking techniques for fiber Bragg grating sensors. *Sensors* **2017**, *17*, 2368. [CrossRef]
23. Bodendorfer, T.; Muller, M.; Hirth, F.; Koch, A. Comparison of different peak detection algorithms with regards to spectrometric fiber Bragg grating interrogation systems. In Proceedings of the International Symposiumon Optomechatronic Technologies (ISOT), Istanbul, Turkey, 21–23 September 2009.
24. De Luca, M.; Ioele, G.; Ragno, G. Cumulative area pre-processing (CAP): A new treatment of UV data for the analysis of complex pharmaceutical mixtures. *J. Pharm. Biomed. Anal.* **2014**, *90*, 45–51. [CrossRef] [PubMed]
25. Tesei, A.; Regazzoni, C.S. HOS-based generalized noise pdf models for signal detection optimization. *Signal Process.* **1998**, *65*, 267–281. [CrossRef]

sensors

MDPI

Communication

Speed of Light in Hollow-Core Photonic Bandgap Fiber Approaching That in Vacuum

Xiaolu Cao [1,2], Mingming Luo [1,2,*], Jianfei Liu [1,2], Jie Ma [1,2], Yundong Hao [3] and Yange Liu [3]

1 School of Electronic and Information Engineering, Hebei University of Technology, Tianjin 300401, China; 202231903062@stu.hebut.edu.cn (X.C.); jfliu@hebut.edu.cn (J.L.); jie.ma@hebut.edu.cn (J.M.)
2 Hebei Key Laboratory of Advanced Laser Technology and Equipment, Tianjin 300401, China
3 Institute of Modern Optics, Nankai University, Tianjin 300350, China; hydnku@163.com (Y.H.); ygliu@nankai.edu.cn (Y.L.)
* Correspondence: mmluo@hebut.edu.cn

Abstract: A Fresnel mirror is introduced at a hollow-core photonic bandgap fiber end by fusion splicing a short single-mode fiber segment, to reflect the light backward to an optical frequency domain reflectometry. The backward Fresnel reflection is used as a probe light to achieve light speed measurement with a high resolution and a high signal-to-noise ratio. Thus, its group velocity is obtained with the round-trip time delay as well as the beat frequency at the reflection peak. Multiple Fresnel peaks are observed from 2180.00 Hz to 13,988.75 Hz, corresponding to fusion-spliced hollow-core fiber segments with different lengths from 0.2595 m to 1.6678 m, respectively. The speed of light in the air guidance is calculated at 2.9753×10^8 m/s, approaching that in vacuum, which is also in good agreement with 2.9672×10^8 m/s given by the numerical analysis with an uncertainty of 10^{-3}. Our demonstration promises a key to hollow-core waveguide characterization for future wide-bandwidth and low-latency optical communication.

Keywords: hollow-core photonic bandgap fiber; Fresnel reflection; optical frequency domain reflectometry; light speed measurement

1. Introduction

As a promising solution to a primary concern in mass data transmission, hollow-core microstructured optical fibers (HC-MOFs) are expected to break through the capacity limit of commercial single-mode fibers (SMFs) [1–3]. The HC-MOF is considered as an ideal combination of a free-space light transmission and optical waveguide, for its low latency, low dispersion, and high threshold and flexibility [4–6], which shows prospects in high-capacity signal and energy transfer [7,8].

Due to the strong confinement over 99.8% in the air core, the light propagates in HC-MOFs at a group velocity approaching that in vacuum [6]. However, achieving hollow-core fiber leakage, scattering, and micro-bending loss levels comparable to or lower than conventional fibers has proved to be a difficult challenge until recently [9–11]. Moreover, subject to the distortion in nano-scaled structures during the manufacturing, the intrinsic characteristics of the optical waveguide are not convincing enough just with those results from a numerical analysis. As an example, light speed in HC-MOFs becomes a crucial and tough issue to be addressed, where plenty of smart schemes and impressive techniques were taken [12–20].

The direct measurement of group velocity is recognized as time delay evaluation using modulated probe light, which is improved but still limited in accuracy by loss, distances, response time, and dispersion [12–14]. For the poor signal-to-noise ratio (SNR) caused by higher loss in HC-MOFs, the performances degenerate in the time-of-flight technique [14], re-circulating loop [15], and optical network tester [16]. From the grating Bragg wavelength, the group velocity can be determined, but there are errors due to the uncertainty in pitch of the phase masks used for grating inscription [17]. In the low-coherent interferometry method, the group velocity can be derived by measuring the group velocity dispersion (GVD) over a continuous spectral range. However, the dynamic range is limited by dispersion with a low power and excessive bandwidth source. Also, the spatial alignment is hard to achieve at a high precision from its complexity and instability [18–20]. Due to the high output light intensity, the methods abovementioned result in a poor SNR using forward transit light. Thus, an all-fiber optical frequency domain reflectometry (OFDR) stands out using backward Fresnel reflection as the probe light, whose light intensity is 3-4 orders higher than that of the background Rayleigh scattering [21]. It brings a high-resolution and a high-SNR characterization method for mode identification and evolution in microstructured optical fibers (MOFs) [22–24]. Moreover, group birefringence in MOFs can also be measured and regulated using OFDR, where the polarized Fresnel reflection between the silica core and free space is easy to identify at the fiber end face [25,26]. Due to the absence of the refractive index (RI) gradient at the fiber end, however, direct group velocity measurement with OFDR is not applicable in HC-MOFs when compared with solid core fibers.

In this paper, the backward reflection at the man-made Fresnel mirror in a hollow-core photonic bandgap fiber (HC-PBF) is harvested with the OFDR, in which the group velocity can be obtained with the related time delay and the beat frequency. Several hollow-core fiber segments with different lengths from 0.2595 m to 1.6678 m are fusion-spliced to a single-mode fiber end face, leading to their beat-frequency differences from 2180.00 Hz to 13,988.75 Hz, respectively. Thus, considering the distortion between the actual HC-PBF and its ideal model, the averaged group velocity in the HC-PBF is calculated as 2.9753×10^8 m/s, which is close to 2.9672×10^8 m/s given by the numerical analysis. Our demonstration provides a universal and flexible method for light speed measurement in air guidance, which is achieved just within a few meters of a fiber segment and applicable to all kinds of HC-MOFs as well.

2. Principle and Experimental Setup

The spliced SMF-HCPBF joint is introduced as a Fresnel mirror for group velocity measurement as shown in Figure 1a. Notably, the Fresnel mirror is deliberately manufactured at the HC-PBF end to reflect the light backward to the OFDR, for the absence of the RI gradient between the air core and free space. To avoid the excitation of the surface super-modes in HC-PBF, the pig-tail SMF1 in OFDR is mechanically aligned with its 3D adjustment in the fusion splicer.

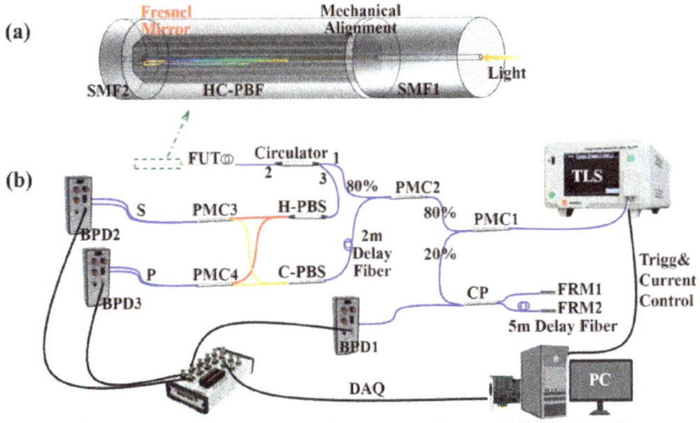

Figure 1. (**a**) A perspective view of a HC-PBF sample with a Fresnel mirror. (**b**) The schematic setup of the dual-polarization coherent OFDR system.

Figure 1b shows a schematic diagram of our experimental setup for group velocity measurement in HC-PBF based on OFDR [22]. Frequency-swept light from the tunable laser source (TLS) is split by a 20/80 polarization-maintaining coupler (PMC), PMC1. The 20% optical power goes into an auxiliary interferometer consisting of a 1:1 coupler (CP) and two Faraday Rotation Mirrors (FRMs), which generates a sine reference collected by a balanced photo detector (BPD), BPD1, for sweeping nonlinearity calibration. In the main interferometer, the remaining 80% is split once again by another 20/80 PMC2. Most of the optical power enters the fiber under testing (FUT) from port 1 to port 2 through a circulator. Meanwhile, the rest enters a reference path with a 2 m delay fiber to make up the optical path difference between the reference and test paths. Both the backward reflection from port 3 and the local light from the reference path split into S/P polarizations by two polarization beam splitters (PBSs), respectively. The polarized beams interfere at PMC3 and PMC4 according to their polarizations, and the interferences are then collected by BPD2 and BPD3 separately. The bandpass signals from the three BPDs are acquired by a 4-channel digital data acquisition (DAQ), for post processing once the frequency sweep begins.

Figure 2 shows an extremely small group dispersion at 10^{-6} of the fundamental mode (LP_{01} mode) in HC-PBF, with its ideal and actual transverse cross-sections shown as the insets. A 7-cell hollow core is surrounded by 8 rings of cladding holes to form the multilayer reflection microstructures and photonic band gaps. The phase refractive index is cosine projection along the axial direction and lower than that in vacuum, effectively confining the light transmission in the air core. In the numerical analysis, the dimensions of the cross-section model are adjusted according to the microscopic image, where the pitch (Λ, distance between two adjacent holes) is 4.1 μm, the core radius (R) is 7.9 μm, and the cladding air hole diameter-to-pitch ratio (d/Λ) is 0.968. The hexagonal air hole circular angle diameter-to-pitch ratio (dc/Λ) is 0.3, the core filet diameter-to-pitch ratio (Dc/Λ) is 0.5, and t refers to the wall thickness of the central air core. The material RI of the fusion silica background is expressed by the Sellmeier equation [27]. Thus, the wavelength-dependent phase RI curve is depicted with data from Finite Element Method (FEM) COMSOL Multiphysics, and the group RI of the LP_{01} mode is further calculated with Equation (1).

$$n_g(\lambda) = n_{eff}(\lambda) - \lambda \cdot dn_{eff}(\lambda)/d\lambda, \tag{1}$$

where n_g and n_{eff} represent the group RI and phase effective RI, respectively, and λ is the wavelength. Subsequently, the group velocity v_g and the time delay τ can be calculated with Equation (2).

$$v_g = \frac{L}{\tau} = c\left(\frac{d\beta}{dk}\right)^{-1} = \frac{c}{n_g} = \frac{c}{n_{eff} - \lambda \cdot dn_{eff}/d\lambda}, \tag{2}$$

where L is the fiber length, c is the speed of light in vacuum, β is the propagation constant, and k is the wavevector in vacuum. Accordingly, the group velocity of the LP_{01} mode is theoretically obtained at 2.9672×10^8 m/s with the time delay 3.3702 μs/km. Considering the expansion and distortion of the hollow core and innermost air holes, the actual light speed in the air guidance might be faster than that in simulation. Thus, the group velocity increases 1.6862‰ of the air guidance when the diameter of the hollow core expands by 1%.

When frequency-swept light reaches the HCPBF-SMF2 joint, significant reflection occurs at the Fresnel mirror due to the large RI gradient. Thus, the group velocity v_g in hollow-core fibers is obtained with the time delay τ and the beat frequency f_{Beat} at the Fresnel reflection peak with Equation (3).

$$v_g = 2L/\tau = 2L\gamma/f_{Beat}, \tag{3}$$

where $2L$ refers to the round-trip propagation in HC-PBF, and γ is the scanning rate of the laser.

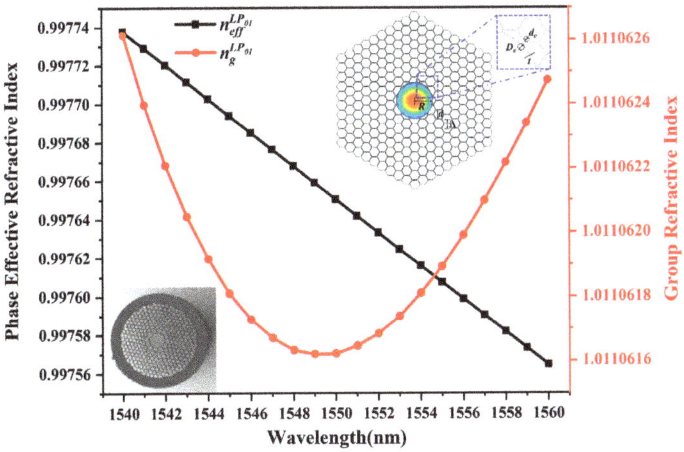

Figure 2. The phase and group dispersion curves of the LP_{01} mode in HC-PBF with the microscopic image, model of the cross-section, and LP_{01} mode profile as insets.

For an actual OFDR system, however, the total time delay includes delay τ_{sys} in the OFDR system, τ_{S1} in the SMF "pig-tail" of length L_{S1}, τ_{HC} in HC-PBF of length L_{HC}, and τ_{S2} in Fresnel mirror SMF of length L_{S2}. Thus, the total beat frequency related to time delay can be expressed with Equation (4).

$$\begin{aligned} f_{Beat} &= f_{Beat}^{sys} + f_{Beat}^{S1} + f_{Beat}^{HC} + f_{Beat}^{S2} \\ &= \gamma(\tau_{sys} + \tau_{S1} + \tau_{HC} + \tau_{S2}) \\ &= \gamma(\tau_{sys} + 2L_{S1}/v_g + 2L_{HC}/v_g^{LP_{mn}} + 2L_{S2}/v_g). \end{aligned} \tag{4}$$

Figure 3 shows the beat-frequency differences from 2184 Hz to 14,037 Hz corresponding to HC-PBFs with different lengths applied in the experiment from 0.2595 m to 1.6678 m. By numerical simulation based on the OFDR principle, the beat-frequency difference also shows a linear relation to the fiber length, providing a method for group velocity measurement in air-guidance hollow-core fibers with Equation (3).

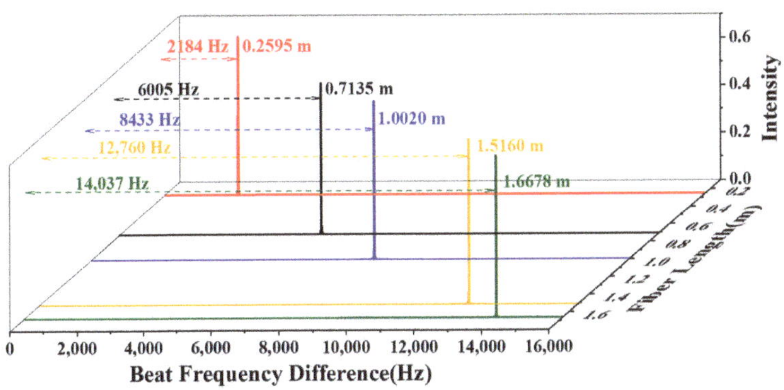

Figure 3. The numerical simulation of the beat-frequency difference using HC-PBF segments with different lengths based on the OFDR principle.

3. Results and Discussion

The Santec tunable semiconductor laser TSL-570 is set to 10 mW at a scanning rate of 10 nm/s (γ = 1.2487 THz/s), within a spectral range from 1540 nm to 1560 nm. As the

Fresnel mirror mentioned above in Figure 1b, a short SMF segment is fusion-spliced to the HC-PBF with different lengths. The other side of the HC-PBF is mechanically aligned with the SMF pig-tail using a fusion splicer, Fujikura 80s, Japan, to ensure the reflection back to the OFDR. The multi-component interferences are expanded in the frequency domain by the Fast Fourier Transform (FFT) method, where the reflections at different positions are clearly distinguished in Figure 4a. The external port of the OFDR is located at 1483.13 Hz, and the next on the right is the mechanical alignment between the SMF1 and HC-PBF at 16,218.13 Hz. The Fresnel mirror presents itself at the HCPBF-SMF2 joint of 24,626.25 Hz, where the end face of the SMF2 shows at 26,118.13 Hz. By local zooming at the HCPBF-SMF2 joint in Figure 4b, the Fresnel peak of the LP_{01} mode is observed with remaining energy of the surface super-modes or high-order modes on the right.

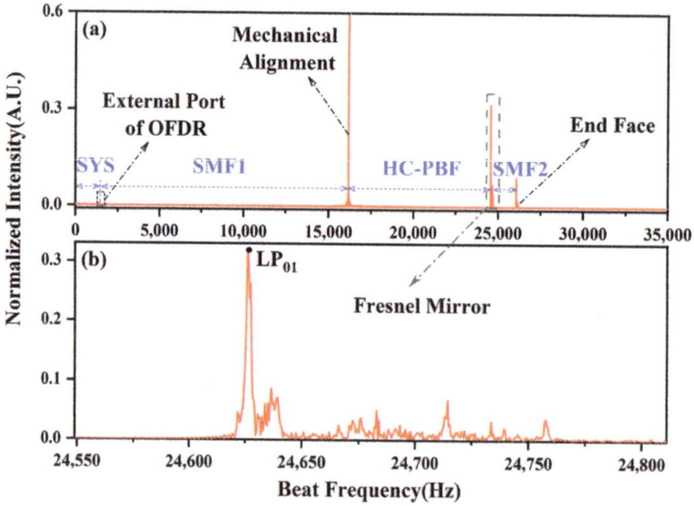

Figure 4. The observation of the Fresnel reflection peak in the frequency domain with 1.0020 m long HC-PBF as an example in (**a**) the whole frequency range and (**b**) the detailed frequency range.

Several HC-PBF segments with different lengths are further tested in the same way. The beat-frequency difference is calculated with the positionings N_1 at the mechanical alignment and N_2 at the Fresnel mirror, from the beginning to end of the HC-PBF expressed by Equation (5).

$$f_{Beat}^{HC} = (N_2 - N_1)\Delta f = (N_2 - N_1)f_s/N, \tag{5}$$

where Δf represents the OFDR frequency resolution of 0.62 Hz, and f_s and N are the sampling rate of 2 M Hz and sampling point of 3.2 M separately set by DAQ.

The beat-frequency differences show a linear relation to their lengths of the HC-PBFs, as the black scatters with error bars shown in Figure 5. The measured group velocity and time delay of the LP_{01} mode are averaged at 2.9753×10^8 m/s and 3.3610 μs/km in HC-PBF separately, which are in good agreement with those in the numerical analysis. Subject to deformation during the manufacturing and the deviation in fiber length measurement, the deviation in $v_g^{LP_{01}}$ becomes larger along the increasing fiber length shown as the blue error bars and listed in the inset table of Figure 5. Thus, an uncertainty of the light speed in HC-PBF is 10^{-3} of 2.9753×10^8 m/s with a good linearity.

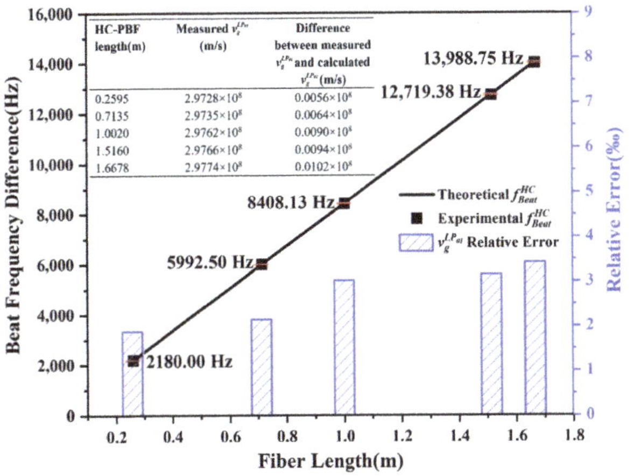

Figure 5. The uncertainty analysis between the numerical analysis and experimental results.

4. Conclusions

To measure the speed of light guided in an air core approaching that in vacuum, a Fresnel mirror is introduced at the hollow-core fiber end to reflect the light back to the OFDR. Multiple Fresnel reflection peaks are observed at different beat frequencies from 2180.00 Hz to 13,988.75 Hz, corresponding to different round-trip time delays over fiber segments with different lengths from 0.2595 m to 1.6678 m. Finally, considering the deformation of the cross-section and deviation in length measurement, the group velocity is calculated as 2.9753×10^8 m/s approaching that in vacuum. Our demonstration promises an accurate and flexible method for group velocity measurement, which is beneficial to understand the waveguide characteristics of specific optical fibers. This OFDR-based instrument is easy to operate and applicable just within a few meters of hollow-core fiber for future mass data optical communication.

Author Contributions: Conceptualization, X.C. and M.L.; methodology, X.C.; software, X.C. and Y.H.; writing—original draft preparation, X.C.; writing—review and editing, X.C. and M.L.; supervision, J.L., J.M. and Y.L. All authors have read and agreed to the published version of the manuscript.

Funding: This research was funded by the National Natural Science Foundation of China, grant number 6210031560; Natural Science Foundation of Hebei Province, grant number F2023202016; Natural Science Foundation of Tianjin City, grant number 21JCQNJC00780; and Spring Sunshine Program of the Ministry of Education, grant number HZKY20220259.

Institutional Review Board Statement: Not applicable.

Informed Consent Statement: Not applicable.

Data Availability Statement: Data underlying the results presented in this paper are not publicly available at this time but may be obtained from the authors upon reasonable request.

Acknowledgments: The authors thank Nankai University and Shenzhen Tongsheng Optoelectronics Co., Ltd. for providing the hollow-core photonic bandgap fiber.

Conflicts of Interest: The authors declare no conflicts of interest.

References

1. Zhang, X.; Feng, Z.; Marpaung, D.; Fokoua, E.N.; Sakr, H.; Hayes, J.R.; Poletti, F.; Richardson, D.J.; Slavík, R. Low-loss microwave photonics links using hollow core fibres. *Light. Sci. Appl.* **2022**, *11*, 213. [CrossRef] [PubMed]
2. Sakr, H.; Chen, Y.; Jasion, G.T.; Bradley, T.D.; Hayes, J.R.; Mulvad, H.C.H.; Davidson, I.A.; Fokoua, E.N.; Poletti, F. Hollow core optical fibres with comparable attenuation to silica fibres between 600 and 1100 nm. *Nat. Commun.* **2020**, *11*, 6030. [CrossRef] [PubMed]

3. Gao, S.; Wang, Y.; Ding, W.; Hong, Y.; Wang, P. Conquering the Rayleigh Scattering Limit of Silica Glass Fiber at Visible Wavelengths with a Hollow-Core Fiber Approach. *Laser Photonics Rev.* **2019**, *14*, 1900241. [CrossRef]
4. Cregan, R.F.; Mangan, B.J.; Knight, J.C.; Birks, T.A.; Russell, P.S.J.; Roberts, P.J.; Allan, D.C. Single-mode photonic band gap guidance of light in air. *Science* **1999**, *285*, 1537–1539. [CrossRef] [PubMed]
5. Knight, J.C. Photonic crystal fibres. *Nature* **2003**, *424*, 847–851. [CrossRef]
6. Poletti, F.; Wheeler, N.V.; Petrovich, M.N.; Baddela, N.; Fokoua, E.N.; Hayes, J.R.; Gray, D.R.; Li, Z.; Slavík, R.; Richardson, D.J. Towards high-capacity fibre-optic communications at the speed of light in vacuum. *Nat. Photonics* **2013**, *7*, 279–284. [CrossRef]
7. Hong, Y.; Bradley, T.D.; Taengnoi, N.; Bottrill, K.R.H.; Hayes, J.R.; Jasion, G.T.; Poletti, F.; Petropoulos, P.; Richardson, D.J. Hollow-Core NANF for High-Speed Short-Reach Transmission in the S + C + L-Bands. *J. Light. Technol.* **2021**, *39*, 6167–6174. [CrossRef]
8. Huang, L.; Wang, Y.; Zhang, Y.; Cheng, T.; Wang, L.; Jiang, H. High-efficiency 6-hole structure anti-resonant hollow-core fiber 2.79 μm Cr, Er:YSGG high-energy pulse laser transmission system. *Opt. Laser Technol.* **2024**, *175*, 110743. [CrossRef]
9. Fokoua, E.N.; Mousavi, S.A.; Jasion, G.T.; Richardson, D.J.; Poletti, F. Loss in hollow-core optical fibers: Mechanisms, scaling rules, and limits. *Adv. Opt. Photon.* **2023**, *15*, 1–85. [CrossRef]
10. Osório, J.H.; Amrani, F.; Delahaye, F.; Dhaybi, A.; Vasko, K.; Melli, F.; Giovanardi, F.; Vandembroucq, D.; Tessier, G.; Vincetti, L.; et al. Hollow-core fibers with reduced surface roughness and ultralow loss in the short-wavelength range. *Nat. Commun.* **2023**, *14*, 1146. [CrossRef]
11. Jasion, G.T.; Sakr, H.; Hayes, J.R.; Sandoghchi, S.R.; Hooper, L.; Fokoua, E.N.; Saljoghei, A.; Mulvad, H.C.; Alonso, M.; Taranta, A.; et al. 0.174 dB/km hollow core double nested antiresonant nodeless fiber (DNANF). In Proceedings of the 2022 Optical Fiber Communications Conference and Exhibition (OFC), San Diego, CA, USA, 10 March 2022; pp. 1–3.
12. Danielmeyer, H.G.; Weber, H.P. Direct measurement of the group velocity of light. *Phys. Rev. A* **1971**, *3*, 1708. [CrossRef]
13. Danielson, B.L. Precise length measurements in multimode optical fibers. *Appl. Opt.* **1991**, *30*, 3867–3872. [CrossRef] [PubMed]
14. Shen, W.; Du, J.; Sun, L.; Wang, C.; Zhu, Y.; Xu, K.; Chen, B.; He, Z. Low-Latency and High-Speed Hollow-Core Fiber Optical Interconnection at 2-Micron Waveband. *J. Light. Technol.* **2020**, *38*, 3874–3882. [CrossRef]
15. Hong, Y.; Bottrill, K.R.H.; Bradley, T.D.; Sakr, H.; Jasion, G.T.; Harrington, K.; Poletti, F.; Petropoulos, P.; Richardson, D.J. Low-Latency WDM Intensity-Modulation and Direct-Detection Transmission over >100 km Distances in a Hollow Core Fiber. *Laser Photonics Rev.* **2021**, *15*, 2100102. [CrossRef]
16. Kuschnerov, M.; Mangan, B.J.; Gong, K.; Sleiffer, V.A.J.M.; Herrmann, M.; Nicholson, J.W.; Fini, J.M.; Meng, L. Transmission of commercial low latency interfaces over hollow-core fiber. *J. Light. Technol.* **2016**, *34*, 314–320. [CrossRef]
17. Cruz, J.L.; Barmenkov, Y.O.; Díez, A.; Andres, M.V. Measurement of phase and group refractive indices and dispersion of thermo-optic and strain-optic coefficients of optical fibers using weak fiber Bragg gratings. *Appl. Opt.* **2021**, *60*, 2824–2832. [CrossRef]
18. Mori, K.; Morioka, T.; Saruwatari, M. Group velocity dispersion measurement using supercontinuum picosecond pulses generated in an optical fibre. *Electron. Lett.* **1993**, *29*, 987–989. [CrossRef]
19. Arosa, Y.; Fuente, R.D.L. Refractive index spectroscopy and material dispersion in fused silica glass. *Opt. Lett.* **2020**, *45*, 4268–4271. [CrossRef]
20. Zhang, X.; Gao, S.; Wang, Y.; Ding, W.; Wang, X.; Wang, P. 7-cell hollow-core photonic bandgap fiber with broad spectral bandwidth and low loss. *Opt. Express* **2019**, *27*, 11608–11616. [CrossRef]
21. Michaud-Belleau, V.; Fokoua, E.N.; Bradley, T.D.; Hayes, J.R.; Chen, Y.; Poletti, F.; Richardson, D.J.; Genest, J.; Slavik, R. Backscattering in antiresonant hollow-core fibers: Over 40 dB lower than in standard optical fibers. *Optica* **2021**, *8*, 216–219. [CrossRef]
22. Luo, M.; Liu, J.; Tang, C.; Wang, X.; Lan, T.; Kan, B. 0.5 mm spatial resolution distributed fiber temperature and strain sensor with position-deviation compensation based on OFDR. *Opt. Express* **2019**, *27*, 35823–35829. [CrossRef] [PubMed]
23. Xu, X.; Luo, M.; Liu, J.; Gao, Q.; Li, C.; Guo, H.; Liu, Y.-G.; Ge, C. Observation and characterization of the high order modes in a six-mode fiber using an OFDR method. *Opt. Express* **2022**, *30*, 26341–26347. [CrossRef] [PubMed]
24. Ahn, T.-J.; Kim, D.Y. High-resolution differential mode delay measurement for a multimode optical fiber using a modified optical frequency domain reflectometer. *Opt. Express* **2005**, *13*, 8256–8262. [CrossRef] [PubMed]
25. Gao, Q.; Luo, M.; Liu, J.; Luan, N.; Xu, X.; Li, C.; He, Y. Group Birefringence Regulation and Measurement with Twin Zeros in a Selectively Infiltrated Microstructured Optical Fiber Based on OFDR. *J. Light. Technol.* **2022**, *40*, 308–312. [CrossRef]
26. Meng, X.; Luo, M.; Liu, J.; Zhao, S.; Zhou, R. Birefringence characterization in a dual-hole microstructured optical fiber using an OFDR method. *Appl. Opt.* **2024**, *63*, 772–776. [CrossRef]
27. Malitson, I.H. Interspecimen comparison of the refractive index of fused silica. *J. Opt. Soc. Am.* **1965**, *55*, 1205–1209. [CrossRef]

sensors

Article

Scenarios for Optical Encryption Using Quantum Keys †

Luis Velasco [1,*], Morteza Ahmadian [2], Laura Ortiz [3], Juan P. Brito [3], Antonio Pastor [4], Jose M. Rivas [4], Sima Barzegar [1], Jaume Comellas [1], Vicente Martin [3] and Marc Ruiz [1]

1 Advanced Broadband Communications Center (CCABA), Universitat Politècnica de Catalunya (UPC), 08034 Barcelona, Spain
2 Electrical Engineering, Chalmers University of Technology, 41296 Gothenburg, Sweden
3 Industrial Engineering School, Universidad Politécnica de Madrid, 28040 Madrid, Spain
4 Telefonica Innovación Digital, 28050 Madrid, Spain
* Correspondence: luis.velasco@upc.edu
† This manuscript is the extension of a conference paper: Ahmadian, M.; Vicente, R.; Brito, J.; Lopez-Garcia, A.; Pastor, A.; Rivas, J.; Comellas, J.; Ruiz, M.; Martín, V.; Velasco, L. Experimental Demonstration of Optical Encryption Using Quantum Keys: Two Scenarios. In Proceedings of the Optical Fiber Communications Conference and Exhibition (OFC), San Diego, CA, USA, 24–28 March 2024.

Abstract: Optical communications providing huge capacity and low latency remain vulnerable to a range of attacks. In consequence, encryption at the optical layer is needed to ensure secure data transmission. In our previous work, we proposed LightPath SECurity (LPSec), a secure cryptographic solution for optical transmission that leverages stream ciphers and Diffie–Hellman (DH) key exchange for high-speed optical encryption. Still, LPSec faces limitations related to key generation and key distribution. To address these limitations, in this paper, we rely on Quantum Random Number Generators (QRNG) and Quantum Key Distribution (QKD) networks. Specifically, we focus on three meaningful scenarios: In Scenario A, the two optical transponders (Tp) involved in the optical transmission are within the security perimeter of the QKD network. In Scenario B, only one Tp is within the QKD network, so keys are retrieved from a QRNG and distributed using LPSec. Finally, Scenario C extends Scenario B by employing Post-Quantum Cryptography (PQC) by implementing a Key Encapsulation Mechanism (KEM) to secure key exchanges. The scenarios are analyzed based on their security, efficiency, and applicability, demonstrating the potential of quantum-enhanced LPSec to provide secure, low-latency encryption for current optical communications. The experimental assessment, conducted on the Madrid Quantum Infrastructure, validates the feasibility of the proposed solutions.

Keywords: optical encryption; Quantum Random Number Generator; Quantum Key Distribution; Post-Quantum Cryptography

1. Introduction

The high capacity of optical communications brought by higher-order modulation schemes, like Quadrature Amplitude Modulation (QAM), and their inherent low latency make them ideal for supporting 5G and beyond communications [1] from the transport core to the access network [2]. However, in standard optical communications, data are generally transmitted as plain text because implementing cryptographic methods to run at line speeds of hundreds of Gb/s is challenging, and their implementation might impact on the performance of the supported services, especially on the added transmission delay [1]. In addition, optical networks are themselves vulnerable to a variety of attacks such as eavesdropping, physical infrastructure attacks, interception, and jamming [3]. Therefore, quantum communications, providing superior security, continue their development and can finally provide the required capacity and performance [4] to replace classical optical communications, services must be individually secured at the packet layer using standard stream ciphers, like Advanced Encryption Standard [5] and ChaCha [6]. In parallel, quantum computing will soon be able to perform relevant cryptographic operations and break public cryptography. Therefore, advancing in the deployment of more secure communication infrastructures should be prioritized to secure the optical layer, so as to avoid the risks of building a secure platform on top of an unsecure one.

In our previous works [7,8], we proposed *LightPath SECurity* (LPsec), a secure cryptographic solution for point-to-point and point-to-multipoint optical transmission (e.g., 16-QAM) that: (*i*) extends optical transponders (Tp) with a fast bit stream Data Encryption Mechanism (DEM) using stream ciphers; (*ii*) implements key exchange based on the Diffie–Hellman (DH) [9] protocol orchestrated by the Software-Defined Networking (SDN) controller; and (*iii*) distributes symmetric keys through the encrypted optical channel. LPSec has shown its ability to provide data stream encryption at line speed, while introducing noticeable low processing delay.

Nonetheless, two main limitations can be found on LPSec: (*Lim-A*) *key generation*. This limitation refers to the use of a secure Pseudo-Random Number Generator (PRNG) to generate keys, which produces not truly random random-number sequences; and (*Lim-B*) *key distribution*. This limitation comes from the use of the previous key to encrypt the new key, which creates key dependency that can be exploited if one key is compromised. These limitations make LPSec to be not secure against the quantum threat.

To address these issues, in this paper, we propose using Quantum technologies, specifically: (*i*) Quantum Random Number Generators (QRNG) [10] to produce true-random numbers coming from inherently non-deterministic quantum processes, which solves *Lim-A*; and (*ii*) Quantum Key Distribution (QKD) networks [11,12], which provide an Information Theoretically Secure (ITS) protocol for distributing symmetric keys, thus facing *Lim-B*. In fact, although QRNG can be used on its own for classical processes, they are usually used inside the QKD devices. Hence, a QKD network can be understood as a network to generate and distribute truly random keys between any two points in the network. The combination of LPSec and quantum cryptography represents a step forward for the security of optical communications, as it allows upgrading technologies currently deployed in operators network with security resistant to attacks from quantum computing, thus extending their lifespan. The maturity of QKD devices and their commercial availability have allowed the deployment of experimental infrastructures (see, e.g., [13,14]). Furthermore, efforts devoted to the standardization of application interfaces for key retrieval (e.g., ETSI QKD 004 [15]), have eased QKD integration and enabled a wide-range of use cases.

However, although QKD can share infrastructure with classical optical communications, deploying a QKD network infrastructure is expensive and has distance limitations until quantum repeaters are commercially available. Therefore, in the application of QKD to LPSec, it will be very common that only one of the Tps of the optical connection is within the reachable perimeter of the QKD network, e.g., on optical connections between core/metro and access/aggregation sites. This situation entails reducing the security level in order to extend the optical network, since although standalone QRNGs can be used for key generation (thus solving *Lim-A*), key distribution mechanism still relies on LPSec, which keeps unsolved *Lim-B*.

A classical possibility to overcome the distance limitations and preserve resistance quantum attacks by Shor's algorithm [16] is to use Post-Quantum Cryptography (PQC) algorithms, especially those providing a Key Encapsulation Mechanism (KEM). Since PQC algorithms are primarily based on the computational difficulty of solving certain mathematical problems, *Lim-B* can be unlocked. Recently, the National Institute of Standards and Technology (NIST) has published FIPS 203 [17], FIPS 204 [18], and FIPS 205 [19], which specify algorithms derived from CRYSTALS-Dilithium, CRYSTAL-KYBER and SPHINCS+ [20]. For our implementation, we have selected the Bit Flipping Key Encapsulation (BIKE) scheme [21]. BIKE is one of the earliest code-based cryptosystems and has proven resistant to quantum attacks by the Shor's algorithm due to its NP-hardness in decoding random linear codes [22].

QKD and PQC are the solutions against cryptographic attacks via quantum computing. However, they both are complementary; QKD offers ITS, while PQC facilitates scalable authentication. In consequence, some previous works have proposed solutions involving QKD and PQC. Just to cite some very recent examples, the authors in [23] presented a quantum-secure architecture with PQC and QKD tailored explicitly for mobile networks and provided use cases that emphasize the need for advanced protection measures. An authenticated hybrid key exchange protocol that incorporates PQC and QKD and provides both forward and post-compromise security was introduced in [24]. The authors in [25] aimed to establish a secure system for key exchange by using PQC schemes in the classical channel of QKD to address authentication and encryption critical security challenges, so as to ensure reliable communication across quantum and classical channels, while the authors in [26] use PQC to provide encryption between KMS in two different domains. QKD and PQC hybridizations have been showcased experimentally, e.g., the authors in [27] proposed different methods of interconnecting three QKD testbeds in Europe using hybridizations of QKD and PQC algorithms.

In this paper, we explore the outlined scenarios for optical encryption using quantum keys where keys are: (*A*) retrieved from a QKD network, assuming that both Tps are inside the security perimeter of the QKD network; (*B*) retrieved from a QRNG in the case that only one of the Tps is inside that security perimeter, and distributed using LPSec to the other Tp; and (C) retrieved from a QRNG as in (B) but encapsulated using BIKE as KEM and exchanged to the other Tp using LPSec. In all cases, the retrieved keys leverage the true-random number properties of quantum physics. Such keys are then expanded to the desired line rate using a secure PRNG. For all the aforementioned scenarios, we assume that an optical connection between two Tps is setup using a SDN controller and both Tps connect to a local key manager (KM) in the case of QKD or just one of the Tps connect to a local key server in the case of relying on a QRNG. The workflows have been designed in the Horizon Europe ALLEGRO project [28] and their experimental assessment has been carried out on the Madrid Quantum Infrastructure (MadQCI) [14]. The rest of this paper is organized as follows. Section 2 first briefly introduces LPSec and presents in depth the three application scenarios that are the focus of this work. Section 3 describes the workflows for implementing the scenarios in practice and Section 4 presents the experiments carried out for the assessment of the solutions. Finally, Section 5 draws the main conclusions of this work.

2. Optical Encryption Using Quantum Keys

2.1. Brief Summary of LPSec

LPsec [7] extends the standard coherent Tp with tailored solutions for both key exchange and encryption/decryption (see Figure 1). Cryptographic blocks operate at line speeds and therefore, optical encryption in LPSec is based on simple operations that can be easily implemented inside the optical Tp. For the key exchange, LPSec includes a mechanism based on the DH key exchange, where: (*i*) the initial public keys of the two end parties, i.e., the Transmitter (Tx) and the Receiver (Rx), are exchanged via the SDN controller; and (*ii*) keys are periodically updated through the optical channel to enhance the security level. As for the DEM, LPSec relies on two nested ciphers: (*i*) a stream cipher ($E_1(\cdot)$) that encrypts input data chunks m of predefined size with a key kx. Key kx is the result of expanding an input key k using a cryptographically secure PRNG. Encryption of m is performed is implementing as bitwise XOR operation, i.e., $c_1 = E_1(m, kx) = m \oplus kx$, where kx and m have the same length; and (*ii*) a substitution cipher ($E_2(\cdot)$) for scrambling symbols. A Lookup Table (LUT) is used to create a ciphered gray map constellation through LUT permutations of incoming bits. The output ciphertext c_2 is produced by the combination of the inner stream cipher $E_1(\cdot)$ and the outer substitution cipher $E_2(\cdot)$, i.e., $c_2 = E_2(E_1(m, kx), k)$.

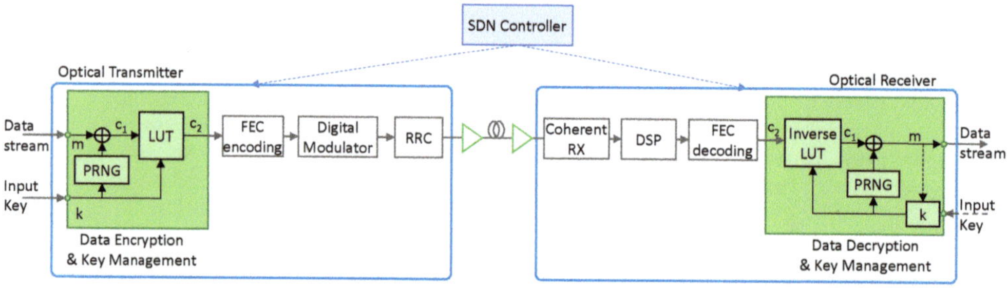

Figure 1. Optical coherent system implementing LPsec (extended from [7]).

Each cipher has its drawbacks, but when combined they provide the required security level to encrypt data at 100 s of Gb/s exhibiting negligible transmission delay. For illustrative purposes, let us assume that input keys k are generated using the Xoshiro256+ PRNG, which is able to generate 256-bit keys in less than 1 ns using an Intel Core i7 [29]. After the input data are encrypted using stream cipher $E_1(\cdot)$, the substitution cipher $E_2(\cdot)$ adds more complexity to the brute-force attack, e.g., key spaces of 44 bits are produced using 16-QAM. That combination results in 300 bits key sizes, thus raising the effective computation required by the *birthday* paradox to over 2^{150}, which can be considered safe enough for the foreseeable future [30]. This makes that LPsec can be considered secure against exhaustive search attacks.

However, (*i*) the sequence of stream keys generated by the PRNG from a given input key k cannot be infinite as this would reduce the security level; and (*ii*) the LUT permutation needs to be periodically regenerated to minimize vulnerabilities. As a result, the lifetime of keys is limited, e.g., to 1 s., which entails new keys being periodically made available at the two Tps. To facilitate communication between the two Tps, a special frame named Key exchange Frame (KxF) is used. A KxF is generated by the Tx and sent to the Rx periodically. The KxF includes a header of a fixed size that allows the Rx to detect its arrival.

2.2. Application Scenarios

In this work, we focus on the encryption of an optical signal, e.g., a 16-QAM signal, using the keys retrieved from a quantum key generation source. Retrieved keys are used in LPSec as input to the key expansion mechanism based on a secure PRNG to encrypt the input data stream at line rate. Then, three scenarios are analyzed depending on how keys are made available at the Rx side.

In the first scenario (*Scenario A*), both TpA1 and TpB are in sites covered by the QKD network (e.g., in metro/core sites), so keys k can be retrieved from the local KMs using a standard interface and used as input of the PRNG at both sides. In this scenario, (*i*) QKD is in charge of the generation and the distribution of truly random quantum keys; and (*ii*) the LPSec KxF header is used for synchronization purposes between the two Tps. Note that this is still needed to be sure that new keys are used for encryption and decryption at the right times. Figure 2a presents *Scenario A*, where the two Tps are co-located with QKD systems, each including its respective KM exposing an interface for key retrieval. An SDN controller is used for lightpath setup and LPSec configuration. Finally, two TP agents are responsible for configuring the local Tps and communicate with the SDN controller. *Scenario A* extends case 1 "undefined KSID in a single link scenario" defined in [15], to be used for LPSec connection setup.

In this scenario, the ITS security provided by the QKD network is reduced by the high expansion rate needed to expand retrieved key to the line rate, as QKD systems provide throughputs from hundreds of kb/s to few Mb/s, depending on the system.

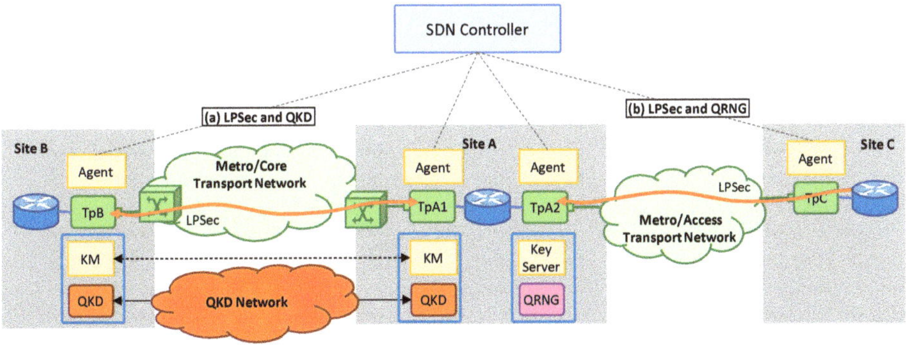

Figure 2. Scenarios under analysis: (**a**) both Tps are inside QKD security perimeter (e.g., in metro/core sites), and (**b**) one Tp is not within the security perimeter (e.g., in an access site) (**b**).

To achieve the highest possible security level, both Tps need to be within the security perimeter of the QKD network, which is not always possible, e.g., when the optical connection involves the access network. In such cases, LPSec can be used to extend farther the reach of the security perimeter of the quantum network (*Scenario B*). Figure 2b depicts the scenario where only one of the TPs (TpA2) is in a metro/core site covered by the QKD network and a single QRNG is used as the entropy source. In this case, keys *k* can be retrieved by the local Tp from a local QRNG that provides a vendor-proprietary interface for key retrieval. Such random keys are used to encrypt the input data stream and exchanged to the remote Tp using the KxF header encrypted with the previous key. Typically, QRNGs have very high throughput (about 1 Gb/s), so moderate expansion is needed in this case to reach the desired line rate.

Note that in Scenario B, keys are used to encrypt the next keys that are distributed within the next KxF to be used during the next period. This creates a dependency that can be exploited to decrypt all the transmission if one of the keys is compromised. To avoid that, new keys, retrieved from a local QRNG, are used as part of key encapsulation (*Scenario C*). To that end, we modified the BIKE scheme to use the QRNG keys as seeds thus, replacing the PRNG in the encapsulation function. Random seeds together with the public key of TpC are inputs to generate the session key *ks* and the cyphertext *C*. Session keys *ks* are kept secret and used for data encryption, whereas cyphertexts *C* are exchanged with TpC that decapsulates them with its private key to retrieve *ks* and used for data decryption. Note that by doing so, we are ensuring higher randomness and increasing the overall security of the system. Figure 3 presents the modified optical coherent Tx and Rx that include both KEM and LPSec DEM.

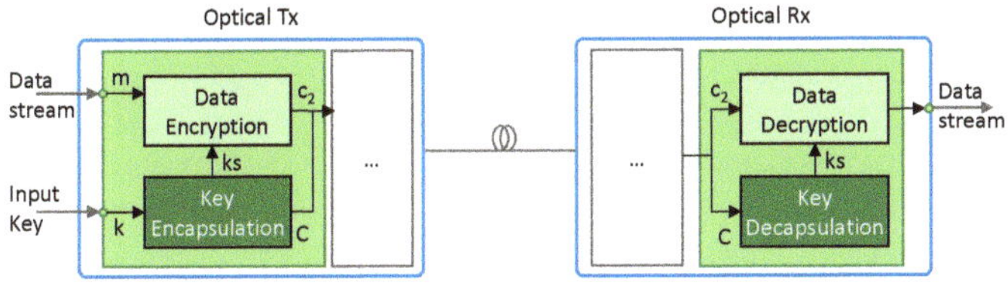

Figure 3. Extension of the optical system implementing LPsec DEM with BIKE KEM for Scenario C.

3. Workflows

This section details the workflows to be implemented for each of the identified scenarios. The workflows start after the lightpath has been established over the optical network (step 0 in the workflows). LPSec requires an initial public key exchange to be carried out through the SDN controller as part of the provisioning process; the keys are used to generate the particular KxF header pattern that will be used on the optical channel between the two Tps. It is worth noting that details of the workflows have been intentionally omitted for the sake of simplicity; please refer to [7] for details related to LPSec, e.g., the use of public/private keys, KxF, and LUT.

3.1. Scenario A: LPSec DEM and QKD

In Scenario A (Figure 4), the SDN collects the public key and the ID of TpB (labeled 1 in Figure 4) and sends them to the agent of TpA1 together with the details of the local KM (2). The agent in TpA1 connects to the local KM using the OPEN_CONNECT function and indicating the ID of TpA1 as source and that of TpB as destination. The KM replies with the *Key stream ID* (KSID) to be used for retrieving quantum keys (3). The KM in site A coordinates internally with the KM in site B to create the association KSID in both sites (4). The KSID needs to be used in site B, so TpA1 agent replies the SDN controller with it together with its public key, the ID of the local Tp and the KxF header pattern encrypted with the public key of TpB (5). The SDN controller, in turn, sends them to TpB agent (6). The agent in site B uses the KSID together with the IDs of the local Tp to connect to the local KM using the OPEN_CONNECT function (7) and reports to the SDN controller. At this time, both ends are ready to use the optical connection with encryption.

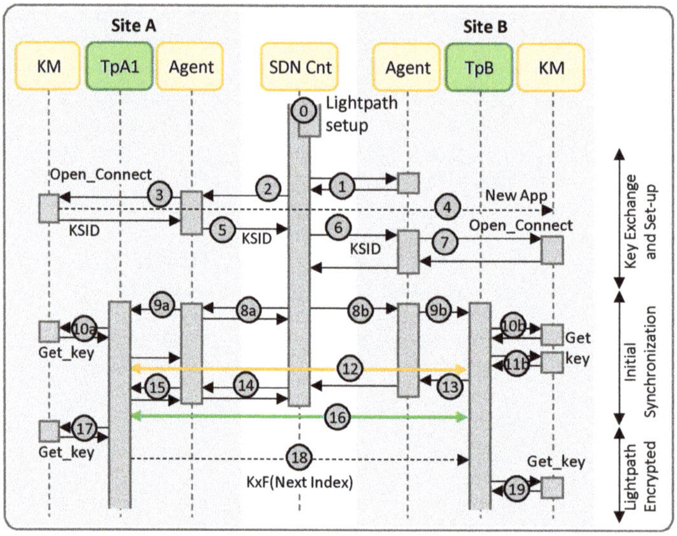

Figure 4. Workflow for Scenario A.

The next step is to synchronize the two Tps so they can start using the right keys to encrypt the data stream. Such initial synchronization is triggered by the SDN controller, which requests both ends to start the process (8a, 8b). In the case of the agent of TpA1, it acknowledges the request and asks the local Tp to get the first key (9a). TpA1 gets the key using the GET_KEY function with the obtained KSID (10a), uses it to encrypt a known bit stream and inserts the KxF header periodically including in the header the index of the retrieved key (12). Meanwhile, on site B, the agent sends the KSID and the ID of TpA1 to the local Tp (9b), which retrieves two keys (10b, 11b): the first key will be used during the synchronization period and the second once that period finishes and the real data stream is received. When TpB is able to successfully decrypt the known bit stream and finds the KxF header with the same index as the one got from the local KM, it notifies its agent that reports to the SDN controller (13). The SDN controller requests TpA1 agent to start (14), which in turns notifies the local Tp (15). At this time, both ends are ready to start with the real data encryption (16).

Once the Tps are synchronized, keys are updated periodically and synchronization is carried out using the KxF header. TpA1 gets the next key from the local KM (17), sends the index inside the KxF header (18) and starts encrypting the incoming data stream with the new key. Note that TpB already had the key for decrypting the data, so it uses that immediately, and gets the next key from the local KM (19). Steps 17–19 repeat at every time interval until the lightpath is torn down, when both agents use the KSID to terminate the association with the KMs.

3.2. Scenario B: LPSec DEM and QRNG

As in Scenario A, an initial public key exchange is carried out and used to generate the particular KxF header pattern that will be used on the optical channel between the two Tps for synchronization and key distribution. The SDN first collects the public key of TpC (labeled 1 in Figure 5) and sends it to TpA2 via its agent (2). The agent requests TpA2 to retrieve a key from the local key server that will be used for data encryption (3–4). As in Scenario A, TpA2 uses the retrieved key to encrypt a known bit stream and inserts the KxF header periodically (5). In addition, the key is sent to the SDN controller encrypted using the public key of TpC, together with the generated KxF header pattern, which are sent to TpC via its agent (6–7).

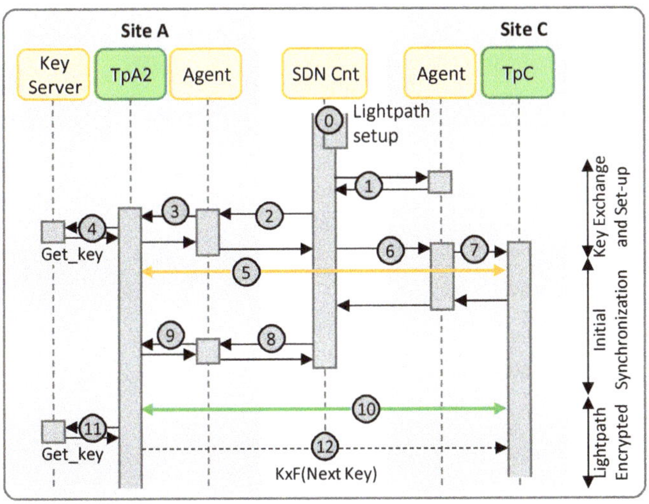

Figure 5. Workflow for Scenario B.

Once TpC is able to get synchronized with TpA2 through the optical channel, it replies the agent, which in turn notifies the SDN controller. Then, the SDN controller sends the confirmation to the agent in TpA2 (8), which notifies TpA2 (9) that the encryption of the incoming data stream can start (10).

Periodically, TpA2 retrieves a new key from the local key server (11), encrypts it using the previously retrieved key and sends it in the next KxF header (12). Steps 11–12 repeat at every time interval until the lightpath is torn down.

3.3. Scenario C: LPSec DEM, QRNG and PQC KEM

In this scenario, quantum keys are used in the PQC KEM to produce the session keys *ks* and the cyphertext C (Figure 6). Session keys *ks* are used as input to the DEM, while cyphertexts C are distributed between the Tps using the KxF header, which increases the security of the system. As in scenarios A and B, an initial public key exchange is carried out and used to generate the particular KxF header pattern. However, in this scenario, TpC generates its public-private keys specifically for the current lightpath. In addition, TpA2 needs the public key of TpC to produce the session key *ks* and the cyphertext *C*.

Therefore, when the SDN starts the workflow to collect the public key of TpC (labeled 1 in Figure 6), a new pair of public-private keys is generated by TpC (2) to be used to generate the KxF header pattern and the public key is collected by the agent that sends back in response to the SDN controller, which sends it to the agent of TpA2 (3). The agent requests TpA2 to retrieve a key from the local key server that will be used as a seed for data encryption (4–5). Specifically, TpA2 uses the retrieved seed to generate the session key and the cyphertext (6). The session key is used to encrypt a known bit stream and the cyphertext is added to the KxF header that is inserted periodically (7). In addition, the generated KxF header pattern and the cyphertext are sent to the SDN controller, which sent them to TpC via its agent (8–9). The cyphertext is then decapsulated using the private key of TpC to get the session key to decrypt the incoming optical signal (10).

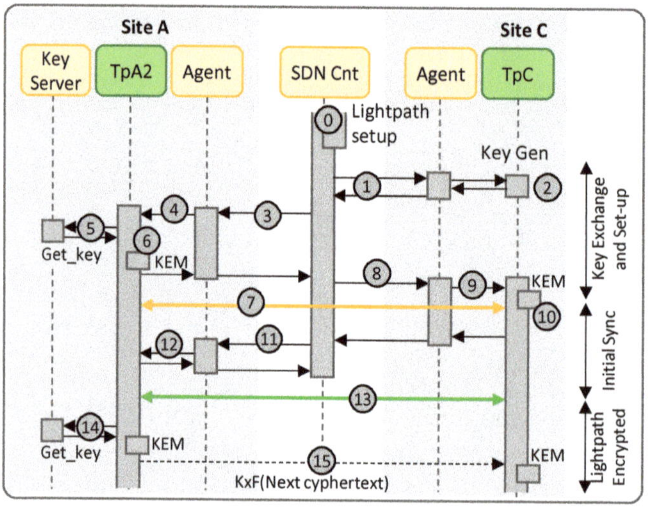

Figure 6. Workflow for Scenario C.

The rest of the workflow is similar to that of Scenario B. Once TpC is able to get synchronized with TpA2 through the optical channel, it replies the agent, which in turn notifies the SDN controller. Then, the SDN controller sends the confirmation to the agent in TpA2 (11), which notifies TpA2 (12) that the encryption of the incoming data stream can start (13). Periodically, TpA2 retrieves a new key from the local key server (14) and uses it to generate a new session key and cyphertext. The new session key will be used to encrypt the data stream in the next period, while the cyphertext is sent in the next KxF header encrypted using the previous session key (15). Steps 14–15 repeat at every time interval until the lightpath is torn down.

4. Experimental Assessment

In this section, the setup used for the experimental assessment is first described and then, the workflows of the application scenarios described in the previous section are assessed.

4.1. Setup for the Experiments and Security Analysis

The setup in Figure 2 has been experimentally assessed using real QKD and QRNG systems. Specifically, a QKD link using experimental Huawei continuous-variable QKD devices [31] are used for Scenario A. The link connects two Telefonica's facilities in Madrid, Spain; it spans 15 km and it has 10.2 dB losses to provide secret key rate of 8.4 Kb/s through an ETSI GS QKD 004 interface. In addition, a QuSIDE QRNG system [32] deployed in UPM premises is used for Scenarios B and C as quantum entropy source. The QRNG implements a proprietary phase-diffusion quantum random number generation technology and has embedded randomness metrology capabilities [33,34] to produce very high-quality random bits at 4 Gb/s, thus enabling a noticeable reduction in key expansion. The system exposes a proprietary REST API interface to deliver the random bits.

Using these systems to generate input keys k noticeably increases the security of LPSec. To illustrate this, Table 1 presents the total key size and the effective computation required by the brute-force attack for the plain LPSec using the Xoshiro256+ PRNG as key generator, and when LPSec uses keys from the Quantum systems described above. Note that the use of quantum systems makes the size space added by the substitution cipher $E_2(\cdot)$ to be not significant and thus, it can be removed without affecting the security level of LPSec. This would noticeably simplify, and thus reduce the cost, of the optical transponders implementing LPSec.

Table 1. Size of the keys and effective computation required by the brute-force attack.

	Size of k (bits)	Total Key Size ($E_1(\cdot) + E_2(\cdot)$)	Effective Computation Required (*birthday* Paradox)
Plain LPSec	256	300	2^{150}
Keys from QKD	8602	8646	$2^{4,322}$
Keys from QRNG	4,194,304	4,194,348	$2^{2,097,174}$

The optical transmission, LPSec, including encryption and key exchange, as well as the modified BIKE scheme are implemented in a simulator running in UPC premises in Barcelona, Spain. The simulator has been developed in Python 3.10 and it is based on the architecture in [35], where several modules, including algorithms and interfaces communicate through a Redis DB in publish-subscribe mode. The setup was deployed inside Docker containers on an Ubuntu Server 22.04 LTS as operating system.

Finally, a video streaming service was implemented in Python 3.10 using the OpenCV library [36] to demonstrate the connectivity and data encryption. Due to the software implementation, the video streaming service was configured to run at 30 fps, which resulted in a smooth visualization in all the scenarios.

4.2. Assessment of the Application Scenarios

Figure 7 reproduces the capture of messages exchanged between the different systems under Scenario A. Message exchange between modules implemented in the simulator is supported by the Redis serialization protocol (RESP). Actually, every message between two modules in the capture in Figure 7 is a summarization of two exchanged messages between the originating module and the Redis DB for publishing the message on a defined topic and between the Redis DB and the second module (which is subscribed to the topic) for message delivery. The numbering of the messages follows that of the workflow in Figure 4 to facilitate their identification. The details of messages exchanged between TpA1 and the KM in site A for OPEN_CONNECT (message 3) and GET_KEY (messages 10a/b, etc.) are also shown in Figure 7. The OPEN_CONNECT function requires as inputs the IDs of the application source (TpA1) and destination (TpB), as well as some Quality-of-Service parameters (in our tests, we configured the length of the keys to 8 bytes) and it returns a KSID. The subsequent GET_KEY calls get the obtained KSID together with the index of the key to retrieve.

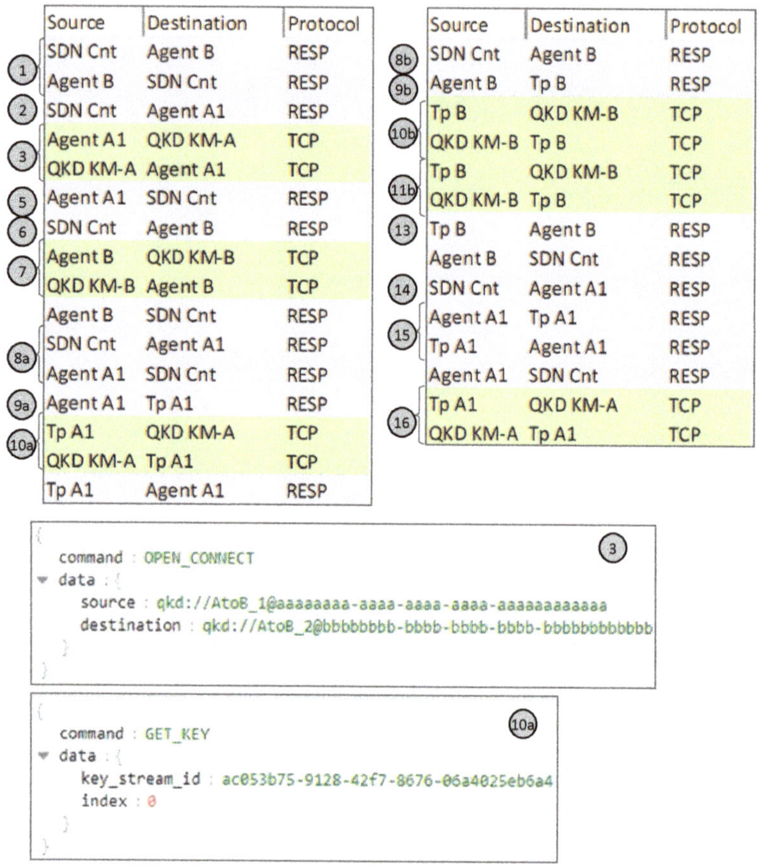

Figure 7. Capture of messages exchanged in the workflow for Scenario A and details of key messages.

Figure 8 shows the log text in the terminals executing TpA1 and TpB. The keys and their indexes are printed when they are retrieved from the KMs. In this scenario, TpA1 obtains a KSID for the lightpath and shared it with TpB, so both Tps can get the keys from their local KM. TpA1 gets a key with the next index from the local KM and sends that index in the next KxF header. TpB gets two keys, one being the current and the other being the next one to be used as soon as a new KxF header is received specifying the next index in message 18. Then, a key with the next index is retrieved from the local KM. The capture shows the first three iterations.

Figure 9 presents the capture of the exchanged messages under Scenario B, as well as the log text on terminals running the Tps under Scenario B. In this scenario, the keys (8 bytes) are retrieved by TpA2 from the local QRNG and exchanged encrypted through the optical channel using the KxF header (messages 5 and 12). In addition, the first key is shared with TpC thought the control plane for the initial synchronization (message 7).

Finally, Figure 10 shows the captures for Scenario C. A modified version of the BIKE scheme is used in: (*i*) both TpA2 and TpC to generate the pairs of public-private keys that will be used for the lightpath; (*ii*) in TpA2 to encapsulate the key retrieved from the QRNG and produce *ks* (8 bytes) and *C* (1549 bytes); and (*iii*) in TpC to decapsulate the received cyphertext and obtain *ks*. The first cyphertext is exchanged though the control plane (message 10) and used for the initial synchronization. In addition, TpA2 sends the initial cyphertext (message 7) and the subsequent ones (messages 15) within the KxF header though the optical channel.

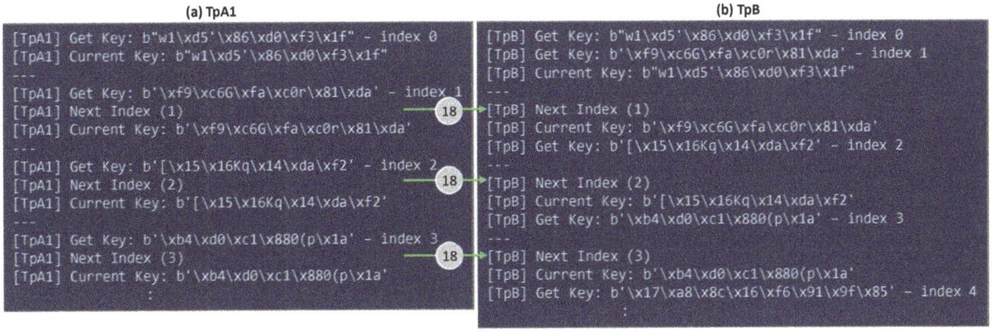

Figure 8. Outputs on terminals running TpA1 and TpB for Scenario A.

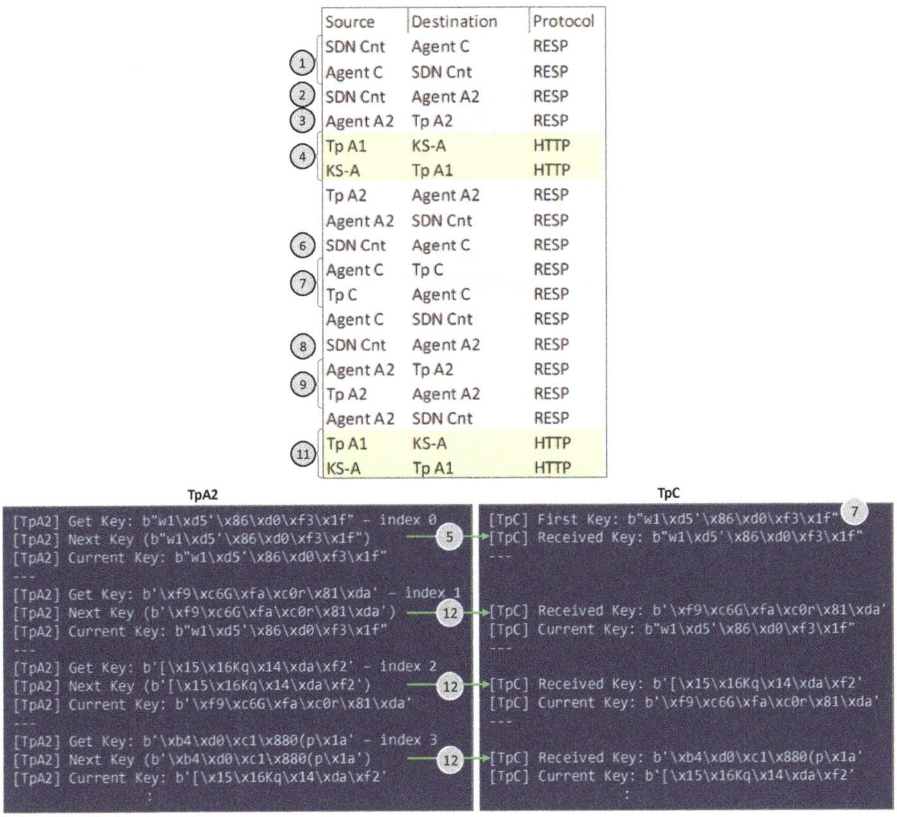

Figure 9. Capture of messages exchanged in the workflow and outputs on terminals running Tps for Scenario B.

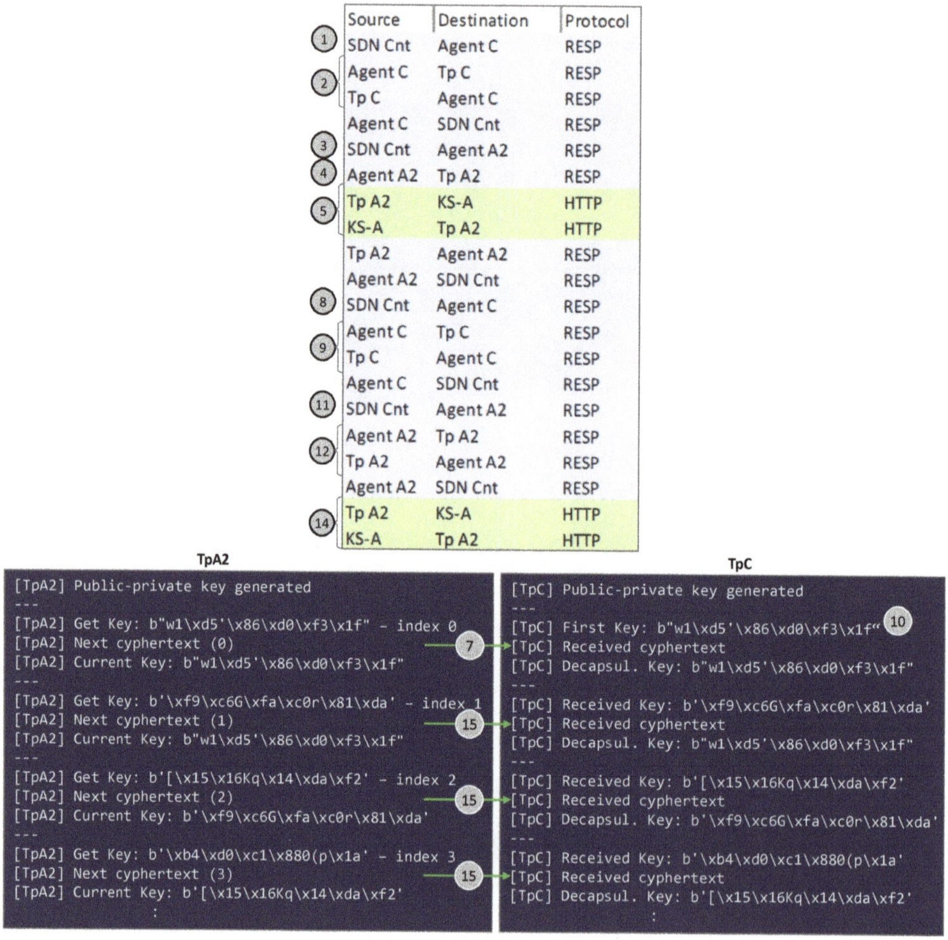

Figure 10. Capture of messages exchanged in the workflow and outputs on terminals running Tps for Scenario C.

5. Concluding Remarks

In this paper, we have explored the use of quantum solutions for improving the security of classical high-speed optical communications. In particular, we focused on improving LPSec, which has shown its ability to provide data stream encryption at line speed, while introducing noticeable low processing delay. However, as any classical cryptographic, LPSec is vulnerable to quantum attacks. Specifically, LPSec shows limitations in the way keys are generated and distributed. For that very reason, LPSec has been extended in this paper with truly random keys generated by quantum devices. The combination of LPSec and quantum security represents a clear benefit for network operators as it allows extending the lifespan of the optical technologies currently deployed in their networks, and therefore the proposed solutions have great potential for its deployment in current transport networks.

Three application scenarios have been experimentally assessed, differing in how the quantum keys are managed and distributed between the two Tps involved in the optical transmission. In Scenario A, the two Tps are within the security perimeter of the QKD network. In Scenario B, only one of the Tps is within the QKD network, so keys are retrieved from a QRNG and distributed using LPSec. Finally, Scenario C extends Scenario B by implementing a PQC KEM to secure key exchanges. A summary of the analysis is presented in Table 2. The table presents the main advantages and disadvantages of each application scenario, also highlighting their key functionalities and requirements in order to provide some insight of the applicability and practical application of the different proposals.

Table 2. Summary of the scenarios and comparative analysis.

Scenario	Description	Advantages	Disadvantages	Security Level
A	Both Tps are in the QKD network. Keys are retrieved directly from local KMs and expanded using a PRNG.	• The highest security as keys are distributed using the QKD network. • Direct key retrieval from KMs.	• Large key expansion to match optical line speed (100 s Gb/s).	Highest
B	Only one Tp is within the QKD network. Keys are generated by a local QRNG and transmitted encrypted using the KxF header.	• Extends the reach of the security perimeter of the quantum network. • High throughput with QRNG. Moderate key expansion required.	• Key dependency vulnerability if one key is compromised.	Medium/High
C	Only one Tp is within the QKD network. A modified BIKE scheme is used for KEM, ensuring higher randomness. Session keys are generated and cyphertexts are securely exchanged.	• Extends the reach of the security perimeter of the quantum network. • Increased overall security by removing key dependency vulnerability.	• Complexity of PQC KEM.	Very High

We have illustrated the scenarios with optical connections between metro/core sites that are assumed to be inside the Quantum security perimeter, and between a metro/core site and an access site, where the latter is assumed to be outside the security perimeter. Of course, the applicability of the solutions investigated in this paper is not restricted to those scenarios but it can be generalized to any other scenario depending on the availability of Quantum systems in one or the two end sites of the optical connection.

Overall, the investigated application scenarios show a practical and efficient way to combine classical and quantum optical technologies for enhancing the security of high-speed optical transmission. This represents a noticeable achievement of the ALLEGRO project [28], which aims at designing and validating a novel end-to-end sliceable, reliable, and secure architecture for next-generation optical networks, achieving high transmission/switching capacity.

Author Contributions: Conceptualization, L.V., J.C. and M.A.; methodology, L.V. and M.R.; investigation, L.V., M.A. and J.P.B.; writing—original draft, L.V. and M.A.; writing—review and editing, L.V., J.C., S.B., M.R., M.A., L.O., J.P.B., V.M., A.P. and J.M.R.; supervision, A.P., J.M.R., V.M. and L.V. All authors have read and agreed to the published version of the manuscript.

Funding: This work has received funding from the European Union's Horizon Europe research and innovation program under grant agreement No. 101092766 (ALLEGRO Project) and from the ICREA Institution.

Institutional Review Board Statement: Not applicable.

Informed Consent Statement: Not applicable.

Data Availability Statement: Data are contained within the article.

Conflicts of Interest: Authors Antonio Pastor and Jose M. Rivas were employed by the company Telefonica Innovación Digital. The remaining authors declare that the research was conducted in the absence of any commercial or financial relationship that could be construed as a potential conflict of interest.

References

1. Shariati, B.; Velasco, L.; Pedreno-Manresa, J.-J.; Dochhan, A.; Casellas, R.; Muqaddas, A.S.; De Dios, O.G.; Canto, L.; Lent, B.; López De Vergara, J.E.; et al. Demonstration of Latency-Aware 5G Network Slicing over Edge Computing Enabled Optical Metro Network. *J. Opt. Commun. Netw.* **2022**, *14*, A81–A90. [CrossRef]
2. Velasco, L.; Wright, P.; Lord, A.; Junyent, G. Saving CAPEX by Extending Flexgrid-based Core Optical Networks towards the Edges. *J. Opt. Commun. Netw.* **2013**, *5*, A171–A183. [CrossRef]
3. Fok, M.; Wang, Z.; Deng, Y.; Prucnal, P. Optical Layer Security in Fiber-Optic Networks. *IEEE Trans. Inf. Forensics Secur.* **2011**, *6*, 725–736. [CrossRef]
4. Iqbal, M.; Velasco, L.; Costa, N.; Napoli, A.; Pedro, J.; Ruiz, M. Investigating Imperfect Cloning for Extending Quantum Communication Capabilities. *Sensors* **2023**, *23*, 7891. [CrossRef]
5. *FIPS-197*; Specification for the Advanced Encryption Standard (AES). National Institute of Standards and Technology (NIST): Gaithersburg, MD, USA, 2001.
6. Bernstein, D. ChaCha, a variant of Salsa20. *Workshop Rec. SASC* **2008**, *8*, 3–5.
7. Iqbal, M.; Velasco, L.; Costa, N.; Napoli, A.; Pedro, J.; Ruiz, M. LPsec: A Fast and Secure Cryptographic System for Optical Connections. *J. Opt. Commun. Netw.* **2022**, *14*, 278–288. [CrossRef]
8. Iqbal, M.; Velasco, L.; Ruiz, M.; Costa, N.; Napoli, A.; Pedro, J.; Comellas, J. Supporting Heterogenous Traffic on top of Point-to-Multipoint Light-Trees. *Sensors* **2023**, *23*, 2500. [CrossRef]
9. Diffie, W.; Hellman, M. New directions in cryptography. *IEEE Trans. Inf. Theory* **1976**, *IT-22*, 644–654. [CrossRef]
10. Pironio, S.; Acín, A.; Massar, S.; Boyer, A.; Matsukevich, D.; Maunz, P.; Olmschenk, S.; Hayes, D.; Luo, L.; Manning, T.; et al. Random numbers certified by Bell's theorem. *Nature* **2010**, *464*, 1021–1024. [CrossRef]
11. Martin, V.; Martinez-Mateo, J.; Peev, M. *Introduction to Quantum Key Distribution*; Wiley: Hoboken, NJ, USA, 2017.
12. Ahmadian, M.; Ruiz, M.; Comellas, J.; Velasco, L. Cost-effective ML-powered polarization-encoded quantum key distribution. *J. Light. Technol.* **2022**, *40*, 4119–4128. [CrossRef]
13. Aguado, A.; Lopez, D.; Pastor, A.; Lopez, V.; Brito, J.; Peev, M.; Poppe, A.; Martin, V. Quantum cryptography networks in support of path verification in service function chains. *J. Opt. Commun. Netw.* **2020**, *12*, B9–B19. [CrossRef]
14. Martin, V.; Brito, J.; Ortiz, L.; Mendez, R.; Buruaga, J.; Vicente, R.; Sebastián-Lombraña, A.; Rincon, D.; Perez, F.; Sanchez, C.; et al. MadQCI: A heterogeneous and scalable SDN QKD network deployed in production facilities. *npj Quantum Inf.* **2024**, *10*, 80. [CrossRef]
15. *Quantum Key Distribution (QKD); Application Interface*; ETSI GS QKD 004 v.2.1.1; ETSI: Sophia Antipolis, France, 2020.
16. Shor, P. Algorithms for quantum computation: Discrete logarithms and factoring. In Proceedings of the 35th Annual Symposium on Foundations of Computer Science, Santa Fe, NM, USA, 20–22 November 1994.
17. *NIST FIPS 203*; Module-Lattice-Based Key-Encapsulation Mechanism Standard. National Institute of Standards and Technology: Gaithersburg, MD, USA, 2024. [CrossRef]
18. *NIST FIPS 204*; Module-Lattice-Based Digital Signature Standard. National Institute of Standards and Technology: Gaithersburg, MD, USA, 2024. [CrossRef]
19. *NIST FIPS 205*; Stateless Hash-Based Digital Signature Standard. National Institute of Standards and Technology: Gaithersburg, MD, USA, 2024. [CrossRef]
20. NIST Post-Quantum Cryptography. Available online: https://csrc.nist.gov/Projects/post-quantum-cryptography/ (accessed on 10 October 2024).
21. Official Web Page of BIKE Suite. Available online: https://bikesuite.org (accessed on 10 October 2024).
22. Nosouhi, M.; Shah, S.; Pan, L.; Doss, R. Bit Flipping Key Encapsulation for the Post-Quantum Era. *IEEE Access* **2023**, *11*, 56181–56195. [CrossRef]
23. Hoque, S.; Aydeger, A.; Zeydan, E. Exploring Post Quantum Cryptography with Quantum Key Distribution for Sustainable Mobile Network Architecture Design. In Proceedings of the 4th Workshop on Performance and Energy Efficiency in Concurrent and Distributed Systems, Pisa, Italy, 3–4 June 2024; pp. 9–16.
24. Garms, L.; Paraiso, T.; Hanley, N.; Khalid, A.; Rafferty, C.; Grant, J.; Newman, J.; Shields, A.; Cid, C.; O'Neill, M. Experimental Integration of Quantum Key Distribution and Post-Quantum Cryptography in a Hybrid Quantum-Safe Cryptosystem. *Adv. Quantum Technol.* **2024**, *7*, 2300304. [CrossRef]
25. Ghashghaei, F.; Ahmed, Y.; Elmrabit, N.; Yousefi, M. Enhancing the Security of Classical Communication with Post-Quantum Authenticated-Encryption Schemes for the Quantum Key Distribution. *Computers* **2024**, *13*, 163. [CrossRef]
26. Rivas, J.; Melgar, A.; Morales, J.; Pastor, A.; Folgueira, J.; Poti, L.; Krilakis, K.; Syvridis, D.; Mandilara, A.; Velasco, L.; et al. A Security Plane Architecture for Ultra-low-energy, High-capacity Optical Transport Networks. In Proceedings of the 2024 International Conference on Quantum Communications, Networking, and Computing (QCNC), Kanazawa, Japan, 1–3 July 2024.

27. Brauer, M.; Vicente, R.; Buruaga, J.; Méndez, R.; Braun, R.; Geitz, M.; Rydlichkowski, P.; Brunner, H.; Fung, F.; Peev, M.; et al. Linking QKD Testbeds across Europe. *Entropy* **2024**, *26*, 123. [CrossRef]
28. HORIZON-CL4-2022 "Agile Ultra-Low Energy Secure Networks" (ALLEGRO). Available online: https://www.allegro-he.eu/ (accessed on 10 October 2024).
29. Blackman, D.; Vigna, S. Scrambled Linear Pseudorandom Number Generators. *arXiv* **2019**, arXiv:1805.01407. [CrossRef]
30. Ferguson, N.; Schneier, B.; Kohno, T. *Cryptography Engineering: Design Principles and Practical Applications*; Wiley Publishing: Hoboken, NJ, USA, 2010.
31. Brunner, H.; Fung, C.; Peev, M.; Méndez, R.; Ortiz, L.; Brito, J.; Martín, V.; Rivas-Moscoso, J.; Jiménez, F.; Pastor, A.; et al. Demonstration of a switched CV-QKD network. *EPJ Quantum Technol.* **2023**, *10*, 38. [CrossRef]
32. Quside. Available online: https://quside.com/ (accessed on 10 October 2024).
33. Mitchell, M.; Abellán, C.; Amaya, W. Strong experimental guarantees in ultrafast quantum random number generation. *Phys. Rev. A* **2015**, *91*, 012314. [CrossRef]
34. Abellán, C.; Amaya, W.; Mitrani, D.; Pruneri, V.; Mitchell, M. Generation of fresh and pure random numbers for loophole-free Bell tests. *Phys. Rev. Lett.* **2015**, *115*, 250403. [CrossRef]
35. Velasco, L.; González, P.; Ruiz, M. Distributed Intelligence for Pervasive Optical Network Telemetry. *J. Opt. Commun. Netw.* **2023**, *15*, 676–686. [CrossRef]
36. OpenCV. Available online: https://pypi.org/project/opencv-python/ (accessed on 10 October 2024).

sensors

MDPI

Article

Compact Silicon-Arrayed Waveguide Gratings with Low Nonuniformity

Chengkun Yang [1,*], Zhonghao Zhou [1,*], Xudong Gao [1], Zhengzhu Xu [1], Shoubao Han [1], Yuhua Chong [1], Rui Min [2], Yang Yue [3] and Zongming Duan [1]

[1] Center for East China Research Institute of Electronics Engineering, Hefei 230036, China; gaoxudong311@126.com (X.G.); zhenzhupearl@163.com (Z.X.); shoubao_cetc@126.com (S.H.); chchyh@163.com (Y.C.); duanzongm@163.com (Z.D.)

[2] Center for Cognition and Neuroergonomics, State Key Laboratory of Cognitive Neuroscience and Learning, Beijing Normal University, Zhuhai 519087, China; rumi@doctor.upv.es

[3] School of Information and Communications Engineering, Xi'an Jiaotong University, Xian 710049, China; yueyang@xjtu.edu.cn

* Correspondence: yangchk1225@163.com (C.Y.); zhouzhonghao0901@163.com (Z.Z.)

Abstract: Array waveguide gratings (AWGs) have been widely used in multi-purpose and multi-functional integrated photonic devices for Microwave photonics (MWP) systems. In this paper, we compare the effect of output waveguide configurations on the performance of AWGs. The AWG with an output waveguide converging on the grating circle had larger crosstalk and lower nonuniformity. We also fabricated a 1 × 8 AWG with an output waveguide converging onto the SOI's grating circle, whose central operation wavelength was around 1550 nm. The fabricated AWG has a chip size of 500 μm × 450 μm. Experimental results show that the adjacent channel crosstalk is −12.68 dB. The center channel insertion loss, as well as 3 dB bandwidth, are 4.18 dB and 1.22 nm at 1550 nm, respectively. The nonuniformity is about 0.494 dB, and the free spectral range is 19.4 nm. The proposed AWG is expected to play an important role in future MWP systems given its good nonuniformity and insertion loss level.

Keywords: microwave photonics; array waveguide gratings; integrated photonic devices

1. Introduction

Microwave photonics (MWP), where photonic devices and techniques are required to generate, manipulate, transport, and measure highspeed radiofrequency (RF) signals, has attracted growing research interest [1–3]. Conventional MWP systems consist of relatively expensive discrete fiber optic components, which have the disadvantages of bulky size and susceptibility to external perturbations. The advent of the advanced photonic integration platform, which allows the fabrication of multi-purpose and multi-functional circuits on a photonic chip [4], offers the required framework for MWP technologies. By these means, stable, cost-effective, and scalable integrated photonic devices can be developed.

Recently, multi-purpose and multi-functional integrated photonic devices have been developed for many MWP applications, such as optical beamformers [5,6], optical true time delay [7,8], millimeter wave signal generation [9], and wavelength interrogation [10], etc. Most integrated photonic devices consist of discrete integrated photonic components, especially array waveguide gratings (AWGs). AWGs are one kind of angular dispersion passive device where multi-beam interference is exploited to introduce certain optical path differences through the array waveguide and interferometrically image different wavelengths into the slab waveguide, which is received by different output waveguides. Owing to their distinguished advantages, such as high resolution, multi-channel, high integration, low crosstalk, and low loss, AWGs have been widely applied in multi-spectral imaging [11,12], spectrum analysis [13–15], dense-wavelength-division multiplexing [16,17], astronomy [18], biomedical applications [19], and optical sensing [20,21].

AWGs have been developed based on various materials such as silica [22–24], indium phosphide (InP) [25,26], polymer [19,27], silicon nitride [28–30], thin-film lithium niobate (TFLN) [31–33], and silicon [34–36]. Silica-based AWGs have been successfully commercialized due to their minimal loss and mature processing technology. However, their large footprint, caused by silica's low refractive index, hinders high-density integration. InP-based AWGs exhibit excellent optoelectronic properties and waveguide integration capabilities, allowing for efficient optical coupling and transmission. Although they benefit from relatively mature processing technology, they are hindered by high production costs, poor thermal stability, and a low damage threshold. Polymer-based AWGs offer low cost, flexibility, and ease of manufacture. However, they typically exhibit relatively low thermal stability and damage thresholds. Silicon nitride-based AWGs provide a moderately high refractive index and low absorption loss. However, they also have disadvantages, including stress issues and limited thermo-optic effects. TFLN-based AWGs use the electro-optic effect to fast-tune their operating wavelength. However, their manufacturing process is complex, and they suffer from low mechanical strength, poor thermal stability, and high costs. Low-index materials have managed to demonstrate excellent performance and commercial maturity. Furthermore, highly integrated compact silicon photonics circuits can provide high refractive index contrast in silicon nanowire waveguides, reducing the waveguide bending radius and miniaturizing the device size. Silicon photonics coms-compatible technology is also suitable for fabricating silicon-based AWGs, showing good feasibility in mass production at low cost.

AWGs can have various structural forms, with the main differences lying in the arrangement of the phase array. The common structures are S-shaped [37,38], arc-shaped [11,39,40], and horseshoe-shaped [41–43]. Horseshoe-shaped AWGs have a constant length for all array arms, which contributes to better performance and is straightforward from a design perspective. Arc-shaped AWGs minimize the number of waveguide junctions, but junction losses and mode conversion at these junctions can degrade performance. S-shaped AWGs exhibit lower dispersion values due to S-bend waveguide arrays, where the dispersion of one curved section is counteracted by a second section with opposite curvature [44]. Other novel structures have been proposed to improve the performance of arrayed waveguide gratings, such as overlapped AWGs [45], reflection-shaped AWGs [46], etc.

However, silicon AWGs do not perform well in terms of crosstalk and insertion loss characteristics. It should be noted that a few factors deteriorate the optical transmission characteristics of AWGs, including the coupling between single-mode waveguides and reflections induced by mode mismatching between the single-mode waveguide and the free propagation region (FPR). To solve these issues, tapers are usually used to establish efficient connections between single-mode waveguides and FPRs. The tapers in an AWG should be long enough to lower the optical loss level and widen the operation bandwidth of each channel. Different types of tapers have been developed to improve AWG performance. Linear tapers proved efficient in lowering propagation loss but ineffective in suppressing crosstalk between different channels [46,47]. Double-etched tapers can solve both propagation loss and crosstalk issues, but the fabrication procedure is complicated, where two steps of etching are normally required [34,48]. Compact parabolic tapers fabricated based on simple procedures have been employed in AWGs to reduce insertion loss and suppress crosstalk at the user terminal [43,49]. The loss of uniformity directly affects its high-speed modulation performance, tolerable wavelength drift, and the performance of cascaded systems. Several methods have been investigated to improve the loss uniformity, such as 2×1 multimode interference couplers [50], mode-field converters [51], auxiliary waveguides [52], etc.

In this study, we designed and fabricated a 1×8 arc-shaped AWG on a SOI platform with a central operation wavelength around 1550 nm. Parabolic tapers were introduced to bridge the single-mode waveguides and FPRs for better insertion loss and crosstalk performances while maintaining chip compactness. We compared the effect of output waveguide configurations on the performance of AWGs. Then, we numerically simulated the impacts of the array waveguide number on its optical transmission characteristics to optimize the primary parameters of AWGs. We fabricated AWGs based on the simulation results, and experimental results indicated that the AWGs outperform in insertion loss and nonuniformity.

2. Theory and Design

An AWG consists of input/output waveguides, input/output star coupler, and array waveguides. Array waveguides connect input/output star couplers with identical Rowland circle designs through a constant path length difference between neighboring waveguides. AWG principles are described as follows: When wideband light is launched into the input star coupler from the central waveguide, it diffracts and forms a Gaussian optical field. This field diverges into each channel of array waveguides. Then, light propagates in the arrayed waveguides and arrives at an output aperture with a phase difference of $n_{eff}k_0\Delta L$ between the neighboring waveguides, where $k_0 = 2\pi/\lambda$ is the wavenumber and n_{eff} is the effective refractive index of the fundamental mode propagating in each waveguide. After transmitting through array waveguides, the light output diverges in the output coupler and shifts along the image plane, with the wavelength deviating from the central wavelength λ_c. The output waveguides are arranged over this image plane and each collects one section of the spatially separated spectrum. The input and output waveguides satisfy the equation:

$$n_s d_a \sin\theta_i + n_{eff}\Delta L + n_s d_a \sin\theta_o = m\lambda \quad (1)$$

For central input, θ_i and θ_o tend to 0, the constant path length difference between adjacent array waveguides satisfies the equation:

$$\Delta L = m\lambda_c/n_{eff} \quad (2)$$

The pitch of output waveguides is provided by the equation:

$$d_o = \frac{\Delta\lambda D m}{d_a}\frac{n_g}{n_{eff}n_s} \quad (3)$$

where d_a is the distance between the adjacent array waveguides; $\Delta\lambda$ is the channel spacing; D is the diameter of the Rowland circle; m is the diffraction order; n_g is the group index of the fundamental mode of the arrayed waveguide; n_s is the effective index of the FPR; θ_i and θ_o are the angles between the input/output waveguide and the central channel waveguide, respectively.

The structural design of the taper between the single-mode waveguide and FPR is an important issue affecting AWG performance. Compact parabolic tapers reduce insertion loss and receiver crosstalk in AWGs. In this study, the parabolic taper was optimized to improve AWG performance. The profile of the parabolic taper was designed using the following equation [11]:

$$y = A(Lt - x)^m + B \quad (4)$$

where A and B are constant, as determined by the waveguide width and taper width; m is the power of the exponent and Lt is the total taper length.

Since the waveguide and receiver aperture designs already determine the input and output waveguide widths, only two parameters need optimization: the length of the taper and its geometry, which is defined by the exponent m. In the study [11], various taper geometries with m ranging from 1 to 4 were simulated. The authors found that the parabolic taper with m = 2 achieved the most efficient adiabatic coupling, as shown in Figure 1a. Figure 1c shows the transmission of the parabolic taper was investigated for different taper lengths. In these simulations, light was defined to propagate from the wide to the narrow waveguide. The mode conversion efficiency increased with increments in taper length and a 5.0 μm long parabolic taper achieved transmission over 97%. The simulated field intensity for a parabolic taper length of 5 μm at a wavelength of 1550 nm is shown in Figure 1b.

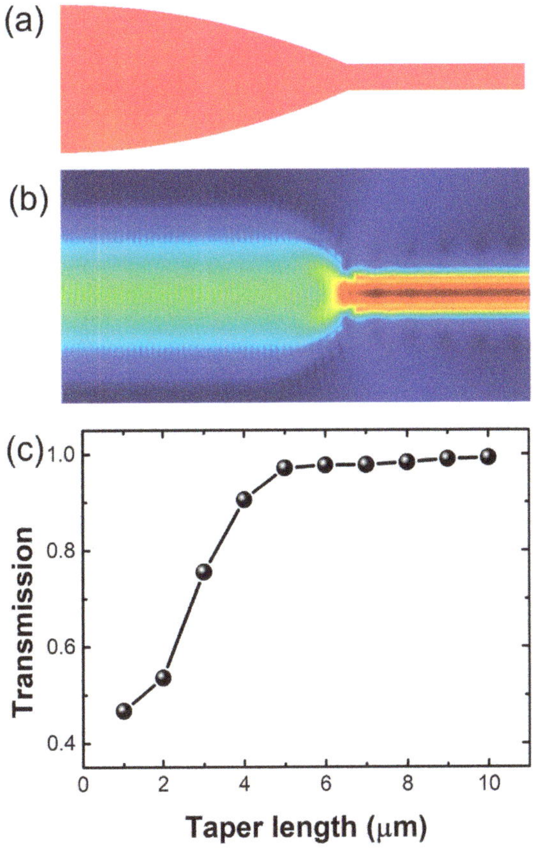

Figure 1. (**a**) Schematic diagram of a parabolic taper; (**b**) simulated field intensity for a taper length of 5 μm at a wavelength of 1550 nm; (**c**) transmission of tapers as a function of taper length at a wavelength of 1550 nm.

To investigate the effect of output waveguide configurations on the performance of AWGs, two AWG types were designed, as shown in Figure 2. Both include Rowland circle designs and have the same parameters except for the output waveguide configurations. In the Type-1 design, the input/output waveguides all point toward the center of the Rowland circle with a radius of D/2, as shown in Figure 2a. In the Type-2 design, the input/output waveguides lie on the Rowland circle and converge on the grating circle, as shown in Figure 2b. The array waveguides of two AWG types lie on the grating circle with a radius of D.

Channel spacing and the number of output waveguides are specified by $\Delta\lambda$ and N_o, respectively. The widths of the input waveguide, arrayed waveguides, and output waveguides are set to the same value. The distance between the adjacent array waveguides (d_a) is determined by the fabrication resolution of the lithography. Other parameters of the AWG are calculated based on the theoretical model in [28]. The optimal parameters of the AWG are shown in Table 1. The AWG transmission spectrum is simulated based on the 2.5-dimensional finite–difference time–domain (FDTD) method, showing far-field projection across the star coupler and phase errors through the array waveguide.

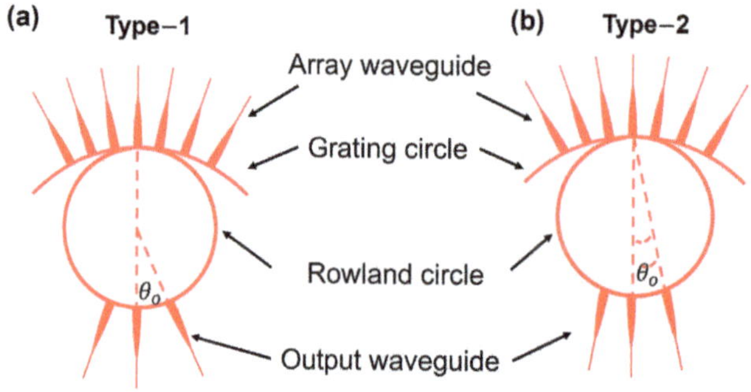

Figure 2. Schematic layout of output star coupler of the AWG structure. (**a**) Type-1 output waveguides all point toward the center of the Rowland circle; (**b**) Type-2: output waveguide converges on the grating circle.

Table 1. Primary design parameters of the AWG.

Parameter	Symbol	Value	Units
Channel spacing	$\Delta\lambda$	1.6	nm
Central wavelength	λ_c	1549.3	nm
Length increment	ΔL	28.29	µm
Number of input/output channels	N_i/N_o	1/8	
Spacing of input/output waveguide	d_i/d_o	2.7	µm
Spacing of array waveguides	d_a	1.7	µm
Diffraction order	m	43	
Gap between adjacent tapers	g_a	0.2	µm
Length of the tapers	L_t	5	µm
Diameter of Rowland circle	D	103	µm
Waveguide width	w	0.45	µm

The number of array waveguides impacts insertion loss, adjacent channel crosstalk, nonuniformity, and overall device size. Adjacent channel crosstalk is defined as the difference between the peak insertion loss within the channel passband and the minimum insertion loss within the passband of adjacent channels. Conversely, non-adjacent channel crosstalk is the difference between the peak insertion loss within the channel passband and the minimum insertion loss within the passband of all non-adjacent channels. According to the International Telecommunication Union standard ITU-T G.671, the channel passband width is defined as a quarter of the AWG channel spacing. As shown in Figure 3, with increasing array waveguides, the insertion loss of AWGs decreases and the adjacent channel crosstalk of the AWG increases. The reason for this is that the light field at the receiving end intensifies with increasing arrayed waveguide increments, thereby lowering AWG insertion loss to some extent. However, with the increased number of array waveguides, the phase error introduced through the fabrication process may also increase the adjacent channel crosstalk of the device. Upon comparing AWG to two output waveguide configurations, output waveguide configurations affect crosstalk and nonuniformity. Type-2 has a lower nonuniformity. Both configurations do not affect the insertion loss. Simulated spectral responses for the AWG with 20 array waveguides are shown in Figure 4. From Figure 4a,b, it can be concluded that the adjacent channel crosstalk is −14.69 dB. The insertion loss and 3 dB bandwidth of the central channel are 3.25 dB and 1.09 nm at 1550 nm, respectively. The transmission nonuniformity is about 1.96 dB, and the free spectral range is 19.42 nm. From Figure 4c,d, it can be concluded that the adjacent channel crosstalk is −15.28 dB.

The insertion loss and 3 dB bandwidth of the central channel are 3.17 dB and 1.08 nm at 1550 nm, respectively. The transmission nonuniformity is about 0.627 dB, and the free spectral range is 19.42 nm.

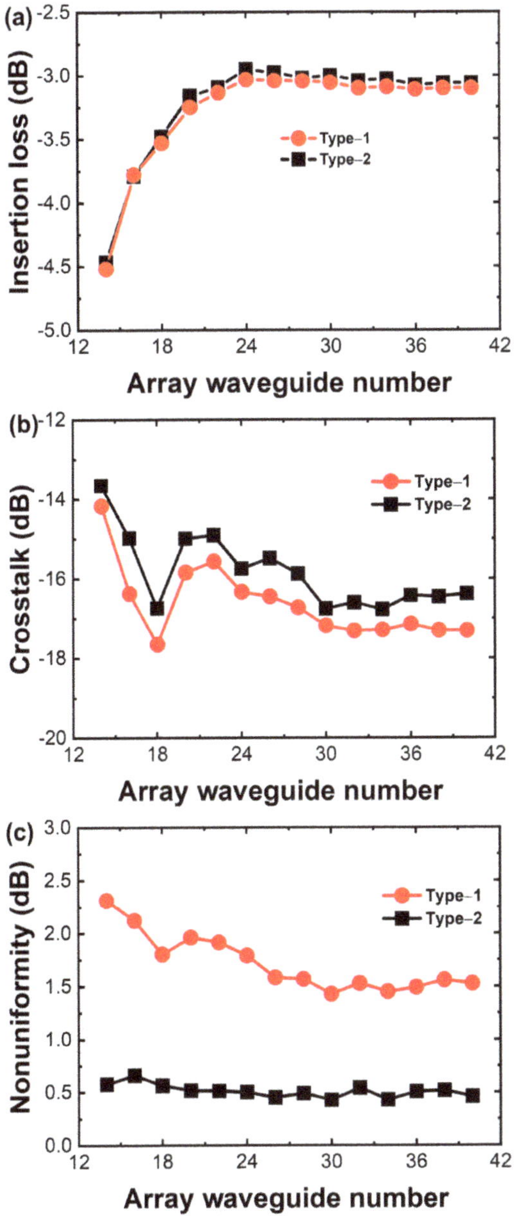

Figure 3. Simulated AWG performance as a function of array waveguides. (**a**) Insertion loss; (**b**) adjacent channel crosstalk; (**c**) nonuniformity.

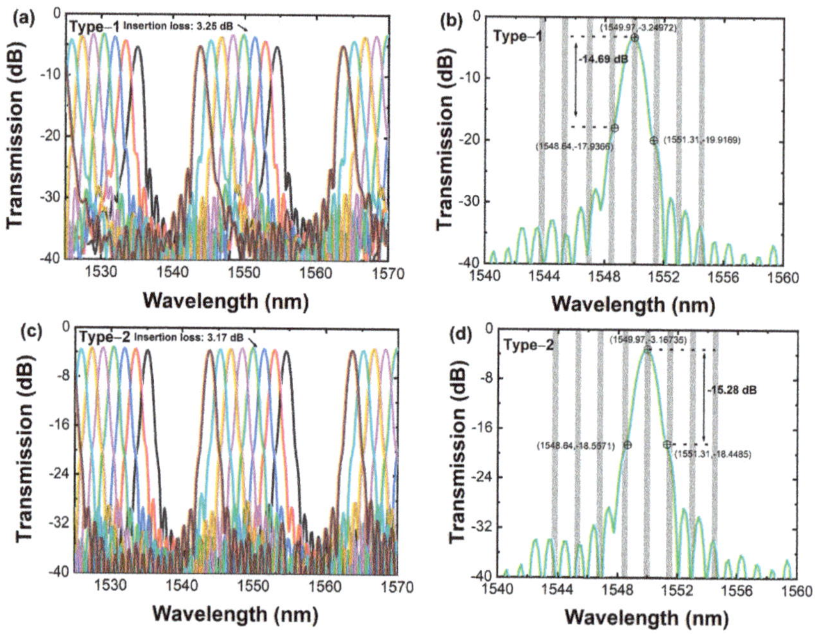

Figure 4. Type-1: (**a**) simulated spectral responses for the AWG with 20 array waveguides; (**b**) simulated spectral responses for the central output port. Type-2: (**c**) simulated spectral responses for the AWG with 20 array waveguides; (**d**) simulated spectral responses for the central output port.

3. Measurements and Discussion

For comparison, the AWG with output waveguides converging on the grating circle (Type-2) was chosen for fabrication. The optimally designed AWGs were fabricated by Chongqing United Microelectronics Center (CUMEC, Chongqing, China) using the CSiP180AL technology platform. This platform provides MPW services based on standard SOI with 2 μm BOX and 220 nm top silicon. Its silicon photonic chip manufacturing processes are compatible with standard CMOS processes, including deep UV lithography and inductively coupled plasma dry etching. The microscopic image of the AWG and enlarged microscope images of the output star coupler and parabolic tapers are shown in Figure 5. The total footprint of the fabricated AWG is 500 μm × 450 μm. A coupling platform was used to test the performance of the fabricated device. A broadband light from a superluminescent diode (SLD) was launched into a single-mode fiber and then coupled to the input waveguide via a grating coupler. The light from the output waveguide was connected to an optical spectrum analyzer (OSA, AQ6370C, Yokogawa, Japan) with single-mode fibers to collect transmission light.

The transmission spectrum of the AWG is obtained by subtracting measurements of the same grating coupler taken back-to-back, as shown in Figure 6. As shown in Figure 6b, the adjacent channel crosstalk is −12.68 dB and the non-adjacent channel crosstalk is −14.66 dB. The center channel insertion loss and 3 dB bandwidth are 4 dB and 1.22 nm for the central output port, respectively. For the edge output port, the adjacent channel crosstalk is −12.15 dB and the non-adjacent channel crosstalk is −15.32 dB, as shown in Figure 6c. The nonuniformity is about 0.494 dB and the free spectral range is 19.4 nm.

The nonuniformity matches well with the simulation results while the channel central wavelength shifts by 0.79 nm, the insertion loss deteriorates by 1.01 dB and the channel crosstalk varies by 2.6 dB, are shown in Table 2. The AWG's crosstalk and insertion loss is worse than mainstream AWGs. The bending loss of the arrayed waveguides is not considered in the simulation process because the array waveguides normally experience large bending. Fabrication error is the main reason accounting for the slight deviation in measurement results from the originally designed values. Due to fabrication errors, the sidewall of the waveguides is no longer smooth, resulting in large scattering losses. The roughness of the waveguides is the largest factor that deteriorates the insertion loss feature. Additionally, the crosstalk is more sensitive to the sidewall roughness. The incremental overlap between the waveguide guiding mode and the sidewall of the waveguide leads to higher crosstalk. In the designed AWG, the width of the bent waveguide is narrower than the waveguide width of mainstream AWGs. As the waveguide becomes narrower, the guiding mode in the bent waveguide tends to strongly overlap with the sidewall of the waveguides, causing larger phase errors and higher crosstalk. Due to fabrication errors, the refractive index of the waveguide may deviate from the design parameters, causing the experimentally central wavelength to shift from the original designed value.

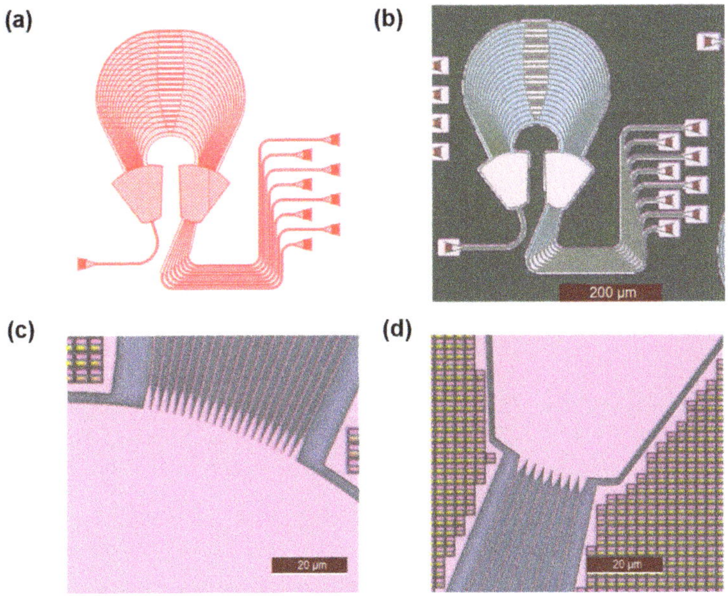

Figure 5. Illustrations of the fabricated AWGs. (**a**) A snippet from the GDSII mask layout file of an AWG; (**b**) microscopic picture of an AWG; (**c**) microscopic picture of the parabolic tapers between the FPR and array waveguides; (**d**) microscopic picture of the parabolic tapers between the output waveguides and FPR.

Table 2. Simulation and measurement results of the AWG.

Parameter	Simulation	Measurement	Units
Central wavelength	1549.3	1550.09	nm
Insertion loss	3.17	4.18	dB
Crosstalk	−15.28	−12.68	dB
Nonuniformity	0.627	0.494	dB
3-dB bandwidth	1.08	1.22	dB
FSR	19.42	19.4	nm

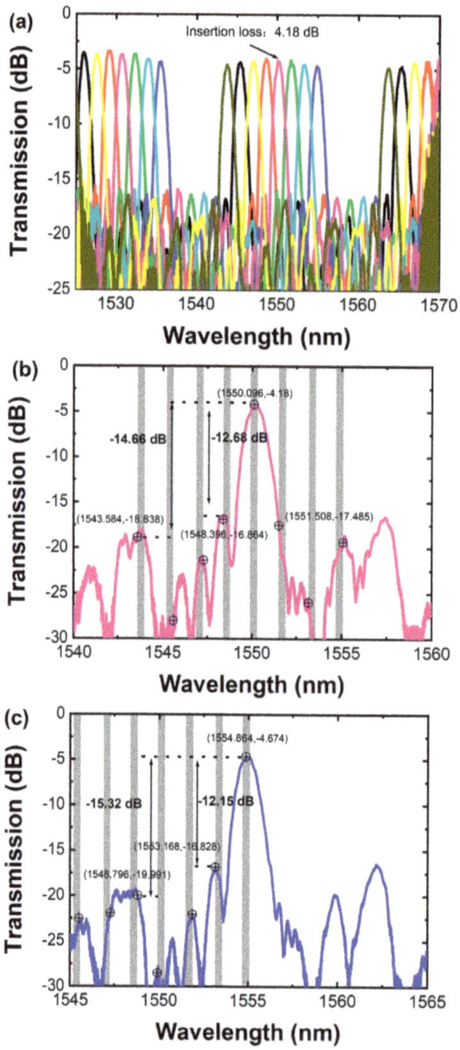

Figure 6. (**a**) Transmission spectrum for all output ports; (**b**) the central output port; (**c**) the edge output port.

4. Conclusions

In conclusion, a 1 × 8 AWG on a SOI platform with a 1550 nm operation wavelength was designed and fabricated. The output waveguide configurations of AWGs affect crosstalk and nonuniformity. The AWG with an output waveguide converging on the grating circle had larger crosstalk and lower nonuniformity. The fabricated AWG with an output waveguide converged on the grating circle had an area of 500 μm × 450 μm. Experimental results indicated that the single-channel crosstalk was −12.68 dB, and the center channel insertion loss and 3 dB bandwidth were 4.18 dB and 1.22 nm at 1550 nm, respectively. The nonuniformity was about 0.494 dB, and the free spectral range was 19.4 nm. For MWP systems, such as optical beamformers, optical true time delay, millimeter wave signal generation, and wavelength interrogation, silicon AWGs were expected to have the lowest crosstalk and insertion loss. The specific requirements for crosstalk and insertion loss performance in silicon AWGs depend on the system parameters, which are still under investigation and will be reported in the near future.

Author Contributions: Conceptualization, C.Y. and Z.Z.; methodology, C.Y., Z.Z., X.G., Z.X. and S.H.; validation, C.Y., Z.Z. and Y.C.; formal analysis, C.Y., Z.Z. and Y.C.; investigation, C.Y., Z.Z. and X.G.; data curation, C.Y. and Z.Z.; writing—original draft preparation, C.Y.; writing—review and editing, C.Y., R.M. and Y.Y.; visualization, C.Y.; supervision, Z.D.; project administration, Z.X., S.H. and Z.D.; funding acquisition, Z.X., S.H. and Z.D. All authors have read and agreed to the published version of the manuscript.

Funding: This study was financed by the Key R&D Program of Anhui Province (2022a05020027, 202104a05020052) and the China National Natural Science Foundation of China (62075247).

Institutional Review Board Statement: Not applicable.

Informed Consent Statement: Not applicable.

Data Availability Statement: Data underlying the results presented in this paper are not publicly available at this time but may be obtained from the authors upon reasonable request.

Conflicts of Interest: The authors declare no conflicts of interest.

References

1. Seeds, A.J.; Williams, K.J. Microwave photonics. *J. Light. Technol.* **2006**, *24*, 4628–4641. [CrossRef]
2. Minasian, R.A. Photonic signal processing of microwave signals. *IEEE Trans. Microw. Theory Tech.* **2006**, *54*, 832–846. [CrossRef]
3. Marpaung, D.; Roeloffzen, C.; Heideman, R.; Leinse, A.; Sales, S.; Capmany, J. Integrated microwave photonics. *Laser Photonics Rev.* **2013**, *7*, 506–538. [CrossRef]
4. Munoz, P.; Domenech, J.D.; Artundo, I.; den Bested, J.H.; Capmany, J. Evolution of fabless generic photonic integration. In Proceedings of the 2013 15th International Conference on Transparent Optical Networks (ICTON), Cartagena, Spain, 23–27 June 2023; pp. 1–3.
5. Doménech, J.; Muñoz, P.; Capmany, J. Discretely tunable microwave photonics beamformer based on ring resonators and arrayed waveguide gratings. In Proceedings of the Silicon Photonics and Photonic Integrated Circuits II, Brussels, Belgium, 12–16 April 2010; pp. 373–380.
6. Hu, G.; Cui, Y.; Yang, Y.; Yang, J.; Tian, D.; Yun, B.; Zhang, R. Optical beamformer based on diffraction order multiplexing (DOM) of an arrayed waveguide grating. *J. Light. Technol.* **2019**, *37*, 2898–2904. [CrossRef]
7. Vidal, B.; Corral, J.L.; Piqueras, M.A.; Marti, J. Optical delay line based on arrayed waveguide gratings' spectral periodicity and dispersive media for antenna beamforming applications. *IEEE J. Sel. Top. Quantum Electron.* **2002**, *8*, 1202–1210. [CrossRef]
8. Wang, X.; Zhou, L.; Lu, L.; Wang, X.; Chen, J. Integrated optical delay line based on a loopback arrayed waveguide grating for radio-frequency filtering. *IEEE Photonics J.* **2020**, *12*, 1–11. [CrossRef]
9. Carpintero, G.; Rouvalis, E.; Ławniczuk, K.; Van Dijk, F.; Chtioui, M.; Renaud, C.C.; Leijtens, X.J.; Bente, E.A.; Seeds, A. Millimeter-wave signal generation by optical heterodyne of two channels from an arrayed waveguide grating-based multi-wavelength laser. In Proceedings of the 2012 IEEE International Topical Meeting on Microwave Photonics, Noordwijk, The Netherlands, 11–14 September 2012; pp. 51–54.
10. Guo, H.; Xiao, G.; Mrad, N.; Albert, J.; Yao, J. Wavelength interrogator based on closed-loop piezo-electrically scanned space-to-wavelength mapping of an arrayed waveguide grating. *J. Light. Technol.* **2010**, *28*, 2654–2659. [CrossRef]
11. Zhang, Z.; Wang, Y.; Tsang, H.K. Ultracompact 40-channel arrayed waveguide grating on silicon nitride platform at 860 nm. *IEEE J. Quantum Electron.* **2019**, *56*, 1–8. [CrossRef]
12. Akca, B.; Považay, B.; Alex, A.; Wörhoff, K.; De Ridder, R.; Drexler, W.; Pollnau, M. Miniature spectrometer and beam splitter for an optical coherence tomography on a silicon chip. *Opt. Express* **2013**, *21*, 16648–16656. [CrossRef] [PubMed]
13. Zhang, Z.; Wang, Y.; Wang, J.; Yi, D.; Chan, D.W.U.; Yuan, W.; Tsang, H.K. Integrated scanning spectrometer with a tunable micro-ring resonator and an arrayed waveguide grating. *Photonics Res.* **2022**, *10*, A74–A81. [CrossRef]
14. Zhang, Z.; Wang, Y.; Tsang, H.K. High resolution silicon nitride spectrometer by integrating micro-rings and arrayed waveguide gratings in tandem. In Proceedings of the 2019 IEEE 16th International Conference on Group IV Photonics (GFP), Singapore, 28–30 August 2019; pp. 1–2.
15. Ismail, N.; Choo-Smith, L.-P.; Wörhoff, K.; Driessen, A.; Baclig, A.; Caspers, P.; Puppels, G.; De Ridder, R.; Pollnau, M. Raman spectroscopy with an integrated arrayed-waveguide grating. *Opt. Lett.* **2011**, *36*, 4629–4631. [CrossRef] [PubMed]
16. Bogaerts, W.; Selvaraja, S.K.; Dumon, P.; Brouckaert, J.; De Vos, K.; Van Thourhout, D.; Baets, R. Silicon-on-insulator spectral filters fabricated with CMOS technology. *IEEE J. Sel. Top. Quantum Electron.* **2010**, *16*, 33–44. [CrossRef]
17. Zou, J.; Ma, X.; Xia, X.; Wang, C.; Zhang, M.; Hu, J.; Wang, X.; He, J.-J. Novel wavelength multiplexer using (N+ 1)×(N+ 1) arrayed waveguide grating and polarization-combiner-rotator on SOI platform. *J. Light. Technol.* **2021**, *39*, 2431–2437. [CrossRef]

18. Stoll, A.; Zhang, Z.; Haynes, R.; Roth, M. High-Resolution Arrayed-Waveguide-Gratings in Astronomy: Design and Fabrication Challenges. *Photonics* **2017**, *4*, 30–32. [CrossRef]
19. Li, H.; Wang, Y.; Lin, Z.; Zhang, S.; Zhang, Z.; Zhang, C.; Tang, C.; Li, E.; Prades, J.D. Design, fabrication and characterization of arrayed waveguide grating devices based on different polymers. *Opt. Commun.* **2023**, *530*, 129139. [CrossRef]
20. Weng, S.; Yuan, P.; Zhuang, W.; Zhang, D.; Luo, F.; Zhu, L. SOI-Based Multi-Channel AWG with Fiber Bragg Grating Sensing Interrogation System. *Photonics* **2021**, *8*, 214. [CrossRef]
21. Li, S.; Yuan, P.; Li, T.; Li, K.; Zhu, L. SOI-based 15-channel arrayed waveguide grating design for fiber Bragg grating interrogator. *Photonics Nanostruct.-Fundam. Appl.* **2023**, *53*, 101113. [CrossRef]
22. Stoll, A.; Madhav, K.; Roth, M. Performance limits of astronomical arrayed waveguide gratings on a silica platform. *Opt. Express* **2020**, *28*, 39354–39367. [CrossRef] [PubMed]
23. Li, S.; Zhang, J.; Wang, L.; You, J.; Wang, Y.; Yin, X.; Chen, J.; Sun, B.; An, J.; Wu, Y. The O-band 20-channel 800 GHz Arrayed Waveguide Grating based on silica platform for 1 Tb/s or higher-speed communication system. *Opt. Laser Technol.* **2022**, *156*, 108475. [CrossRef]
24. Seyringer, D.; Serecunova, S.; Uherek, F.; Seyringer, H.; Chovan, J. 256-Channel 10-GHz AWG demultiplexer for ultra-dense WDM. In Proceedings of the 2023 23rd International Conference on Transparent Optical Networks (ICTON), Bucharest, Romania, 2–6 July 2023; pp. 1–4.
25. Bissessur, H.; Gaborit, F.; Martin, B.; Pagnod-Rossiaux, P.; Peyre, J.-L.; Renaud, M. 16channel phased array wavelength demultiplexer on InP with low polarisation sensitivity. *Electron. Lett.* **1994**, *30*, 336–337. [CrossRef]
26. Schindler, A.; Garten, A.-B.; Boerma, H.; Ganzer, F.; Runge, P.; Schell, M. Low crosstalk INP-Based arrayed waveguide grating PIC. In Proceedings of the 2022 IEEE Photonics Conference (IPC), Vancouver, BC, Canada, 13–17 November 2022; pp. 1–2.
27. Takenobu, S.; Kuwana, Y.; Takayama, K.; Sakane, Y.; Mitsufumi, O.; Hideki, S.; Norbert, K.; Brinker, W.; Huihai, Y.; Zawadzki, C. All-polymer 8x8 AWG wavelength router using ultra low loss polymer optical waveguide material (CYTOPTM). In Proceedings of the Optical Fiber Communication Conference, San Diego, CA, USA, 24–28 February 2008; p. JWA32.
28. Caut, A.; Girardi, M.; Torres-Company, V.; Larsson, A.; Karlsson, M. Channel Scalability of Silicon Nitride (De-) multiplexers for Optical Interconnects at 1 µm. *J. Light. Technol.* **2023**, *42*, 276–286. [CrossRef]
29. Shang, K.; Pathak, S.; Qin, C.; Yoo, S.B. Low-loss compact silicon nitride arrayed waveguide gratings for photonic integrated circuits. *IEEE Photonics J.* **2017**, *9*, 1–5. [CrossRef]
30. van Wijk, A.; Doerr, C.R.; Ali, Z.; Karabiyik, M.; Akca, B.I. Compact ultrabroad-bandwidth cascaded arrayed waveguide gratings. *Opt. Express* **2020**, *28*, 14618–14626. [CrossRef]
31. Tu, H.; Zhang, Y.; Guo, W. Arrayed Waveguide Grating Based on Z-cut Lithium Niobate Platform. In Proceedings of the 2023 Opto-Electronics and Communications Conference (OECC), Shanghai, China, 2–6 July 2023; pp. 1–3.
32. Yu, Y.; Yu, Z.; Zhang, Z.; Tsang, H.K.; Sun, X. Wavelength-division multiplexing on an etchless lithium niobate integrated platform. *ACS Photonics* **2022**, *9*, 3253–3259. [CrossRef]
33. van Wijk, A.C.; Doerr, C.R.; Akca, I.B. Custom Arrayed Waveguide Gratings with Improved Performance. *Adv. Photonics Res.* **2023**, *4*, 2300198. [CrossRef]
34. Wang, J.; Sheng, Z.; Li, L.; Pang, A.; Wu, A.; Li, W.; Wang, X.; Zou, S.; Qi, M.; Gan, F. Low-loss and low-crosstalk 8 × 8 silicon nanowire AWG routers fabricated with CMOS technology. *Opt. Express* **2014**, *22*, 9395–9403. [CrossRef] [PubMed]
35. Zou, J.; Le, Z.; Hu, J.; He, J.-J. Performance improvement for silicon-based arrayed waveguide grating router. *Opt. Express* **2017**, *25*, 9963–9973. [CrossRef] [PubMed]
36. Xie, C.; Zou, X.; Zou, F.; Zhang, Y.; Yan, L.; Pan, W. High-performance 32-channel silicon arrayed waveguide grating with 100 GHz spacing. *IEEE Photonics Technol. Lett.* **2023**, *35*, 369–372. [CrossRef]
37. Ismail, N.; Sun, F.; Sengo, G.; Wörhoff, K.; Driessen, A.; De Ridder, R.M.; Pollnau, M. Improved arrayed-waveguide-grating layout avoiding systematic phase errors. *Opt. Express* **2011**, *19*, 8781–8794. [CrossRef] [PubMed]
38. Adar, R.; Henry, C.H.; Dragone, C.; Kistler, R.C.; Milbrodt, M.A. Broad-band array multiplexers made with silica waveguides on silicon. *J. Light. Technol.* **1993**, *11*, 212–219. [CrossRef]
39. Cheung, S.; Su, T.; Okamoto, K.; Yoo, S. Ultra-compact silicon photonic 512 × 512 25 GHz arrayed waveguide grating router. *IEEE J. Sel. Top. Quantum Electron.* **2013**, *20*, 310–316. [CrossRef]
40. Gatkine, P.; Veilleux, S.; Hu, Y.; Bland-Hawthorn, J.; Dagenais, M. Arrayed waveguide grating spectrometers for astronomical applications: New results. *Opt. Express* **2017**, *25*, 17918–17935. [CrossRef] [PubMed]
41. Li, L.; Xiong, H.-N.; Li, X.; Chen, X.; Wang, C.; Le, Z.; Wang, X.; Ma, X.; Zou, J. Crosstalk reduction for Arrayed waveguide gratings on Silicon-on-Insulator platform. *Opt. Laser Technol.* **2024**, *175*, 110817. [CrossRef]
42. Fu, X.; Zhang, L.; Yang, L. Experimental Demonstration of a Horseshoe-shaped 16-channel Arrayed Waveguide Grating (De) multiplexer. In Proceedings of the 2019 PhotonIcs & Electromagnetics Research Symposium-Spring (PIERS-Spring), Rome, Italy, 17–20 June 2019; pp. 2189–2192.
43. Ye, T.; Fu, Y.; Qiao, L.; Chu, T. Low-crosstalk Si arrayed waveguide grating with parabolic tapers. *Opt. Express* **2014**, *22*, 31899–31906. [CrossRef]
44. Smit, M.K.; Van Dam, C. PHASAR-based WDM-devices: Principles, design and applications. *IEEE J. Sel. Top. Quantum Electron.* **1996**, *2*, 236–250. [CrossRef]

45. Dai, D.; Liu, L.; Wosinski, L.; He, S. Design and fabrication of an ultrasmall overlapped AWG demultiplexers based on α-Si nanowire waveguides. *Electron. Lett.* **2006**, *42*, 400–402. [CrossRef]
46. Okamoto, K.; Ishida, K. Fabrication of silicon reflection-type arrayed-waveguide gratings with distributed Bragg reflectors. *Opt. Lett.* **2013**, *38*, 3530–3533. [CrossRef] [PubMed]
47. Fu, X.; Dai, D. Ultra-small Si-nanowire-based 400 GHz-spacing 15 × 15 arrayed-waveguide grating router with microbends. *Electron. Lett.* **2011**, *47*, 1. [CrossRef]
48. Pathak, S.; Vanslembrouck, M.; Dumon, P.; Van Thourhout, D.; Verheyen, P.; Lepage, G.; Absil, P.; Bogaerts, W. Effect of mask discretization on performance of silicon arrayed waveguide gratings. *IEEE Photonics Technol. Lett.* **2014**, *26*, 718–721. [CrossRef]
49. Fu, J.; Chen, Y.; Sun, S.; Han, Y.; Huang, Y.; Lin, S.; Yang, L.; Yu, J. A Low-loss 4-channel Arrayed Waveguide Grating Using Parabolic Tapered Waveguides. In Proceedings of the Asia Communications and Photonics Conference, Beijing, China, 24–27 October 2020; p. M4A. 83.
50. Kamei, S.; Ishii, M.; Kaneko, A.; Shibata, T.; Itoh, M. NXN Cyclic-Frequency Router With Improved Performance Based on Arrayed-Waveguide Grating. *J. Light. Technol.* **2009**, *27*, 4097–4104. [CrossRef]
51. Sakamaki, Y.; Kamei, S.; Hashimoto, T.; Kitoh, T.; Takahashi, H. Loss uniformity improvement of arrayed-waveguide grating with mode-field converters designed by wavefront matching method. *J. Light. Technol.* **2009**, *27*, 5710–5715. [CrossRef]
52. Sheng, Z.; Dai, D.; He, S. Improve channel uniformity of an Si-nanowire AWG demultiplexer by using dual-tapered auxiliary waveguides. *J. Light. Technol.* **2007**, *25*, 3001–3007. [CrossRef]

sensors

MDPI

Article

Enhancing Security in Visible Light Communication: A Tabu-Search-Based Method for Transmitter Selection

Ge Shi [1,*], Wei Cheng [2], Xiang Gao [2], Fupeng Wei [1], Heng Zhang [2] and Qingzheng Wang [1,*]

1 The School of Information Engineering, North China University of Water Resources and Electric Power, Zhengzhou 450046, China
2 The School of Electronics and Information, Northwestern Polytechnical University, Xi'an 710129, China; pupil_119@nwpu.edu.cn (W.C.); 1486812648@mail.nwpu.edu.cn (X.G.)
* Correspondence: shige@ncwu.edu.cn (G.S.); wangqingzheng@ncwu.edu.cn (Q.W.)

Abstract: In this paper, we explore the secrecy performance of a visible light communication (VLC) system consisting of distributed light-emitting diodes (LEDs) and multiple users (UEs) randomly positioned within an indoor environment while considering the presence of an eavesdropper. To enhance the confidentiality of the system, we formulate a problem of maximizing the sum secrecy rate for UEs by searching for an optimal LED for each UE. Due to the non-convex and non-continuous nature of this security maximization problem, we propose an LED selection algorithm based on tabu search to avoid getting trapped in local optima and expedite the search process by managing trial vectors from previous iterations. Moreover, we introduce three LED selection strategies with a low computational complexity. The simulation results demonstrate that the proposed algorithm achieves a secrecy performance very close to the global optimal value, with a gap of less than 1%. Additionally, the proposed strategies exhibit a performance gap of 28% compared to the global optimal.

Keywords: physical layer security; visible light communication; LED selection; secrecy rate

1. Introduction

Visible light communication (VLC) is an emerging wireless technology that uses visible light to transmit data. VLC works by modulating the intensity of optical signals from light sources to encode and transmit information. VLC has gained significant attention recently due to its potential for fast, secure, and energy-efficient data transmission [1]. It can operate where radio frequency (RF) signals are restricted, such as in hospitals and aircraft. VLC also offers transmission security since the signals do not penetrate walls and are not susceptible to electromagnetic interference. VLC has various applications, including indoor positioning [2], smart lighting, and high-speed internet access [3].

VLC systems have a broadcasting characteristic, which allows multiple users (UEs) to access the information transmitted from every light-emitting diode (LED) that is inside their detective areas. A malicious UE can potentially gather confidential data not intended for them. Traditional upper techniques may not effectively safeguard VLC systems since each transmission requires private key exchanges and management. Fortunately, physical layer security (PLS) has appeared as a promising method to enhance wireless security by utilizing wireless channel characteristics. In PLS, the performance is determined by the transmission performance gap between a legitimate channel and an eavesdropper (Eve) channel [4]. Therefore, to enhance the security of VLC systems, strategies must be designed to improve the signal-to-interference-plus-noise ratio (SINR) of the legitimate UE while degrading that of the Eve. This approach has been extensively investigated in RF systems, but PLS in VLC systems requires special investigation due to their differing characteristics.

Beamforming-based techniques demonstrate superior performance as they enable transmitters to steer the direction of optical signals and adjust their power levels [5,6]. By dynamically controlling these parameters, beamforming-based methods can focus the transmitted light toward specific UEs or areas of interest. This targeted transmission improves signal strength, enhances coverage, avoids eavesdropping, and reduces interference from other sources, leading to a superior performance in terms of signal reception and reliability. However, beamforming-based methods may require additional equipment, such as lenses and signal processing circuits, to adjust the amplitude and direction of the optical signals to create a beam targeting UEs. VLC systems are typically intensity-modulation/direct-detection systems, which use incoherent light and cannot directly control the phase of the light wave. Beamforming devices that are designed for coherent optical communication systems, such as those using infrared radiation lasers [7], cannot be directly transplanted into VLC systems.

Another promising technique for improving the performance of VLC systems is LED selection. While LED selection methods may not achieve the same level of performance as beamforming-based methods, they offer a compromise between performance and complexity [8]. The computational complexity comparison between these two methods can be analyzed based on two factors: the solution space and the problem dimension. Beamforming-based methods refine solutions in a continuous space, while LED selection methods search for solutions in a discrete space. Due to its finer-grained solution space, the former methods generally require more iterations compared to the latter methods. In addition, the problem dimension for beamforming-based methods is M times larger than that for LED selection methods, where M represents the number of UEs. In beamforming-based methods, the optimization variables for a secrecy-rate maximization problem are M beamforming vectors that allocate the coefficients from LEDs to each UE, with a size of $1 \times K$, where K represents the number of LEDs. On the other hand, in LED selection methods, the optimization variable is one vector that includes the indices of LEDs, with a size of $1 \times K$.

To enhance the transmission quality in LED selection methods, the optimal criterion is often to associate an LED–UE pair that has the shortest Euclidean distance. This is because the useful signal strength transmitted from the LED towards the UE is the strongest in this case. Fangxin et al. proposed an LED selection algorithm based on the support vector machine of a VLC system to improve the bit error rate, where the Euclidean distance are selected as the key metric of the system to construct the label vector of the training samples [9]. Yitian et al. designed an Euclidean distance-based LED selection scheme to improve the bit error rate by reducing the effect of channel correlation [10]. Yang et al. proposed a hybrid dimming scheme to maximize the throughput of a multiple-user, multiple-cell VLC system [11]. The LED selection subproblem was formulated as a mixed-integer problem and solved using a penalty method. For security purposes, the main idea behind LED selection is to choose the best combinations of LEDs and UEs that can maximize the data rates of UEs while minimizing that of Eves. Cho et al. proposed an LED selection scheme that chooses the nearest LED for UEs to reduce the secrecy outage probability of a VLC system, providing a practical and near-optimal solution [12]. Motivated by these results, a TS-based LED selection scheme is proposed to maximize the sum secrecy rate of VLC systems in this paper, which has not been investigated yet. A brief summary of the above work is given in Table 1.

Table 1. Main features of related works and comparison with this paper.

Paper	Method	Main Contribution
Fangxin et al. [9]	Support vector machine	Improved bit error rate
Yang et al. [10]	Penalty method	Improved throughput
Yitian et al. [11]	Euclidean distance-based grouping strategy	Improved bit error rate
Cho et al. [12]	Euclidean distance-based strategy	Reduced secrecy outage probability
This paper	Tabu search	Improved secrecy rate

In this paper, we investigate a multi-UE VLC channel that comprises multiple LEDs with the objective of transmitting confidential messages to UEs in the presence of an active Eve. Assuming that UEs and the active Eve send feedback through uplink media, the CSI of UEs and Eve is known by the center control unit. In the center control unit, a UE selects an LED based on an algorithm or strategy. The primary goal of the study is to improve the secrecy performance by assigning a distinct LED to each UE while taking the presence of an Eve into account. To achieve this goal, we formulate a sum secrecy rate maximization problem with integer variables, which is non-convex in nature, and introduce a tabu search (TS)-based algorithm that provides a sub-optimal solution with low complexity. To explore the solution spaces in this algorithm, we design a neighboring vector whose element selects a random LED within the UE's receiving area. Furthermore, we design three simple LED selection strategies to reduce the complexity further. Simulation results and comparisons of the proposed algorithm with three simple strategies and global search are provided to demonstrate the effectiveness of the proposed algorithm.

The rest of the paper is organized as follows: In Section 2, we describe the system model for the downlink transmission of the multi-UE VLC system. Section 3 presents the TS-based LED selection algorithm and three strategies while providing a complexity and convergence analysis. In Section 4, we offer numerical simulation results and comparisons. Finally, we conclude the paper in Section 5.

2. System Model

There are M UEs and K LEDs in an interior setting with the presence of an Eve. We assume that a UE selects one LED to connect, and the UE-to-LED connection is a one-to-one pair. $\mathbf{x} = [k_1, k_2, k_m, \cdots, k_M]$ is an index vector that indicates the connection between LEDs and UEs, where $k_m \in \mathbf{A}$ is the index of the LED that is associated with the m-th UE, and $\mathbf{A} = \{1,\ 2,\ k, \cdots,\ K\}$ denotes a set that contains the index of every LED. Once the index vector $\mathbf{x}$ is determined, each UE receives its confidential signal from a specific LED. LEDs that are not selected by any UE are only used for illumination, not for transmitting signals. However, due to the broadcasting nature of VLC, a UE also receives signals from nearby LEDs which are selected by other UEs.

For example, as shown in Figure 1, the m-th UE (magenta triangle) selects the k_m-th LED (blue fork) and the j-th UE selects the k_j-th LED. The m-th UE receives two different signals in its reception range. The reception ranges of UEs are denoted as magenta dashed circles. The signal from the k_m-th LED is confidential information, whereas the signal from the k_j-th LED is considered user interference. The useful links and interference links are represented by blue lines and red lines, respectively. When an Eve (green solid circle) intends to eavesdrop on the m-th UE, the signal from the k_j-th LED is regarded as interference. The reception ranges of the Eve are denoted as a green dashed circle. In the meantime, the j-th UE only receives its signal from the k_j-th LED since no LED selected by other UEs is within its reception range.

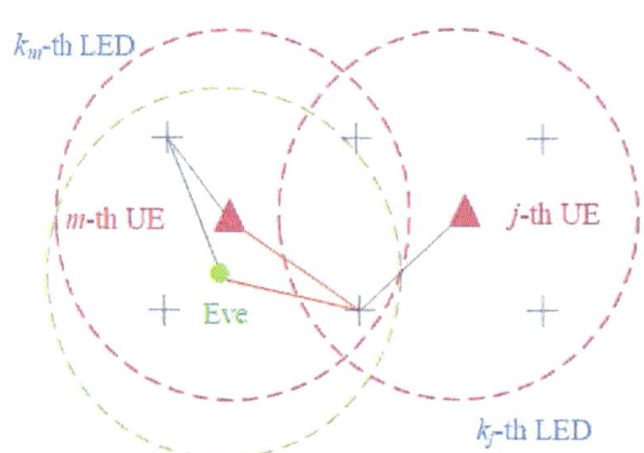

Figure 1. System model.

In the general case, assuming the index vector $\mathbf{x}$ is fixed, the signal received at the m-th UE is formulated as follows:

$$y_m = h_{k_m,m} s_m + \sum_{j=1,\, j\neq m}^{M} h_{k_j,m} s_j + n_m, \tag{1}$$

where $\mathbb{E}\{s_m^2\} = \mathbb{E}\{s_j^2\} = 1$ and $\mathbb{E}\{s_m\} = \mathbb{E}\{s_j\} = 0$. The channel gain from the k_m-th LED to the m-th UE is represented as $h_{k_m,m}$; s_m is the information signal for the m-th UE; the channel gain from the k_j-th LED to the m-th UE is $h_{k_j,m}$; and s_j is the information signal for the j-th UE and part of the interference signals for the m-th UE. If the k_j-th LED falls outside the m-th UE's reception range, the channel gain is zero, i.e., $h_{k_j,m} = 0$. The noise source inside the receivers' circuit is mainly dominated by thermal noise and shot noise. This is modeled as additive zero-mean Gaussian noise, i.e., $n_m \sim \mathcal{N}(0, \xi^2)$. The channel gain from the k_m-th LED to the m-th UE can be expressed as [13]:

$$h_{k_m,m} = \frac{(\gamma+1)A_R}{2\pi d_{k_m,m}^2} \cos(\phi_{k_m,m})^{\gamma} \cos(\varphi_{k_m,m}) f(\varphi_{k_m,m}) g_{\mathrm{of}}, \tag{2}$$

where $\gamma = \frac{-1}{\log_2 \cos\phi_{1/2}}$ denotes the Lambertian emission order, $\phi_{1/2}$ and A_R are the half-intensity radiation angle and the area of receiver's photodiode (PD), respectively, $d_{k_m,m}$ is the Euclidean distance from the k_m-th LED to the m-th UE, $\phi_{k_m,m}$ and $\varphi_{k_m,m}$ are the irradiance angle and incidence angle, respectively, g_{of} is the gain of the optical filter, and $\mathrm{f}(\varphi_{k_m,m})$ denotes the gain of the optical concentrator, which is given by [13]:

$$\mathrm{f}(\varphi_{k_m,m}) = \begin{cases} \frac{q^2}{\sin^2(\varphi_{k_m,m})} & 0 \leq \varphi_{k_m,m} \leq \Theta, \\ 0 & \varphi_{k_m,m} \geq \Theta, \end{cases} \tag{3}$$

where q and Θ are the refractive index and the field of view (FoV) of the PD used at the m-th UE's side, respectively. The extent of the FoV dictates the reception range for stationary UEs. The SINR of the m-th UE is calculated as follows:

$$\mathrm{R}_m = \frac{1}{2}\log_2\left(1 + \frac{e}{2\pi}\frac{\left|h_{k_m,m}\right|^2}{\sum\limits_{j=1,\, j\neq m}^{M}\left|h_{k_j,m}\right|^2 + \xi^2}\right). \tag{4}$$

Assuming that the Eve overhears the confidential signal for the m-th UE, which is transferred from the k_m-th LED, the signal received by the Eve is formulated as:

$$\mathrm{ye} = g_{k_m} s_m + \sum_{j=1,\, j\neq m}^{M} g_{k_j} s_j + n_e, \tag{5}$$

where g_{k_m} is the channel gain from the k_m-th LED to the Eve; s_m is the confidential signal for the m-th UE and also useful signal for the Eve; g_{k_j} is the channel gain from the k_j-th LED to the Eve; and s_j is the confidential signal for the j-th UE and part of interference signal for the Eve. Thus, the SINR of the Eve is calculated as follows:

$$\mathrm{Re} = \frac{1}{2}\log_2\left(1 + \frac{e}{2\pi}\frac{\left|g_{k_m}\right|^2}{\sum\limits_{j=1,\, j\neq m}^{M}\left|g_{k_j}\right|^2 + \xi^2}\right). \tag{6}$$

Based on the lower bound of the channel capacity considered in [14], the achievable secrecy rate of the m-th UE is given by substituting (4) and (6) into the following equations:

$$\mathrm{SR}_m = [\mathrm{R}_m - \mathrm{Re}]_+. \tag{7}$$

The sum secrecy rate maximization problem can be modeled as

$$\max \sum_{m=1}^{M} \mathrm{SR}_m, \tag{8}$$

$$\text{s.t. } \mathbf{x} = [k_1, k_2, k_m, \cdots, k_M] \in \mathbb{N}^M, k_m \in \mathbf{A}, \ \forall m, \tag{9}$$

The above problem is an integer programming problem, and it involves searching for an optimal index vector $\mathbf{x}$. The discrete nature of the solution space and the non-convexity of the problem are two of the main challenges in solving this problem. The former makes it impossible to find the optimal solution through simple interpolation or approximation, while the latter means there are many local optimal solutions, making it challenging to find the global optimal solution.

3. Proposed Algorithm and Strategies

3.1. Tabu Search-Based LED Selection Algorithm

LED selection can involve large search spaces due to the number of possible combinations of LEDs and UEs. Decision tree algorithms explore multiple branches and evaluate potential splits at each level, which can lead to an increased computational complexity and longer processing times [15]. Genetic algorithms, with their population-based approach and genetic operations, may have a higher computational complexity as they evaluate and evolve multiple solutions in parallel [16]. However, tabu search's local search approach and memory-based mechanisms can efficiently explore the neighborhood and avoid revisiting local optima, making it suitable for handling large search spaces [17]. The algorithm was first introduced in [18] and has since been extensively studied and applied in various fields [19,20]. Whei-Min developed an improved TS algorithm for a nonlinear economic dispatch problem that classic Lagrange-based algorithms failed to solve [19]. Srinidhi proposed a layered detection approach in conjunction with TS and showed that it works very well in terms of both performance and complexity in multiple-input, multiple-output systems with a large number of antennas [20].

The TS algorithm begins with an initial solution, defines a set of neighboring solutions based on a neighborhood criterion, and moves to the best one among the neighboring solutions. This process has a certain number of iterations, after which the algorithm is stopped and the best solution of all the iterations is declared as the final result. In defining the neighborhood of the solution in a given iteration, the algorithm attempts to avoid cycling by prohibiting the moves to the solutions of the past few iterations. The solutions of the past few iterations are recorded in the 'tabu list' and parameterized by the 'tabu period', which is changed depending on the number of repetitions of the solutions that are marked in the search path. The performance of TS is highly dependent on the selection of its parameters, such as definitions of the neighborhood, the tabu list, and the tabu period.

Neighborhood Definition : Let $\mathbf{x}^{(l)} = \left[k_1^{(l)}, k_2^{(l)}, \cdots, k_M^{(l)}\right]$ denote an index vector belonging to the solution space in the l-th iteration, where $k_m \in \mathbf{A}$ is the index of the LED that is selected by the m-th UE. The neighborhood of the k_m-th LED for the m-th UE is chosen based on $\mathbf{B}_m$. $\mathbf{B}_m$ is a fixed subset of $\mathbf{A}$, and it contains the indices of every reachable LED for the m-th UE. As long as the m-th UE does not move, $\mathbf{B}_m$ is a fixed subset. The cardinality and members of this set are not the same for different UEs, but depend on their FoV and position. Note that the maximum and minimum values of the cardinality are K and 1, respectively. We refer to a neighbor vector $\mathbf{z}^{(l)}(q) = \left[z_1^{(l)}(q), z_2^{(l)}(q), \cdots, z_M^{(l)}(q)\right]$ as the q-th neighbor of $\mathbf{x}^{(l)}$, where the m-th element $z_m^{(l)}(q)$ is the q-th neighbor of $k_m^{(l)}$,

$q = 1 \cdots M$. The $z_m^{(l)}(q)$ -th LED is the q-th neighbor of the $k_m^{(l)}$-th LED for the m-th UE, the formula of which is written as follows:

$$z_m^{(l)}(q) = \omega(k_m^{(l)}), \forall q = 1 \cdots M. \tag{10}$$

where $\omega(k_m^{(l)})$ represents that $z_m^{(l)}(q)$ is randomly selected from $\mathbf{B}_m$ but does not include $k_m^{(l)}$, i.e., $z_m^{(l)}(q) \in \mathbf{B}_m \backslash k_m^{(l)}$. We choose M neighbor vectors that differ from a given vector $\mathbf{x}^{(l)}$ in the solution space. An operation on $\mathbf{x}^{(l)}$ which gives $\mathbf{x}^{(l+1)}$ belonging to the neighborhood of $\mathbf{x}^{(l)}$ is referred to as a move. The algorithm is said to execute a move if $\mathbf{x}^{(l+1)} = \mathbf{z}^{(l)}(q)$. We note that the number of candidates to be considered for a move is equal to M in any of the iterations.

Tabu List : A tabu list $\mathbf{T}$ of size $M \times K$ is a matrix whose entries denote the tabu values of moves. The tabu value of a move means that the move cannot be considered for that many successive iterations. There are M rows in $\mathbf{T}$, where each row represents the indices of UEs from 1 to M. The K columns of the $\mathbf{T}$ matrix correspond to the indices of LEDs from 1 to K. In other words, the (m, k_m)-th entry of the tabu list corresponds to the connection between the k_m-th LED and the m-th UE. The entries of the tabu list are updated in each iteration, and they are used to determine the direction in which the search proceeds.

Tabu Period: The tabu period, a non-negative integer parameter, is defined as follows. If combinations of LEDs and UEs are accepted as the next move in an iteration, these combinations will be restricted for several subsequent iterations. The number of restricted iterations is denoted by the tabu period.

TS Algorithm: Let $\boldsymbol{\alpha}^{(l)}$ be a solution that has the best sum secrecy rate found till the l-th iteration. The algorithm starts with an initial solution vector $\mathbf{x}^{(0)}$. Set $\boldsymbol{\alpha}^{(0)} = \mathbf{x}^{(0)}$. Note that different initial solutions can lead to different results. All the entries of the tabu list are set to zero. The following steps (1) to (3) are performed in each iteration. Consider the l-th iteration and the current solution $\mathbf{x}^{(l)}$ in the algorithm, $l \geq 0$.

Step (1): The sum secrecy rate of the neighbor vector $\mathrm{SR}(\mathbf{z}^{(l)}(q))$ is computed. Let

$$\mathbf{z}^{(l)}(q^*) = \arg\max_q \mathrm{SR}(\mathbf{z}^{(l)}(q)), \forall q = 1 \cdots M. \tag{11}$$

The neighbor vector $\mathbf{z}^{(l)}(q^*)$ is accepted as the next move if any one of the following two conditions is satisfied:

$$\mathrm{SR}(\mathbf{z}^{(l)}(q^*)) > \mathrm{SR}(\boldsymbol{\alpha}^{(l)}), \tag{12}$$

$$\mathbf{T}(m, {z_m}^{(l)}(q^*)) = 0, \forall m = 1 \cdots M. \tag{13}$$

The first equation indicates that the secrecy performance of the neighbor vector $\mathbf{z}^{(l)}(q^*)$ surpasses that of the best index vector $\boldsymbol{\alpha}^{(l)}$ up till the l-th iteration. The second formula denotes that the $(m, {z_m}^{(l)}(q^*))$-th entry in the tabu list is marked as zero, which means that the connection between the ${z_m}^{(l)}(q^*)$-th LED and the m-th UE is not forbidden. If this neighbor is not accepted as the next move (i.e., either one of the conditions above is not satisfied), the neighbor vector $\mathbf{z}^{(l)}(q^{**})$ with the second best sum secrecy rate is found such that

$$\mathbf{z}^{(l)}(q^{**}) = \arg \max_{q: q^{**} \neq q^*} \mathrm{SR}(\mathbf{z}^{(l)}(q)). \tag{14}$$

Next, check for acceptance of $\mathbf{z}^{(l)}(q^{**})$ by substituting it in (12) and (13). If this also cannot be accepted, the procedure is repeated for the neighbor vector $\mathbf{z}^{(l)}(q^{***})$ with the third-best sum secrecy rate, and so on. If any unrestricted neighbor is accepted as the next move, proceed to Step (2). Otherwise, advance to Step (4). Continue this process until an acceptable move is made.

Step (2): Let $\mathbf{z}^{(l)}(\hat{q})$ be a neighbor vector for which the move is permitted. Make

$$\mathbf{x}^{(l+1)} = \mathbf{z}^{(l)}(\hat{q}), \tag{15}$$

where the m-th element $z_m{}^{(l)}(\hat{q})$ of the neighbor vector $\mathbf{z}^{(l)}(\hat{q})$ represents the index of the $\hat{q}$-th neighbor LED for the k_m-th LED. It should be noted that the best permissible neighbor vector is selected as the solution vector for the subsequent iteration. Next, if this neighbor vector satisfies the first condition (12), it is assigned to the best solution $\boldsymbol{\alpha}^{(l+1)}$ in the $(l+1)$-th iteration.

Step (3): The connection between the $z_m{}^{(l)}(\hat{q})$-th LED and the m-th UE should be prohibited for several upcoming iterations to prevent cycling. Thus, the $(m, z_m{}^{(l)}(\hat{q}))$-entry of the tabu list is assigned by the tabu period, denoted as M:

$$\mathbf{T}(m, z_m{}^{(l)}(\hat{q})) = M, \ \forall m = 1 \cdots M, \tag{16}$$

Step (4): All the non-zero entries of the tabu list are reduced by one, which is achieved as follows:

$$\mathbf{T}(m, k_m) = \max\{\mathbf{T}(m, k_m) - 1, 0\}, \ \forall k_m = 1 \cdots K, \tag{17}$$

The algorithm terminates in Step 4 if the following stopping criterion is satisfied; otherwise, it goes back to Step (1).

Stopping Criterion: The algorithm described above is stopped if the maximum number of iterations Max_l is reached. Also, if the total number of repetitions of the current solution is greater than a threshold Max_t, the algorithm is stopped. The average complexity of the algorithm is $O(MK)$. The procedure of the algorithm is further illustrated in Figure 2.

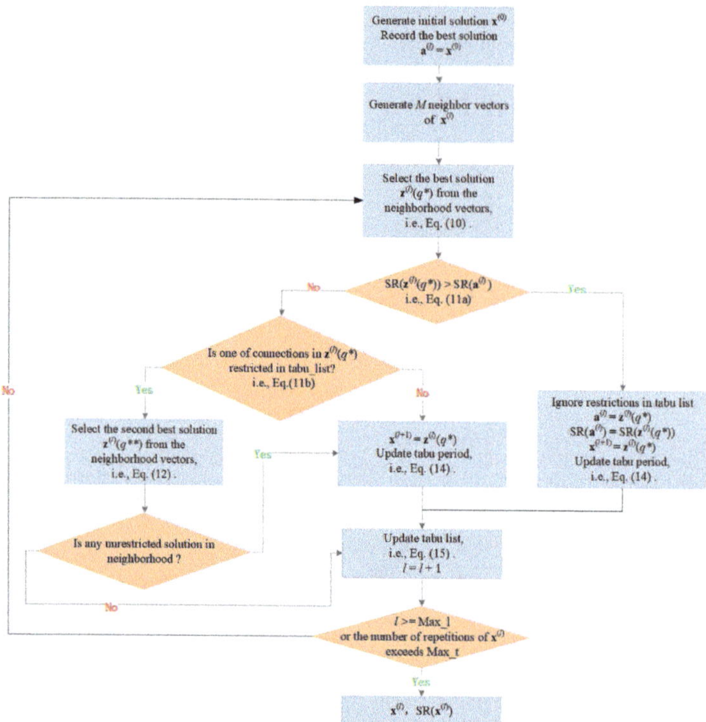

Figure 2. The procedure of the proposed algorithm (Orange diamond and blue rectangle are depicted for the decision and process in the algorithm, respectively).

3.2. Simple LED Selection Strategies

Three simple LED selection strategies are proposed as benchmarks in this scenario, which require no iterations and have a lower complexity.

Random strategy: In this strategy, each UE is assigned to a random LED within its reception range. The solution vector, denoted as $\mathbf{x} = [k_1, k_2, k_m, \cdots, k_M]$, represents the assignment of LEDs to UEs, where k_m is the index of the LED that the m-th UE randomly chooses from the set $\mathbf{B}_m$. Note that the random strategy does not consider any optimization criteria or system constraints. If any LED is already connected to a UE, then it cannot be selected as the transmitter for other UEs.

Channel gain-based strategy: In this strategy, each UE is assigned to the LED with the highest channel gain, which is similar to the LED selection method proposed in [12]. $k_m \in \mathbf{B}_m$ is the index of the LED that is associated with the m-th UE.

Eve's channel gain-based strategy: In this strategy, each UE is allocated to the LED with the highest channel gain that is greater than Eve's channel gain. If a UE's channel gain is less than Eve's, the corresponding LED is banned from being selected by that UE. Thus, the indices of the banned LEDs are eliminated from the set $\mathbf{B}_m$. Assuming that β is the set containing the indices of banned LEDs, $k_m \in \mathbf{B}_m \backslash \beta$ is the index of the LED that is associated with the m-th UE.

Since the main procedure of the above strategies involves sorting channel gains, the average complexity of the above three strategies is $O(K)$.

3.3. Global Search

The result of the global search is used to measure the optimality of the proposed algorithm and strategies. To achieve this, a list that contains all possible solutions is created, and the best solution is traversed by substituting every solution vector $\mathbf{x} = [k_1, k_2, k_m, \cdots, k_M]$ into (8). Thus, the average complexity of the algorithm is $O(K^M)$.

3.4. Convergence Analysis

Figure 3 illustrates the iteration times of the proposed algorithm in comparison to global search. The proposed algorithm and global search achieve the same optimal value of 2.18 bit/s. The global search terminated at the 384th iteration. The optimal value was determined after an ergodic examination of all reasonable combinations between 5 UEs and 25 LEDs. Two kinds of scenarios need to be excluded from the examination: multiple UEs connecting to the same LED, and one LED connecting to one UE not inside its illumination area. In comparison, the proposed algorithm terminated at the 50th iteration when it reached the stopping criterion (that is, the repetition number of the best solution is greater than the threshold). As shown in Figure 3, the algorithm reached the best value four times. The best solution was first obtained in the 21st iteration, and it was also the optimal value.

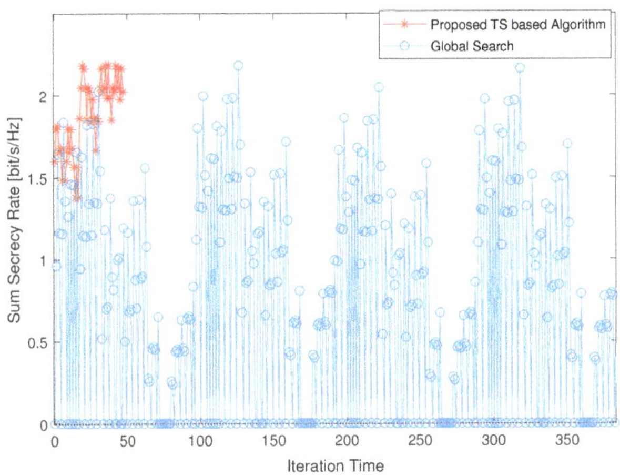

Figure 3. Comparison of the proposed algorithm and global search regarding convergence performance.

4. Numerical Results

In this section, we study the secrecy performance of the proposed VLC system in a 10 m × 10 m × 3 m indoor setting. Figure 4 demonstrates the geometrical configuration of 25 LEDs within an indoor room. Note that the simulations below employ a Monte Carlo-based approach, and the secrecy performance is determined by averaging the results from 500 different instances. In each instance, the coordinates of the UEs and the Eve are randomly generated using a uniform distribution within the interval of $[0, 10]$. The parameters used in the following simulations are shown in Table 2.

Table 2. Simulation parameters.

Parameter	Value
Room size ($W \times L \times H$)	$(10 \times 10 \times 4)$ m^3
Average electrical ambient noise (ξ)	−98 dBm
Lambertian emission order (β)	1
Half-intensity radiation angle ($\theta_{1/2}$)	60°
PD surface area (A_R)	1 cm^2
Optical filter gain (g_{of})	1
Maximum power of an LED	23 dBm
PD FoV (Θ)	50°
Refractive index (q)	1.5

Figure 4 shows a layout example of five UEs and one Eve. In this case, the coordinates of the UEs are (1.34, 1.59, 0.8), (5.28, 6.05, 0.8), (8.66, 4.19, 0.8), (2.48, 5.89, 0.8), and (7.07, 6.00, 0.8), while the coordinates of the Eve are (2.13, 4.25, 0.8). Blue forks denote the coordinates of the LEDs. The positions of the UEs are marked with magenta triangles, and magenta dashed circles denote their respective coverage areas. The green circle represents Eve's location, and the green dashed circles denote its coverage areas.

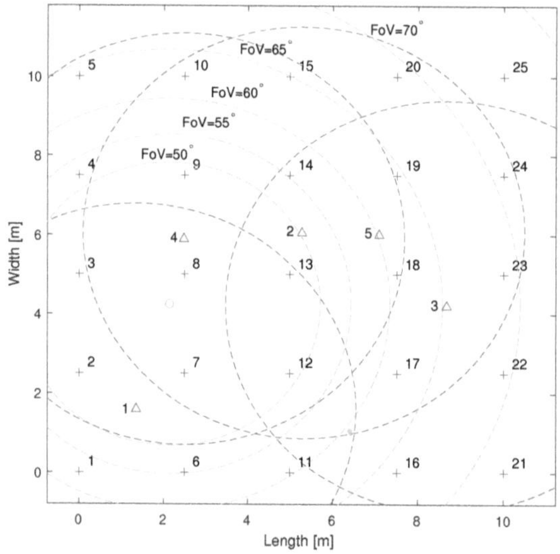

Figure 4. Placement of LEDs (blue) and example of a layout for UEs (magenta) and an Eve (green).

Figure 5 displays the changes in the sum secrecy rate as a function of the maximum power of each LED. The improvement in the secrecy rate is attributed to the increased signal strength transmitted by the LED. When the maximum power of the LED increases, the m-th UE can receive a stronger confidential signal from its intended LED. If there is no interference signal from nearby LEDs, the increased power directly contributes to an enhancement in the achievable data rate at the m-th UE. Furthermore, the results of the global search represent the optimal performance in this scenario, and the proposed algorithm's performance is very close to it. The proposed algorithm outperforms three simple strategies, and the performance gap between them becomes more significant with increasing power. Notably, the performance of the Eve's channel-gain-based strategy is superior to that of the channel-gain-based strategy. This is because the latter does not take into account the position of the Eve, whereas the former allows UEs to reject the connection of LEDs that provide a better channel quality to the Eve.

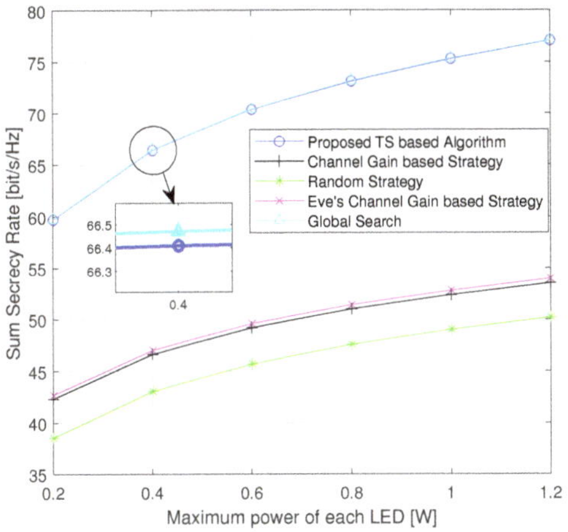

Figure 5. Average sum secrecy rate with respect to the maximum power of each LED (randomly generated positions of UEs and Eves for each instance).

Figure 6 illustrates the variation in the sum secrecy rate with respect to the number of UEs. As the number of UEs increases, the sum secrecy rate initially experiences a rapid growth. This is due to the increased potential for multi-user diversity. UEs that are in favorable positions, for instance, those far away from the Eve, contribute to a higher sum secrecy rate. However, the growth rate slows down as the number of UEs continues to increase. This is attributed to the increased inter-user interference and the limited number of LEDs that can be selected for UEs.

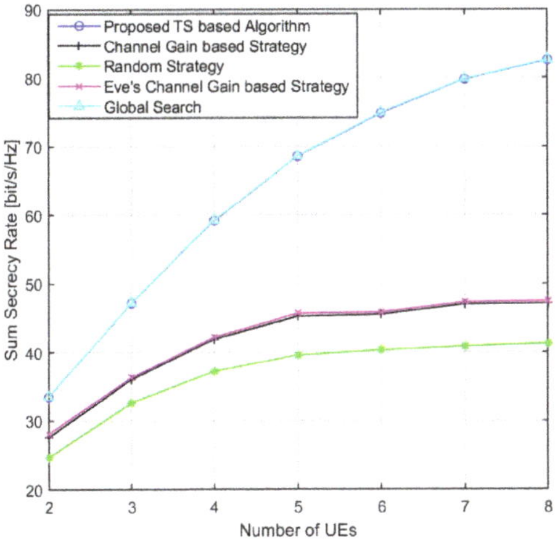

Figure 6. Average sum secrecy rate with respect to the total number of randomly placed UEs.

Figure 7 illustrates the performance of the sum secrecy rate as a function of the Eve's FoV. The FoV of the UEs is 50 degrees. The secrecy performance exhibits a slightly increasing trend with an increasing FoV. This is because both interference and useful signals grow on the Eve's side. As depicted in Figure 4, the green circles represent the detectable areas of the Eve. As the FoV increases, the green circles become larger, encompassing more UEs and LEDs, which allows the Eve to receive useful signals from more UEs. For example, at a FoV of 70 degrees, the Eve begins to receive signals from the third UE. Therefore, compared with the case where its FoV is smaller than 70 degrees, the Eve's achievable rate becomes greater than zero when overhearing the third UE. However, if the Eve overhears other UEs, their achievable rates become lower due to interference from the third UE.

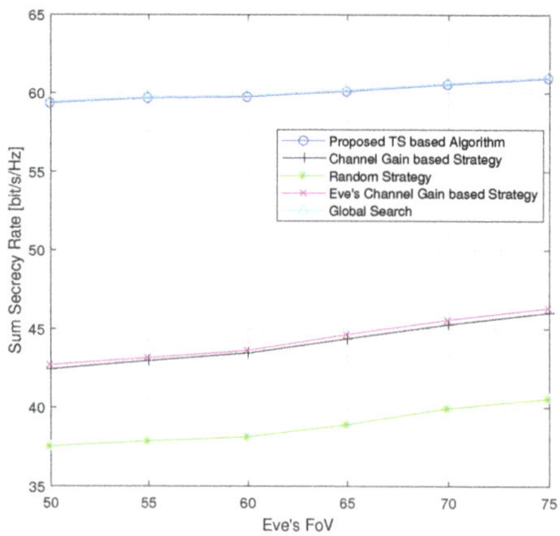

Figure 7. Average sum secrecy rate with respect to the Eve's FoV (randomly generated positions of UEs and Eves for each instance).

Figure 8 displays the changes in the sum secrecy rate as a function of the FoV of the UEs, while the Eve has a fixed FoV of 50 degrees. As the FoV of the UEs increases, the downward trend in the secrecy performance becomes less pronounced. This is because with a wider FoV, UEs detect signals from more LEDs and receive a stronger interference signals. Once the FoV becomes large enough to connect all LEDs, the interference remains constant and hence the secrecy performance stabilizes.

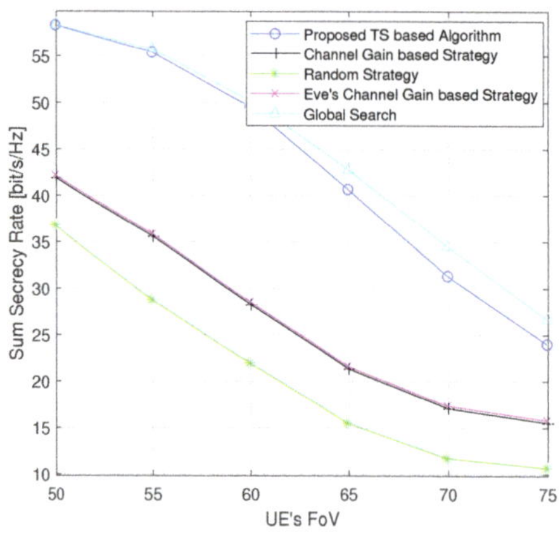

Figure 8. Average sum secrecy rate with respect to the UEs' FoVs (randomly generated positions of UEs and Eves for each instance).

Figure 9 shows that the accuracy of the Eve's position influences the performance of the proposed algorithm. Both the secrecy performance of the proposed algorithm and the proposed Eve's channel-gain-based strategy gradually deteriorate as the localization error of the Eve increases. However, the performance of the channel-gain-based strategy remains stable, as this strategy assigns the LEDs to the UEs without considering the position of the Eve.

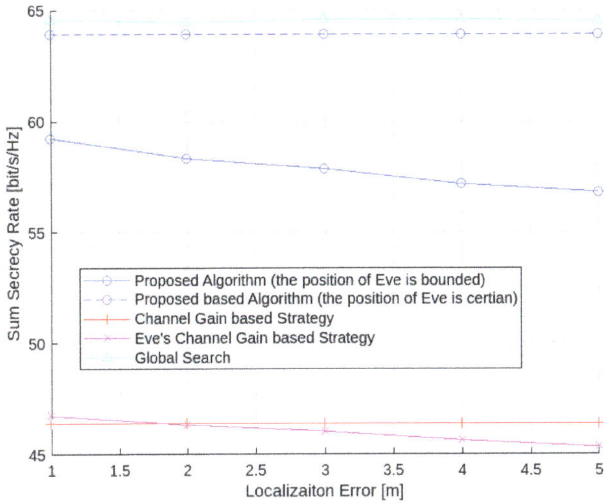

Figure 9. Average sum secrecy rate with respect to the localization error of the Eve (randomly generated positions of UEs and Eves for each instance).

5. Conclusions

In this paper, we present a multi-LED VLC system incorporating multiple UEs and a single Eve. Depending on the perfect CSI of the UEs and the Eve, we propose an LED selection algorithm based on a TS to enhance the secrecy performance of the system. The simulation results demonstrate that the proposed algorithm offers a near-optimal solution, with a performance gap of less than 1%, while it achieves faster convergence in comparison to the global search. Additionally, we design three simple LED selection strategies to reduce the computational complexity, which all exhibit a performance gap of around 28% compared to the global search. The Euclidean distance-based criterion studied in previous work [12] has the same performance as the proposed channel-gain-based strategy (since in VLC, the channel gain is proportional to the distance). It is easy to implement in practical experiments since the distances between LEDs and the UE can be measured by using the received signal strength technique. However, the proposed algorithm achieves a near-optimal performance with a slight increase in computational complexity. With more hardware resources, such as adding more RAM to store the tabu list and using a camera to detect the position of the Eve, the proposed algorithm should be a better choice to improve the secrecy performance in a VLC system. Furthermore, the simulation results show the impact of various system parameters on the secrecy performance. The results indicate that increasing the number of UEs and the maximum power of each LED and reducing the UEs' FoV improves the secrecy performance. On the other hand, augmenting the FoV of the Eve has a negligible impact on the secrecy performance.

Author Contributions: Conceptualization, G.S. and W.C.; methodology, G.S.; software, F.W.; validation, H.Z., G.S. and X.G.; formal analysis, H.Z.; investigation, F.W.; resources, X.G.; data curation, F.W.; writing—original draft preparation, G.S.; writing—review and editing, Q.W.; visualization, W.C.; supervision, W.C.; project administration, G.S.; funding acquisition, G.S. All authors have read and agreed to the published version of the manuscript.

Funding: This research was funded by the Key Scientific Research Projects of Colleges and Universities in Henan Province under grant number 23A520031; the National Natural Science Foundation of China under grant number 61702185; and the Key Scientific Research Project of Colleges and Universities in Henan Province under grant number 18A520034.

Institutional Review Board Statement: Not applicable.

Informed Consent Statement: Not applicable.

Data Availability Statement: Data are contained within the article.

Conflicts of Interest: The authors declare no conflicts of interest.

References

1. Aboagye, S.; Ndjiongue, A.R.; Ngatched, T.M.N.; Dobre, O.A.; Poor, H.V. RIS-Assisted Visible Light Communication Systems: A Tutorial. *IEEE Commun. Surveys Tuts.* **2023**, *25*, 251–288. [CrossRef]
2. Shi, G.; Li, Y.; Cheng, W.; Dong, L.; Yang, J.; Zhang, W. Accuracy analysis of indoor visible light communication localization system based on received signal strength in non-line-of-sight environments by using least squares method. *Opt. Eng.* **2019**, *58*, 1–11. [CrossRef]
3. Naser, S.; Sofotasios, P.C.; Bariah, L.; Jaafar, W.; Muhaidat, S.; Al-Qutayri, M.; Dobre, O.A. Rate-Splitting Multiple Access: Unifying NOMA and SDMA in MISO VLC Channels. *IEEE Open J. Veh. Technol.* **2020**, *1*, 393–413. [CrossRef]
4. Khisti, A.; Wornell, G.W. Secure Transmission With Multiple Antennas I: The MISOME Wiretap Channel. *IEEE Trans. Inf. Theory* **2010**, *56*, 3088–3104. [CrossRef]
5. Mostafa, A.; Lampe, L. Physical-Layer Security for MISO Visible Light Communication Channels. *IEEE J. Sel. Areas Commun.* **2015**, *33*, 1806–1818. [CrossRef]
6. Shi, G.; Aboagye, S.; Ngatched, T.M.N.; Dobre, O.A.; Li, Y.; Cheng, W. Secure Transmission in NOMA-Aided Multiuser Visible Light Communication Broadcasting Network with Cooperative Precoding Design. *IEEE Trans. Inform. Foren. Sec.* **2022**, *17*, 3123–3138. [CrossRef]
7. Morant, M.; Trinidad, A.; Tangdiongga, E.; Koonen, T.; Llorente, R. Experimental Demonstration of mm-Wave 5G NR Photonic Beamforming Based on ORRs and Multicore Fiber. *IEEE Trans. Microw. Theory Tech.* **2019**, *67*, 2928–2935. [CrossRef]
8. Cho, S.; Chen, G.; Coon, J.P. Zero-Forcing Beamforming for Active and Passive Eavesdropper Mitigation in Visible Light Communication Systems. *IEEE Trans. Inf. Forensics Secur.* **2021**, *16*, 1495–1505. [CrossRef]
9. Zhang, F.; Wang, F.; Zhang, J.; Zuo, T. SVM aided LEDs selection for generalized spatial modulation of indoor VLC systems. *Opt. Commun.* **2021**, *497*, 127161. [CrossRef]
10. Yang, Y.; Yang, Y.; Chen, M.; Feng, C.; Xia, H.; Cui, S.; Poor, H.V. Joint LED Selection and Precoding Optimization for Multiple-User Multiple-Cell VLC Systems. *IEEE Internet Things J.* **2022**, *9*, 6003–6017. [CrossRef]
11. Chen, Y.; Gao, S.; Tu, G.; Chen, D. Group-Based LED Selection for Generalized Spatial Modulation in Visible Light Communication. *IEEE Commun. Lett.* **2021**, *25*, 3022–3026. [CrossRef]
12. Cho, S.; Chen, G.; Coon, J.P. Securing Visible Light Communication Systems by Beamforming in the Presence of Randomly Distributed Eavesdroppers. *IEEE Trans. Wirel. Commun.* **2018**, *17*, 2918–2931. [CrossRef]
13. Obeed, M.; Salhab, A.M.; Alouini, M.; Zummo, S.A. On Optimizing VLC Networks for Downlink Multi-User Transmission: A Survey. *IEEE Commun. Surv. Tuts.* **2019**, *21*, 2947–2976.
14. Yapıcı, Y.; Güvenç, I. NOMA for VLC Downlink Transmission with Random Receiver Orientation. *IEEE Trans. Commun.* **2019**, *67*, 5558–5573. [CrossRef]
15. Browne, C.B.; Powley, E.; Whitehouse, D.; Lucas, S.M.; Cowling, P.I.; Rohlfshagen, P.; Tavener, S.; Perez, D.; Samothrakis, S.; Colton, S. A Survey of Monte Carlo Tree Search Methods. *IEEE Trans. Comput. Intell. AI Games* **2012**, *4*, 1–43. [CrossRef]
16. Zheng, G.; Gong, C.; Xu, Z. Constrained Partial Group Decoding with Max–Min Fairness for Multi-Color Multi-User Visible Light Communication. *IEEE Trans. Commun.* **2019**, *67*, 8573–8584. [CrossRef]
17. Lai, X.; Fu, Z.H. A Tabu Search Approach with Dynamical Neighborhood Size for Solving the Maximum Min-Sum Dispersion Problem. *IEEE Access* **2019**, *7*, 181357–181368. [CrossRef]
18. Glover, F. Tabu Search—Part I. *ORSA J. Comput.* **1989**, *1*, 190–206. [CrossRef]
19. Lin, W.M.; Cheng, F.S.; Tsay, M.T. An improved tabu search for economic dispatch with multiple minima. *IEEE Trans. Power Syst.* **2002**, *17*, 108–112. [CrossRef]
20. Srinidhi, N.; Datta, T.; Chockalingam, A.; Rajan, B.S. Layered Tabu Search Algorithm for Large-MIMO Detection and a Lower Bound on ML Performance. *IEEE Trans. Commun.* **2011**, *59*, 2955–2963. [CrossRef]

sensors

MDPI

Article

A High-Capacity Optical Metro Access Network: Efficiently Recovering Fiber Failures with Robust Switching and Centralized Optical Line Terminal

Rahat Ullah [1,2,3], Sibghat Ullah [4,*], Ahmad Almadhor [5], Hathal Salamah Alwageed [6], Abdullah A. Al-Atawi [7], Jianxin Ren [1,2,3] and Shuaidong Chen [1,2,3]

1 Institute of Optics and Electronics, Nanjing University of Information Science & Technology, Nanjing 210044, China
2 Jiangsu Key Laboratory for Optoelectronic Detection of Atmosphere and Ocean, Nanjing University of Information Science & Technology, Nanjing 210044, China
3 Jiangsu International Joint Laboratory on Meterological Photonics and Optoelectronic Detection, Nanjing University of Information Science & Technology, Nanjing 210044, China
4 School of Electronic Science and Engineering, Southeast University, Nanjing 210096, China
5 Department of Computer Engineering and Networks, College of Computer and Information Sciences, Jouf University, Sakaka 72388, Saudi Arabia
6 College of Computer and Information Sciences, Jouf University, Sakaka 72388, Saudi Arabia
7 Department of Computer Science, Applied College, University of Tabuk, Tabuk 47512, Saudi Arabia
* Correspondence: sibghat@bupt.edu.cn

Abstract: This study proposes and presents a new central office (CO) for the optical metro access network (OMAN) with an affordable and distinctive switching system. The CO's foundation is built upon a novel optical multicarrier (OMC) generation technique. This technique provides numerous frequency carriers that are characterized by a high tone-to-noise ratio (TNR) of 40 dB and minimal amplitude excursions. The purpose is to accommodate multiple users at the optical network unit side in the optical metropolitan area network (OMAN). The OMC generation is achieved through a cascaded configuration involving a single phase and two Mach Zehnder modulators without incorporating optical or electrical amplifiers or filters. The proposed OMC is installed in the CO of the OMAN to support the 1.2 Tbps downlink and 600 Gbps uplink transmission, with practical bit error rate (BER) ranges from 10^{-3} to 10^{-13} for the downlink and 10^{-6} to 10^{-14} for the uplink transmission. Furthermore, in the OMAN's context, optical fiber failure is a main issue. Therefore, we have proposed a possible solution for ensuring uninterrupted communication without any disturbance in various scenarios of main optical fiber failures. This demonstrates how this novel CO can rapidly recover transmission failures through robust switching a and centralized OLT. The proposed system is intended to provide users with a reliable and affordable service while maintaining high-quality transmission rates.

Keywords: optical communication; optical metro access network; wavelength division multiplexing; optical ring network; optical multicarrier generation

1. Introduction

Metro access networks must provide a high data rate and bandwidth to keep pace with escalating bandwidth demands. The demand for triple-play services necessitates using a high number of distributed laser diodes at both the transmitter and receiver. There is growing evidence that an optical metro access network (OMAN) is a viable option for next-generation optical networks due to its low operating costs, big capacity, low latency, and high transmission efficiency [1–3]. Similarly, the convergence of wireless and optical technologies will facilitate the provision of large bandwidth for high data rates, improving service quality [4]. Different techniques are introduced to converge both the optical and

wireless networks [5,6]. As a meeting point between service providers and operators, OMANs are particularly well suited for adapting to the extreme acceleration of dynamics in optical systems, using efficient components like optical nodes and transponders [7,8]. This means that these networks can quickly adjust to rapid changes in network conditions like traffic volume. Access and conventional convergence layers can be integrated into the metro access layer of the OMAN [9]. One accepted trend in the optical access network is merging metro and access structures under the cost-effectiveness paradigm to support many users [10]. However, new optical line terminals should be designed to support a large number of users. This allows for transparent reach between the metro network and the end-user. However, bridging optical access and metro networks requires a careful estimation of the deployment costs and power budgeting to facilitate the end-user with the guaranteed facilities [11]. Many researchers have widely deployed the ring topology in OMANs due to its self-healing characteristics and reliability for access networks [12–14]. X. Li et al. deployed a dual fiber ring structure for protection measurements and multiple transmitters at the optical line terminal side (OLT) [15]. Similarly, a ring-based OMAN network is deployed with two OLTs in [16] for protection purposes.

The number of frequencies in an OMAN is rising in parallel with the country's increasing need for high-speed data transmission. Because of this, many light sources are needed to accommodate many users [15]. The OLT's price tag is critical in determining the final bill. It is possible to generate a wide range of frequencies from a single laser source using an optical multicarrier (OMC) generator [17–19]. The cascaded arrangement of a different or a single modulator is one of several recent ways of generating OMCs [20–23]. The OMC is used in the optical access network [22,24,25], free space optics [26], in the OMAN [17,27], and for other applications [28]. By boosting efficiency and simplifying maintenance, the OMC generators deployed in an OMAN can efficiently deliver a variety of frequencies. A spectrum-efficient network with an add–drop multiplexer and a centralized OMC source was utilized for the OMAN. Similarly, the carrier wavelength reuse concept was presented using an OMC-based ring network and a reconfigurable optical add–drop multiplexer [29].

In the provided references, some papers have proposed OMC-based sources as the best solution for an OMAN but did not provide detailed proposals of their OLTs [27,30,31]. In the OMAN, each user needs a specific wavelength, which increases the OLT's cost. A similar add–drop multiplexer has been used on the ONU side, but it is unable to drop many frequencies at once and is hence unable to handle the OMC source [31–33]. In the OMAN, in a core network, a ring architecture is an ideal topology for connecting different nodes, and an OLT is the main source for data handling/transmission to the ONUs. Therefore, for a ring-based architecture, two or more OLTs are tested in case a fiber failure occurs in the core network [16]. Hence, deploying more OLTs increases network costs. Most importantly, the data rate in the cited work is not sufficiently high to fulfill the demands of applications for the next-generation access network.

OpEx and CapEx constraints and the development of metro network traffic have led to a search for architectures with reduced energy consumption, smaller footprints, and cheaper costs. For this reason, a centralized OLT in the OMAN using WDM technology is configured in a ring architecture at the feeder fibers. At the same time, star topology is set up at the user end. Deploying OFCs in an OMAN is important for enhancing data transmission by providing high bandwidth to each user with high-frequency spacing. This could ensure a high data rate transmission efficiently with superb scalability, provide low latency, and fulfil the urban network demands.

An OMCG-based OLT, which can efficiently transmit 20 Gbps duo-binary modulated data at a frequency spacing of 20 GHz, is presented in this study. At each OMAN remote node (RN), optical switches (OSws) can support downlink and uplink transmission. Our OMC spectrum distribution divides the entire wavelengths into six RNs, with ten wavelengths each. Depending on the end-user's location, this system can increase or decrease the number of RNs and ONUs. At each RN, the ONUs are connected through star-topology dis-

tribution cables. Overall, 60 ONUs are connected to the central office, enabling a broadcast of 1200 Gpbs, with 20 Gpbs being allocated to each user at varying distances. WDM-PON assigns each subscriber a distinct wavelength. By eliminating ROADMs, OADMs, and various light sources, a cost-effective and efficient OMAN is presented and supported by the results. The main contributions of this paper are summarized here for easy understanding by the readers:

- This study proposes a new central office for an OMAN based on an OMC generator, focusing on affordability and an efficient switching system.
- The OMCG is based on a cascaded configuration of a phase modulator (PM) and two Mach Zehnder modulators (MZMs), driven by an RF signal controlling the carriers' frequency spacing. Over sixty frequencies are generated with a high tone-to-noise ratio (TNR) of 40 dB and minimal amplitude difference among the generated carriers.
- An efficient switching mechanism is introduced in the proposed OMAN to handle the transmissions during the fiber failures in the main core fiber in the ring architecture. This failure could be a single fiber failure or a multistage fiber failure, which are also considered and presented for proper maintenance/management of the transmission. The proposed CO design can cope with different scenario-based problems without compromising the transmission quality.

2. Architecture of the Proposed Optical Metro Access Network

Figure 1 shows the OMAN's dual-fiber ring construction. The metro access network consists of one or two central offices and M RNs [9,15,16,29,30], with a ring feeder fiber design. When building an OMAN, redundant fibers are buried [14]. A feeder fiber is utilized for network protection; when one fiber is broken, transmission can be shifted to the second fiber. Dormant access ring fibers can be utilized for protection. A dual-fiber ring connects the CO and RNs in this network. One central office and six RNs can distribute multiple wavelengths to access nodes. It can increase (RNs→N) based on the access node location. Providing multiple wavelengths (λ1→λm) can also enhance ONUs. In the CO, the OMCG OLT offers 61 stable and healthy frequency tones.

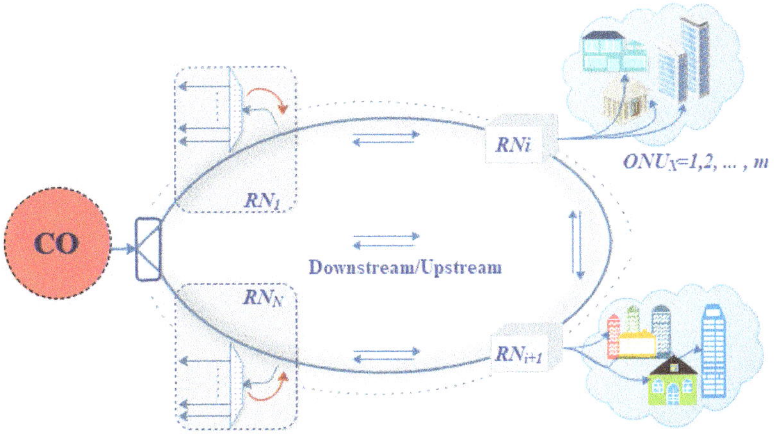

Figure 1. The proposed metro access network. CO: central office; RN: remote node; ONU: optical network unit.

2.1. Schematic of the CO

The schematic diagram of the proposed CO is shown in Figure 2a. It is usually more favorable for the OMAN to have a great number of frequency channels. This can be achieved by either using laser arrays or frequency combs. We have utilized an

OMC generation scheme, shown in Figure 2b, to provide several carriers to facilitate over 60 end-users with high data rates.

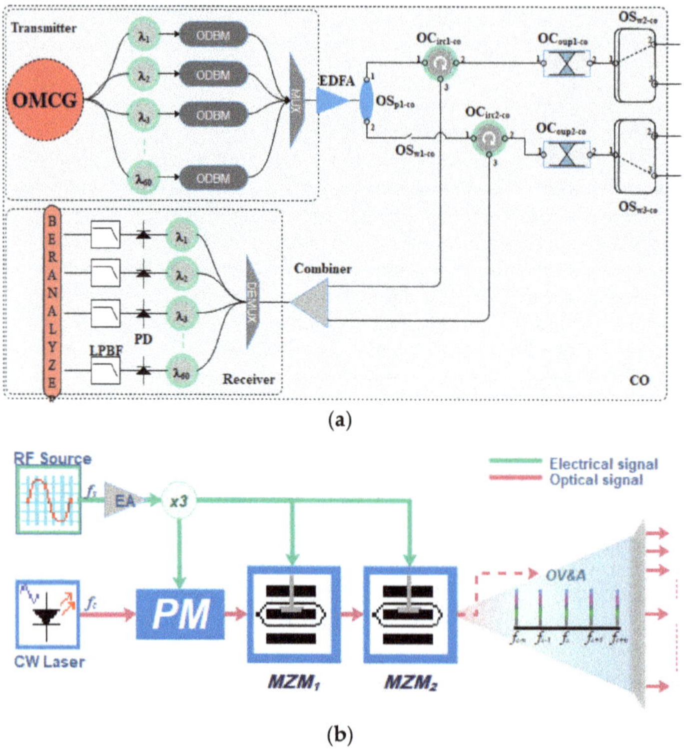

Figure 2. (**a**) Schematic diagram of the proposed central office. OMCG: optical multicarrier generator; ODBM: optical duo-binary modulator; EDFA: erbium doped fiber amplifier; optical switch: optical switch; OC: optical coupler; PD: photodetector; LPBF: low bandpass Bessel filter (**b**) Schematic of the proposed optical multicarrier generator. EA: electrical amplifier; PM: phase modulator; MZM: Mach–Zehnder Modulator; OV and A: optical visualizer and analyzer.

The proposed technique in this study can handle higher data rates with increased frequency spacing and practically double the bandwidth. The proposed OLT, placed at the CO of the metro network, is shown in Figure 2a. It essentially eliminates the need for a single laser diode per single subscriber. The end-face of the multicarrier generator is connected to a 1xN optical fork (a fork with N outputs) arrangement, followed by a set of optical filters to separate the frequencies. The filtered carriers are injected into an optical duo-binary modulator ~20 Gbps (ODBM). An optical multiplexer combines the modulated signals into one (MUX). A 1 × 2 optical splitter (OSp1-co) amplifies and splits the multiplexed signal, which is then followed by two optical circulators (OCircs-co), optical couplers (OCoups-co), and optical switches (OSws-co). The OSp1-co's first port is coupled to the OCirc1-co's input port1. The OCirc1-co's port 2 is linked to the OCoup1-co's port 1. To address the demand for network expansion, the OSp1-co's second port is connected to an optical switch (OSw1-co). Following the switch is OCirc2-co, which has the same structure as OCirc1-co. The OCoup1-co and OCoup2-co end-ports are connected to OSw2-co and OSw3-co, which direct the signal to either the operating mode fiber or the protection fiber. The protection mechanism section explains the needs and working mechanism of OSw2-co and OSw3-co. This arrangement of OCirc-co, OCoup-co, and OSw-co is for normal working conditions, and one is kept reserved for the protection mode for the primary feeder fiber, both for uplink and downlink transmission. The switch manages the uplink and downlink transmissions in normal conditions or when a fault is detected, it switches the transmission into the reserved/protected fiber cable.

Similarly, the CO contains another section of uplink transmission after receiving the data during the normal operation via OSw2-co or during the protection mode through OSw3-co. The data are passed from the OCoups-co to the OCircs-co via port2 and forwarded to the power combiner via port 3 of the OCircs-co. The outputs of these two OCircs-co are combined to receive the signal from both modes directly. The receiver section is further discussed in the coming sections.

2.2. The Principal of the OMC Generator

The OMC generator is based on the cascaded configuration of a single phase modulator (PM) and two Mach Zehnder modulators (MZMs). This proposed technique uses electro-optic modulators driven by an RF source that produces a multiharmonic signal, separated by a fixed frequency termed frequency spacing that can be modified using the RF signal. The optical input of the PM is connected to the CW laser source, whereby the first MZM (MZM1)'s optical input is driven by the PM's output and the second MZM (MZM2) by the output of the first MZM.

The phase modulator (PM) changes the injected optical signal in response to the provided amplified RF signal. Consider the optical continuous wave (CW) with electric field E_0 and angular frequency ω_0. The CW's signal before entering the PM can be given as $E_{cw}(t) = E_0 e^{j\omega_0 t}$, where E_0 represents the amplitude, and ω_0 represents the angular frequency of the laser. The RF amplified sinusoidal signal can be described as $V_{RF}(t) = V_{\max}(\cos(\omega_{RF}t + \phi)$, where V_{max} and ω_{RF} represent the amplitude and angular frequency of the RF source. The phase modulator produces a phase shift that is proportional to the provided RF signal, which can be given as ϕ. Therefore, the phase-modulated optical signal can be given as

$$E_{PM}(t) = E_{cw}(t)\exp(jG\beta\sin(\omega_{RF}t + \phi)) \tag{1}$$

where G is the gain in the RF signal, and we will consider it a constant value ~1 for ease of understanding the further equations, β is the modulation index of the PM, ϕ is the phase of the RF signal, and t represents the time. Therefore, the term $\beta\sin(\omega_{RF} + \phi)$ represents the phase modulation introduced by the PM. After expanding the above equation, it can be given as

$$E_{PM}(t) = E_0\exp(j\omega_0 t)\exp(j\beta\sin(\omega_{RF}t + \phi)) \tag{2}$$

Equation (2) can be expanded further by using the well-known Jacobi–Anger expansion to express the phase-modulated signal in terms of the Bessel functions, as given below:

$$E_{PM}(t) = E_0\exp(j\omega_0 t)\sum_{n=-\infty}^{n=\infty} J_n(\beta)\exp(jn(\omega_{RF}t + \phi)) \tag{3}$$

where J_n is the n-th order of the Bessel function of the first kind. This expression shows the creation of sidebands with frequencies $\omega_0 \pm n\omega_{RF}$, with their amplitude being determined by the Bessel function $J_n(\beta)$, and ω_{RF} represents the frequency spacing between the comb lines. This equation captures the essence of phase modulation under the impact of an RF signal. The Bessel functions represent the energy distribution created due to modulation among various sidebands. The output of the PM is shown in Figure 3a. The output of the PM is connected in cascade to the first MZM. It modulates the intensity of the optical signal whose output can be given as

$$\begin{aligned} E_{MZM1}(t) &= E_{PM}(t).[1 + m\cos(\omega_{RF}t)] \\ &= E_0\exp(j\omega_0 t)\sum_{n=-\infty}^{n=\infty} J_n(\beta)\exp(jn(\omega_{RF}t + \phi)).[1 + m\cos(\omega_{RF}t)] \end{aligned} \tag{4}$$

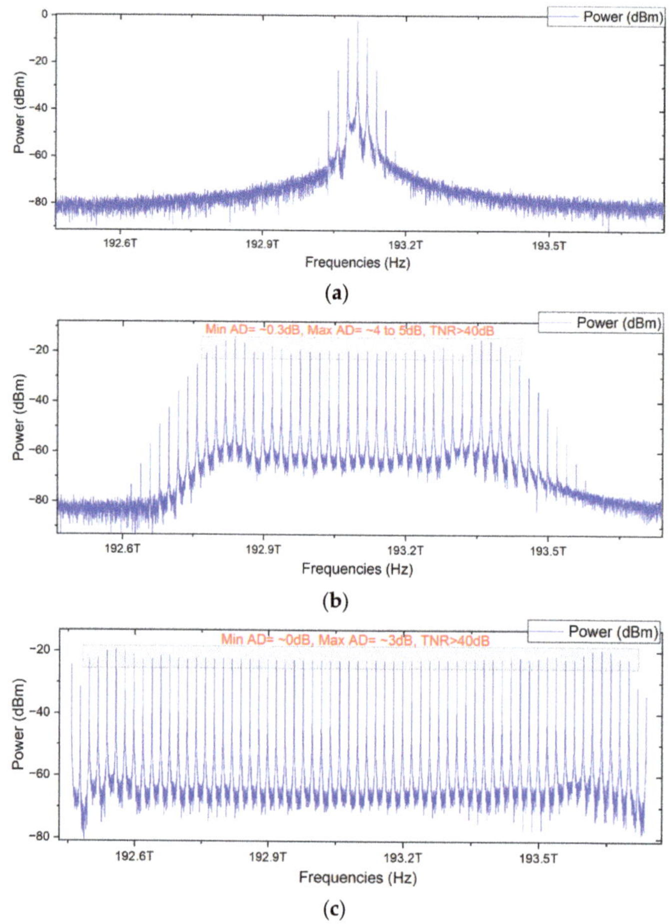

Figure 3. Generation of OMC using three modulators; (**a**) output of the phase modulator; (**b**) output of MZM1; (**c**) output of MZM2.

As $\cos(\omega_{RF}t) = \frac{1}{2}[\exp(j\omega_{RF}t) + \exp(-j\omega_{RF}t)]$, therefore, after putting this in the above Equation (4), it becomes

$$E_{MZM1}(t) = E_0 \exp(j\omega_0 t) \sum_{n=-\infty}^{n=\infty} J_n(\beta) \exp(jn(\omega_{RF}t + \phi)). \\ \left[1 + \frac{m}{2}\{\exp(j\omega_{RF}t) + \exp(-j\omega_{RF}t)\}\right] \tag{5}$$

Here, m shows the modulation depth of the MZM that controls the additional sidebands generated by the MZM. The above expression shows a sum of frequency components that is generated by the PM and then further modulated by the first MZM, where every term corresponds to a separate frequency tone at both sides of the central frequency ~$j_0(\beta)$. This comb is practically finite, because the Bessel function decreases with increasing of the "n". After passing this modulated OFC through the second MZM, more carriers are added to the OFC, which can be presented as

$$E_{MZM1}(t) = E_0 \exp(j\omega_0 t) \sum_{n=-\infty}^{n=\infty} J_n(\beta) \exp(jn(\omega_{RF}t + \phi)). \\ \left[1 + \frac{m}{2}\{\exp(j\omega_{RF}t) + \exp(-j\omega_{RF}t)\}\right].\left[1 + \frac{m'}{2}\{\exp(j\omega_{RF}t) + \exp(-j\omega_{RF}t)\}\right] \tag{6}$$

If both MZMs are identical, then it can be given as

$$E_{MZM2}(t) = E_0 \exp(j\omega_0 t) \sum_{n=-\infty}^{n=\infty} J_n(\beta) \exp(jn(\omega_{RF}t + \phi)). \\ \left[1 + \tfrac{m}{2}\{\exp(j\omega_{RF}t) + \exp(-j\omega_{RF}t)\}\right]^2 \quad (7)$$

This equation is mathematically complex because of the double modulation effect, where each frequency tone from the first MZM is further increased. Hence, we developed a large bandwidth in the proposed scheme.

An amplified 1×3 RF signal is used to modify the electrical inputs of these modulators. The RF source frequency is kept constant at 20 GHz, the amplitude and DC biasing of the RF are 1 a.u., and the sample rate is 1.28×10^{12}. The frequency of the CW laser is 193.1 THz as a center frequency, with a linewidth of 10 MHz and dynamic noise of 3 dB. The PM converts the signal into a wideband signal composed of ultra-short optical pulses that are broadened by the MZM1. The output signal is further expanded by employing the MZM2, which achieves over 60 ultra-flat optical carriers. Figure 3 depicts the output of the three modulators. The resulting optical spectrum, shown in the image, has a tone-to-noise ratio of more than 40 dB and amplitude differences of 0–3 dB. The scheme has a total bandwidth of 1.2 THz. The proposed network for OFC generation has successfully generated over sixty carriers which is higher than or equal in numbers to most of the cited works [19,20,25,28,34].

3. Downlink and Uplink Transmission Operations

During the regular mode operation, the downlink and uplink data are transmitted in the feeder fibers (working fibers) from the CO to RNs and vice versa. The CO contains an OMC source as an OLT that generates X number of frequencies, utilized for downlink transmission, which can be received by Y number of receivers (X = Y) back into the CO after an uplink transmission. The X number of transmitted wavelengths is received by the M number of RNs, where each RN passes Z number of frequencies. A DeMUX can separate the desired tones for the dedicated number of ONUs. As demonstrated in Figure 2a, port 1 of the OSw1-co is connected to the working fiber. The schematic of RN and ONU is shown in Figure 4, where OSw1-RNx followed by a 1×2o-fork receives the signal. The 1×2o-fork performs the following two tasks: (i) it extends the traffic towards the end-users through 1xNo-fork, or by using DeMUX to separate the frequency tones, where N represents the number of users connected to the fork/DeMUX, and (ii) it extends the signal from output port 2 towards the second switch (OSw2-RNj), which extends the fiber in the clockwise direction towards the next RN (OSw-RNj1), where the same procedure is repeated and a ring network is formed. Consequently, if "M" RNs are connected in the ring, and at each RN, 1:No-forks are connected across the ring, the total number of subscribers in the network can be given by M × N. The end-face of each 1:No-fork or the DeMUX terminates in the subscriber premises through a distribution fiber (DF). The DF is connected to a bandpass filter that works as a DeMUX arrangement and recovers the desired carrier frequency for the intended subscriber, which is split in two by a 1×2 OSp; half of it is used for demodulating the signal. In contrast, half of the signal's power is used for colorless uplink transmission.

In the proposed setup, the multiplexed signal that is transmitted from the CO is received by six RNs. Therefore, the overall spectrum of 1.2 THz is divided into six groups of frequencies with a range from 192.50 THz to 193.70 THz. The frequency range dedicated to the RN1 is from 192.52 to 192.70 THz. Similarly, RN2 is from 192.72 to 192.90 THz, and so on. At each RN, ten tones are dedicated to the end-users. The summary of the frequency distributions among the different RNs is given in Table 1.

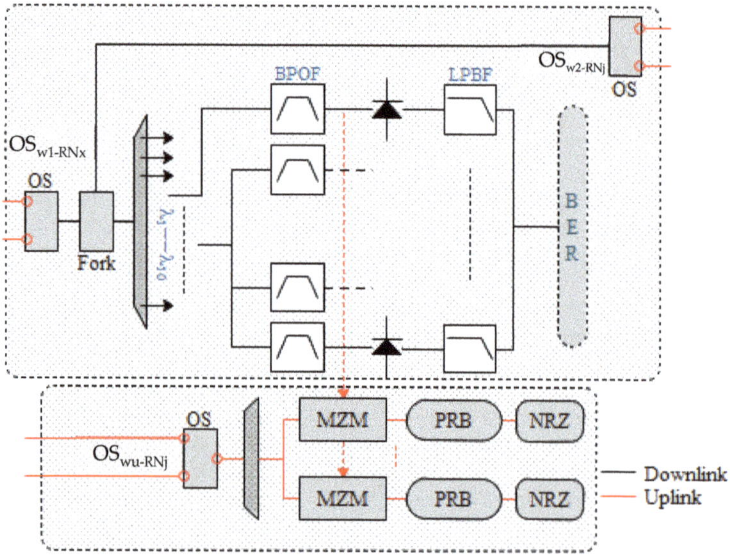

Figure 4. Schematic diagram of the RN + ONU.

Table 1. Summary of the frequency distribution.

RNX	Frequencies per RN	Frequency Ranges (20 GHz Frequency Spacing)
RN1	F_{01} to F_{10}	192.52 THz→192.7 THz
RN2	F_{11} to F_{20}	192.72 THz→192.9 THz
RN3	F_{21} to F_{30}	192.92 THz→193.1 THz
RN4	F_{31} to F_{40}	193.12 THz→193.3 THz
RN5	F_{41} to F_{50}	193.32 THz→193.5 THz
RN6	F_{51} to F_{60}	193.52 THz→193.7 THz

For downlink transmission among the six RNs, we simulated five RNs, and for each RN, two frequencies were selected for carrying the 20 Gbps ODBM-based modulated data. As mentioned in Table 1, 60 subscribers can be facilitated at six different RNs, where each RN is at the span of 5 Km feeder fiber, while each user is at a distance of 2 km from the RN. In this huge network, we have provided the simulation results for two users per RN for downlink and uplink transmission. The multiplexed signal received at 5 km (at RN1), 10 km (at RN2), and up to 25 km (at RN5), each with a fiber span of 5 km, is demonstrated in Figure 5a, which shows the drop/loss of power from one ONU to the other. The power loss across each RN can be observed clearly in the figure. To test the applicability of the scheme, a variable optical attenuator (VOA) is used at the output of the $OS_{w1\text{-}co}$ and $OS_{w2\text{-}co}$. The VOA attenuates the signal's power level and distorts the signal, and we can calculate the desired BER values. The value of the VOA is set in an iterative manner, where the min value is 0, and the maximum is restricted to 9. Each iteration advances the value by an addition of 1. The simulation parameters are listed in Table 2.

The achieved low bit error rate of 10^{-13} to 10^{-3} is shown in Figure 5b, which shows a min. BER of approximately 10^{-13} for 20 Gbps downlink transmission. Hence, the scheme guarantees efficient performance for 20 Gbps downlink transmission. The graph shows the power loss across the channel and other passive components.

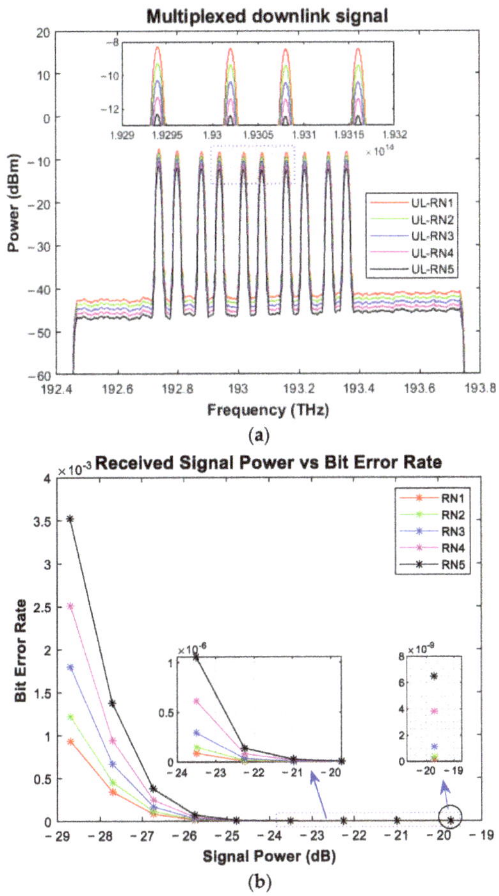

Figure 5. (**a**) The multiplexed signal received at each RN: (**b**) bit error rate (BER) vs. received signal power.

Table 2. Simulation parameters.

Parameter	Value	Parameter	Value
Global Parameters		Optical Fiber	
Bit rate	20×10^9 bit/s	Reference wavelength	1552 nm
Sequence length	256 bits	Length	5 km
Samples per bit	128	Attenuation	0.2 dB/km
Symbol rate	10×10^9	Dispersion	16.75 ps/nm/km
CW Laser		Dispersion slope	0.075 ps/nm^2/km
Frequency	193.1 THz	Differential group delay	0.2 ps/km
Power	5 dBm	Effective area	80 um^2
Linewidth	0.1 MHz	Fract. Raman contribution	0.18

Upon receiving the signal at $OS_{w1\text{-}RNj}$, as shown in Figure 4, which is split after the identification via an optical filter, part of it is reused for uplink transmission. The signal is identified using a Gaussian optical filter and fed into an intensity modulation on-off keying (IM-OOK) for data modulation. For uplink transmission, the data rate of the simulation environment is reduced to 10 GHz. The use of a duobinary modulation scheme in the downlink transmission proved to be useful to re-modulate the downstream signal into the OOK upstream signal at the RN [35]. A symmetrical operation at 10 Gbps over a 25 km reach has already been proposed and demonstrated experimentally [36]. Even a symmetric 10 Gbps WDM-PON has been proven to reach distances of 80 km without using any remote amplification [37]. After modulation, the signals of both ONUs are combined and sent back to the optical fiber. Before injecting the signal into the fiber, $OS_{wu\text{-}RNj}$ is used to direct the signal to the working fiber or protection fiber. The OSw1-co and OSw2-co are utilized at the central office to identify the signal mode. If the signal is received from the working fiber, the $OS_{w1\text{-}co}$ directs the signal to the $OC_{oup1\text{-}co}$, which is then forwarded to the $OC_{irc1\text{-}co}$. If the signal is received from the protection fiber, the $OS_{w2\text{-}co}$ directs the signal to the $OC_{oup2\text{-}co}$, which is then forwarded to the $OC_{irc2\text{-}co}$. The output of both circulators is combined using an ideal Optical Combiner, and the output is fed into a 1 × 10 WDM De-MUX. A Gaussian optical filter is utilized at the De-MUX to identify each wavelength. After wavelength identification, a direct detection optical receiver (DDOR) receives the signal. The DDOR is based on a PIN photodiode and a low-pass Bessel filter, and the results are calculated using BER spectrum analyzers. The received channels promise a good performance, and the results can be seen in Figure 6a, where the multiplexed signal shows a continuous downfall in the received optical power of around 1 dBm. The same VOA is used to test the performance of the scheme. When the value of the VOA is at its minimum, the minimum BER is exceptionally low, whereas the scheme offers a minimum BER of slightly below 10×10^{-3} for the high attenuation value, as shown in Figure 6b, which greatly supports the applicability of the proposed scheme.

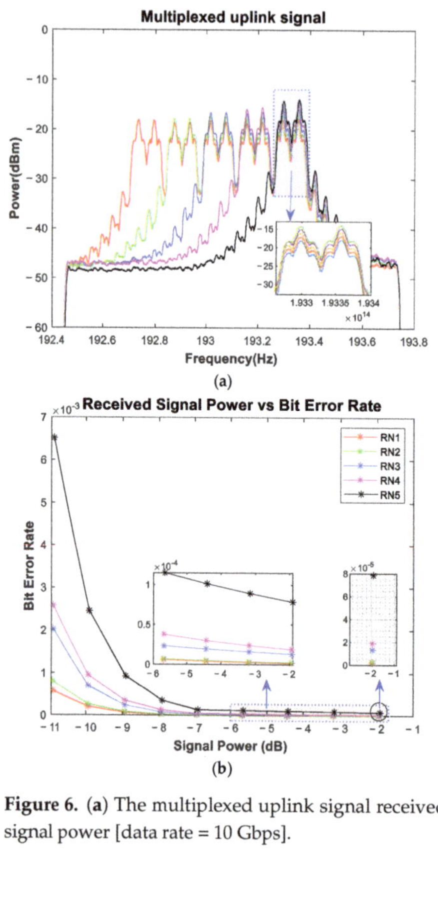

Figure 6. (**a**) The multiplexed uplink signal received from five RNs: (**b**) the BER vs. the received signal power [data rate = 10 Gbps].

4. Protection Mechanism

The protection mechanism of the proposed scheme can be characterized by two categories: fault-A and fault-B. The fault-A protection mechanism can be considered when a single working feeder fiber fails between any two RNs, i.e., RN_j and RN_{j+1}; then, the $OS_{w1\text{-}co}$ or the $OS_{w2\text{-}RNj}$ *at* RN_j transfers the load to the second port that is connected with the protection fiber. Similarly, the optical switch at RNj+1 transfers the load to the switch's second port, which accepts the data from the protection fiber. Therefore, normal operation can be easily restored after repairing the working feeder fiber by utilizing the protection fiber. This platform can work even for numerous single-feeder fiber faults, as it easily transfers the transmissions to the protected fiber. As shown in Figure 7, three cases, I, II, and III, show the faults at three single feeder fibers at different positions. The simulation results of these cases can be studied in Figure 8. The receiving powers across the downlink transmissions show us the power loss during the transmission and during the fiber failure cases.

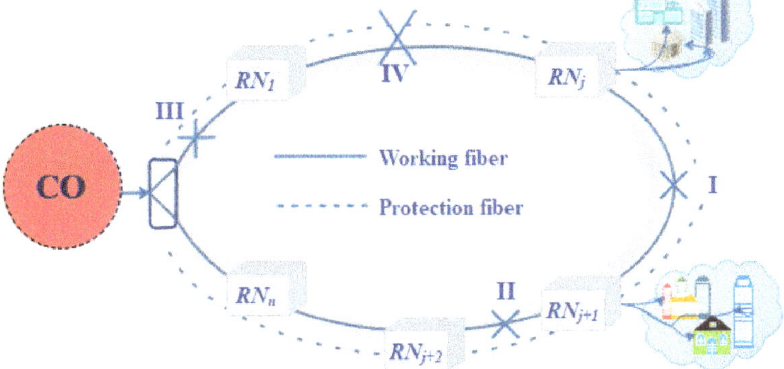

Figure 7. Fiber failures at different positions.

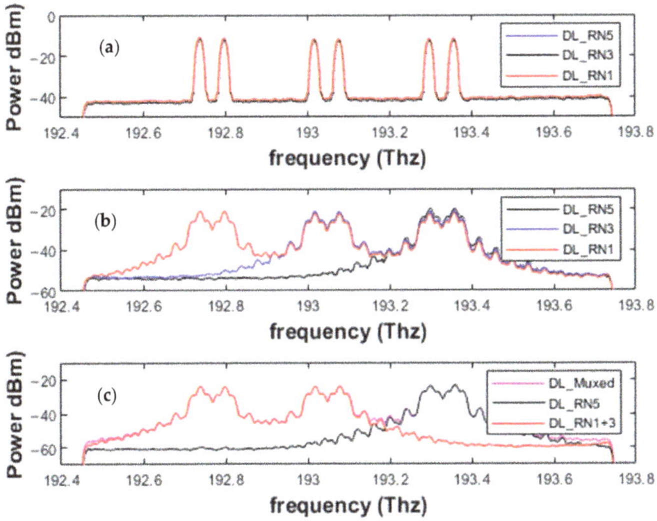

Figure 8. Simulation results: (**a**) multiplexed downlink signal at each RN, (**b**) received signal from each RN during class-A failures, and (**c**) received signal from each RN during a class-B failure.

The fault-B protection mechanism is used when the feeder fiber fails and $OS_{w2\text{-}co/RNj}$ cannot restore normal operation. In the situation that occurred at I and II, the fault protection mechanism works as we described earlier. For the fault at IV, the state of the binary switch $OS_{w1\text{-}co}$ is changed before the $OC_{irc2\text{-}co}$ is turned on, and the signal is transmitted across the OC_{irc}. This signal is connected to the $OS_{w2\text{-}co}$. Here, the direction of the signal is changed, and the signal at $OS_{w1\text{-}co}$ is directly received by the RN1, whereas the RN1 and RN5 are no longer connected due to a failure of the feeder and protection fiber, and the $OS_{w2\text{-}co}$ changes the flow and transmits the signal to RN5. The working mechanism of the upload transmission is the same for RN1; however, the uplink signal generated by RN5 can no longer connect to RN1. Therefore, the uplink signal is also received by the $OS_{w2\text{-}co}$. All the switches in the CO ($OS_{w1,2,3\text{-}co}$) and ONUs realize the working conditions in normal and protection mode.

Similarly, the proposed model can also work in the dual feeder fiber failure, i.e., if both the normal mode fiber and the protection fibers fails, as shown in Figure 7 by IV between the RN1 and RNj. In this scenario, the switch in the CO (OSCO-1) can manage both the downlink and uplink transmission, and both the branches will work in the two tree networks for the time being, until the repairing of the feeder fibers occurs.

5. Conclusions

This paper introduces a centralized single OLT to support many users in a WDM ring architecture-based OMAN. A traditional OLT of the OMAN is replaced with a cost-effective OMC source that offers numerous frequency tones to carry the bidirectional information across different spans of fibers. These fibers are connected in a ring architecture with different RNs, and ten ONUs can be connected to each RN based on the star topology. At each RN, part of the downlink signal, which can be considered colorless, is utilized for uplink transmission. The OMC source generates more than sixty comb lines with measurable parameters, making it more suitable to deploy in an OMAN that is meant for entertaining many subscribers while facilitating heavy data rate support. Across each carrier of the generated OFC, 20 Gbps data are transmitted based on optical duobinary modulation. At the ONU side, VOAs are utilized to calculate the BERs of the received signal, which ranges from 10^{-3} to 10^{-13} for the downlink transmission. Similarly, an OOK is utilized for the uplink transmission to test the feasibility of the proposed OFC in the OMAN system. The BER during the uplink transmission ranges from 10^{-6} to 10^{-14}. Furthermore, the detailed analyses are discussed and explained with their respective results for the various cases of main optical fiber failures in the optical metro access network, along with the system's robust switching mechanism.

Author Contributions: Methodology, R.U., S.U., A.A.A.-A. and J.R.; Software, R.U., A.A.A.-A., J.R. and S.C.; Validation, R.U., S.U. and S.C.; Formal analysis, A.A. and H.S.A.; Investigation, A.A. and H.S.A.; Resources, H.S.A. and J.R.; Data curation, R.U., S.U. and S.C.; Writing—original draft, S.U., H.S.A. and S.C.; Writing—review & editing, A.A. and A.A.A.-A. All authors have read and agreed to the published version of the manuscript.

Funding: National Key Research and Development Program of China (No. 2018YFB1801703); National Natural Science Foundation of China (No. 61835005, 61822507, 61522501, 61475024, 61675004, 61705107, 61727817, 61775098, 61720106015, 61875248); Beijing Young Talent (No. 2016000026833ZK15); BUPT Excellent Ph.D. Students Foundation (No. CX2020301); Open Fund of State Key Laboratory of Information Photonics and Optical Communications (Beijing University of Posts and Telecommunications); Jiangsu talent of innovation and entrepreneurship; Jiangsu team of innovation and entrepreneurship; Postgraduate Research & Practice Innovation Program of Jiangsu Province (No. SJKY19_0972).

Institutional Review Board Statement: Not applicable.

Informed Consent Statement: Not applicable.

Data Availability Statement: Data are contained within the article.

Conflicts of Interest: The authors declare no conflict of interest.

References

1. Pan, B.; Xue, X.; Guo, X.; Calabretta, N. Precise time distribution and synchronization for low latency slotted optical metro-access networks. *J. Light. Technol.* **2022**, *40*, 2244–2253. [CrossRef]
2. Hall, M.N.; Foerster, K.-T.; Schmid, S.; Durairajan, R. A survey of reconfigurable optical networks. *Opt. Switch. Netw.* **2021**, *41*, 100621. [CrossRef]
3. Zhang, C.; Zhang, Q.; Chen, C.; Jiang, N.; Liu, D.; Qiu, K.; Liu, S.; Wu, B. Metro-access integrated network based on optical OFDMA with dynamic sub-carrier allocation and power distribution. *Opt. Express* **2013**, *21*, 2474–2479. [CrossRef] [PubMed]
4. Filgueiras, H.R.D.; Lima, E.S.; Cunha, M.S.B.; Lopes, C.H.D.S.; De Souza, L.C.; Borges, R.M.; Pereira, L.A.M.; Brandão, T.H.; Andrade, T.P.V.; Alexandre, L.C. Wireless and Optical Convergent Access Technologies Toward 6G. *IEEE Access* **2023**, *11*, 9232–9259. [CrossRef]
5. Yang, H. Optical and wireless convergence network based on blockchain. In *Security Issues in Fog Computing from 5G to 6G: Architectures, Applications and Solutions*; Springer: Berlin/Heidelberg, Germany, 2022; pp. 131–143.
6. Xia, J.; Li, T.; Cheng, Q.; Glick, M.; Crisp, M.; Bergman, K.; Penty, R. A future proof reconfigurable wireless and fixed converged optical fronthaul network using silicon photonic switching strategies. *J. Light. Technol.* **2022**, *41*, 1610–1618. [CrossRef]
7. Feng, N.; Ma, M.; Li, Z.; Li, S.B. Flexible Media Access Control Layer Key Technologies for Elastic Optical Access Network. In Proceedings of the 2023 21st International Conference on Optical Communications and Networks (ICOCN), Qufu, China, 31 July–3 August 2023; IEEE: Piscataway, NJ, USA, 2023; pp. 1–3.
8. Layec, P.; Dupas, A.; Verchère, D.; Sparks, K.; Bigo, S. Will metro networks be the playground for (true) elastic optical networks? *J. Light. Technol.* **2017**, *35*, 1260–1266. [CrossRef]
9. Li, X.; Gan, C.; Yan, Y.; Qiao, H. Grid architecture of a metro-access optical network to support discretionary peer-to-peer intracommunication and intercommunication between ONUs. *J. Opt. Commun. Netw.* **2019**, *11*, 130–139. [CrossRef]
10. Ruffini, M.; Achouche, M.; Arbelaez, A.; Bonk, R.; Di Giglio, A.; Doran, N.J.; Furdek, M.; Jensen, R.; Montalvo, J.; Parsons, N.; et al. Access and metro network convergence for flexible end-to-end network design [invited]. *J. Opt. Commun. Netw.* **2017**, *9*, 524–535. [CrossRef]
11. Pato, S.; Borges, N.; Pedro, J. Capacity prospects of future high density metro-access networks. In Proceedings of the 2013 15th International Conference on Transparent Optical Networks (ICTON), Cartagena, Spain, 23–27 June 2013.
12. Gou, K.; Gan, C.; Zhang, X.; Zhang, Y. A tangent-ring optical TWDM-MAN enabling three-level transregional reconfigurations and shared protections by multipoint distributed control. *Opt. Commun.* **2018**, *410*, 855–862. [CrossRef]
13. Bianco, A.; Bonald, T.; Cuda, D.; Indre, R.M. Cost, power consumption and performance evaluation of metro networks. *J. Opt. Commun. Netw.* **2013**, *5*, 81–91. [CrossRef]
14. Guo, Y.; Gan, C.; Zhan, N. A novel expandable intersection-ring architecture supporting optical network units' discretionary peer-to-peer communication in metro-access network. *Int. J. Commun. Syst.* **2022**, *35*, e5079. [CrossRef]
15. Li, X.; Gan, C.; Chen, Y.; Qiao, H.; Yan, Y. Dual-Fiber-Ring Architecture Supporting Discretionary Peer-to-Peer Intra-Communication and Bidirectional Inter-communication in Metro-Access Network. *IEEE Access* **2019**, *7*, 52360–52370. [CrossRef]
16. Lafata, P. Protection of passive optical network by using ring topology. In Proceedings of the 2012 35th International Conference on Telecommunications and Signal Processing, Prague, Czech Republic, 3–4 July 2012.
17. Kaszubowska-Anandarajah, A.; Ahmad, S.T.; Roeloffzen, C.G.H.; van Dijk, P.W.L.; Sharma, A.; Srivastava, M.; Lakshmijayasimha, P.D.; Pascual, M.D.G.; Smyth, F.; Braddell, J. Reconfigurable photonic integrated transmitter for metro-access networks. *J. Opt. Commun. Netw.* **2023**, *15*, A92–A102. [CrossRef]
18. Lakshmijayasimha, P.D.; Kaszubowska-Anandarajah, A.; Martin, E.P.; Landais, P.; Anandarajah, P.M. Expansion and phase correlation of a wavelength tunable gain-switched optical frequency comb. *Opt. Express* **2019**, *27*, 16560–16570. [CrossRef] [PubMed]
19. Zhou, J.; Ban, X.; Zhang, J.; Li, J.; Yang, J.; Jia, D.; Ge, C.; Wang, Z. Ultra-flat multicarrier light source based on recirculating frequency shift loop driven by a parity–time symmetric optoelectronic oscillator. *Appl. Opt.* **2023**, *62*, 6857–6863. [CrossRef] [PubMed]
20. Das, B.; Mallick, K.; Mandal, P.; Dutta, B.; Barman, C.; Patra, A.S. Flat optical frequency comb generation employing cascaded dual-drive Mach-Zehnder modulators. *Results Phys.* **2020**, *17*, 103152. [CrossRef]
21. Ullah, R.; Ullah, S.; Khan, G.Z.; Mao, Y.; Ren, J.; Zhao, J.; Chen, S.; Li, M.; Khan, J. Ultrawide and tunable self-oscillating optical frequency comb generator based on an optoelectronic oscillator. *Results Phys.* **2021**, *22*, 103849. [CrossRef]
22. Ullah, R.; Bo, L.; Ullah, S.; Yaya, M.; Tian, F.; Xiangjun, X. Cost effective OLT designed from optical frequency comb generator based EML for1. 22 Tbps wavelength division multiplexed passive optical network. *Opt. Fiber Technol.* **2018**, *43*, 49–56. [CrossRef]
23. Ullah, R.; Ullah, U.; Ullah, S.; Al-atawi, A.A.; Alwageed, H.S.; Chen, S.; Ren, J. Dual Electro-Optic Comb Generator by Heterodyning Two Combs for Extending the Optical Bandwidth. *IEEE Photonics Technol. Lett.* **2023**, *35*, 1186–1189. [CrossRef]
24. Lakshmijayasimha, P.D. Reconfigurable Multi-Carrier Transmitters and Their Application in Next Generation Optical Networks. Ph.D. Thesis, Dublin City University, Dublin, Ireland, 2022.

25. Llorente, R.; Fito, V.; Morant, M. Optical combs and multicore fiber as technology enablers for next-generation datacenter infrastructure. In *Proceedings of the Metro and Data Center Optical Networks and Short-Reach Links V*; SPIE: Philadelphia, PA, USA, 2022; Volume 12027, pp. 54–63.
26. Ullah, R.; Ullah, S.; Imtiaz, W.A.; Khan, J.; Shah, P.M.A.; Kamran, M.; Ren, J.; Chen, S. High-Capacity Free Space Optics-Based Passive Optical Network for 5G Front-Haul Deployment. *Photonics* **2023**, *10*, 1073. [CrossRef]
27. Jin, W.; Zhang, C.; Chen, C.; Zhang, Q.; Qiu, K. Scalable and reconfigurable all-optical VPN for OFDM-based metro-access integrated network. *J. Light. Technol.* **2014**, *32*, 318–325. [CrossRef]
28. Ionin, A.A.; Kinyaevskiy, I.O.; Klimachev, Y.M.; Sagitova, A.M.; Andreev, Y.M. CO laser sum-frequency comb for atmosphere sensing. *Infrared Phys. Technol.* **2019**, *100*, 62–66. [CrossRef]
29. Cai, Y.; Cheng, J.; Yan, Y. Spectrum-efficient optical drop-add-drop network with a centralized multi-carrier light source. *Photonic Netw. Commun.* **2014**, *27*, 89–98. [CrossRef]
30. Matsuura, M.; Oki, E. Multi-carrier distributed WDM ring network based on reconfigurable optical drop-add-drop multiplexers and carrier wavelength reuse. In Proceedings of the IEEE International Conference on Communications, Cape Town, South Africa, 23–27 May 2010.
31. Pan, B.; Yan, F.; Xue, X.; Magelhaes, E.; Calabretta, N. Latency Study of Fast Optical Add-Drop Multiplexer Based Metro Access Network with Edge Computing for 5G Applications. In Proceedings of the OECC/PSC 2019—24th OptoElectronics and Communications Conference/International Conference Photonics in Switching and Computing 2019, Fukuoka, Japan, 7–11 July 2019.
32. Pan, B.; Yan, F.; Xue, X.; Magelhaes, E.; Calabretta, N. Performance assessment of a fast optical add-drop multiplexer-based metro access network with edge computing. *J. Opt. Commun. Netw.* **2019**, *11*, 636–646. [CrossRef]
33. Miao, W.; De Waardt, H.; Calabretta, N. Optical Label Switching Add-Drop Multiplexer for High-Performance Metro Networks. *IEEE Photonics Technol. Lett.* **2016**, *28*, 1065–1068. [CrossRef]
34. Lin, C.; Wang, Y.; Wang, Y.; Wang, D.; Xiong, F.; Zhang, D.; Zhang, J.; Wang, A.; Du, P.; Peng, X. Multi-carrier broadband signal generation based on a mutually dual-domain mode-locked optoelectronic oscillator. *Opt. Commun.* **2023**, *529*, 129094. [CrossRef]
35. Muciaccia, T.; Gargano, F.; Passaro, V.M.N. Passive optical access networks: State of the art and future evolution. *Photonics* **2014**, *1*, 323–346. [CrossRef]
36. Chiuchiarelli, A.; Proietti, R.; Presi, M.; Choudhury, P.; Contestabile, G.; Ciaramella, E. Symmetric 10 Gb/s WDM-PON based on a cross wavelength-reusing scheme to avoid rayleigh backscattering and maximize band usage. In Proceedings of the 2009 IEEE LEOS Annual Meeting Conference Proceedings, Belek-Antalya, Turkey, 4–8 October 2009; IEEE: Piscataway, NJ, USA, 2009; pp. 555–556.
37. Cho, K.-Y.; Hong, U.-H.; Jung, S.-P.; Takushima, Y.; Agata, A.; Sano, T.; Horiuchi, Y.; Suzuki, M.; Chung, Y.-C. Long-reach 10-Gb/s RSOA-based WDM PON employing QPSK signal and coherent receiver. *Opt. Express* **2012**, *20*, 15353–15358. [CrossRef]

sensors

MDPI

Article

Optimizations of Double Titanium Nitride Thermo-Optic Phase-Shifter Heaters Using SOI Technology

Eylon Eliyahu Krause and Dror Malka *

Faculty of Engineering, Holon Institute of Technology (HIT), Holon 5810201, Israel; eylon1909@gmail.com
* Correspondence: drorm@hit.ac.il

Abstract: A commercial thermo-optic phase shifter (TOPS) is an efficient solution to the imbalance problem in the fabrication process of Mach–Zehnder modulator (MZM) arms. The TOPS consumes electrical power and transforms it into thermal energy, which changes the real part of the effective refractive index at the waveguide and adjusts the MZM transfer function to work in the linear region. The common model being used today is constructed with only one heater; however, this solution requires more electrical power, which can increase the transmitter system cost. To reduce the system energy cost, we propose a pioneering optimal double titanium nitride heater model under forward biasing at 1550 nm wavelength using the standard silicon-on-insulator technology. Numerical investigations were carried out on the key relative geometrical parameters, heat distribution at the silicon layer, thermal crosstalk, and laser wavelength drift. Results show that the optimal TOPS design can function with a low electrical power of 19.1 mW to achieve a π-phase shift, with a low thermal crosstalk of 0.404 and very low optical losses over 1 mm length. Thus, the proposed device can be used for improving the imbalance problem in MZMs with low electrical power consumption and low losses. This functionality can be utilized to obtain better performances in transmitter systems for data centers and long-range optical communication system applications.

Keywords: titanium nitride; SOI; silicon; phase shifter; MZM

1. Introduction

The integration of thermo-optic phase shifters (TOPSs) using silicon-on-insulator (SOI) technology presents an appealing solution for various applications, including sensors, high-speed communications, filters, and broadband electro-optic modulators. In these applications, TOPSs play a crucial role in encoding light through intensity or phase modulation [1–5]. TOPSs are being created monolithically, usually with silicon (Si) as the waveguide (wg), other dioxides semiconductors like silica (SiO_2) as the covering material, and other metals/doping semiconductors for the heaters. TOPSs are being integrated into a carrier depletion-based Mach–Zehnder modulator (MZM), a device that modulates a signal by either difference in length or a change in the effective refractive index of the transverse electric (TE) mode due to heating [2]. This phenomenon arises from the depletion carrier effect or thermal changes around the MZM. Heating charges create differences in voltage potentials, which start a charge movement inside the MZM and consequently change the operation voltage [6,7]. The relation between the operating voltage and the optical signal at the output is co-sinusoidal. The linear area is defined as an approximation around the bias quadrature points, which occurs firstly at $\pi/2$. TOPSs heat the MZM in a uniform pattern and thus keep the operating voltage at the linear region, which allows us to transmit signals with a bandwidth as wide as 25 Gbit/s and beyond, and maximize the signal-to-noise ratio [8–11]. The TOPS requirement comes from the imprecise fabrication of the MZM; the two lines that split the optical signal, in which the signal propagates through each one of them and then combines the two for phase modulation, are not identical in length, and a difference in their lengths causes a difference in the phase from each path. A phase difference, unless intentionally applied (like the TOPS itself), needs to be avoided. Otherwise, it would create non-in-phase interference at the end of the MZM, which would damage the signal, and in addition, in an integrated feedback loop, it could cause destabilization with positive feedback [12–15]. TOPSs are an efficient solution for the imbalance problem [4], yet there has not been a use case for more than one heater in TOPSs. This paper commences with the optimization of a dual metal-based titanium nitride (TiN) TOPS design with thermal simulation based on finite difference equation (FDE) and optical simulations based on the finite element method (FEM). The metal-based heaters were chosen due to the TiN advantage over doped semiconductors in attributes such as low specific heat (598 J/kgK) versus (711 J/kgK) in Si, while having roughly the same thermal conductivity (28 W/mK) as Si based phosphorus doping (25 W/mK) [4]. Its separation from the wg has negligible attenuation coefficient, making it a superior solution for long range communication systems. The simulation in the Lumerical suite was performed to the thermal crosstalk between the two heaters of an MZI switch for both TOPS types, for various heater powers, and for various distances between the heaters. This paper thus comprehensively reports on double TOPS design optimizations and crosstalk mitigation in a standard SOI process, which, to the best of our knowledge, had not been presented in the literature [16,17] and can be an efficient solution with low power consumption for optical communication systems.

2. Design of the Device and Theoretical Aspects

Figure 1 shows the Si strip waveguide structure in the x-y plane, while the light propagates on the z-axis. The colors in the figure are defined as follows: yellow is air, light blue is the TiN heaters, light grey is SiO_2, and purple is Si wg. The dimensionality of the components are as follows: the height of the air is 5 μm; the height of the SiO_2 (h_{SiO_2}) is 2 μm, which also called buried oxide (BOX); the height of the Si wg (h_{wg}) is 220 nm, and its width (w_{wg}) is 500 nm. The dimensionality is required to obtain a fundamental TE mode solution to the operated wavelength (λ_0) of 1550 nm (which can be used in TM mode as well). The device length (along the z axis) is 1 mm. The metal used in the double heaters is TiN, and their locations in the structure were optimized to obtain better performances; the optimization parameters are the distance between the two heaters (d_{TiN}) as well as their center of mass distance from the Si wg (d). The TiN dimensions were set to be suitable to the fab capability fabrication conditions that exist today. Hence, each TiN heater has the same height of 220 nm (h_{TiN}) and width of 2 μm (w_{TiN}).

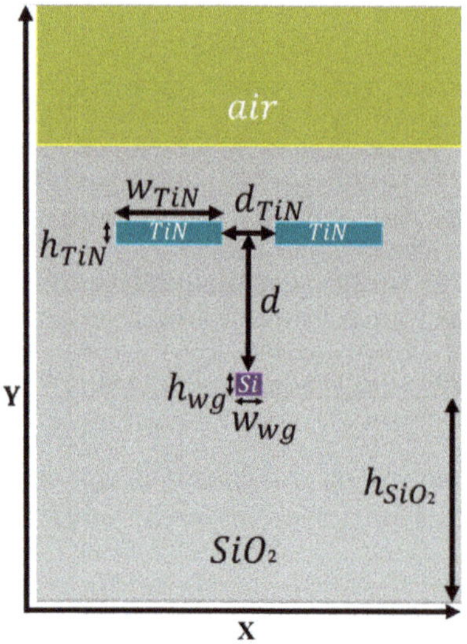

Figure 1. Schematic sketch of double TiN in Si strip wg structure in the x–y plane.

In the following study case, there is an optimization of a model that uses a dual heater structure based on TiN. The heating units keep the temperature constant across the model, which changes the electron distribution and approximately keeps the biased voltage in the linear region. The crucial input, aside from the materials and geometrical aspects, is the electrical power. To validate and use the model with a driven voltage, the relation between Ohm's law, the electrical power of the resistor, and the formula for solid resistance will yield the following:

$$V_\pi = \sqrt{P_\pi \frac{A}{\sigma l}} \tag{1}$$

where P_π is the total power needed for a π-phase shift, σ is the electrical conductivity, A is the surface area of the resistor, and l is the length of the resistor. Since the geometrical and material aspects are constant, we can see a square root relation between the operating voltage and the total power. In the simulation, we utilize the Heat tool, a simulation for studying semiconductor devices based on a discrete approximation of thermodynamics and its relation to Maxwell's equation. The results are both graphic, by heat distribution plotting, and numerical, relating the consumed electrical power to a coordinate on the media. The heat transport equation is given by:

$$\rho c_\rho \frac{\partial T}{\partial t} - \nabla(k\nabla T) = Q \tag{2}$$

where T denotes the temperature at each media coordinate, ρ represents mass density, c_ρ denotes specific heat, k stands for thermal conductivity, and Q represents the rate of applied heat energy transfer. By incorporating the electrical conductivity σ, the electric current equation (Ohm's law) can be concurrently solved with the heat transport equation. Internally, in the media, the charge density may change its distribution at each point, over time, due to the external voltage applied. By plugging the continuity equation, we will be able to determine the charge density both in the media space and its change over time.

$$\frac{\partial \rho}{\partial t} = -\nabla J \tag{3}$$

The Poisson equation describes the relation between charge distribution and the external voltage in the media. To avoid unnecessary oscillatory behavior in the distribution, we solve for the DC steady-state power consumption:

$$-\nabla(\varepsilon\nabla V) = \rho \tag{4}$$

The DC step response, which contributes to heat dissipation, is stable since it generates an exponential decay in the charge density, which corresponds to a left pole.

$$\frac{\partial \rho}{\partial t} + \frac{\sigma}{\varepsilon}\rho = 0 \tag{5}$$

The decay process is defined by the relaxation time $\tau = \varepsilon/\sigma$. When examining changes in the system over time, scales much greater than τ ($t >> \tau$), such as the quasi-static approximation, becomes relevant, resulting in a negligible value for $\partial\rho/\partial t$. It should be emphasized that in steady-state conditions, the rate of change of the charge density ($\partial\rho/\partial t$) is intentionally disregarded. In the case that the charge density changes much slower for the relaxation time, the solution becomes much simpler by being homogenous [18–21]:

$$\nabla(\sigma E) = 0 \tag{6}$$

Simply by using Ohm's law once again, we can directly relate the emerging change in the media's electric field and the given electrical power consumption:

$$P = JE = \sigma E^2 \tag{7}$$

The dissipated power resulting from Ohmic loss is accounted for in the heat transport equations as the transfer rate of heat energy, where the entire power consumption is converted into heat. The HEAT tool discretizes and solves either the heat transport equation alone or the coupled heat transport and electric current equations using a finite-element mesh on a media's surface or throughout its volume. The simulation region is divided into multiple domains along material boundaries, each possessing distinct physical characteristics. The accuracy of the simulation can be changed by the number of points in the media, that are being taken into account when solving the equations. Accuracy itself is a tradeoff with the runtime of the simulation. The materials utilized in the simulation range from insulators to conductors and even fluids. Each material type has its own parameters; for some, they are already given, and for others, they can be implemented or changed by the user. The system of equations addressed in a heat transport simulation accommodates both steady-state and time-varying solutions. Since the simulation works in both ways, now, we can set initial conditions for the temperature and have a constraint to keep it constant over time:

$$\frac{\partial T}{\partial t} = 0 \tag{8}$$

We will now have to solve a special case for the heat equation with its restriction:

$$-\nabla(k\nabla T) = Q \tag{9}$$

Steady-state simulations allow for the examination of the system's performance under a consistent operating condition. In contrast, by specifying an initial temperature (and electric field if required), the equations can be solved at desired time periods. This enables the evaluation of the component's time-dependent behavior, facilitating a direct analysis of its response in the time domain, particularly for signals with long-time oscillation analysis [22–28].

Increasing the voltage show a respective increase in the Ohmic heating. The factor that needs to be taken into account for heating is thermal crosstalk, which is defined as the ratio between the increment in temperature at the passive component and the active component. In our case, the passive component is the Si wg, and the active is the double TiN heaters pair above the wg [29]. The cross talk is a good approximation to its analog measurement given that in the simulation, we view the crosstalk on the surface of XY plane, as seen in Figure 1, as a matrix, with each point being summed over is a location on the matrix:

$$\text{Normalized} - \text{thermal} - \text{crosstalk} = \frac{\sum\limits_{n,m} Y_{nm} - Y_{ambient}}{\sum\limits_{q,r} X_{qr} - Y_{ambient}} \tag{10}$$

where Y_{nm} is a sample of passive components surface at location n,m; X_{qr} is a sample of active components surface at location q,r; and $Y_{ambient} = X_{ambient} = 300$ K is the TOPS at ambient temperature. The ohmic heating itself, as mentioned above, is causing a phase shift, which we are using to stabilize the imbalance problem. The temperature-dependent phase change in the heated waveguide is:

$$\Delta\phi = \frac{2\pi L}{\lambda_0}\frac{dn}{dT}\Delta T \tag{11}$$

where L is the phase shifter length, λ_0 is the operated wavelength of 1550 nm with a Si thermo-optic coefficient of $dn/dT \sim 1.8 \times 10^{-4}\ K^{-1}$, T = 300 K, used as the room temperature and the respective temperature change ΔT. The temperature difference limit will occur at our maximum phase shift ($\Delta\phi = \pi$):

$$\Delta T_{\pi} = \frac{\lambda_0}{2L dn/dT} \tag{12}$$

It is crucial to consider that the device length scale typically ranges in the hundreds of micrometers. This is because the resistance of the heaters increases proportionally with their length. Consequently, higher voltages would be required to drive the system, leading to reduced power efficiency. Additionally, the phase-shift calibration should be limited to a maximum of π-phase shift. Beyond this point, there is no significant advantage gained in terms of additional phase shifting; rather, it would result in unnecessary additional power consumption [4]. The heat dissipation is dependent on the attributes of the active and passive components. The time constant of the TOPS (τ) is defined by the amount of time for a change of 1×10^{-1} from the total change in the steady state and is given by:

$$\tau = \frac{H}{GA} \tag{13}$$

where G denotes the thermal conductance between the heated Si wg and the Si substrate, and H denotes the heat capacity of the TOPS.

The figure of merit (FOM), which is the total power consumption multiplied by the heat dissipation, is given by:

$$FOM = P_{\pi}\tau \tag{14}$$

To examine our TOPS efficiency and to select the optimal design, a calculation of the FOM has been performed to evaluate the TOPS performances.

3. Simulation Results

The TOPS was analyzed and solved using HEAT solver and MODE tools based on Lumerical suite software. The optical simulations were performed using the MODE solver tool, which is based on the finite difference eigenmode (FDE) numerical method. The heat simulations were performed using the HEAT solver tool, which solves the system of equations describing the distribution temperature as a function of power. In our simulations, all height and width values are taken into the constraints of the fab fabrication ability, where the smallest width of the heaters allowed after fabrication is 2 μm. The heat diffusion equation in an area of 8 μm width × 3.5 μm height. Key parameters such as thermal conductivity are taken into account.

Figure 2a–d shows the thermal profile of the TOPS with different distance between the two TiN heaters. The high peek temperature of the TiN heaters (339–336 K) can be noticed in Figure 2a–d, and the heaters location was set above 1.1 μm in order to avoid the possibility of light coupling from the Si wg to the heaters, which is a key feature to obtain the very low attenuation coefficient.

After obtaining the index delta data from the HEAT solver, the fundamental mode of the Si wg-based TOPS is being solved by mode solver for single TE mode at 1550 nm wavelength. Figure 3a–e shows the fundamental mode solution profile of the double TiN heaters at different distances between the heaters, and from this solution, the change of the effective index can be obtained for each design. Thus, combined with Equation (1), the voltage for obtaining phase change of π can be found, as shown in Figure 3a–e.

Figure 4 shows the phase shift as a function of the total heaters power for five different distances between the heaters. It can be noticed from Figure 4 that more electrical power for obtaining a phase shift of π is needed in cases of the distance between the two heaters is over 1 μm. Hence, the study of the tradeoff between the electrical power and the thermal crosstalk needs to be investigated to select the optimal design.

To choose the optimal design, a lower thermal crosstalk must be obtained, such that the contribution of one heater unit would not overheat the other unit and affect the heating of the Si wg area. Overheating can cause unwanted phase-shift deviation, which can easily lead to more optical losses [30–37]. Figure 5 shows the thermal crosstalk versus the distance between the two heaters. The results from Figures 2–5 have been analyzed in order to select the optimal design. From Figure 5, it can be observed that a higher normalized thermal crosstalk over 0.45 is obtained in cases of short distance between the heaters which is less than 0.85 μm. Thus, the best design must have less than 0.45 normalized thermal crosstalk. Moreover, the shorter the distance between the heaters, the more efficient the power consumption will be, which means that the consumption will be lower to obtain a π-phase shift, but at the same time, the tradeoff with thermal crosstalk will be greater. Therefore, the optimal design consists of a 1 µm distance between the two heaters. In this case, the power for obtaining a π-phase shift is considerably lower, with 18.95 mW total power and a low normalized thermal crosstalk of 0.404.

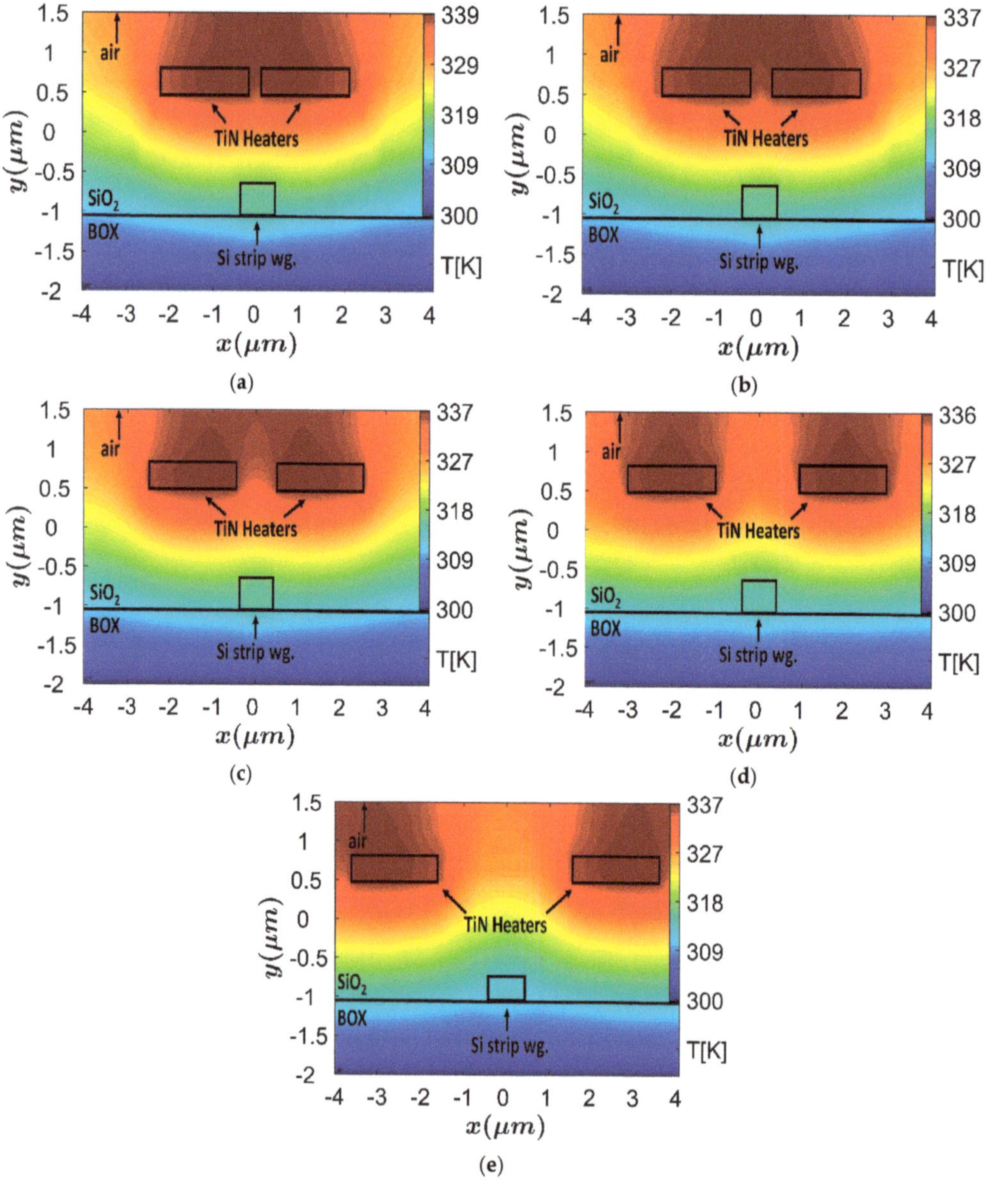

Figure 2. The heat distribution at the XY plane, each plot represents the change in the distance between the heaters: (**a**) d_{TiN} = 0.5 µm, (**b**) d_{TiN} = 0.75 µm, (**c**) d_{TiN} = 1 µm, (**d**) d_{TiN} = 2 µm, (**e**) d_{TiN} = 3 µm.

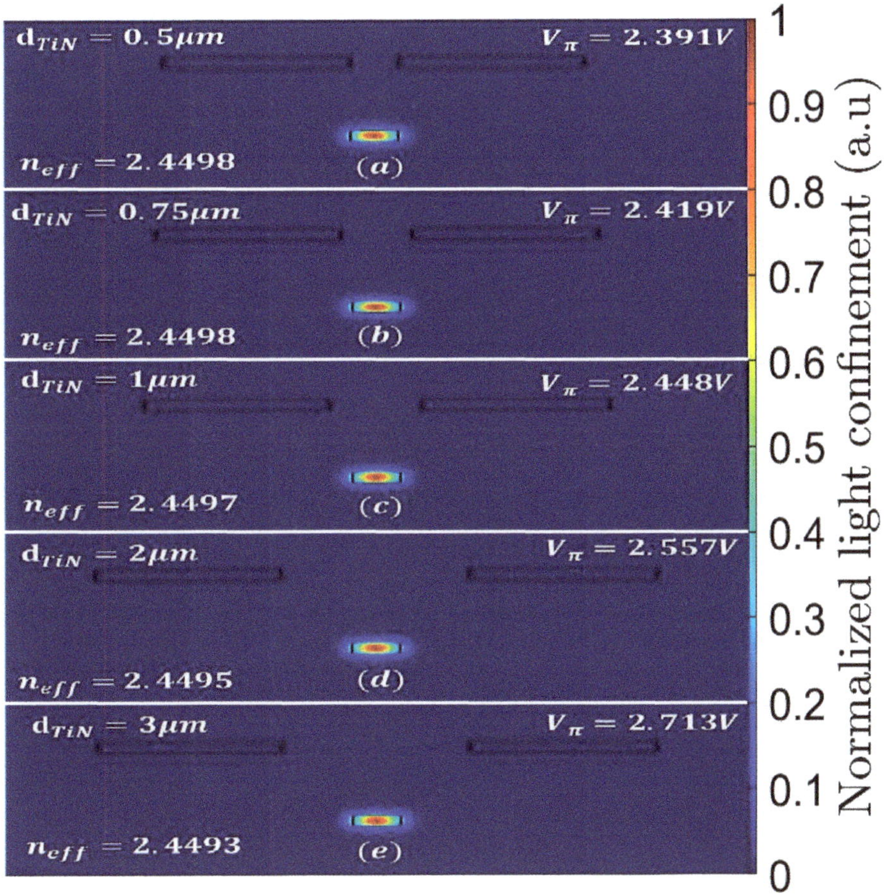

Figure 3. Solution of the fundamental TE mode at 1550 nm wavelength inside the Si strip: (**a**) d_{TiN} = 0.5 μm, (**b**) d_{TiN} = 0.75 μm, (**c**) d_{TiN} = 1 μm, (**d**) d_{TiN} = 2 μm, (**e**) d_{TiN} = 3 μm.

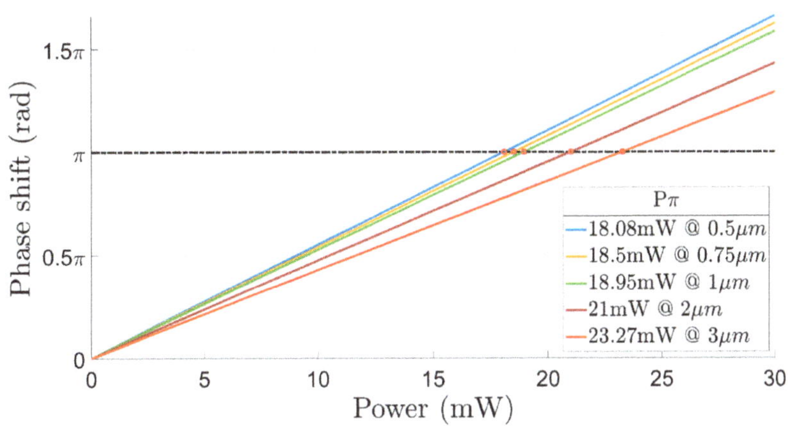

Figure 4. Phase shift as a function of power at different distances between the heaters.

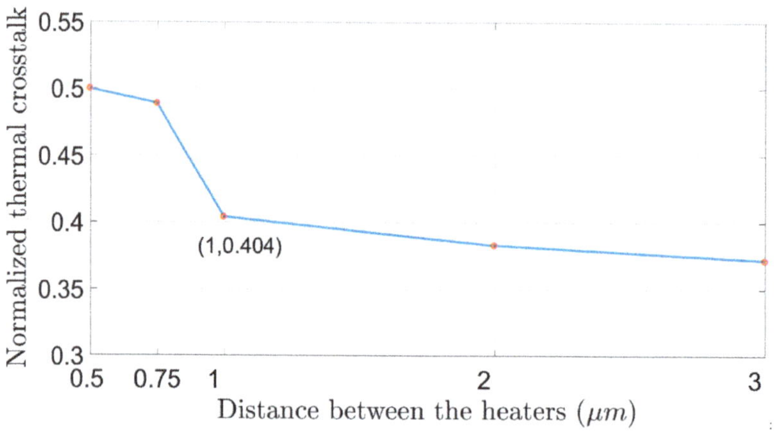

Figure 5. Normalized thermal crosstalk as a function of distance between the heaters.

Another important characteristic is the height distance between the double TiN heaters and the Si wg. The TiN heaters have a positive imaginary refractive index that can increase the optical losses in case the heaters are located close to the Si wg. In addition, in this case, less electrical power is needed to obtain a phase shift of π. Thus, an optimization on the d parameter is needed to be achieved in order to understand the tradeoff between optical losses and power consumption. Figure 6a–d shows the thermal profile of the TOPS with a different d value (0.5–2 µm). The resultant data were analyzed by solving the TE mode and extracting the power for the phase shift of π and the absorption coefficient (attenuation). In these figures, it can be noticed that a strong thermal change over the Si wg area is obtained at short distances of d, as shown in Figure 6a,b.

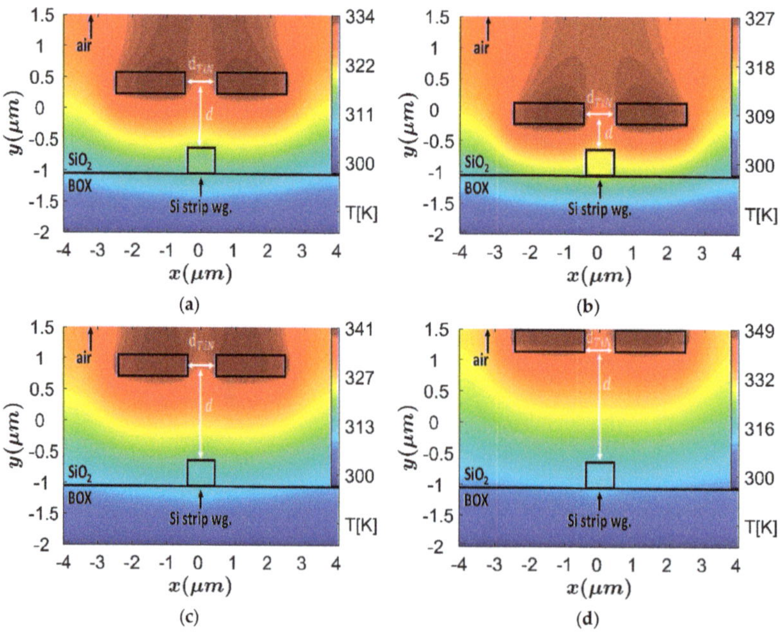

Figure 6. Heat distribution of the TOPS under different height distances between the heaters centered to the top of the Si strip wg: **(a)** d = 0.5 µm, **(b)** d = 1 µm, **(c)** d = 1.5 µm, **(d)** d = 2 µm.

Figures 7 and 8 shows the optimization of the d value, which includes how much power is needed to obtain the phase shift of π at different values of d and the attenuation (losses) for each case. From the results shown in Figures 7 and 8, it is clear that the optimal d value is 1 µm, with a low power of 19.1 mW and without suffering losses. Thus, selecting this distance ensure that the light of the TE fundamental mode does not has an overlap with the double TiN areas.

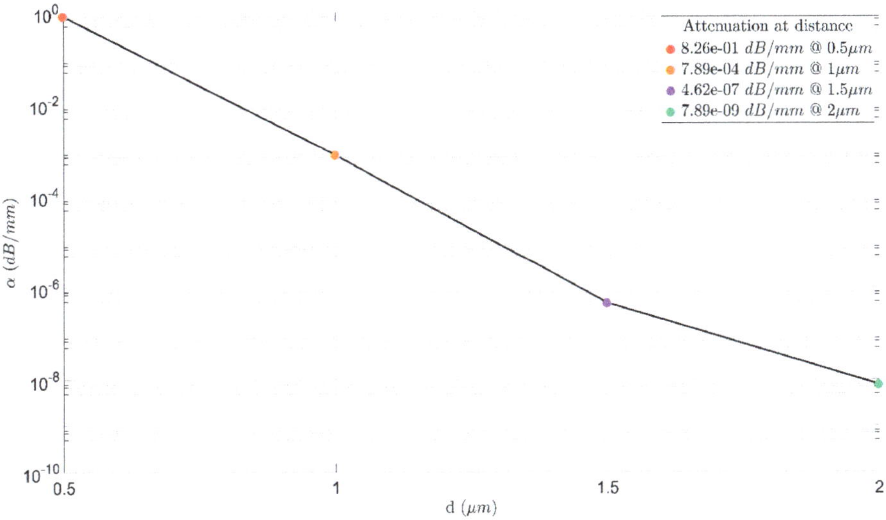

Figure 7. Attenuation as a function of parameter d.

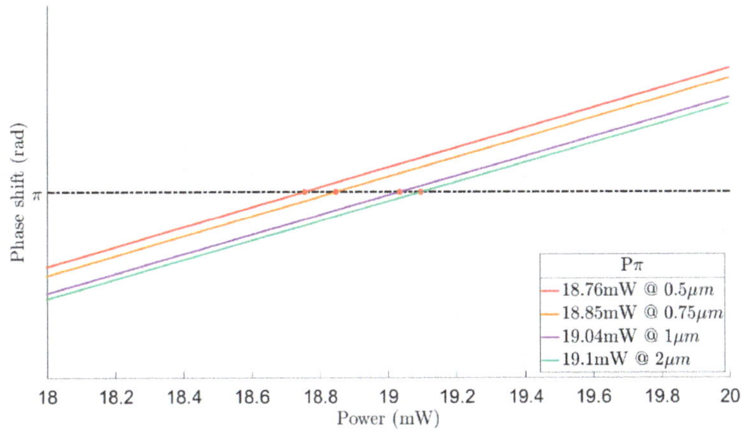

Figure 8. Total power consumption for π-phase shift as a function of parameter d. The results in Figure 8 correspond by color to the results in Figure 7.

Figure 9 shows the tolerance sensitive of the total π-phase shift power consumption for the wavelength drift effect over the C-band spectrum. This investigation is crucial because the wavelength in which the laser operates is not constant and has a shift-drift effect around the desired wavelength due to the heating process of the laser lifetime. Hence, a shift of around 0.1 nm per degree Celsius can happen for Si wg [38], and usually, a commercial laser can cause a heat increase of over 50 degrees from room temperature. It can be observed from Figure 9 that the proposed TOPS has a good sensitive to the laser drift effect, with only a minor power shift of 60 μW due to a temperature increase of 50 Celsius degrees due to laser heating and an overall good power consumption stability over the C-band spectrum, with only a small power shift of ±0.25 mW from the operating wavelength.

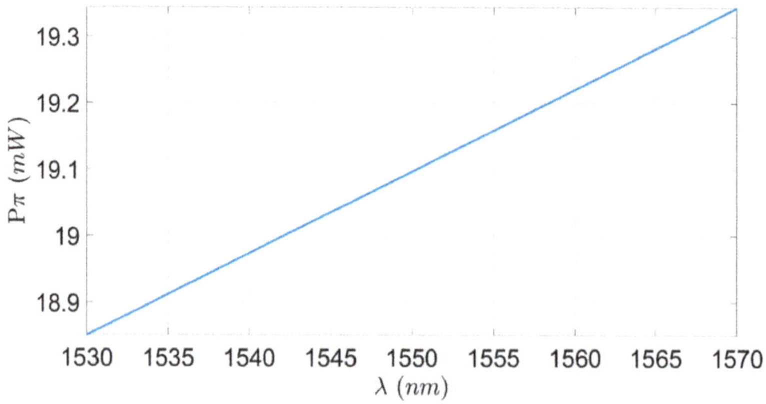

Figure 9. P_π as a function of the wavelengths over the C-band spectrum.

The rise time of the optimal TOPS design was found to be 2.31 μs, which is the time it takes the heat to dissipate (heat diffusion). This time constant can be theoretically improved by fabricating TiNs with a shorter width. In our case, the width was selected to be 2 μm because of the fabrication limit. To understand the advances of this new design, a comparison between the main properties of the proposed TOPS and other state-of-the-art TOPS components was performed, where all TOPS are operated under forward bias operation mode. The main properties are TOPS type, TOPS length (L), optical losses from the heaters, voltage for π-phase shift ($V\pi$), power for π-phase shift ($P\pi$), rise time (τ), FOM, and the year of publication. The results in Table 1 show that our double TiN heater design is superior to other TOPS designs in all aspects. Our design has the best power electric consumption, 19.1 mW, with very low optical loss due to the heaters and fast switch time (rise time) using a PS with a length of 1 mm. Another advance of this technology compared to the doped Si TOPS design is the ability to design a PS with a large length without consuming too much electrical power. Thus, for designing an MZM with a large RF line, it will be better to use a double TiN heater design to reduce energy costs.

Table 1. A comparison between different types of TOPS.

TOPS Type	L (μm)	Losses (dB)	V_π (v)	P_π (mW)	τ (μs)	FOM $P_\pi\ \tau$ (mWμs)	Year of Publication
Doped Si (P) [7]	61.6	0.23	4.36	24.8	2.7	66.9	2014
Doped Si (N) [39]	320	0.08	36.4	19.1	2.471	47.2	2022
TiN [4]	320	<0.4	2.33	21.4	5.6	119.8	2018
Double TiN	1000	7.89×10^{-4}	3.07	19.1	2.31	44.12	Our work

4. Conclusions

A new TOPS design based on double TiN heaters using a commercial 220 nm SOI strip wg under forward biased has been studied in this research. This study shows how to design TOPS with a high energy efficiency that can be suitable to fulfill the commercial requirements for designing MZMs that work with RF lines around 1 mm in length. The optimal parameters of the TOPS design are a distance between the two TiN heating units of 1 μm, a distance between the center of the heaters mass and the upper surface of the Si strip wg of 1 μm, and a PS length of 1 mm. The new TOPS design shows good stability to the thermal laser drift effect, with only a minor power shifting of ± 0.25 mW across the C-band spectrum. The results show that the proposed TOPS can be used for shifting a π-phase shift using low electrical power (19.1 mW) over 1 mm PS length, with an excellent normalized thermal crosstalk of 0.40 between the two TiN heater units, a fast rise time of 2.31 µs, very low optical losses of 7.72×10^{-9} dB/mm, and FOM of 44.12 mWµs. Thus, the new TOPS design grants us groundbreaking energy efficiency compared to other TOPS designs, such as doped Si or only using one TiN heater unit. This new TOPS can be easily integrated into MZMs that work under high-speed conditions, which usually require a large RF line (around 1 mm) to solve the imbalance arms issue, with a low cost of electrical energy. From a commercial view, this new design is very attractive because of the ability to reduce the driver voltage level for achieving phase shift control in the MZM. Thus, this ability from a system view can be utilized for reducing the transmitter costs.

Author Contributions: D.M. conceptualized the project; D.M. offered guidance and direction; E.E.K. devised the device's design; E.E.K. conducted simulations with the assistance of D.M.; D.M. and E.E.K. authored the paper and response letter, and generated the figures. All authors have read and agreed to the published version of the manuscript.

Funding: This research received no external funding.

Conflicts of Interest: The authors declare no conflict of interest.

References

1. Mynbaev, D.K. What Will Modern Photonics Contribute to the Development of Future Optical Communication Technology. In *Future Trends in Microelectronics: From Nanophotonics to Sensors to Energy*; Wiley-IEEE Press: New York, NY, USA, 2013; pp. 160–172.
2. Okamoto, M.; Mendez-Astudillo, M.; Kita, T. Thermo-optic phase shifter with sub-microsecond switching time and low power consumption. *Jpn. J. Appl. Phys.* **2020**, *59*, SOOA03. [CrossRef]
3. Doerr, C.; Chen, L.; Nielsen, T.; Aroca, R.; Chen, L.; Banaee, M.; Azemati, S.; McBrien, G.; Park, S.Y.; Geyer, J. L band silicon photonics coherent modulator/receiver. In Proceedings of the IEEE Optical Fiber Communications Conference and Exhibition, San Diego, CA, USA, 6–10 March 2016.
4. Tait, A.N.; de Lima, T.F.; Zhou, E.; Wu, A.X.; Nahmias, M.A.; Shastri, B.J.; Prucnal, P.R. Neuromorphic photonic networks using silicon photonic weight banks. *Sci. Rep.* **2017**, *7*, 7430. [CrossRef]
5. Gao, W.; Sun, J.; Sun, X.; Yan, Y.; Gao, L.; Zhang, D. Fabrication of a polymer CPW electro-optic modulator using a strip-loaded waveguide structure. *Microw. Opt. Technol. Lett.* **2009**, *51*, 2097–2100. [CrossRef]
6. Watts, M.R.; Sun, J.; DeRose, C.; Trotter, D.C.; Young, R.W.; Nielson, G.N. Adiabatic thermo-optic Mach Zehnder switch. *Opt. Lett.* **2013**, *38*, 733–735. [CrossRef] [PubMed]
7. Harris, N.C.; Ma, Y.; Mower, J.; Baehr-Jones, T.; Englund, D.; Hochberg, M.; Galland, C. Efficient, compact and low loss thermo-optic phase shifter in silicon. *Opt. Express* **2014**, *22*, 10487–10493. [CrossRef]
8. Jacques, M.; Samani, A.; El-Fiky, E.; Patel, D.; Xing, Z.; Plant, D.V. Optimization of thermo-optic phase-shifter design and mitigation of thermal crosstalk on the SOI. *Platf. Opt. Express* **2019**, *27*, 10456–10471. [CrossRef] [PubMed]
9. Fujisawa, T.; Takahata, K.; Tadokoro, T.; Kobayashi, W. Long-reach 100 Gbit Ethernet light source based on 4 × 25-Gbit/s 1.3-µm InGaAlAs EADFB lasers. *IEICE Trans. Electron.* **2011**, *94*, 1167–1172. [CrossRef]
10. Dahan, D.; Mahlab, U.; Levy, D. Stimulated Brillouin scattering based in-band OSNR monitoring technique for 40 Gbps and 100 Gbps optical transparent networks. *Opt. Express* **2010**, *18*, 15769–15783. [CrossRef]
11. Barthomeuf, S.; Saliou, F.; Anet Neto, L.; Chanclou, P.; Erasme, D. TDM-PON PAM Downstream Transmission for 25 Gbit/s and beyond. *Photonics* **2018**, *5*, 45. [CrossRef]
12. Chattopadhyay, T.; Saw, A.K. Wideband optical comb generation using an injection-locked feedback loop. *Opt. Commun.* **2011**, *284*, 4611–4614. [CrossRef]
13. Doylend, J.K.; Jessop, P.E.; Knights, A.P. Silicon photonic dynamic optical channel leveler with external feedback loop. *Opt. Express* **2010**, *18*, 13805–13812. [CrossRef]
14. Kim, S.; Ahn, S.-H. A Cost-effective Sub-Terahertz Continuous Wave Generation Scheme Using a Broadband Optical Source and An Optical Feedback Loop. *J. Infrared Millim. Terahertz Waves* **2013**, *34*, 109–118. [CrossRef]
15. Kang, G.; Kim, S.-H.; You, J.-B.; Lee, D.-S.Y.H.; Ha, Y.-G.; Kim, J.-H.; Yoo, D.-E.; Lee, D.-W.; Youn, C.-H.; Yu, K.; et al. Silicon-Based Optical Phased Array Using Electro-Optic p-i-n Phase Shifters. *IEEE Photonics Technol. Lett.* **2019**, *31*, 1685–1688. [CrossRef]
16. Helszajn, J. *Microwave Polarizers, Power Dividers, Phase Shifters, Circulators, and Switches*; Wiley-IEEE Press: New York, NY, USA, 2019; pp. 104–108.
17. Tafur Monroy, I.; Tangdiongga, E. *Crosstalk in WDM Communication Networks*; Springer: Berlin/Heidelberg, Germany, 2002; pp. 166–170.

18. Oliver, J. Modelling strong discontinuities in solid mechanics via strain softening constitutive equations. parts 1,2. *Int. J. Numer. Methods Eng.* **1996**, *39*, 3575–3600. [CrossRef]
19. Namazifard, A.; Parsons, I.D. A distributed memory parallel implementation of the multigrid method for solving three-dimensional implicit solid mechanics problems. *Int. J. Numer. Methods Eng.* **2004**, *61*, 1173–1208. [CrossRef]
20. Adams, M.F. Algebraic multigrid methods for constrained linear systems with applications to contact problems in solid mechanics. *Numer. Linear Algebra Appl.* **2004**, *11*, 141–153. [CrossRef]
21. Nguyen-Thoi, T.; Liu, G.R.; Nguyen-Xuan, H. An n-sided polygonal edge-based smoothed finite element method (nES-FEM) for solid mechanics. *Int. J. Numer. Methods Biomed. Eng.* **2010**, *27*, 1446–1472. [CrossRef]
22. Dadabayev, R.; Malka, D. A visible light RGB wavelength demultiplexer based on polycarbonate multicore polymer optical fiber. *Opt. Laser Technol.* **2019**, *116*, 239–245. [CrossRef]
23. Meftah, A.M.; Meftah, A.F.; Hiouani, F.; Merazga, A. Numerical simulation of the defect density influence on the steady state response of a silicon-based p–i–n cell. *J. Phys. Condens. Matter* **2004**, *16*, 2003–2016. [CrossRef]
24. Schmidt, J.A.; Goldie, D.M. Photocurrent decay from the steady-state in thin film hydrogenated amorphous silicon: Numerical simulation analysis of experimental results. *Thin Solid Film.* **2020**, *696*, 137793. [CrossRef]
25. Merazga, A.; Tobbeche, S.; Main, C.; Al-Shahrani, A.; Reynolds, S. Numerical simulation of the steady state photoconductivity in hydrogenated amorphous silicon including localized state electron hopping. *J. Phys. Condens. Matter* **2006**, *18*, 3721. [CrossRef]
26. Meyne, S.; Bernadi, P.; Birtel, P.; David, J.-F.; Jacob, A.F. Large-Signal 2.5-D Steady-State Beam-Wave Interaction Simulation of Folded-Waveguide Traveling-Wave Tubes. *IEEE Trans. Electron Devices* **2016**, *63*, 4961–4967. [CrossRef]
27. Chen, L.; Bagci, H. Steady-state Simulation of Semiconductor Devices using Discontinuous Galerkin Methods. *IEEE Access* **2019**, *8*, 16203–16215. [CrossRef]
28. Willis, K.J.; Hagness, S.C.; Knezevic, I. Terahertz conductivity of doped silicon calculated using the ensemble Monte Carlo/finite-difference time-domain simulation technique. *Appl. Phys. Lett.* **2010**, *96*, 8491. [CrossRef]
29. Lohn, A.J.; Mickel, P.R.; Marinella, M.J. Analytical estimations for thermal crosstalk, retention, and scaling limits in filamentary resistive memory. *J. Appl. Phys.* **2014**, *21*, 115. [CrossRef]
30. Liao, L.; Lim, D.R.; Agarwal, A.M.; Duan, X.; Lee, K.K.; Kimerling, L.C. Optical transmission losses in polycrystalline silicon strip waveguides: Effects of waveguide dimensions, thermal treatment, hydrogen passivation, and wavelength. *J. Electron. Mater.* **2000**, *29*, 1380–1386. [CrossRef]
31. Grant, S.A.; Soufiane, A.; Shirk, G.; Martin, S.W. Degradation-induced transmission losses in silica optical fibers. *Lasers Surg. Med.* **1997**, *21*, 65–71. [CrossRef]
32. Revin, D.G.; Wilson, L.R.; Carder, D.A.; Cockburn, J.W.; Steer, M.J.; Hopkinson, M.; Vicet, A. Measurements of optical losses in mid-infrared semiconductor lasers using Fabry–Pérot transmission oscillations. *J. Appl. Phys.* **2004**, *95*, 7584–7587. [CrossRef]
33. Zinke, T.; Fischer, U.; Splett, A.; Schuppert, B.; Petermann, K. Comparison of optical waveguide losses in silicon-on-insulator. *Electron. Lett.* **1993**, *29*, 2031–2033. [CrossRef]
34. Tran, M.; Huang, D.; Komljenovic, T.; Peters, J.; Malik, A.; Bowers, J. Ultra-Low-Loss Silicon Waveguides for Heterogeneously Integrated Silicon/III-V Photonics. *Appl. Sci.* **2018**, *8*, 1139. [CrossRef]
35. Vlasov, Y.A.; McNab, S.J. Losses in single-mode silicon-on-insulator strip waveguides and bends. *Opt. Express* **2004**, *12*, 1622–1631. [CrossRef] [PubMed]
36. Tresna, W.P.; Putra, A.W.S.; Maruyama, T. Optical-loss measurement of a silicon-slab waveguide. *Curr. Opt. Photonics* **2020**, *4*, 6.
37. Dong, P.; Qian, W.; Liao, S.; Liang, H.; Kung, C.-C.; Feng, N.-N.; Asghari, M. Low loss silicon waveguides for application of optical interconnects. *IEEE Photonics* **2010**, *1*, 191–192.
38. Samoi, E.; Benezra, Y.; Malka, D. An ultracompact 3 × 1 MMI power-combiner based on Si slot-waveguide structures. *Photonics Nanostruct. Fundam. Appl.* **2020**, *39*, 100780. [CrossRef]
39. Sitbon, E.; Ostrovsky, R.; Malka, D. Optimizations of thermo-optic phase shifter [heaters using doped silicon heaters in Rib waveguide structure. *Photonics Nanostruct. Fundam. Appl.* **2022**, *51*, 101052. [CrossRef]

sensors

MDPI

Article

Parabolic-Index Ring-Core Fiber Supporting High-Purity Orbital Angular Momentum Modes

Yuanpeng Liu [1], Yingning Wang [1], Wenpu Geng [1], Wenqian Zhao [1], Hao Zhang [1], Weigang Zhang [1], Zhongqi Pan [2] and Yang Yue [3,*]

[1] Institute of Modern Optics, Nankai University, Tianjin 300350, China
[2] Department of Electrical & Computer Engineering, University of Louisiana at Lafayette, Lafayette, LA 70504, USA
[3] School of Information and Communications Engineering, Xi'an Jiaotong University, Xi'an 710049, China
[*] Correspondence: yueyang@xjtu.edu.cn

Abstract: We design a graded-index ring-core fiber with a GeO_2-doped silica ring core and SiO_2 cladding. This fiber structure can inhibit the effect of spin-orbit coupling to mitigate the power transfer among different modes and eventually enhance the orbital angular momentum (OAM) mode purity. By changing the high-index ring core from the step-index to parabolic graded-index profile, the purity of the $OAM_{1,1}$ mode can be improved from 86.48% to 94.43%, up by 7.95%. The proposed fiber features a flexible structure, which can meet different requirements for mode order, effective mode area, etc. Simulation results illustrate that the parabolic-index ring-core fiber is promising in enhancing the OAM mode purity, which could potentially reduce the channel crosstalk in mode-division-multiplexed optical communication systems.

Keywords: orbital angular momentum; graded index; mode purity; ring fiber

1. Introduction

A key issue in modern optical fiber communication systems is how to meet the rapidly growing traffic demand. To satisfy the explosive requirement for data bandwidth, extensive research has been conducted on multiplex optical signals in diverse photon dimensions, such as time, wavelength, amplitude, phase, polarization, and space [1–5]. According to Shannon's theorem, there is an upper bound to the capacity in single-mode fiber transmission systems, which is a function of the available bandwidth and the signal-to-noise ratio of the link. With the continuous development of wavelength-division, polarization-division, and other multiplexing technologies, the capacity of communication systems has gradually approached the Shannon limit, and researchers have tried to implement other physical dimensions of the data signal to increase the transmission capacity and improve the spectral efficiency. Spatial division multiplexing (SDM) utilizes the degree of freedom of light waves in the physical dimension of space, which can be applied in multi-core fibers, multi-mode fibers, or multi-core multi-mode fibers to further improve the transmission capacity of optical fiber communication systems [6–10].

The SDM technology can be achieved in a multi-core fiber where each core works as a spatially independent channel for transmission, or in a multi-mode fiber where mutually orthogonal modes transmit different data, or in a multi-core multiple-mode fiber where greater transmission capacity in a single fiber is achieved by combining spatial multiplexing with mode multiplexing. Through the multiplexing over space and mode, the SDM technology can increase the communication capacity by one to two orders of magnitude, which has been one of the mainstream research and development directions of ultra-large-capacity optical fiber communication technology [11,12]. Currently, a scalar modal set composed of several eigenmodes, the linearly polarized (LP) modes, are widely used in SDM systems because of the simple excitation condition. However, a random mode coupling among the channels will arise during propagation because of fabrication imperfections or bending, which can introduce unwanted crosstalk between the optical modes. Nevertheless, the detrimental effect of mode coupling can be fully compensated for using a multiple-input multiple-output (MIMO) digital signal processing (DSP) algorithm, which could extremely increase the complexity and cost of the communication system. Different methods to reduce the use of MIMO processing have been investigated. One feasible scheme is mode-group multiplexing, which employs low differential group delays (DGD) and/or DGD management to lower the complexity of MIMO DSP [13–16]. Moreover, polarization-maintaining optical fibers, for example, the elliptical-core fibers, are also adopted for MIMO-less SDM, in which all the degeneracies among the vectorial modes are broken [17,18].

In addition to the above schemes, the multiplexing technique of beams carrying orbital angular momentum (OAM) offers a new direction for simplifying communication systems [19,20]. OAM is a natural property of a beam with a helical phase and, theoretically, has infinite topological charges. The helical phase front of a beam carrying OAM can be described as $e^{il\varphi}$, where φ is the azimuth angle and l is the topological charge [21]. OAM modes with different topological charges are mutually orthogonal, so different OAM modes can provide a new degree of freedom for multiplexing technology. Furthermore, the multiplexing of OAM is independent of polarization and wavelength, which makes it have significant potential in improving the capacity of optical fiber communication systems [22–26]. With proper design, different OAM modes transmitted in the fiber can have a large index separation to mitigate the modal crosstalk. Compared with the traditional LP modes, it will have a smaller inter-mode crosstalk, which means it can be used to lessen the demand on the MIMO system [16,20,27].

To achieve stable transmission of OAM beams, limit the generation of unwanted radially high-order modes, and match their doughnut-shaped optical field distributions, various ring-core fibers (RCFs) have been proposed [28]. In 2009, Ramachandran et al. found that the annular high-refractive-index region in the fiber was adapted to the intensity profile of the vector beam, allowing the vector beam to maintain better stability and purity [29]. In 2013, the same kind of ring-core fiber was used to transmit OAM modes. This kind of vortex fiber was used to simultaneously couple two OAM modes with $l = \pm 1$ and two polarization-multiplexed fundamental modes [20]. In 2014, C. Brunet et al. designed and fabricated a ring fiber that is capable of transmitting 36 OAM modes [30]. In 2015, an air-core fiber was reported to further increase the refractive index difference between eigenmodes, which could increase the number of OAM modes supported in the fiber [31]. However, the large refractive index contrast between the air core and the high-index ring will greatly enhance the spin-orbit coupling (SOC) effect in the fiber, so that the OAM carried by a portion of the transmitted beam will be converted to spin angular momentum (SAM), which means the transmitted vortex light is no longer a pure circularly polarized OAM mode. Thus, the purity of the mode will be reduced, leading to the reduction in the demultiplexing efficiency of the orthogonal OAM modes at the receiving end and the increase in the crosstalk between modes [32–35]. Therefore, finding a proper solution for the problem of the reduced purity of the OAM modes can provide a certain reference for the design and manufacturing of OAM fibers in the future. Ref. [36] proposed a dual-step-index RCF structure to suppress the spin-orbit effect and, thus, improve the OAM mode purity. Similar to Ref. [36], the graded-index profile is also a promising solution, which can suppress the SOC by reducing the abrupt refractive index change. In recent years, some researchers have used the graded-index profile (GIP) in their proposals. Nevertheless, the related fibers in these works either are not the ring-core fiber [37], which cannot match the mode field distribution of the OAM modes well, or have complicated structures that may be challenging for the current fabrication technology [38]. Therefore, it is necessary to systematically analyze the characteristics of the OAM modes in the graded-index ring-core fiber (GIRCF). Conducting this kind of research would, thus, be beneficial to further understand the effect of the refractive index profile on the OAM mode properties.

In this paper, we design a graded-index ring-core fiber and investigate the properties of its supported OAM mode. By adopting a parabolic-index profile to reduce the abrupt index change between the ring core and the cladding, the SOC effect can be restrained at the boundary between the ring core and cladding. The power of the transmitted OAM mode can be well-maintained in its initial mode order, and the corresponding mode purity can be thus improved. At 1550 nm, the GIRCF can support up to a total of 322 OAM modes. Additionally, the purity of the first-order OAM mode in GIRCF is notably higher than that of the step-index ring core fiber, which can be increased from 86.48% to 94.43%.

2. Concept and Fiber Structure

2.1. Fiber Structure

Figure 1a illustrates the concept of multiplexing OAM modes with different orders in a GIRCF, and Figure 1b depicts the cross-section and index profile of the designed GIRCF. As shown in Figure 1b, the GIRCF is composed of a high-index GeO_2-doped silica ring core and SiO_2 background. The refractive index profile in the high-index ring-core region can be expressed as:

$$n(r) = \begin{cases} n_1 \times \sqrt{1 - \frac{n_1^2 - n_2^2}{n_1^2} \times \left(\frac{2|r-r_0|}{\Delta r} \right)^{\alpha}} & ; r_1 < r_2 \\ n_2 & ; 0 \leq r \leq r_1 \; or \; r \geq r_2 \end{cases} \tag{1}$$

where n_1, n_2, r_0, Δr, and α are the material index of the GeO_2-doped ring core, the material index of the background silica, the radius from the fiber center to the ring center, the thickness of the ring-core region, and the shape factor, respectively. The refractive indices of the 5% GeO_2-doped silica and silica (SiO_2) at 1550 nm are 1.4515 and 1.444, respectively, and the corresponding relative refractive index difference (Δ) is 0.52% [39]. Their wavelength dependence is considered by using the Sellmeier equations [40]. In GIRCF, r_0 represents the position of the ring core, which can be adjusted from 7 μm to 47 μm to afford more eigenmodes, and the shape factor α controls the gradient of the refractive index between the high-index ring core and cladding. When α is equal to 2, the refractive index profile of the ring-core region is parabolic. As α increases, the upper and lower edges of the parabola will become steeper. When α approaches infinity, the refractive index profile will become a step-index one. The smaller the α is, the slower the refractive index transition between the ring core and cladding is. It can inhibit the coupling effect of SAM and OAM, which is a form of spin-orbit interaction at the ring-core–cladding boundary. Therefore, the power of the OAM mode transmitted in the designed fiber will not couple to the other-order OAM modes, which is beneficial to enhance the OAM mode purity. The thickness of the ring (Δr) also needs proper design to avoid radially high-order modes [41]. All numerical analyses and simulated calculations of the designed fiber are performed using the finite element method (FEM) in COMSOL Multiphysics 5.4 software.

2.2. Fabrication Feasibility of the GIRCF

During the past decade, researchers have fabricated a variety of ring-core fibers. The mole fraction of the GeO_2 in the fiber changes from 5% to almost 30% [30,42–44]. Highly Ge-doped fiber core with GeO_2 content up to 98 mol% has also been manufactured in practice [45]. Moreover, due to the limitations of the current manufacturing technology, the fabricated step-index ring-core fiber (SIRCF) often presents a GIP [28]; thus, it is feasible to fabricate the investigated GIRCF using modified chemical vapor deposition (MCVD) or plasma chemical vapor deposition (PCVD) [46,47]. The following study comprehensively considers the fabrication possibilities and analyzes the OAM mode characteristics in the GIRCF under different GeO_2 mole fractions.

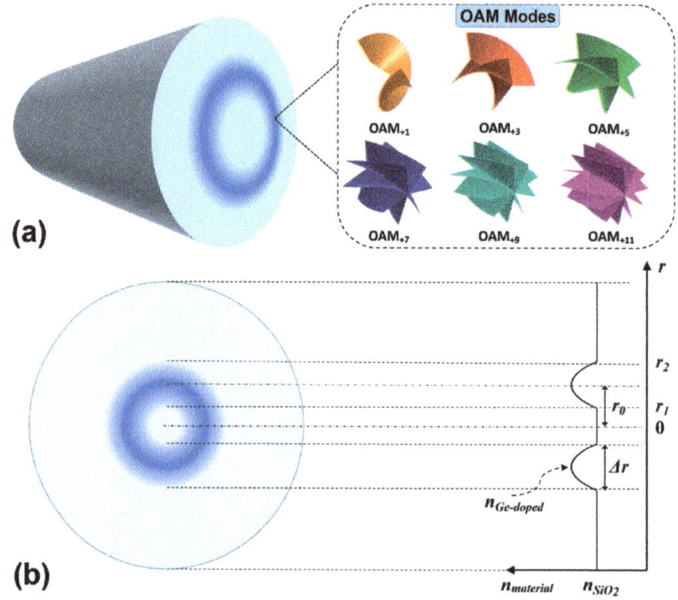

Figure 1. (**a**) Concept of OAM mode-division multiplexing in the proposed GIRCF; (**b**) cross-section view and refractive index profile of the proposed GIRCF.

2.3. OAM Modes Supported in the GIRCF

OAM modes in the GIRCF can be expressed as a superposition of two same-order eigenstates with the same propagation constants:

$$OAM^{\pm}_{\pm l,m} = HE^{even}_{l+1,m} \pm jHE^{odd}_{l+1,m} \tag{2}$$

$$OAM^{\mp}_{\pm l,m} = EH^{even}_{l-1,m} \pm jEH^{odd}_{l-1,m} \tag{3}$$

where l is the topological charge, $|l|$ is the OAM mode order, m is the radial index, and j represents the phase shift of $\pi/2$ between the even and odd modes. The HE and EH in Equations (2) and (3) represent two fundamental eigenstates of the OAM modes. As the radially high-order modes are undesirable, we take $m = 1$ in our designed fiber. The superscripts "$\pm$" represent the right-hand or left-hand circular polarization direction, and the subscripts "$\pm$" represent the right or left rotation direction of the OAM mode phase front, respectively. One of the OAM mode properties is the unlimited topological charge number, so the value of l can theoretically be any integer. In general, the $OAM_{0,1}$ mode composed by $HE_{1,1}$ is not considered as standard OAM because its phase front is not helical.

3. OAM Mode Properties

3.1. Intensity and Phase Distributions

As shown in Figure 2, we compare the normalized intensity and phase distributions of OAM modes ($l = 1, 7, 12$) under different shape factors (α) in RCF. Different OAM modes in the figure are composed of the coherent superposition of the even and odd fiber eigenmodes of $HE_{2,1}$, $HE_{8,1}$, and $HE_{13,1}$. As one can see from the figure, when the refractive index profile evolves from a step to a parabolic shape, the intensity distributions of the corresponding OAM modes are still well-confined in the ring region, and the phase distribution of the $OAM_{l,1}$ mode has a $2l\pi$ change azimuthally.

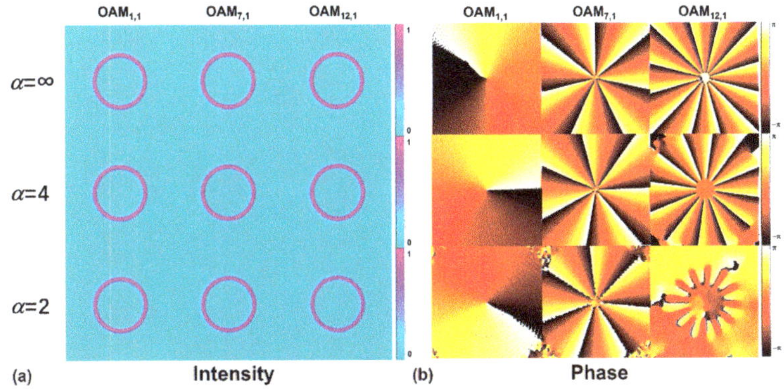

Figure 2. (**a**) Intensity and (**b**) phase distributions of the $OAM_{1,1}$, $OAM_{7,1}$, and $OAM_{12,1}$ modes in the RCF with $\alpha = 2, 4, \infty$, respectively.

3.2. Supported OAM Mode Number and the Effective Refractive Indices

The supported OAM mode number in RCFs can be calculated as [48]:

$$n_{\mathrm{OAM}} = (n_{\mathrm{HE}} - 1 + n_{\mathrm{EH}}) \times 2 \tag{4}$$

where n_{OAM}, n_{HE}, and n_{EH} stand for the OAM mode number, the HE eigenmode number, and the EH eigenmode number supported in RCFs, respectively. Figure 3 displays the maximum number of OAM modes supported in the fiber as a function of the ring-core position and wavelength under different GeO_2 mole fractions when the thickness of the ring-core region is 1.93 μm and the shape factor equals 2. The high refractive index area expands with the increasing radius of the ring-core region (r_0), when the ring-core thickness (Δr) is fixed. As a result, the number of modes supported in GIRCF monotonically increases as the ring-core position moves out, as shown in Figure 3a. When r_0 equals 47 μm and the GeO_2 mole fraction is 5 mol%, the designed GIRCF is capable of supporting up to 34 modes. In addition, Figure 3b indicates that the maximum number of OAM modes supported by GIRCF decreases as the wavelength increases because the higher-order modes will gradually leak out because of the shorter cut-off wavelength.

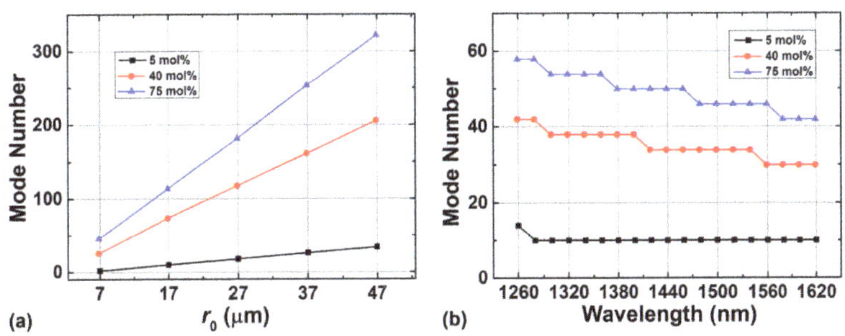

Figure 3. OAM mode number supported in the fiber as a function of (**a**) the position of the ring-core (r_0) and (**b**) wavelength with different mole fractions of GeO_2.

Figure 4a shows the effective refractive indices (n_{eff}) as a function of wavelength for the highest-order $HE_{12,1}$ and lowest-order $HE_{2,1}$ mode supported in the C + L band with 75 mol% GeO_2 in the corresponding ring-core fibers under different shape factors ($\alpha = 2, 4, \infty$) when $r_0 = 7$ μm. The designed GIRCF can support the highest-order $HE_{12,1}$ mode across the C + L band. Compared with the SIRCF, the effective refractive indices of OAM modes in the GIRCF will be lower, which means that some high-order modes (such as $HE_{13,1}$) in the GIRCF will be cut off earlier with increasing wavelength. Under the same condition, Figure 4b illustrates the effective refractive indices of all vector eigenmodes supported in the GIRCF as a function of wavelength, in which the n_{eff} of the eigenmodes decreases with the increasing wavelength. We also calculate the effective index difference Δn_{eff} between the HE eigenmodes and EH eigenmodes, which synthesize the same-order OAM modes. The calculated results indicate that Δn_{eff} is larger than 1×10^{-4} in most cases, except for the ones of $OAM_{8,1}$ and $OAM_{9,1}$. The maximum and minimum Δn_{eff} of the eigenmodes in the designed fiber is 7.78×10^{-4} for $OAM_{2,1}$ and 4.82×10^{-5} for $OAM_{8,1}$, respectively, which suggests a good performance in the modal separation.

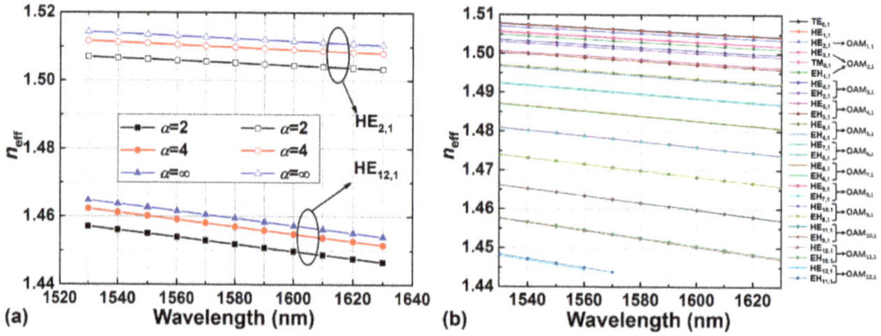

Figure 4. (**a**) Effective refractive indices of the $HE_{2,1}$ and $HE_{12,1}$ modes with α = 2, 4, ∞; (**b**) effective refractive indices of the eigenmodes when α = 2.

3.3. Effective Mode Field Area (A_{eff}) of the OAM Modes

The effective mode area (A_{eff}) as a function of the mode order is shown in Figure 5 with different shape factors (α = 2, 4, ∞) under the radially single-mode condition. A_{eff} is defined as [49]:

$$A_{\text{eff}} = \frac{\left(\iint |E(x,y)|^2 dxdy\right)^2}{\iint |E(x,y)|^4 dxdy} \tag{5}$$

where $E(x, y)$ is the electrical field distribution of the transverse mode field.

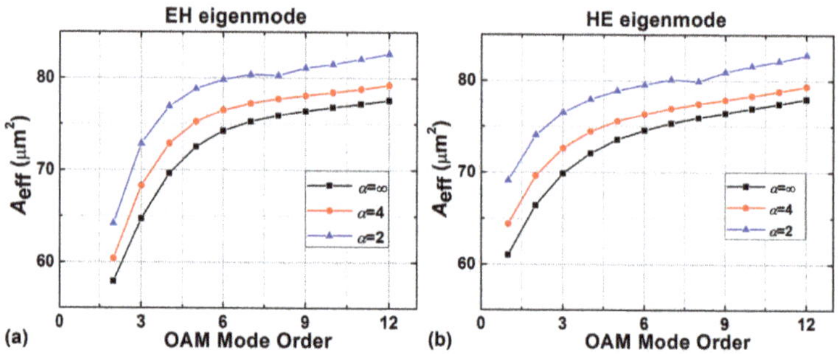

Figure 5. Effective areas of (**a**) the EH-composed OAM modes and (**b**) the HE-composed OAM modes in the ring-core fiber with α = 2, 4, ∞.

As can be seen from Figure 5, the GIRCF has a larger mode effective area than the SIRCF because of the weaker confinement of the mode field in GIRCF, and the effective mode area generally increases with the OAM mode order. Figure 6 depicts the effective mode area of $HE_{2,1}$ and $HE_{8,1}$ in GIRCF with a parabolic index profile and a 1.93 μm ring width as a function of the ring-core position under different mole fractions of GeO_2, respectively. It is clear that the effective mode area increases as the ring-core position moves out and the mole fraction of GeO_2 decreases. The blank area in Figure 6b means that the $HE_{8,1}$ mode has been cut off. To evaluate the ability of the GIRCF to suppress the nonlinear effects, we analyze the nonlinear coefficient γ of the fiber under different GeO_2 mole fractions. The calculation results show that the values of γ are 2.1×10^{-2}/W/km, 1.4/W/km, and 2.8/W/km when the GeO_2 mole fractions are 5 mol%, 40 mol%, and 75 mol%, respectively [50]. According to Ref. [51], the value of γ in traditional single mode fiber is 1.267/W/km. When the mole fraction is up to 40 mol%, the value of γ in GIRCF is still close to the one in SMF, which means they have similar nonlinearity-limited transmission characteristics. Consequently, with a lower GeO_2 mole fraction, the GIRCF could potentially support a longer transmission distance.

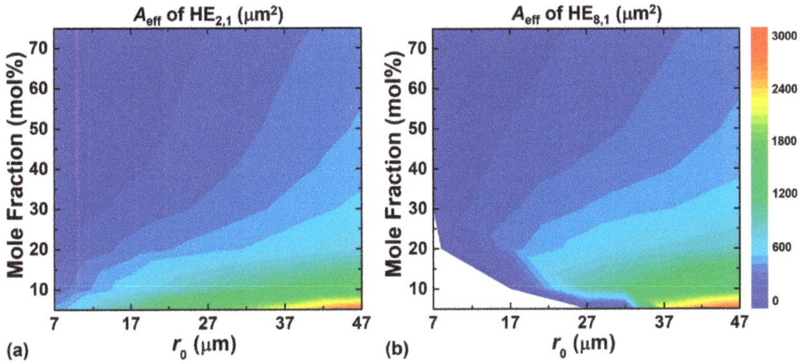

Figure 6. Effective areas of the (**a**) $HE_{2,1}$ and (**b**) $HE_{8,1}$ modes with $\alpha = 2$ under different mole fractions of GeO_2 and the position of the ring-core.

3.4. Mode Purity

In standard optical fibers, the degeneracy of OAM eigenstates is lifted because of the spin-orbit-interaction-induced effective index splitting, which causes the instability of the OAM eigenstates in practice. The ring-core fiber designs can support OAM modes with an order-of-magnitude larger effective refractive index difference, which facilitates stable transmission of a large number of OAM modes. However, the larger refractive index difference also introduces strong SOC. Therefore, the weakly guiding approximation is not considered in the formula of OAM synthesis. In the strong SOC case, the beams propagating in the fiber are the superpositions of the SAM and OAM states [52]. Consequently, the exact solution of the waveguide characteristic equation will not be a single OAM mode, i.e., the synthesized OAM mode is not pure when a large refractive index discontinuity exists at the interface between the ring core and cladding. Part of the power of the lth-order OAM mode transmitted in fiber could be coupled into the other-order OAM modes. Nevertheless, the GIP can greatly alleviate the discontinuous area, thus effectively inhibiting the SOC. Consequently, the GIRCF can enhance the purity of the OAM modes and diminish the intrinsic crosstalk by suppressing the spin-orbit effect at the high-contrast interface. We define the ratio of the dominant component power to the whole power transmitted in the fiber as OAM mode purity, which can be written as [33]:

$$\text{Purity} = \frac{power_{OAM_{major}}}{power_{OAM_{major}} + power_{OAM_{minor}}} \tag{6}$$

Figure 7 illustrates the relationship between the purity of the EH-based and HE-based OAM modes with different shape factors under a radially single-mode condition. The purity of the OAM mode coherently synthesized by the HE eigenmodes is generally higher than that synthesized by the EH eigenmodes with the same order, and higher-order OAM modes have better performance in purity. It should be noted that the GIRCF has higher mode purity than the SIRCF under the same radially single-mode condition. Furthermore, the GIRCF with a smoother refractive index profile has a higher mode purity, which can be attributed to the inhibition of GIP on SOC. In addition, the mode purity of HE-composed $OAM_{1,1}$ in GIRCF reaches up to 94.43%, increasing by 7.95% over the one in the SIRCF, and the purity of EH-composed $OAM_{2,1}$ mode increases by 7.70%. This means that the purity of low-order OAM modes has a significant improvement in the GIRCF. According to Ref. [33], the intrinsic crosstalk can be expressed as:

$$Crosstalk = 10 \times \log_{10}(1 - purity) \tag{7}$$

when the purity is improved from 86.48% to 94.43%, the crosstalk can be reduced from −8.69 dB to −12.54 dB. Moreover, under ideal circumstances, when the OAM mode purity is larger than 96.84%, the system crosstalk can be smaller than −15 dB, which means the transmission system can be a MIMO-free system [20].

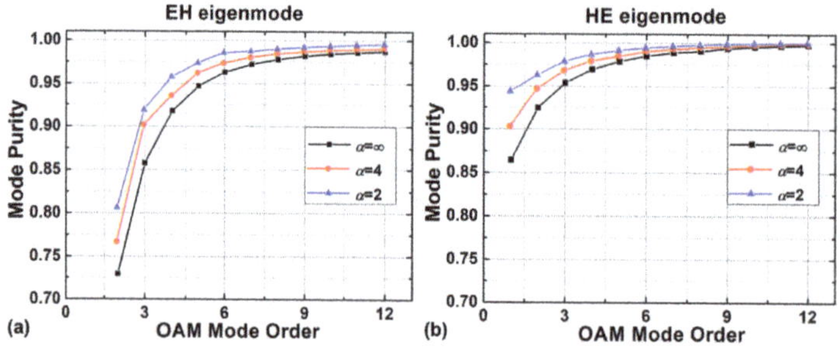

Figure 7. Purity of the (**a**) EH-composed and (**b**) HE-composed OAM modes with different shape factors.

To further observe the effect of spin-orbit interaction and verify the purity enhancement by alleviating the index discontinuity, we calculate and compare the mode purity of HE-composed $OAM_{1,1}$ with different GeO_2 mole fractions and ring-core positions under different shape factors, as shown in Figure 8. As mentioned above, the high GeO_2 mole fraction will introduce strong SOC, which means that the effect of GIP on SOC will be particularly obvious. Therefore, we can see that the purity of the $OAM_{1,1}$ can be promoted significantly under a higher GeO_2 mole fraction.

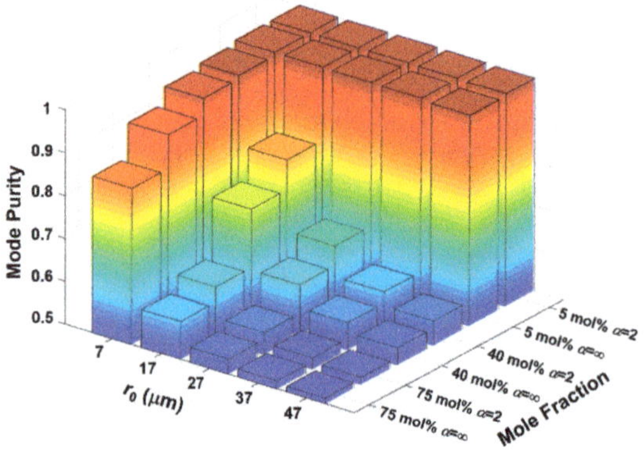

Figure 8. OAM mode purity based on $HE_{2,1}$ eigenmodes as a function of the position of ring-core for different GeO_2 mole fractions.

Considering the defects introduced into the index profile during the fabrication process, a perturbation type of calculation for shape factor α is analyzed when $r_0 = 7$ μm. The purity deviation of the OAM mode (ΔP), as a function of shape factor error ($\Delta\alpha$) from $\alpha = 2$, is displayed in Figure 9. The mode purity for $\alpha = 2$ is used as the reference, and ΔP

represents the deviation between the OAM mode purity under $\alpha + \Delta\alpha$ and the reference. Figure 9a shows that ΔP of $OAM_{1,1}$ varies with $\Delta\alpha$ under three GeO_2 mole fraction cases, and Figure 9b depicts that ΔP of different OAM modes varies with $\Delta\alpha$. As shown in Figure 9a, under the 5 mol% and 40 mol% GeO_2 mole fractions, the deviation of the OAM mode purity could be <0.1% when the shape factor error is within ±0.4, respectively. In addition, the deviation of the OAM mode purity could be <0.2% in high-order OAM modes when the GeO_2 mole fraction is 75 mol%. Therefore, by selecting proper GeO_2 concentration and OAM modes with higher order, the fabrication tolerance could be further increased.

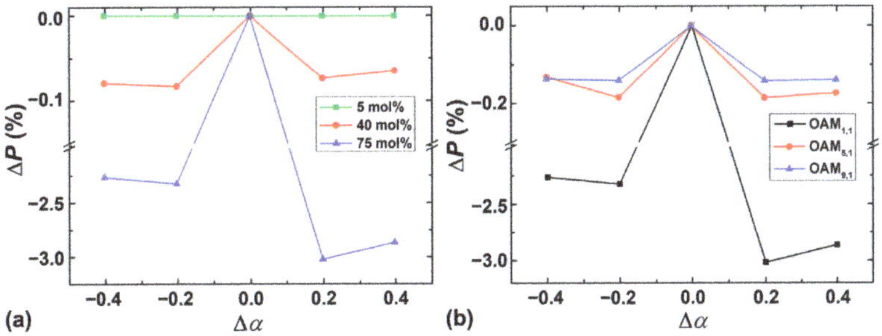

Figure 9. Purity deviation ΔP as a function of $\Delta\alpha$ with (**a**) different mole fractions of GeO_2 for $OAM_{1,1}$ mode, and (**b**) different OAM modes from $OAM_{1,1}$ to $OAM_{9,1}$ for 75 mol% GeO_2 mole fraction.

4. Conclusions and Perspective

In this paper, we design a GIRCF having a GeO_2-doped silica ring core and SiO_2 background. This fiber can restrain the effect of spin-orbit interaction, thus improving the mode purity of OAM modes transmitted in fiber. The GIRCF can support up to a total of 322 OAM modes at 1550 nm. Compared with SIRCF, the modes supported in the GIRCF have different degrees of enhancement in the mode purity, in which the maximum purity improvement of the HE-composed $OAM_{1,1}$ reaches up to 7.95% from 86.48% to 94.43%. Moreover, by changing the position of the ring core and the mole fraction of the GeO_2, the designed GIRCF can satisfy different requirements for SDM. This kind of fiber is promising in the enhancement of OAM mode purity, which can decrease the intrinsic channel crosstalk in optical fiber communication.

Author Contributions: Conceptualization, Y.L., Y.W., W.Z. (Wenqian Zhao) and Y.Y.; methodology, Y.L., Y.W., W.G., W.Z. (Wenqian Zhao) and W.Z. (Weigang Zhang); validation, Y.L. and H.Z.; formal analysis, Y.L., Y.W., H.Z., W.Z. (Weigang Zhang), Z.P. and Y.Y.; investigation, Y.L., Y.W., W.G., W.Z. (Wenqian Zhao) and Y.Y.; data curation, Y.L., Z.P. and Y.Y.; writing—original draft, Y.L.; writing—review and editing, Y.L., Y.W., W.G., W.Z. (Wenqian Zhao), H.Z., W.Z. (Weigang Zhang), Z.P. and Y.Y.; software, Y.Y.; resources, Y.Y.; supervision, Y.Y.; project administration, Y.Y.; funding acquisition, Y.Y. All authors have read and agreed to the published version of the manuscript.

Funding: Natural Science Foundation of Shaanxi Province under grant (2023-JC-YB-587); Key Technologies R & D Program of Tianjin under Grant (20YFZCGX00440); Shaanxi Key Laboratory of Deep Space Exploration Intelligent Information Technology under Grant (2021SYS-04).

Institutional Review Board Statement: Not applicable.

Informed Consent Statement: Not applicable.

Data Availability Statement: Data underlying the results presented in this paper are not publicly available at this time but may be obtained from the authors upon reasonable request.

Conflicts of Interest: The authors declare no conflict of interest.

References

1. Li, G.F.; Bai, N.; Zhao, N.B.; Xia, C. Space-division multiplexing: The next frontier in optical communication. *Adv. Opt. Photonics* **2014**, *6*, 413–487. [CrossRef]
2. Wang, A.; Zhu, L.; Liu, J.; Du, C.; Mo, Q.; Wang, J. Demonstration of hybrid orbital angular momentum multiplexing and time-division multiplexing passive optical network. *Opt. Express* **2015**, *23*, 29457–29466. [CrossRef] [PubMed]
3. Willner, A.E.; Pang, K.; Song, H.; Zou, K.; Zhou, H. Orbital angular momentum of light for communications. *Appl. Phys. Rev.* **2021**, *8*, 041312. [CrossRef]
4. Murshid, S.; Grossman, B.; Narakorn, P. Spatial domain multiplexing: A new dimension in fiber optic multiplexing. *Opt. Laser Technol.* **2008**, *40*, 1030–1036. [CrossRef]
5. Nelson, L.E.; Nielsen, T.N.; Kogelnik, H. Observation of PMD-induced coherent crosstalk in polarization-multiplexed transmission. *IEEE Photon. Technol. Lett.* **2001**, *13*, 738–740. [CrossRef]
6. Richardson, D.J.; Fini, J.M.; Nelson, L.E. Space-division multiplexing in optical fibres. *Nat. Photonics* **2013**, *7*, 354–362. [CrossRef]
7. Kingsta, R.M.; Selvakumari, R.S. A review on coupled and uncoupled multicore fibers for future ultra-high capacity optical communication. *Optik* **2019**, *199*, 163341. [CrossRef]
8. Saitoh, K.; Matsuo, S. Multicore Fiber Technology. *J. Light. Technol.* **2016**, *34*, 55–66. [CrossRef]
9. Saitoh, K.; Matsuo, S. Multicore fibers for large capacity transmission. *Nanophotonics* **2013**, *2*, 441–454. [CrossRef]
10. Rademacher, G.; Luis, R.S.; Puttnam, B.J.; Eriksson, T.A.; Ryf, R.; Agrell, E.; Maruyama, R.; Aikawa, K.; Awaji, Y.; Furukawa, H.; et al. High Capacity Transmission With Few-Mode Fibers. *J. Lightwave Technol.* **2019**, *37*, 425–432. [CrossRef]
11. Wakayama, Y.; Soma, D.; Beppu, S.; Sumita, S.; Igarashi, K.; Tsuritani, T. 266.1-Tbit/s Transmission Over 90.4-km 6-Mode Fiber with Inline Dual C+L-Band 6-Mode EDFA. *J. Lightwave Technol.* **2019**, *37*, 404–410. [CrossRef]
12. Xavier, G.B.; Lima, G. Quantum information processing with space-division multiplexing optical fibres. *Commun. Phys.* **2020**, *3*, 9. [CrossRef]
13. Feng, F.; Jin, X.; O'Brien, D.; Payne, F.; Jung, Y.; Kang, Q.; Barna, P.; Sahu, J.K.; Alam, S.-U.; Richardson, D.J.; et al. All-optical mode-group multiplexed transmission over a graded-index ring-core fiber with single radial mode. *Opt. Express* **2017**, *25*, 13773–13781. [CrossRef] [PubMed]
14. Milione, G.; Ip, E.; Ji, P.; Huang, Y.-K.; Wang, T.; Li, M.-J.; Stone, J.; Peng, G. MIMO-less Space Division Multiplexing with Elliptical Core Optical Fibers. In Proceedings of the Optical Fiber Communications Conference and Exhibition (OFC), Los Angeles, CA, USA, 19–23 March 2017.
15. Zhang, J.; Zhu, G.; Liu, J.; Wu, X.; Zhu, J.; Du, C.; Luo, W.; Chen, Y.; Yu, S. Orbital-angular-momentum mode-group multiplexed transmission over a graded-index ring-core fiber based on receive diversity and maximal ratio combining. *Opt. Express* **2018**, *26*, 4243–4257. [CrossRef]
16. Zhang, X.I.; Chen, S.H.I.; Wang, J.I.A.N. Weakly guiding graded-index ring-core fiber supporting 16-channel long distance mode division multiplexing systems based on OAM modes with low MIMO-DSP complexity. *Opt. Express* **2022**, *30*, 35828–35839. [CrossRef]
17. Liang, J.; Mo, Q.; Fu, S.; Tang, M.; Shum, P.; Liu, D. Design and fabrication of elliptical-core few-mode fiber for MIMO-less data transmission. *Opt. Lett.* **2016**, *41*, 3058–3061. [CrossRef]
18. Ingerslev, K.; Gregg, P.; Galili, M.; Da Ros, F.; Hu, H.; Bao, F.; Castaneda, M.A.U.; Kristensen, P.; Rubano, A.; Marrucci, L. 12 mode, WDM, MIMO-free orbital angular momentum transmission. *Opt. Express* **2018**, *26*, 20225–20232. [CrossRef]
19. Zhang, J.; Liu, J.; Shen, L.; Zhang, L.; Luo, J.; Liu, J.; Yu, S. Mode-division multiplexed transmission of wavelength-division multiplexing signals over a 100-km single-span orbital angular momentum fiber. *Photonics Res.* **2020**, *8*, 1236–1242. [CrossRef]
20. Bozinovic, N.; Yue, Y.; Ren, Y.; Tur, M.; Kristensen, P.; Huang, H.; Willner, A.E.; Ramachandran, S. Terabit-Scale Orbital Angular Momentum Mode Division Multiplexing in Fibers. *Science* **2013**, *340*, 1545–1548. [CrossRef]
21. Allen, L.; Beijersbergen, M.W.; Spreeuw, R.J.; Woerdman, J.P. Orbital angular momentum of light and the transformation of Laguerre-Gaussian laser modes. *Phys. Rev. A* **1992**, *45*, 8185–8189. [CrossRef]
22. Yue, Y.; Yan, Y.; Ahmed, N.; Yang, J.-Y.; Zhang, L.; Ren, Y.; Huang, H.; Birnbaum, K.M.; Erkmen, B.I.; Dolinar, S. Mode properties and propagation effects of optical orbital angular momentum (OAM) modes in a ring fiber. *IEEE Photon. J.* **2012**, *4*, 535–543.
23. Willner, A.E.; Huang, H.; Yan, Y.; Ren, Y.; Ahmed, N.; Xie, G.; Bao, C.; Li, L.; Cao, Y.; Zhao, Z. Optical communications using orbital angular momentum beams. *Adv. Opt. Photonics* **2015**, *7*, 66–106. [CrossRef]
24. Lian, Y.; Yu, Y.; Han, S.; Luan, N.; Wang, Y.; Lu, Z. OAM beams generation technology in optical fiber: A review. *IEEE Sens. J.* **2022**, *22*, 3828–3843. [CrossRef]
25. Wang, J.; Chen, S.; Liu, J. Orbital angular momentum communications based on standard multi-mode fiber. *APL Photonics* **2021**, *6*, 060804. [CrossRef]
26. Wang, J.; Yang, J.-Y.; Fazal, I.M.; Ahmed, N.; Yan, Y.; Huang, H.; Ren, Y.; Yue, Y.; Dolinar, S.; Tur, M.; et al. Terabit free-space data transmission employing orbital angular momentum multiplexing. *Nat. Photonics* **2012**, *6*, 488–496. [CrossRef]
27. Brunet, C.; Rusch, L.A. Optical fibers for the transmission of orbital angular momentum modes. *Opt. Fiber Technol.* **2017**, *35*, 2–7. [CrossRef]
28. Brunet, C.; Ung, B.; Wang, L.; Messaddeq, Y.; LaRochelle, S.; Rusch, L.A. Design of a family of ring-core fibers for OAM transmission studies. *Opt. Express* **2015**, *23*, 10553–10563. [CrossRef]

29. Ramachandran, S.; Kristensen, P.; Yan, M.F. Generation and propagation of radially polarized beams in optical fibers. *Opt. Lett.* **2009**, *34*, 2525–2527. [CrossRef]
30. Brunet, C.; Vaity, P.; Messaddeq, Y.; LaRochelle, S.; Rusch, L.A. Design, fabrication and validation of an OAM fiber supporting 36 states. *Opt. Express* **2014**, *22*, 26117–26127. [CrossRef]
31. Gregg, P.; Kristensen, P.; Ramachandran, S. Conservation of orbital angular momentum in air-core optical fibers. *Optica* **2015**, *2*, 267–270. [CrossRef]
32. Wang, L.; Corsi, A.; Rusch, L.A.; LaRochelle, S. Investigation of Orbital Angular Momentum Mode Purity in Air-Core Optical Fibers. In Proceedings of the IEEE Photonics Society Summer Topical Meeting Series (SUM), Newport Beach, CA, USA, 11–13 July 2016; pp. 203–204.
33. Zhang, Z.; Gan, J.; Heng, X.; Wu, Y.; Li, Q.; Qian, Q.; Chen, D.; Yang, Z. Optical fiber design with orbital angular momentum light purity higher than 99.9%. *Opt. Express* **2015**, *23*, 29331–29341. [CrossRef] [PubMed]
34. Golowich, S. Asymptotic theory of strong spin-orbit coupling in optical fiber. *Opt. Lett.* **2014**, *39*, 92–95. [CrossRef] [PubMed]
35. Guerra, G.; Lonardi, M.; Galtarossa, A.; Rusch, L.A.; Bononi, A.; Palmieri, L. Analysis of modal coupling due to birefringence and ellipticity in strongly guiding ring-core OAM fibers. *Opt. Express* **2019**, *27*, 8308–8326. [CrossRef] [PubMed]
36. Lu, Q.; Tu, J.; Wang, Z.; Gao, S.; Zhou, J.; Xiao, Q.; Shen, L.; Lan, X.; Liu, W.; Li, Z. Spin-orbit coupling suppressed high-capacity dual-step-index ring-core OAM fiber. *Opt. Lett.* **2022**, *47*, 1141–1144. [CrossRef]
37. Rjeb, A.; Ragheb, A.M.; Esmail, M.A.; Fathallah, H.; Machhout, M.; Alshebeili, S.A. Reciprocal Inverse Graded Index Fiber (R-IGIF) for OAM-SDM Networks. *IEEE Photon. J.* **2022**, *14*, 7249211. [CrossRef]
38. Zhang, X.; Ma, Q.; He, W.; Hong, W.; Deng, Z. Graded ring-core photonic crystal fiber for the transmission of OAM modes with high mode purity and quality in C plus L bands. *Opt. Commun.* **2022**, *520*, 128486. [CrossRef]
39. Prajapati, Y.K.; Srivastava, V.K.; Singh, V.; Saini, J.P. Effect of germanium doping on the performance of silica based photonic crystal fiber. *Optik* **2018**, *155*, 149–156. [CrossRef]
40. Fleming, J.W. Dispersion in GeO_2–SiO_2 glasses. *Appl. Opt.* **1984**, *23*, 4486–4493. [CrossRef]
41. Doerr, C.R.; Fontaine, N.; Hirano, M.; Sasaki, T.; Buhl, L.; Winzer, P. Silicon photonic integrated circuit for coupling to a ring-core multimode fiber for space-division multiplexing. In Proceedings of the European Conference and Exposition on Optical Communications, Geneva, Switzerland, 18–22 September 2011; p. Th.13.A.3.
42. Wang, L.; Ruan, Z.; Wang, H.; Shen, L.; Zhang, L.; Luo, J.; Wang, J. Deep learning based recognition of different mode bases in ring-core fiber. *Laser Photonics Rev.* **2020**, *14*, 2000249. [CrossRef]
43. Ge, D.; Gao, Y.; Yang, Y.; Shen, L.; Li, Z.; Chen, Z.; He, Y.; Li, J. A 6-LP-mode ultralow-modal-crosstalk double-ring-core FMF for weakly-coupled MDM transmission. *Opt. Commun.* **2019**, *451*, 97–103. [CrossRef]
44. Zuo, M.; Ge, D.; Gao, Y.; Cui, J.; Huang, S.; Zhou, R.; Guo, Q.; Zhang, Y.; Zhang, D.; Xiao, X.; et al. In 3-Mode Real-Time MDM Transmission Using Single-Mode OTN Transceivers over 300 km Weakly-Coupled FMF. In Proceedings of the Optical Fiber Communication Conference (OFC), San Diego, CA, USA, 6–10 March 2022; p. M4B.4.
45. Deroh, M.; Kibler, B.; Maillotte, H.; Sylvestre, T.; Beugnot, J.-C. Large Brillouin gain in Germania-doped core optical fibers up to a 98 mol% doping level. *Opt. Lett.* **2018**, *43*, 4005–4008. [CrossRef] [PubMed]
46. Dianov, E.A.; Mashinsky, V.A. Germania-based core optical fibers. *J. Light. Technol.* **2005**, *23*, 3500–3508. [CrossRef]
47. Sakaguchi, S.; Todoroki, S. Optical properties of GeO_2 glass and optical fibers. *Appl. Opt.* **1997**, *36*, 6809–6814. [CrossRef] [PubMed]
48. Wang, Y.; Liu, Y.; Zhao, W.; Yang, J.; Geng, W.; Fang, Y.; Bao, C.; Ren, Y.; Wang, Z.; Liu, Y.-g.; et al. Multi-Ring-Air-Core Fiber Supporting Numerous Radially Fundamental OAM Modes. *J. Light. Technol.* **2022**, *40*, 4420–4428. [CrossRef]
49. Wang, W.; Xu, H.-D.; Yang, Q.-H.; Zhou, F.-D.; Li, Z.-R.; Han, Y.; Qi, Y.-F.; Hou, L.-T. Large mode area microstructured fiber supporting 56 super-OAM modes. *Opt. Express* **2019**, *27*, 27991–28008. [CrossRef]
50. Afshar, S.; Monro, T.M. A full vectorial model for pulse propagation in emerging waveguides with subwavelength structures part I: Kerr nonlinearity. *Opt. Express* **2009**, *17*, 2298–2318. [CrossRef]
51. Yue, Y.; Zhang, L.; Wang, J.; Xiao-Li, Y.; Shamee, B.; Karagodsky, V.; Sedgwick, F.G.; Hofmann, W.; Beausoleil, R.G.; Chang-Hasnain, C.J. A "linear" high-contrast gratings hollow-core waveguide and its system level performance. In Proceedings of the Optical Fiber Communication Conference (OFC), San Diego, CA, USA, 21–25 March 2010; p. OTuI5.
52. Gregg, P.; Kristensen, P.; Rubano, A.; Golowich, S.; Marrucci, L.; Ramachandran, S. Enhanced spin orbit interaction of light in highly confining optical fibers for mode division multiplexing. *Nat. Commun.* **2019**, *10*, 4707. [CrossRef]

sensors

MDPI

Article

Face Mask Identification Using Spatial and Frequency Features in Depth Image from Time-of-Flight Camera

Xiaoyan Wang [1], Tianxu Xu [2], Dong An [1], Lei Sun [3], Qiang Wang [4], Zhongqi Pan [5] and Yang Yue [6,*]

1 Institute of Modern Optics, Nankai University, Tianjin 300350, China
2 National Center for International Joint Research of Electronic Materials and Systems, School of Electrical and Information Engineering, Zhengzhou University, Zhengzhou 450001, China
3 Shphotonics, LLC, Tianjin 300450, China
4 Angle AI (Tianjin) Technology Co., Ltd., Tianjin 300450, China
5 Department of Electrical & Computer Engineering, University of Louisiana at Lafayette, Lafayette, LA 70504, USA
6 School of Information and Communications Engineering, Xi'an Jiaotong University, Xi'an 710049, China
* Correspondence: yueyang@xjtu.edu.cn

Abstract: Face masks can effectively prevent the spread of viruses. It is necessary to determine the wearing condition of masks in various locations, such as traffic stations, hospitals, and other places with a risk of infection. Therefore, achieving fast and accurate identification in different application scenarios is an urgent problem to be solved. Contactless mask recognition can avoid the waste of human resources and the risk of exposure. We propose a novel method for face mask recognition, which is demonstrated using the spatial and frequency features from the 3D information. A ToF camera with a simple system and robust data are used to capture the depth images. The facial contour of the depth image is extracted accurately by the designed method, which can reduce the dimension of the depth data to improve the recognition speed. Additionally, the classification process is further divided into two parts. The wearing condition of the mask is first identified by features extracted from the facial contour. The types of masks are then classified by new features extracted from the spatial and frequency curves. With appropriate thresholds and a voting method, the total recall accuracy of the proposed algorithm can achieve 96.21%. Especially, the recall accuracy for images without mask can reach 99.21%.

Keywords: 3D data processing; depth camera; face mask identification

1. Introduction

Face masks are a well-established preventive tool to limit the spread of viruses through droplets and aerosols in the population [1]. With the COVID-19 global pandemic, many countries and regions stipulate that people must wear masks in public places, especially indoor areas and vehicles. In most cases, some staff are assigned to enforce the rule. Nevertheless, the manual monitoring method not only wastes manpower but also can easily cause infection. Contactless mask recognition can avoid the waste of human resources and the risk of exposure, which has been highly valued by many researchers [2–4].

It is important to quickly and accurately identify masks in daily life, especially in areas with a large flow of people. Some researchers have developed integrated systems for rapid screening. Hussain et al. [5] designed an intelligent disinfection screening door that can simultaneously measure the pedestrian's temperature, identify if a mask is worn, and perform the disinfection. The 2D data captured by the camera was first input to the mask recognition module. Then, the neural network was used for classification. In addition to the comprehensive screening system, which can monitor the temperature and detect the mask, there is also a lot of work focused on improving the accuracy of mask recognition. Most mask recognition methods are based on object recognition technology, which automatically identifies whether a person is wearing a mask by combining the camera's input with a computer algorithm. Conventional solutions of object recognition are mainly based on two-dimensional (2D) images, including detecting regions of interest, extracting features, and performing classification [6]. At present, numerous emerging machine learning methods are being used more and more in object recognition. In [7–12], ResNet, Yolo, MobileNet, and other machine learning models are used to recognize face masks. The identification accuracy can exceed 90%. However, the performance of these methods begins to decline significantly in some situations, such as occlusion or variable illumination conditions [13]. In addition, Cao et al. [14] considered night situations and proposed the MaskHunter model, but the accuracy dropped to 71.6%. Even when the additional training module was designed to improve accuracy, the results were not satisfactory. Hence, breaking the trade-offs between accuracy and the adaptability of mask identification is a difficult problem to solve.

In recent years, with the enrichment of information acquisition methods and the development of sensing technology, many three-dimensional (3D) sensors have emerged [15,16], which can not only obtain the shape information of the object but also determine the distance between the object and the camera. Among them, the depth camera, based on the principle of active ranging, has many advantages, such as being insensitive to illumination and unaffected by the contrast of the target. As the popular depth sensors that are commercially available [17–19], the structured light cameras and the time-of-flight (ToF) cameras have different advantages and application scenarios [20]. In most cases, the ToF camera has low complexity and advantages in practical application [21]. Therefore, the ToF camera is more commonly used as a data acquisition device. Since the depth cameras work well in low light and even in dark conditions, object recognition based on obtained 3D images has been used in many scenes, such as human detection, industrial assembly, gesture recognition, and others [22–28]. Luna et al. [29] presented a new method for detecting people only using depth images, and the data was captured by a depth camera in a frontal position. This method ran in real-time using a low-cost CPU platform with high accuracy. Various tasks based on the depth information require different features to be extracted. The introduction of 3D information also increases the computational cost of machine learning models. Moreover, the universal parameters of feature extraction based on 3D information need to be further investigated.

In this paper, we propose and demonstrate a method based on a ToF depth camera to determine whether a person is wearing a mask. The results are divided into three classifications: wearing no mask, wearing a surgical mask, and wearing an N95 mask. The three situations can be easily discriminated by the optimized spatial and frequency features on the facial depth contour. The experimental results show that these characteristics not only can distinguish whether people wear masks but also can determine the types of masks. Unlike the 2D image from the RGB camera, which is easily affected by the environmental illumination, our developed mask identification system based on the ToF camera runs robustly under variable external conditions, especially in the dark environment. Compared with the other 3D imaging sensors, the ToF camera also has the advantage of fast imaging. Through the sampling and dimension reduction of the depth image to obtain the facial contour curve, our method could potentially achieve quick and accurate identification, which is suitable for scenes with illumination variation and rapid identification, such as entrances and exits of the building. In addition, our method can also provide statistical data for epidemiological analysis by monitoring the mask types.

2. Method

Figure 1 illustrates the process of the proposed mask recognition method. In the data acquisition module, the ToF camera is used to collect the facial depth images. The facial contour is then extracted from the depth image as a recognition feature in the next step. The recognition module can be sequentially divided into two sub-modules: one first needs to identify whether a person is wearing a mask. If the output result is yes, the mask type is further differentiated. Otherwise, the module can export the result that there is no mask. The above classification process mainly relies on the feature descriptor, including the spatial and frequency features. The spatial features are extracted from the facial contour, and the frequency features are obtained from the frequency characteristics by using the Fourier transform. There are two popular mask types verified by this method: surgical masks and N95 masks. Surgical masks are the most widely used in daily life, and N95 masks have better prevention abilities and a higher usage rate in hospitals and other risk environments [30]. Thus, the proposed method has great research significance for identifying these two mask types.

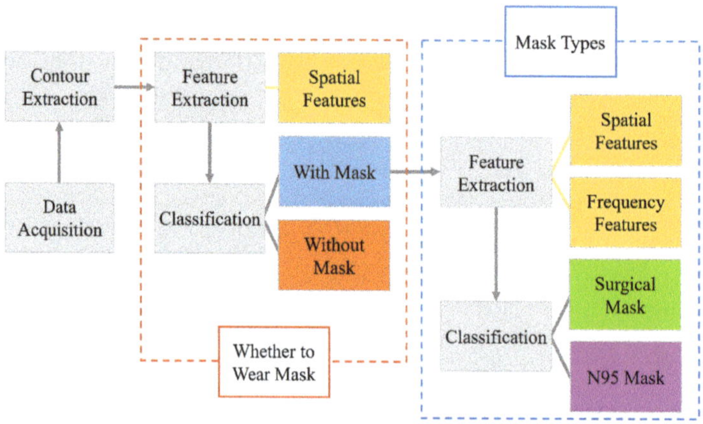

Figure 1. Flow chart of mask recognition based on depth camera.

2.1. Facial Contour Extraction

In order to process the facial data efficiently and save on computational costs, the vertical facial contours are extracted from the 3D facial depth images to achieve rapid recognition. In sequence, spatial features and frequency features are determined from the extracted facial contour. Thus, the precision of the facial contour extraction directly affects the accuracy of the classification. In our method, a contour extraction approach is utilized to obtain appropriate facial contours. In addition to the facial information, there is a lot of redundant information on the obtained depth images. Those useless backgrounds need to be first removed by the clipping method. The designed rhombus template is then applied to reduce the interference of the hair and other noise. The facial contour is ultimately extracted from the rhombic region. The extracted facial contour is smoothed to obtain a contour curve with spatial interpretation.

The above-mentioned process and visualization results of contour extraction are shown in Figure 2. To avoid the influence of background information in the depth image, the human facial depth image is first segmented by the distance filter. With the distance information between the object and the camera being recorded in the depth image, the human face can be easily distinguished from other backgrounds by conditional filtering. By setting the segmented ranges of distance, one can highlight the foreground human face by performing a one-time filtering operation. The camera we used can capture complete and clear faces within the range from 0.1 m to 0.5 m. Therefore, the background with a distance greater than 0.5 m is filtered. The background and face can be completely separated by distance filtering. The images containing only the face can be obtained by setting the background pixel value to infinite and cutting out the pixels with the infinite value. Since the main difference between the images with and without masks is in the bottom part of the face, we directly cut out the top face in the image and only focus on the bottom face with obvious features. As shown in the first column of Figure 2, the whole input image is fully occupied by the face, both horizontally and vertically. Thus, the bottom face can be obtained by clipping from the middle of the image.

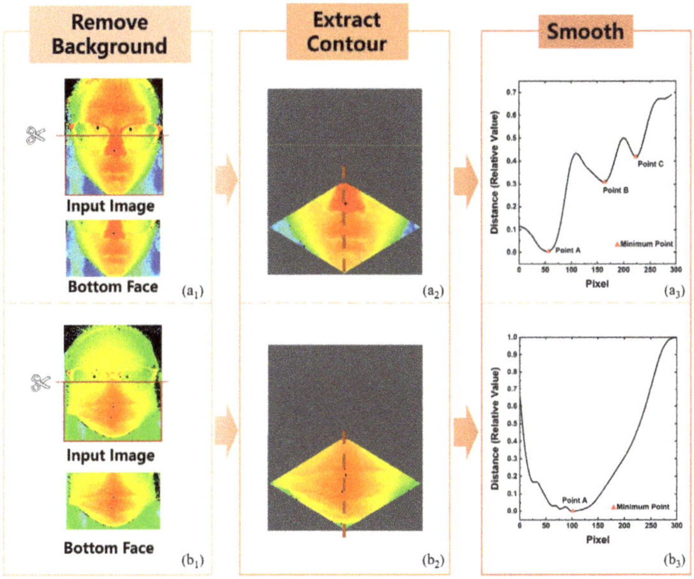

Figure 2. Overview of facial contour extraction. (**a**) Facial contour extraction of the image without a mask and (**b**) facial contour extraction of the image with a surgical mask.

In the extraction process of human facial contour, a rhombus template is designed to filter the redundant hair noise. Through the above steps, we can obtain the bottom face in Figure 2, and the construction of the rhombus is based on the bottom face image. The midpoint of each side in the bottom face image is first picked out. These four points are then taken as the vertices of the rhombus. As shown in the second column of Figure 2, the overlapping part of the rectangular image and rhombus is the target area, and the rest of the rectangular image is removed. The gray part in the figure represents all the filtered parts of the original image, and the contour curve is extracted from the unfiltered color part. According to the geometric characteristics of the human face and depth distance, the facial contour can be obtained by finding the points with the minimum distance in each horizontal line of the rhombus area. The face contour is used to replace the original data, and the feature descriptor is further extracted, which greatly reduces the amount of calculation.

Due to the limited resolution of the depth camera, the extracted facial contour needs to be smoothed so that it can be closer to the original appearance. The extracted facial contour is smoothed with a Gaussian-weighted moving average filter, and the length of the smoothing window is 50. The example of contour smoothness is demonstrated in the third column of Figure 2. The ordinate of the curve is the relative distance between the face and the camera, and the abscissa is the pixel position. The first row (a) displays the contour extraction process of the images without a mask. As can be seen from the curve of a_3, there are three obvious minimum points, A, B, and C, representing the tip of the nose, the top lip, and the bottom lip, respectively. The second row (b) illustrates the contour extraction process of the image with a surgical mask. For contour curves with a mask, point A only represents the global minimum point of the curve, which is caused by the mask shape instead of the nose. Other extreme points of the curve are caused by the mask wrinkles, which have no special significance and are therefore not annotated. All the contour curves need to be normalized, including unifying the length and setting the relative distance of the minimum point to zero.

2.2. Feature Extraction and Classification of Whether to Wear Mask

After obtaining the facial contour curve, the first classification step is deciding whether to wear a mask. The process is shown in Figure 3, and this step identifies wearing or not wearing a mask. If the classification result is with a mask, it is necessary to further identify the mask type. The features used in this classification will be introduced in the following content. Three features are designed from the contour curve to facilitate the classification. The main features are the number of local minimum points on the contour curve, the standard deviation between the farthest points among the local minimum points, and the quadratic coefficient of the expression $Y_1 = a_1x^2 + b_1x + c_1$, which is used to partially fit the contour curve.

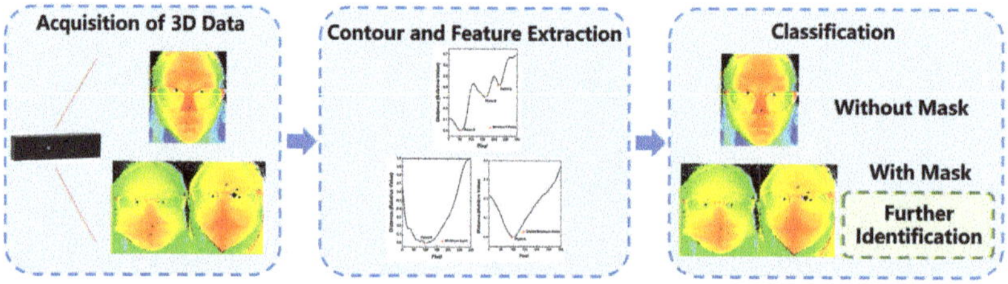

Figure 3. Identification process for whether or not to wear a mask.

Figure 4 depicts the above features of the contour curve without a mask. The local minimum points of the contour curve are labeled in Figure 4a, where point A is the tip of the nose, point B represents the top lip, and point C is the bottom lip. The local minimum points are calculated according to Equation (1):

$$\begin{cases} f(x_j) < f(x_{j-1}) \\ f(x_j) < f(x_{j+1}) \end{cases}, \quad (1)$$

where $f(x_j)$ is the depth value corresponding to point x_j, x is the random point of the contour curve, and the range of j is from 2 to the length of contour curve. When x_j satisfies Equation (1), it is recorded as the local minimum point of the curve. The number of the local minimum point is represented by N.

The standard deviation of the two farthest minimum points is denoted by *SSD* (spatial standard deviation). Point A and point C are the farthest minimum points shown in Figure 4b. *SSD* is calculated as follows:

$$SSD = \sqrt{\frac{\sum_{j=1}^{n_s}(x_j - \overline{x})^2}{n_s - 1}}, \quad (2)$$

where n_s is the number of points between point A and point C, $\overline{x}$ is the average of these points. Especially, when the contour curve has only one minimum point, the value of *SSD* is 0.

Figure 4c explains the partial fitting process. Taking the global minimum of point A of the curve as the center, and the segment curve with the length of $2L$ is then obtained by taking L length from both sides of the curve. The quadratic function $Y_1 = a_1x^2 + b_1x + c_1$ is used to fit this segment curve. According to the properties of the quadratic function, the quadratic coefficient a_1 determines the opening size of the parabola. The value of $|a_1|$ is larger and the opening size of the parabola is smaller.

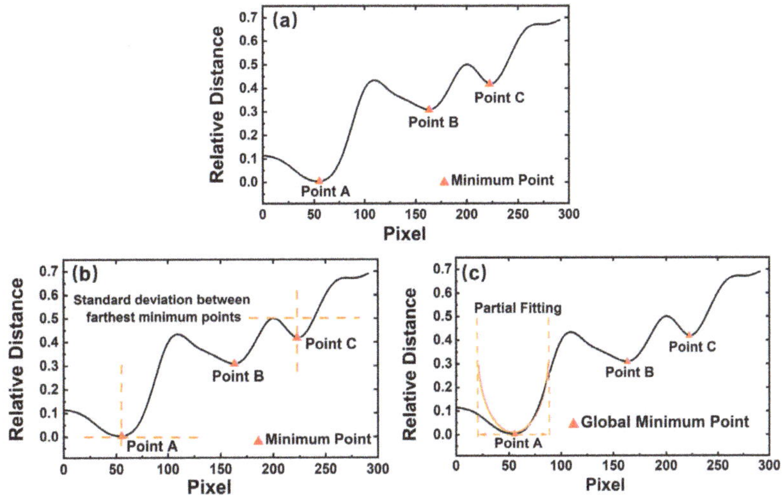

Figure 4. Three features of the image without a mask: (**a**) number of minimum points of the contour curve; (**b**) standard deviation between the farthest minimum points of the contour curve; and (**c**) quadratic coefficient of the partial fitting curve.

After calculating the above features, the type of contour curve can be determined according to the different feature values. Generally, the feature value of the curve without a mask are greater than the feature value of the curve with a mask. Therefore, the threshold values can be set to separate them. For the images without masks, $N > N'$, $SSD > SSD'$, $a_1 > a_1'$, where N', SSD', and a_1' are the thresholds. The voting method is used as classifier to determine whether a mask is worn or not. When more than two of the three features meet the threshold, it is determined that the mask is not worn.

2.3. Feature Extraction and Classification of Mask Types

After identifying that a mask is worn, a further judgement of the mask type is executed. The judgement process is similar to identifying whether to wear a mask or not, which is shown in Figure 5. In our experiment, the mask types are the surgical mask and the N95 mask, respectively. The judgement process is similar to identifying whether to wear a mask. To obtain suitable features to recognize the mask types, three new features are designed from the spatial and frequency levels, including the opening angle of the contour curve, the quadratic coefficient of the expression $Y_1 = a_1x^2 + b_1x + c_1$ (which is used to partially fit the contour curve), and the area of the frequency curve. The corresponding frequency curve is obtained by the Fourier transform.

Figure 5. Identification process for mask types.

Figure 6 depicts the feature visualization of the contour curve with the N95 mask being worn. For contour curves with a mask, point A only represents the global minimum point of the curve, which is caused by the shape of the mask instead of the nose. The global minimum point (x_M, y_M), the left-end point (x_L, y_L), and the right-end point (x_R, y_R) of the contour curve are first located, as shown in Figure 6a. The opening angle, denoted by α, is computed from these three points. Taking the global minimum point as the vertex, the angle between two straight lines is calculated as follows:

$$\alpha = arctan(|\frac{\frac{y_M - y_L}{x_M - x_L} - \frac{y_R - y_M}{x_R - x_M}}{1 + \left(\frac{y_M - y_L}{x_M - x_L}\right) \cdot \left(\frac{y_R - y_M}{x_R - x_M}\right)}|) \quad (3)$$

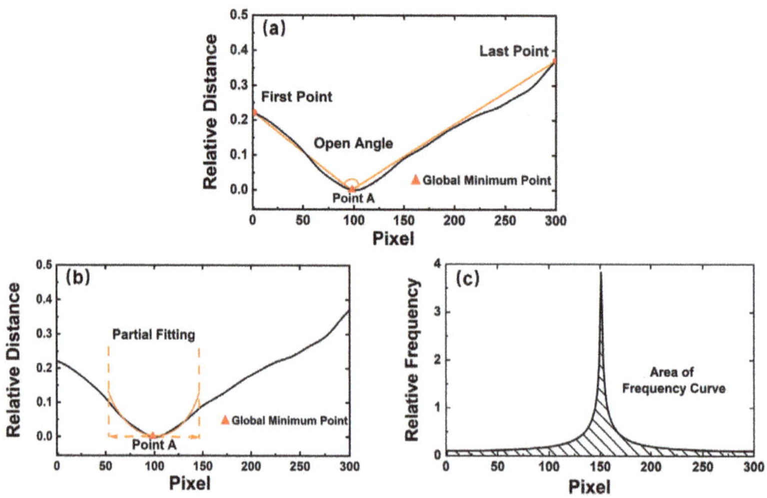

Figure 6. Three features of N95 mask: (**a**) opening angle of the contour curve, (**b**) quadratic coefficient of the partial fitting curve, and (**c**) area of the frequency curve from the Fourier transform of the spatial contour curve.

The process of partial fitting explained in Figure 6b is consistent with Section 2.2. The quadratic function $Y_1 = a_1x^2 + b_1x + c_1$ is used to fit the segment curve. According to the properties of the quadratic function, the quadratic coefficient a1 can also distinguish between the mask types.

After fitting the contour curves with the quadratic function $Y_2 = a_2x^2 + b_2x + c_2$, its frequency curve is obtained by the Fourier transform in Figure 6c. The Fourier transform of the discrete contour curve follows Equation (4):

$$\begin{cases} F(K) = \sum_{j=1}^{n} x(j) W_n^{(j-1)(k-1)} \\ W_n = e^{\frac{-2\pi i}{n}} \end{cases}, \quad (4)$$

where *F(K)* is the frequency curve from the Fourier transform, *n* is the number of the point on the contour curve, and *Wn* is a *n*-th root of unity. The area of the frequency curve is denoted by *FS* and it is calculated as follows:

$$FS = \frac{1}{2} \sum_{j=1}^{n} (f(x_j) + f(x_{j+1})) \quad (5)$$

Due to the different geometric shapes, two mask types have distinct feature values and the appropriate thresholds can separate the mask types. For surgical masks, $\alpha < \alpha'$, $a_1 < a_1'$, $FS < FS'$, where α', a_1', and FS' are the thresholds. The voting method is also used to determine the mask types. When two or more of the above thresholds are met, the mask type is determined to be the surgical mask. Otherwise, the type is the N95 mask.

Figure 7 depicts the feature visualization of the contour curve with the surgical mask being worn. Compared to Figure 6, the partial fitting curve is quite different, and a clearer comparison will be analyzed in the next section.

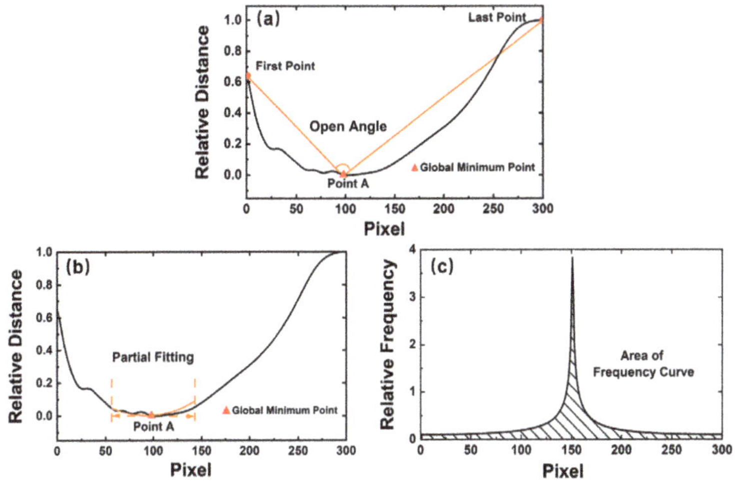

Figure 7. Three features of surgical masks: (**a**) opening angle of the contour curve, (**b**) quadratic coefficient of the partial fitting curve, and (**c**) area of the frequency curve from the Fourier transform of the spatial contour curve.

3. Results

To evaluate the proposed recognition algorithm, a ToF depth camera was used to capture the depth data. The PicoFlexx camera from pmdtechnologies was used for data collection. The PicoFlexx camera is equipped with an IRS1145C Infineon® REAL3™ 3D image sensor based on the principle of ToF. The camera uses an 850-nm Vertical-Cavity Surface-Emitting Laser (VCSEL) as the light source, and the detection range is from 0.1 m to 4 m. We used this camera to collect the front-face image under the indoor environment. For the acquired images, the corresponding resolution is 945×722 pixels, which is expanded from the 244×172 resolution of the IRS1145C sensor by the Software Development Kit (SDK) Royale 3.20.0.62 as a color-coded depth map. These depth images can be divided into three categories: not wearing a mask, wearing a surgical mask, and wearing an N95 mask. The bottom half of the face is obtained by distance filtering and clipping. The 3D view of three situations is shown in Figure 8. The pixel value of the data is normalized to the range from 0 to 1 by the SDK of the ToF camera, where 0 represents the closest distance to the camera and 1 represents the farthest distance detected by the camera. In Figure 8, we flip the meaning of the values for better visualization.

From the 3D view of the bottom face, the face without a mask has more obvious undulations with multiple peaks, and the nose part is conspicuous. The 3D contour of the surgical mask is much smoother in general, but there are more small fluctuations. The shape of the 3D contour for the N95 mask has only one obvious peak.

With the depth image of the bottom face, the facial contour curve is extracted and normalized. In our research, the length of all the contour curves is unified at 300, and the relative distance of the global minimum point is set to zero. Figure 9 shows the 3 types of facial contour curves, and 10 contour curves of each type are selected for overlapping display. When all the contours are normalized, contour curves of the same type show good repeatability and similar geometric features, but different kinds of contour curves have different geometric features. In particular, the mask covers the shape of the human face, so the contour curve with a mask is gentler, while the contour curve without a mask fluctuates more obviously. We can clearly see the characteristics of the bottom face in Figure 9, including the nose, the top lip, and the bottom lip, while they cannot be distinguished from the contour curve with a mask. The contour curves of different mask types are also different. In general, the contour curve of a surgical mask is relatively gentle. However, there are many wrinkle-caused fluctuations in the curve. The N95 mask is made of thicker material, and its depth contour curve is closer to a trapezoidal shape. Depending on the different shapes of the contour curves, the features can be extracted for classification and recognition.

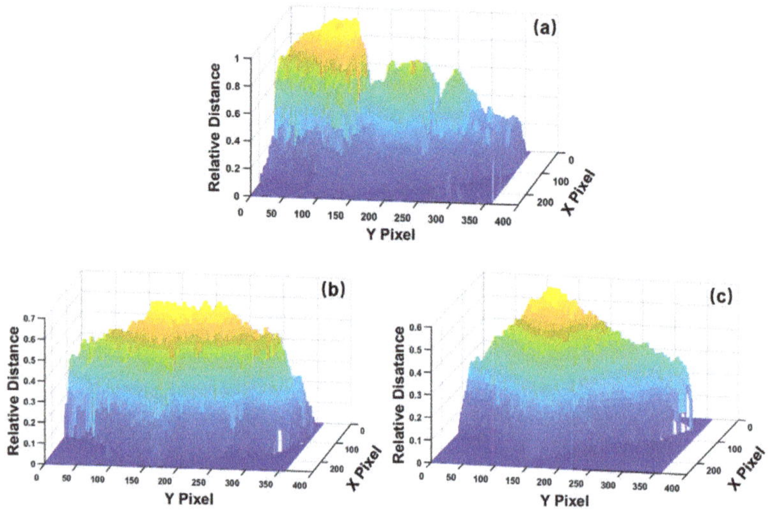

Figure 8. 3D view of the bottom face acquired from the ToF camera: (**a**) without a mask, (**b**) with a surgical mask, and (**c**) with an N95 mask.

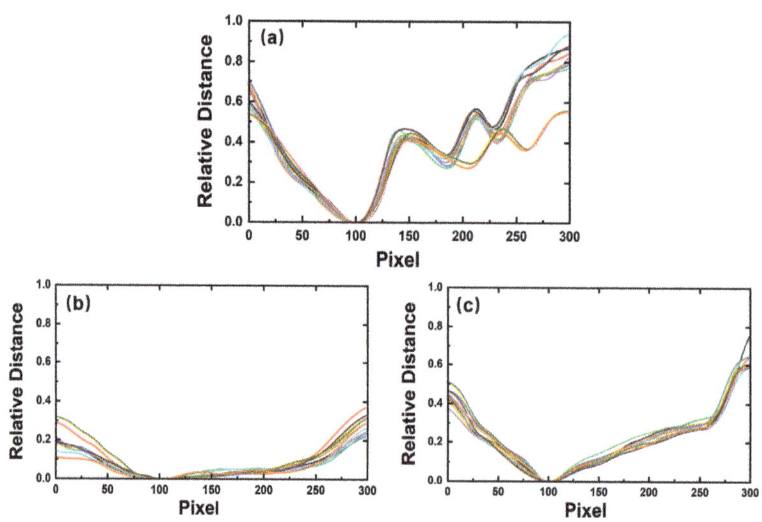

Figure 9. Normalized facial contour curves of 10 samples: (**a**) facial contours without a mask, (**b**) facial contours with a surgical mask, and (**c**) facial contours with an N95 mask.

After obtaining the contour curve, the features N, SSD, and a_1 are calculated to determine whether to wear a mask. A total of 15 images of each type are selected to calculate the three features, which are shown in Figure 10. The black curve represents the feature values of the contour curve without a mask, while the red and blue curves represent the feature values of the contour curve with a surgical mask and an N95 mask, respectively. The minimum number of points is displayed in Figure 10a. One can see that due to the shape of the nose and lips, the number of the minimum points of the contour curve without a mask is always more than three. With the regular shape, the N value of the contour curve with an N95 mask is typically one. However, due to the influence of the folds, the N value of the contour curve with a surgical mask shows great fluctuation, ranging from 1 to 3. Although most N values of the contour curve without mask are greater than those of the contour curves with a mask, there are still overlapping parts. Therefore, a single feature cannot classify the mask wearing condition accurately. The value of N is also not suitable for distinguishing the mask types.

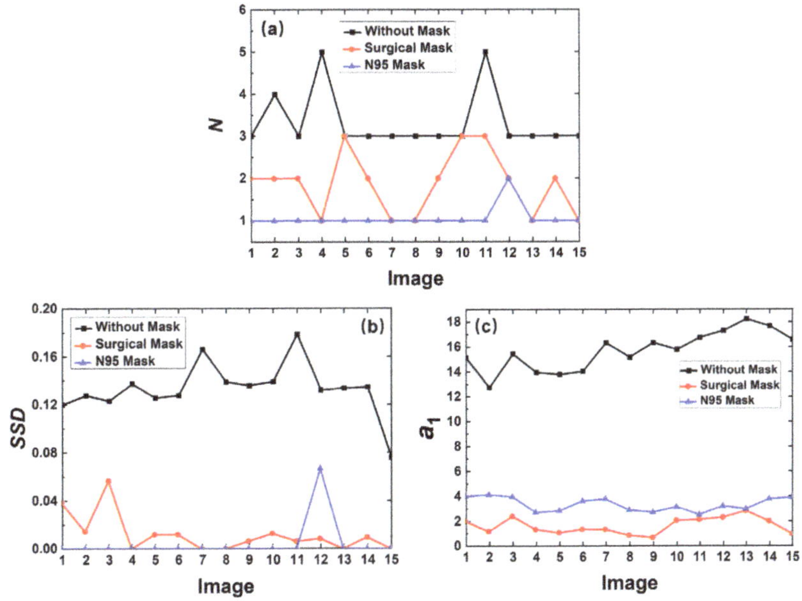

Figure 10. Feature values to identify whether to wear mask captured from 15 images: (**a**) the N value, (**b**) the SSD value, and (**c**) the a_1 value of three types of curves.

Figure 10b shows the standard deviation between the two farthest minimum points. The SSD values of the contour curve without a mask are far greater than those of the contour curve with a mask because of the shape of the human face. The folds of the surgical mask lead to the oscillation of the contour curve with the mask, causing a smaller but unpredictable fluctuation in SSD value. The contour curve with the N95 mask usually has only 1 minimum point, and its SSD is set to 0. However, because the SSD curves of two types of masks also overlap, the SSD value cannot distinguish the mask types well.

Figure 10c demonstrates the quadratic coefficient a_1 of the partial fitting curve. As one can see, the a_1 value of the contour curve without a mask are far greater than those of the contour curve with a mask. For the quadratic function, the value of $|a_1|$ is larger and the opening of the parabola is smaller. The global minimum point is the vertex of the quadratic function. For the contour curve without a mask, the fitting curve represents the nose. For the contour curve with a mask, the fitting curve describes a part of the mask. Therefore, the a_1 value of the contour curve without a mask is greater than that with a mask, which is consistent with the theory. The a_1 values of different mask types also have differentiation, which will be further explained.

In summary, the above three features can distinguish whether or not to wear a mask by setting the thresholds appropriately. The effect of a single feature is limited, while the combination of the three features can achieve better results. The decision conditions are $N > N'$, $SSD > SSD'$, $a_1 > a_1'$, where N', SSD', and a_1' are the thresholds. In our work, N' is 2, SSD' is 0.05, and a_1' is 5. To improve the robustness, the voting method is used for the classification. When more than two evaluation conditions are satisfied, it is considered as without a mask. Otherwise, it is determined to be wearing a mask.

In the identification process of two mask types, the surgical mask and the N95 mask, three features are extracted from the spatial and frequency curves. Figure 8 shows the features to distinguish the mask types from 15 images. The red and blue curves represent the feature values of the contour curve with the surgical mask and the N95 mask, respectively. The opening angle values of these images are illustrated in Figure 11a. It clearly shows that the opening angle of the contour curve with the N95 mask is greater than that with the surgical mask. However, the opening angle values of two types still have overlapping parts and other features are required to be leveraged together.

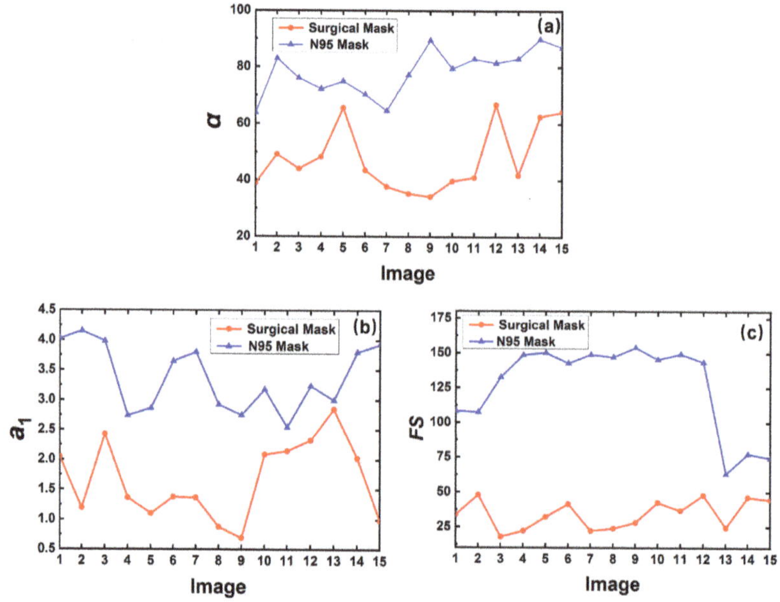

Figure 11. Feature values to identify the mask types captured from 15 images: (**a**) the α value, (**b**) the a_1 value, and (**c**) the FS value of two curve types.

The quadratic coefficient a_1 of the partial fitting curve is shown in Figure 11b. Although the opening angle of the contour curve with an N95 mask is larger than that with a surgical mask, the quadratic coefficient of the partial fitting curve of an N95 mask is smaller, because the chosen part of the N95 mask is sharper together with a more stable shape.

The frequency curve is obtained by the Fourier transform after the quadratic fitting of the contour curve. The area of the frequency curve of the N95 mask is larger than that of the surgical mask in Figure 11c.

The above three features can distinguish whether or not to wear a mask by setting the thresholds appropriately. Due to different wearing habits and soft mask materials, the feature values have fluctuations, and the effect of a single feature is limited. The combination of these three features can achieve better results, and the voting method is also used to classify the mask types. The decision conditions are $\alpha < \alpha'$, $a_1 < a_1'$, $FS < FS'$, where α', a_1', and FS' are thresholds. In our work, α' is 70, a_1' is 2.5, and FS' is 85. When two or more threshold conditions are satisfied, it is determined that the person is wearing a surgical mask. Otherwise, it is judged as an N95 mask.

The algorithm is evaluated from many aspects based on our data. The results can be classified into three categories: *TP* (true positive) means that the results are the same as the real conditions; *FP* (false positive) means samples in the detected category are inconsistent with the real category; and *FN* (false negative) means that the object is detected as the opposite category. Precision is used to measure the proportion of the real cases (*TP*) among all the positive cases (*TP* + *FP*) detected by the algorithm, as shown in Equation (6).

$$Precision = \frac{TP}{TP + FP} \tag{6}$$

Recall is the proportion of detected real cases in all positive cases, which represents the ability of the algorithm to detect the real cases, calculated as follows:

$$Recall = \frac{TP}{TP + FN} \tag{7}$$

Due to the trade-off between the precision and recall, the F_1 score combines them into a single indicator, defined by the harmonic average, as shown in Equation (8).

$$F_1 = 2 \times \frac{P \times R}{P + R} = \frac{TP}{TP + \frac{FN+FP}{2}}, \tag{8}$$

where *P* and *R* represent precision and recall, respectively. If the F_1 score is higher, the test of the algorithm will be more effective.

A total of 369 depth images were captured, including evenly distributed categories: without masks, with surgical masks, and with N95 masks, to test the algorithm. These images were taken by 3 women aged from 23 to 25 over a 1-week timeframe. In order to enrich the sample diversity, the images were taken for people with and without glasses. In addition, it also includes different hairstyles, such as long, straight hair and a ponytail. In the proposed algorithm, there are two classifications: one is whether or not to wear mask and the other is the mask types. The precision, recall, and F_1 score are calculated for the two cases, respectively. For the classification of whether or not to wear mask, the results are listed in Table 1.

Table 1. Detection results of whether or not to wear mask.

Category	Precision	Recall	F_1 Score
Without a mask	0.906	0.992	0.947
With a mask	0.996	0.946	0.970

Although the recall of those without a mask is larger than the recall of those with a mask, the precision and F_1 score of with a mask are larger than that of without a mask. As displayed in Table 1, most images without a mask can be detected correctly. Because of the soft material, surgical masks are more likely to produce wrinkles or reflect the undulation of the face. Therefore, most of the images incorrectly identified as without a mask are actually images with surgical masks.

To evaluate the confusion between the surgical mask and the N95 mask, the classification results on the mask types are shown in Table 2. Since the step is performed after detecting whether or not to wear a mask, the classification of mask types relies on the image detected as with a mask. These two kinds of masks have similar contour curves and feature values, which leads to the drop in accuracy. In addition, the precision and recall of the two types are in a restrictive relation, respectively, and have close F_1 scores.

Table 2. Detection results of whether or not to wear mask.

Category	Precision	Recall	F_1 Score
With a surgical mask	0.847	0.909	0.877
With an N95 mask	0.910	0.842	0.875

The system runs on a regular laptop (Intel Core AMD Ryzen 7 6800H CPU running at 3.2 GHz, and 16 GB of RAM) and MATLAB is used for image processing. The average processing time of the algorithm is 32 ms, which corresponds to 31.55 FPS.

Table 3 presents a comparison of this proposal with the other works. The accuracy of our work is expressed by the average precision. At present, there is less related work to recognize the face mask based on the depth image; most works are based on the RGB images. However, compared with the method based on the RGB image, the depth image captured from the low-cost ToF camera can be less affected by the environment light. Therefore, our accuracy has certain advantages compared with those works. The work [14] considered the RGB data under dark condition; they introduced additional resources to train the network, and the improved accuracy rate in the dark condition was 77.9%. Compared with this work, our proposed method is unaffected by light and the average accuracy can reach 96.9%. In addition, we have also classified the mask types, which are not involved in some works. It can be seen from Jiang et al. [12] that the introduction of more classification situations will affect the accuracy of the method. In the case of sufficient lighting and good image quality, our ToF-camera-based method has close accuracy with the other works of mask recognition based on the RGB images. Most RGB-camera-based works with high identification accuracy are based on mature deep-learning methods. However, compared with most network-based works, our method is more interpretable and with lower computational costs.

Table 3. Comparison of this proposal with other works.

Work	Method	Data	Distinguished Type	Accuracy	Efficiency
Ours	Feature-based	Depth	With/Without Mask Type	96.9% 87.85%	31.55 FPS
Cao et al. [14]	YOLOv4-large	RGB	With/Without Night Time	94% 77.9%	18 FPS
Nagrath et al. [10]	SSDMNV2	RGB	With/Without	92.64%	15.71 FPS
Jiang et al. [12]	SE-YOLOv3	RGB	With/Without/ Correct Wearing	73.7%	15.63 FPS
Walia et al. [7]	YOLOv5	RGB	With/Without	98%	32 FPS
Su et al. [11]	Transfer Learning and Effcient-Yolov3	RGB	With/Without Mask Type	96.03% 97.84%	15 FPS
Yu et al. [8]	YOLO-v4	RGB	With/Without	98.3%	54.57 FPS

4. Conclusions and Discussion

We propose a method to identify whether a person is wearing mask, which can further classify the mask types, such as the surgical mask and the N95 mask. A ToF camera with a simple system and robust data is used to capture the depth image, especially in dark conditions. The designed method can extract facial contours accurately by using the depth information. Facial contour extraction is essentially the sampling and dimensionality reduction of the depth data, which can reduce the amount of data and improve the recognition speed of the algorithm. Additionally, the spatial and frequency features of the facial contour are further extracted. With the appropriate thresholds and the voting method, it can achieve a 99.21% recall accuracy for images without a mask, and a 94.6% recall accuracy for images with a mask. Through reducing the dimension of the data, our method can reach satisfactory accuracy while quickly identifying.

However, there is certain confusion between the two mask types, and the soft material of the mask is the main reason. Therefore, single-depth information is insufficient. In the future, we hope to combine the depth distance with other information to build a better model. Furthermore, people have different habits of wearing masks. The classification category also needs to be further developed, such as the classification of incorrectly worn masks, the recognizable mask types, and the application scenarios. In practical applications, there will be more complex challenges, such as people's movement and a complex crowd environment. When an angle formed between the face and the ToF camera, some depth information will be lost for the occluded part of the face. When the angle between the person and the camera is less than 90°, the facial contour curve can be detected by raw-data rotation, which is also our next research focus. Additionally, the work on larger rotation angles and different directions is also under further study. In addition, the data of multiple people and in motion are also being further considered.

Author Contributions: Conceptualization, Y.Y.; methodology, X.W., L.S. and Y.Y.; software, X.W. and D.A.; validation, X.W., T.X. and D.A.; formal analysis, X.W., L.S. and Y.Y.; investigation, X.W., T.X., D.A. and Y.Y.; resources, X.W. and Y.Y.; data curation, X.W. and D.A.; writing—original draft preparation, X.W.; writing—review and editing, T.X., L.S., Q.W., Z.P. and Y.Y.; visualization, X.W. and T.X.; supervision, Y.Y.; project administration, Y.Y.; funding acquisition, Y.Y. All authors have read and agreed to the published version of the manuscript.

Funding: The National Key Research and Development Program of China (2019YFB1803700); the Key Technologies R&D Program of Tianjin (20YFZCGX00440); and the Shaanxi Key Laboratory of Deep Space Exploration Intelligent Information Technology (2021SYS-04).

Institutional Review Board Statement: Not applicable.

Informed Consent Statement: Not applicable.

Data Availability Statement: Data underlying the results presented in this paper are not publicly available at this time but may be obtained from the authors upon reasonable request.

Conflicts of Interest: The authors declare no conflict of interest.

References

1. Cheng, Y.; Ma, N.; Witt, C.; Rapp, S.; Wild, P.S.; Andreae, M.O.; Pöschl, U.; Su, H. Face Masks Effectively Limit the Probability of SARS-CoV-2 Transmission. *Science* **2021**, *372*, 1439–1443. [CrossRef] [PubMed]
2. Mbunge, E.; Chitungo, I.; Dzinamarira, T. Unbundling the Significance of Cognitive Robots and Drones Deployed to Tackle COVID-19 Pandemic: A Rapid Review to Unpack Emerging Opportunities to Improve Healthcare in Sub-Saharan Africa. *Cogn. Robot.* **2021**, *1*, 205–213. [CrossRef]
3. Goar, V.; Sharma, A.; Yadav, N.S.; Chowdhury, S.; Hu, Y.-C. IoT-Based Smart Mask Protection against the Waves of COVID-19. *J. Ambient Intell. Hum. Comput.* **2022**, 1–12. [CrossRef] [PubMed]
4. Rahim, M.S.A.; Yakub, F.; Hanapiah, A.R.M.; Rashid, M.Z.A. Development of Low-Cost Thermal Scanner and Mask Detection for Covid-19. In Proceedings of the 2021 60th Annual Conference of the Society of Instrument and Control Engineers of Japan (SICE), Tokyo, Japan, 8–10 September 2021.
5. Hussain, S.; Yu, Y.; Ayoub, M.; Khan, A.; Rehman, R.; Wahid, J.A.; Hou, W. IoT and Deep Learning Based Approach for Rapid Screening and Face Mask Detection for Infection Spread Control of COVID-19. *Appl. Sci.* **2021**, *11*, 3495. [CrossRef]
6. Zhao, Z.-Q.; Zheng, P.; Xu, S.; Wu, X. Object Detection with Deep Learning: A Review 2019. *IEEE Trans. Neural Netw. Learn. Syst.* **2019**, *30*, 3212–3232. [CrossRef]
7. Walia, I.S.; Kumar, D.; Sharma, K.; Hemanth, J.D.; Popescu, D.E. An Integrated Approach for Monitoring Social Distancing and Face Mask Detection Using Stacked ResNet-50 and YOLOv5. *Electronics* **2021**, *10*, 2996. [CrossRef]
8. Yu, J.; Zhang, W. Face Mask Wearing Detection Algorithm Based on Improved YOLO-V4. *Sensors* **2021**, *21*, 3263. [CrossRef]
9. Talahua, J.S.; Buele, J.; Calvopiña, P.; Varela-Aldás, J. Facial Recognition System for People with and without Face Mask in Times of the COVID-19 Pandemic. *Sustainability* **2021**, *13*, 6900. [CrossRef]
10. Nagrath, P.; Jain, R.; Madan, A.; Arora, R.; Kataria, P.; Hemanth, J. SSDMNV2: A Real Time DNN-Based Face Mask Detection System Using Single Shot Multibox Detector and MobileNetV2. *Sustain. Cities Soc.* **2021**, *66*, 102692. [CrossRef]
11. Su, X.; Gao, M.; Ren, J.; Li, Y.; Dong, M.; Liu, X. Face Mask Detection and Classification via Deep Transfer Learning. *Multimed. Tools Appl.* **2022**, *81*, 4475–4494. [CrossRef]

12. Jiang, X.; Gao, T.; Zhu, Z.; Zhao, Y. Real-Time Face Mask Detection Method Based on YOLOv3. *Electronics* **2021**, *10*, 837. [CrossRef]
13. Kumar, A.; Kaur, A.; Kumar, M. Face Detection Techniques: A Review. *Artif. Intell. Rev.* **2019**, *52*, 927–948. [CrossRef]
14. Cao, Z.; Shao, M.; Xu, L.; Mu, S.; Qu, H. MaskHunter: Real-time Object Detection of Face Masks during the COVID-19 Pandemic. *IET Image Process* **2020**, *14*, 4359–4367. [CrossRef]
15. Daneshmand, M.; Helmi, A.; Avots, E.; Noroozi, F.; Alisinanoglu, F.; Arslan, H.S.; Gorbova, J.; Haamer, R.E.; Ozcinar, C.; Anbarjafari, G. 3D Scanning: A Comprehensive Survey. *arXiv* **2018**, arXiv:1801.08863.
16. Sansoni, G.; Trebeschi, M.; Docchio, F. State-of-The-Art and Applications of 3D Imaging Sensors in Industry, Cultural Heritage, Medicine, and Criminal Investigation. *Sensors* **2009**, *9*, 568–601. [CrossRef]
17. Zhang, S. High-Speed 3D Shape Measurement with Structured Light Methods: A Review. *Opt. Lasers Eng.* **2018**, *106*, 119–131. [CrossRef]
18. He, Y.; Chen, S. Recent Advances in 3D Data Acquisition and Processing by Time-of-Flight Camera. *IEEE Access* **2019**, *7*, 12495–12510. [CrossRef]
19. Cippitelli, E.; Fioranelli, F.; Gambi, E.; Spinsante, S. Radar and RGB-Depth Sensors for Fall Detection: A Review. *IEEE Sensors J.* **2017**, *17*, 3585–3604. [CrossRef]
20. Zanuttigh, P.; Marin, G.; Dal, M.C.; Dominio, F.; Minto, L.; Cortelazzo, G.M. *Time-of-Flight and Structured Light Depth Cameras: Technology and Applications*, 1st ed.; Springer: Cham, Switzerland, 2016; pp. 199–350.
21. Horaud, R.; Hansard, M.; Evangelidis, G.; Ménier, C. An Overview of Depth Cameras and Range Scanners Based on Time-of-Flight Technologies. *Mach. Vis. Appl.* **2016**, *27*, 1005–1020. [CrossRef]
22. Aggarwal, J.K.; Xia, L. Human Activity Recognition from 3D Data: A Review. *Pattern Recognit. Lett.* **2014**, *48*, 70–80. [CrossRef]
23. Xu, T.; An, D.; Wang, Z.; Jiang, S.; Meng, C.; Zhang, Y.; Wang, Q.; Pan, Z.; Yue, Y. 3D Joints Estimation of the Human Body in Single-Frame Point Cloud. *IEEE Access* **2020**, *8*, 178900–178908. [CrossRef]
24. Bae, J.-H.; Jo, H.; Kim, D.-W.; Song, J.-B. Grasping System for Industrial Application Using Point Cloud-Based Clustering. In Proceedings of the 2020 20th International Conference on Control, Automation and Systems (ICCAS), Busan, Republic of Korea, 13 October 2020.
25. Zengeler, N.; Kopinski, T.; Handmann, U. Hand Gesture Recognition in Automotive Human–Machine Interaction Using Depth Cameras. *Sensors* **2018**, *19*, 59. [CrossRef] [PubMed]
26. Gai, J.; Xiang, L.; Tang, L. Using a Depth Camera for Crop Row Detection and Mapping for Under-Canopy Navigation of Agricultural Robotic Vehicle. *Comput. Electron. Agric.* **2021**, *188*, 106301. [CrossRef]
27. Eric, N.; Jang, J.-W. Kinect Depth Sensor for Computer Vision Applications in Autonomous Vehicles. In Proceedings of the 2017 Ninth International Conference on Ubiquitous and Future Networks (ICUFN), Milan, Italy, 4–7 July 2017.
28. Procházka, A.; Schätz, M.; Vyšata, O.; Vališ, M. Microsoft Kinect Visual and Depth Sensors for Breathing and Heart Rate Analysis. *Sensors* **2016**, *16*, 996. [CrossRef] [PubMed]
29. Luna, C.A.; Losada-Gutiérrez, C.; Fuentes-Jiménez, D.; Mazo, M. Fast Heuristic Method to Detect People in Frontal Depth Images. *Expert Syst. Appl.* **2021**, *168*, 114483. [CrossRef]
30. Das, S.; Sarkar, S.; Das, A.; Das, S.; Chakraborty, P.; Sarkar, J. A Comprehensive Review of Various Categories of Face Masks Resistant to Covid-19. *Clin. Epidemiol. Glob. Health* **2021**, *12*, 100835. [CrossRef]

sensors

MDPI

Review

Revolutionizing Free-Space Optics: A Survey of Enabling Technologies, Challenges, Trends, and Prospects of Beyond 5G Free-Space Optical (FSO) Communication Systems

Isiaka A. Alimi [1,*] **and Paulo P. Monteiro** [1,2]

[1] Instituto de Telecomunicações, University of Aveiro, 3810-193 Aveiro, Portugal; paulo.monteiro@ua.pt
[2] Department of Electronics, Telecommunications and Informatics, University of Aveiro, 3810-193 Aveiro, Portugal
* Correspondence: iaalimi@ua.pt; Tel.: +35-123-437-7900; Fax: +35-123-437-7901

Abstract: As the demand for high-speed, low-latency communication continues to grow, free-space optical (FSO) communication has gained prominence as a promising solution for supporting the next generation of wireless networks, especially in the context of the 5G and beyond era. It offers high-speed, low-latency data transmission over long distances without the need for a physical infrastructure. However, the deployment of FSO systems faces significant challenges, such as atmospheric turbulence, weather-induced signal degradation, and alignment issues, all of which can impair performance. This paper offers a comprehensive survey of the enabling technologies, challenges, trends, and future prospects for FSO communication in next-generation networks, while also providing insights into the current mitigation strategies. The survey explores the critical enabling technologies such as adaptive optics, modulation schemes, and error correction codes that are revolutionizing FSO communication and addressing the unique challenges of FSO links. Also, the integration of FSO with radio frequency, millimeter-wave, and Terahertz technologies is explored, emphasizing hybrid solutions that enhance reliability and coverage. Additionally, the paper highlights emerging trends, such as the integration of FSO with artificial intelligence-driven optimization techniques and the growing role of machine learning in enhancing FSO system performance for dynamic environments. By analyzing the current trends and identifying key challenges, this paper emphasizes the prospects of FSO communication in the evolving landscape of 5G and future networks. In this regard, it assesses the potential of FSO to meet the demands for high-speed, low-latency communication and offers insights into its scalability, reliability, and deployment strategies for 5G and beyond. The paper concludes by identifying the open challenges and future research directions critical to realizing the full potential of FSO in next-generation communication systems.

Keywords: 5G; 6G; adaptive optics; advanced modulation schemes; aperture averaging; artificial intelligence; atmospheric turbulence; cooperative relaying; dual-hop transmission schemes; error correction codes; free-space optical communications; hybrid FSO system; machine learning; millimeter wave; pointing errors; spatial diversity; terahertz communications

1. Introduction

Free-space optical (FSO) communication systems are gaining significant attention due to their potential to offer high data rates, secure communication, and immunity to electromagnetic interference. Unlike traditional radio frequency (RF) systems, FSO utilizes light to transmit data through the atmosphere, making it an attractive solution for a variety of applications. The growing demand for higher bandwidth and the limitations of existing wireless communication technologies drive the exploration and development of FSO systems [1,2]. This section provides an overview of the fundamental principles, key advantages, and challenges associated with FSO communication, setting the stage for a deeper understanding of its potential and the ongoing research in this field.

1.1. Background and Motivation

With the growing demand for high-speed, low-latency communication in the era of 5G and beyond, and the huge capital expenditures associated with public works (such as trench digging and site restoration) limiting the widespread installation of optical fiber infrastructure, FSO communication has emerged as a potent technology poised to meet these demands [3]. FSO communication, which involves the transmission of data through the atmosphere using optical beam, offers an attractive alternative to traditional RF communication due to its potential for low latency, noise immunity, electromagnetic interference immunity, spatial reusability, ease of deployment, being license free, high security, reduced spectrum congestion, and the capability to deliver transmissions at extremely high data rates between two terminals [4,5], whether they are just a few meters apart or separated by thousands of kilometers (long operational range) [6]. FSO systems are especially advantageous for line-of-sight (LoS) communication, bridging gaps where traditional fiber optics or RF systems may not be feasible. For instance, FSO systems utilize narrow laser beams with directional properties. This spatial confinement allows for a high degree of spatial reuse and makes eavesdropping difficult, inherently enhancing security. Utilizing light as the carrier in FSO communications ensures immunity to electromagnetic interference. Additionally, FSO systems are easy to deploy and can be reinstalled without incurring the costs associated with dedicated fiber optic connections. Moreover, the frequency used by FSO technology exceeds 300 GHz, which is unlicensed globally, thus eliminating the need for license fees [1,2,7].

Frequencies in FSO communications are significantly higher than those in RF communications. Consequently, FSO communications can achieve much higher data rates compared to RF communications, and they require antennas that take up less space. Additionally, the coherence of laser light in FSO links can decrease geometrical loss, allowing the transmission of high data rates over long distances [7]. Figure 1 illustrates various FSO applications along with their respective communication ranges. The capability to transmit data over long distances without the need for physical infrastructure makes FSO a suitable candidate for supporting next-generation wireless networks, including 5G and beyond [8]. In this context, the absence of a need for optical fibers between communication nodes allows for flexible application across various use cases and high-speed applications including unmanned aerial vehicles (UAVs), intra- and inter-datacenter connections, inter-building links, high-speed trains, 5G fronthaul and backhaul for wireless cellular networks, indoor and outdoor local- and wide-area networks, metropolitan area network extensions, urban connectivity, last-mile connectivity, inter-chip connectivity, fiber backup, satellite communications, deep-space communications, underwater communication, disaster recovery, video surveillance, and redundant link [2,9,10]. Additionally, it is regarded as a promising solution for addressing the significant traffic demands of sixth-generation (6G) systems [11,12]. These use cases can be classified into two main categories, terrestrial and outer space FSO communications, as illustrated in Figure 2. Given its benefits, FSO is being explored as a solution for last-mile connectivity in situations where fiber deployment is challenging and RF systems fall short of meeting bandwidth needs. Additionally, these advantages position FSO as a highly promising solution for space communications. In this context, by leveraging its strengths, FSO can address many of the challenges associated with traditional communication methods, providing a more efficient and reliable means of data transmission between spacecraft, satellites, and ground stations [13].

Although FSO technology offers undeniable advantages and has a broad range of potential applications, significant technical challenges impede its widespread adoption [14]. For terrestrial applications, atmospheric conditions such as rain, fog, and turbulence can significantly degrade the signal, leading to reduced performance. Furthermore, for general applications, alignment issues and the need for precise beam pointing add to the complexity of FSO system deployment. Addressing these challenges requires the development of enabling technologies and innovative strategies to ensure the reliable performance of FSO systems in diverse environments [2,15].

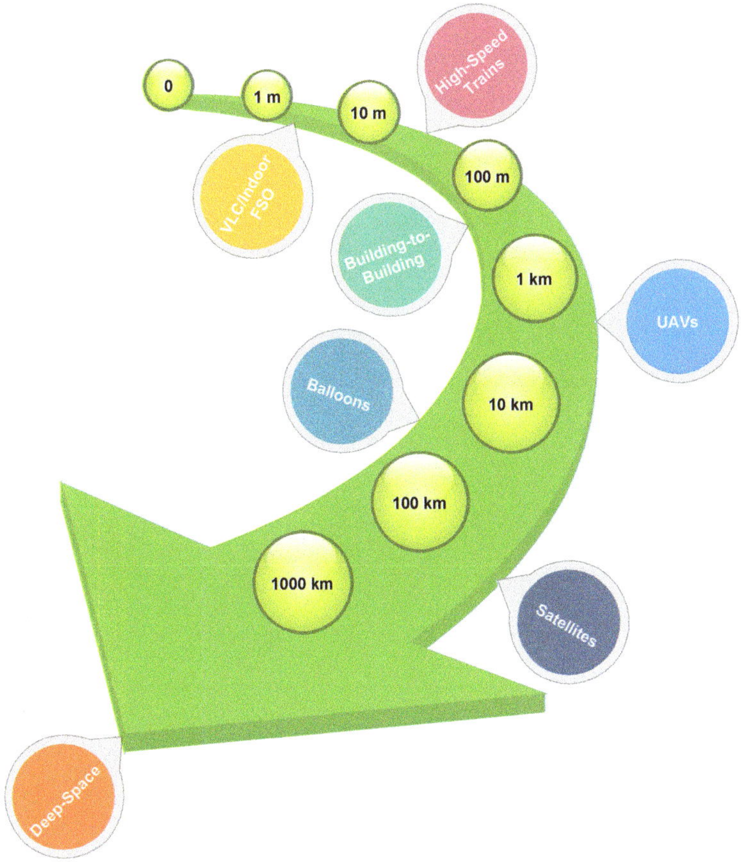

Figure 1. Potential applications of FSO systems based on the operational distance between transceivers.

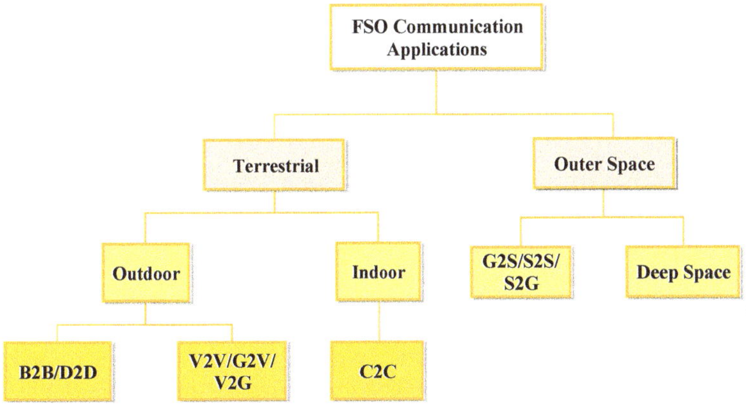

Figure 2. Categorization of FSO communication systems based on their applications. B2B: Building-to-Building, D2D: Device-to-Device, C2C: Chip-to-Chip, V2V: Vehicle-to-Vehicle, V2G: Vehicle-to-Ground, G2V: Ground-to-Vehicle, S2S: Satellite-to-Satellite, S2G: Satellite-to-Ground, and G2S: Ground-to-Satellite.

1.2. Objectives and Scope of the Paper

This paper aims to provide a comprehensive survey of the enabling technologies, challenges, trends, and future prospects of FSO communication in the context of 5G and beyond. The key objectives include the following:

(i) Analysis of Current Challenges: examination of the major challenges faced by FSO systems, including atmospheric turbulence (AT), weather-induced signal degradation, and alignment issues.
(ii) Review of Enabling Technologies: exploration of critical technologies such as adaptive optics (AO), advanced modulation schemes, aperture averaging, spatial diversity, cooperative relaying, hybrid FSO system, and error correction codes (ECCs) that are instrumental in overcoming the challenges associated with FSO communication.
(iii) Exploration of Emerging Trends: discussion of emerging trends, such as the integration of FSO with RF, millimeter-wave (mmWave), and Terahertz (THz) technologies, as well as the use of artificial intelligence (AI) and machine learning (ML) to optimize FSO performance.
(iv) Prospects for 5G and Beyond: assessment of the potential of FSO communication to meet the demands of next-generation networks, including its scalability, reliability, and deployment strategies.
(v) Future Research Directions: identification of open challenges and future research directions necessary to realize the full potential of FSO in the evolving landscape of communication technologies.

The aforementioned objectives highlight the involvement of numerous phenomena in FSO communication, which must be meticulously analyzed to ensure a reliable communication system.

1.3. Contributions

There have been considerable contributions in the literature to FSO communications. Nevertheless, the usual space constraints for publications prevented comprehensive research work in which assiduous efforts towards FSO communications development can be presented. Therefore, an all-inclusive survey in which the trends in FSO communication systems can be studied is highly imperative. Based on this, apart from being a good complement to contributions of the existing work, this paper also offers in-depth clarifications regarding different related concepts and architectures. In this regard, the main contributions of this tutorial paper are as follows:

(i) Comprehensive Survey: The paper provides an extensive review of the enabling technologies critical to advancing FSO communication systems, including AO, modulation schemes, and error correction codes. This thorough examination aids in understanding how these technologies address the unique challenges of FSO links.
(ii) Identification of Challenges: It identifies and discusses significant challenges that hinder the deployment of FSO systems, such as atmospheric turbulence, weather-induced signal degradation, and alignment issues. This highlights the need for targeted solutions and informs future research directions.
(iii) Integration of Hybrid Solutions: The paper explores the integration of FSO with radio frequency, millimeter-wave, and Terahertz technologies, emphasizing how hybrid approaches can enhance reliability and coverage in next-generation networks.
(iv) Emerging Trends: The research highlights emerging trends in FSO communication, particularly the integration of artificial intelligence and machine learning techniques. This focus on AI-driven optimization provides insights into how these technologies can enhance the performance of FSO systems in dynamic environments.
(v) Assessment of Future Prospects: The paper evaluates the potential of FSO to meet the increasing demands for high-speed, low-latency communication in the context of 5G and beyond. It provides a forward-looking perspective on scalabil-

ity, reliability, and deployment strategies that can support the evolution of FSO communication systems.

(vi) Research Directions: Finally, the paper identifies open challenges and outlines future research directions critical to maximizing the effectiveness of FSO technology. This contributes to the broader understanding of the role FSO can play in the future of wireless communication systems.

1.4. Article Structure

The paper is organized into sections to provide a comprehensive overview of FSO communication systems in the context of 5G and beyond. Having introduced the background and motivation along with the objectives and scope of the paper, Section 2 delves into the fundamentals of FSO communication systems, detailing their components and performance metrics. Section 3 addresses the various challenges that FSO systems face, including atmospheric turbulence, weather-induced signal degradation, and alignment issues. Section 4 explores the enabling technologies crucial for advancing FSO communication, such as AO, modulation schemes, and error correction codes, as well as the integration of hybrid systems. Section 5 highlights emerging trends, including the use of artificial intelligence and machine learning to optimize system performance. Section 6 discusses the open challenges and future prospects, providing insights into potential research directions and long-term applications. Finally, Section 7 concludes the paper by summarizing the key findings and emphasizing the importance of continued innovation in the field. The structure and organization of the manuscript adhere to the framework illustrated in Figure 3. Each section is meticulously arranged to follow the sequence and logic depicted in the figure, ensuring a coherent and comprehensive presentation of the content.

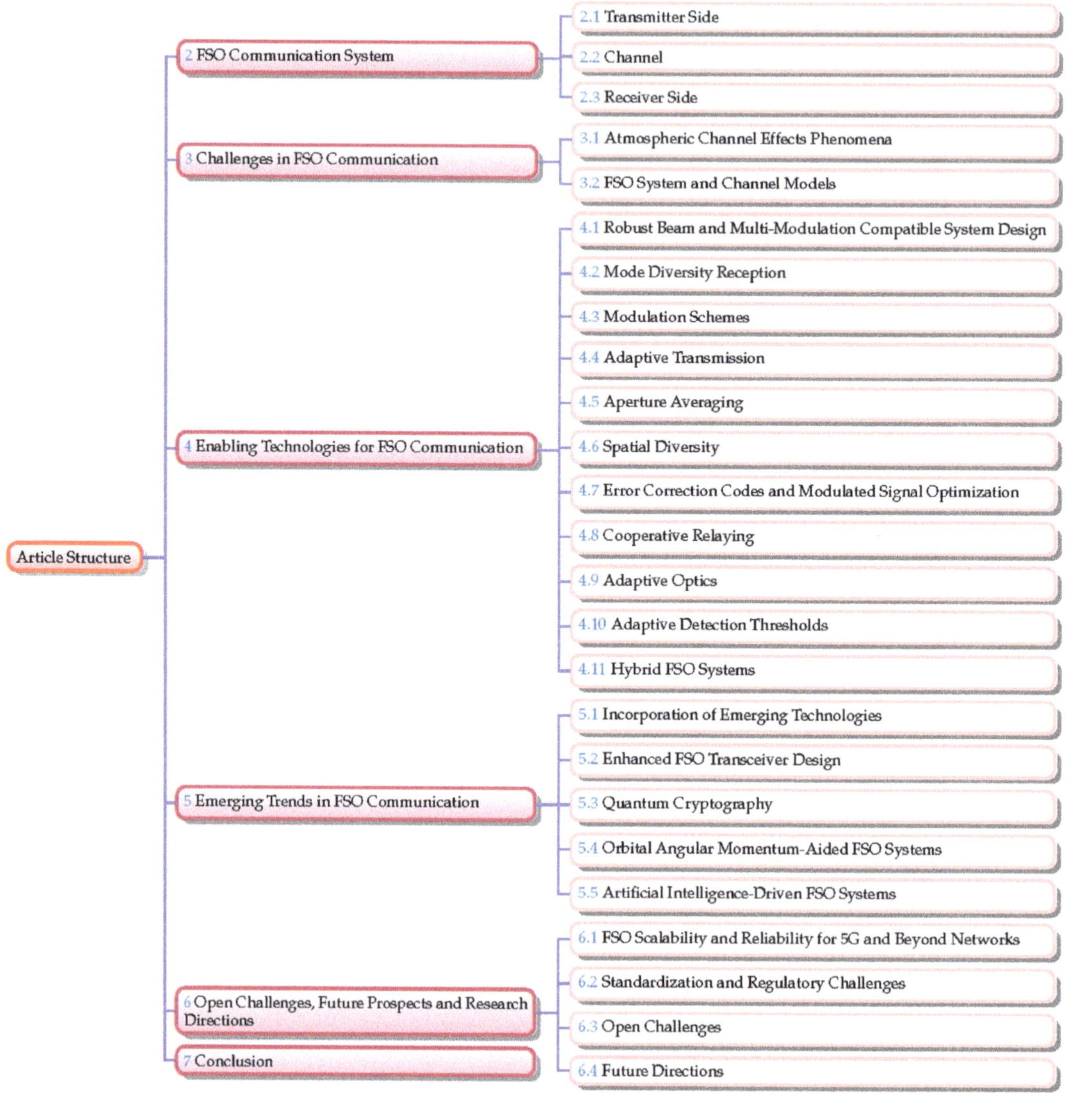

Figure 3. Outline of the article structure.

2. FSO Communication System

FSO communication systems are a vital part of modern telecommunication technologies, offering a LoS data transmission method through the atmosphere [16] and space telecommunications [17]. As illustrated in Figure 4, an FSO system primarily consists of three fundamental components: the transmitter (Tx-source), the channel, and the receiver (Rx-destination). Each of these components plays a critical role in ensuring the successful transmission of optical signals across free space. For instance, information waveforms are generated by a source and then modulated onto an optical carrier. This optical signal is transmitted through the atmosphere, or free-space, towards a remote destination. Along the way, the optical power emitted by the Tx is influenced by various factors before it reaches the Rx. As a result, the information received might not be an exact copy of the originally transmitted information. At the Rx, the optical signal is collected and converted into an electrical current by a photodetector. The Rx then processes this electrical current to retrieve the originally transmitted information [1,3]. So, the FSO communication system's efficiency and performance hinge on the seamless integration and alignment of these three components. The Tx initiates the communication by sending out the optical signal, the channel serves as the medium through which the signal travels, and the Rx completes the process by decoding the transmitted information at the destination. This section outlines the various processes involved in transmitting signals from the source to the destination in an FSO communication system.

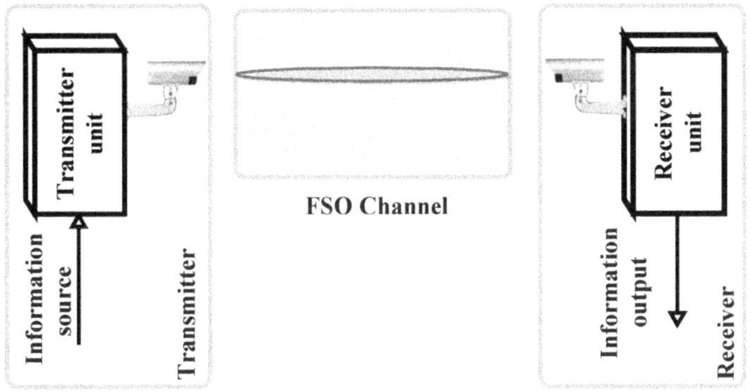

Figure 4. A typical block diagram of an FSO system.

2.1. Transmitter Side

The Tx is responsible for generating and encoding the optical signal that carries the data. It typically comprises a light source, such as a laser diode (LD) or Light-Emitting Diode (LED), which converts electrical data signals into optical signals. The Tx also includes optics to focus and direct the light beam towards the Rx. The FSO Tx operates by encoding and modulating data, amplifying the modulated signal if necessary, and then precisely directing the laser beam through beam-forming optics for transmission across the optical channel.

Figure 5 provides a depiction of a typical FSO Tx. This system comprises several essential components: an optical source, a modulator (in case of external modulation), an optical amplifier (used as needed), and beam-forming optics. Initially, the data bits generated from the information source undergo channel coding, an optional step performed before the modulation process. This coding step enhances the reliability and performance of the communication by adding redundancy to the data. Following this, the encoded data are modulated onto an optical carrier generated by the optical source. This modulated signal is a laser beam that carries the information through the optical channel. To ensure the signal's strength and maintain the quality of transmission over long distances or through adverse conditions, the modulated laser beam may be amplified using an optical amplifier. This amplification process boosts the optical intensity of the beam, making it more robust against attenuation and interference. Once the laser beam is sufficiently amplified, it is then passed through the beam-forming optics. These optics play a critical role in collecting the amplified light and refocusing it into a precise and well-collimated beam. This process ensures that the light beam maintains its integrity and alignment as it travels through free space to reach the Rx [1,3,18].

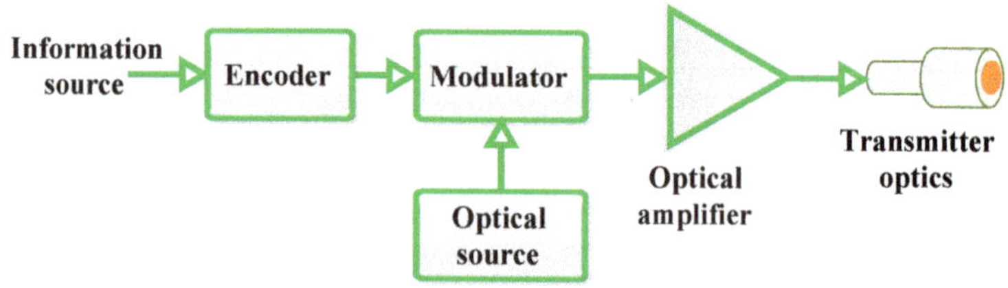

Figure 5. A typical block diagram of an FSO Tx.

2.1.1. Transmission Wavelength

FSO systems are predominantly designed to function in the near-infrared (near-IR) spectrum, specifically within the 750 to 1600 nm wavelength range. While the clear atmosphere is generally highly transparent to near-IR wavelengths, certain atmospheric molecules can cause significant absorption at specific wavelengths within this range. Also, there are particular wavelength windows around 850 nm, 1060 nm, 1250 nm, and 1550 nm, where the atmospheric attenuation is notably low, measuring less than 0.2 dB/km. Notably, the 850 nm and 1550 nm wavelengths align with the standard transmission windows used in fiber optic communication systems [16]. This compatibility allows FSO systems to leverage existing fiber optic components, which are readily available on the market, thereby reducing costs and improving system integration [1,3,18–20].

In addition to near-IR wavelengths, recent research has explored other spectral regions for FSO applications. Wavelengths around 10 μm have been identified for their superior performance in foggy conditions, offering better transmission characteristics. Ultraviolet (UV) wavelengths have also been considered due to their robustness against pointing errors (PEs) and beam blockages. Furthermore, UV wavelengths exhibit lower sensitivity to solar and other background interferences, making them a promising alternative for certain FSO applications [1,21,22]. A detailed discussion of these promising wavelengths is provided in Section 5.2.

2.1.2. Transmission Safety

In FSO communication systems, the primary optical source is typically a semiconductor LD. However, some manufacturers opt for high-power LEDs equipped with beam collimators as an alternative. The chosen optical source must be capable of delivering relatively high optical power across a wide range of temperatures. Additionally, it should feature a long mean time between failures (MTBF), ensuring reliable performance over extended periods. The components associated with the optical source should also have a small physical footprint and consume minimal power [1,3,18]. To meet these stringent requirements, Vertical-Cavity Surface-Emitting Lasers (VCSELs) are predominantly used for operations around the 850 nm wavelength. For operations at the 1550 nm wavelength, Fabry–Pérot (FP) and Distributed Feedback (DFB) lasers are commonly employed. These types of lasers are selected for their efficiency, reliability, and suitability for the specified wavelengths, ensuring optimal performance of the FSO systems [1,18].

Safety considerations are critical when dealing with laser Txs, with the primary concern being the potential risk to the eyes from laser exposure. To mitigate these risks, various safety standards have been established that regulate the transmitted optical power of lasers. These standards take into account factors such as the intensity of the laser beam, its wavelength, proximity to the laser source, peak power in a pulse, and the average power over extended periods [1,3,18,23].

Lasers emit light at different wavelengths, and only certain wavelengths, particularly those in the near-IR (between approximately 400 nm and 1550 nm), have the potential to penetrate the eye with enough energy to harm the retina. Light outside this range is generally absorbed by the cornea and lens before reaching the retina, making it less likely to cause damage. The eye's tolerance to laser power varies with wavelength, primarily due to the differential absorption of light by water, which is a major component of the eye. Specifically, the absorption coefficient at the front of the eye is significantly higher for longer wavelengths (greater than 1400 nm) compared to shorter wavelengths. Consequently, lasers operating at 1550 nm can have a higher allowable transmission power, which makes them suitable for long-distance applications [1,3,18,19,23].

Additionally, the risk associated with laser exposure differs between high-power pulses and low-power continuous beams. A single high-power laser pulse, even if it lasts less than a microsecond, can cause irreversible damage if it directly enters the eye. In contrast, lower-power continuous beams are hazardous primarily with prolonged exposure. Furthermore, the risk of eye damage diminishes with increased distance from the laser source, as the power density of the laser beam decreases with distance [23]. So, when selecting the operating wavelength for FSO, it is crucial to consider eye-safety regulations and standards. This means that the chosen wavelength must ensure that the optical signals used in the FSO system do not pose a risk to human eyes, either through direct exposure or accidental reflections. Compliance with eye-safety requirements helps prevent potential health hazards and ensures the safe deployment and operation of the FSO system in various environments [21].

2.2. Channel

In the FSO communication system, the channel is defined as the free space or atmospheric pathway through which the optical signal is transmitted. Unlike conventional fiber optic systems that use physical cables to guide the light signals, FSO systems depend on a direct LoS between the transmitting and receiving devices. This LoS requirement makes the FSO channel highly susceptible to various environmental influences.

The quality and reliability of the FSO channel can be significantly impacted by several external factors. For terrestrial applications, weather conditions such as fog, rain, and snow can cause attenuation of the signal, reducing its strength as it travels through the atmosphere. Additionally, AT, which includes variations in air density and temperature, can cause the light beam to scatter, further degrading signal quality. Physical obstacles that obstruct the LoS can also interrupt the signal transmission. These environmental factors pose considerable challenges to the stability and performance of FSO communication systems. A comprehensive discussion on these challenges, including detailed analyses of how different conditions affect the FSO channel, is provided in Section 3. The section elaborates on the mechanisms of signal degradation and the strategies employed to mitigate these adverse effects, ensuring more reliable FSO communication.

2.3. Receiver Side

The role of the Rx is to capture the transmitted optical signal and convert it back into an electrical signal. It typically includes a photodetector, which senses the incoming light, and various electronic components to amplify and process the received signal. The Rx must be aligned precisely with the Tx to ensure accurate signal detection and minimize data loss. FSO systems are generally classified into two main types based on their detection mechanisms: non-coherent and coherent systems. Although coherent systems typically offer enhanced performance by effectively rejecting background noise, reducing the effects of AT, and achieving higher sensitivity at the Rx, Intensity Modulation with Direct Detection (IM/DD) systems are more prevalent in terrestrial FSO applications. This is due to their simpler design and lower cost. This section will delve into the details of both IM/DD and coherent FSO systems. Also, alternative solutions to conventional coherent Rxs are presented, involving simpler transmission schemes based on direct detection (DD) systems.

2.3.1. Non-Coherent FSO Systems

Non-coherent FSO systems, also known as IM/DD systems, rely on the intensity of light to transmit information. At the Rx, changes in light intensity are directly detected by a photodetector without the need for a local oscillator (LO) to mix the optical fields. The Rx front-end in IM/DD FSO systems, depicted in Figure 6b, comprises optical filters and a lens. The primary function of the lens is to gather and focus the incoming light beam onto the photodiode (PD). Commercial FSO systems predominantly use solid-state PDs made from materials such as Indium Gallium Arsenide (InGaAs), Silicon (Si), or Germanium (Ge). These materials are chosen due to their high quantum efficiency and their sensitivity to the commonly utilized wavelengths in FSO systems. Furthermore, these PDs possess very short transit times, resulting in high bandwidth and fast response times, making them ideal for FSO applications [1,18].

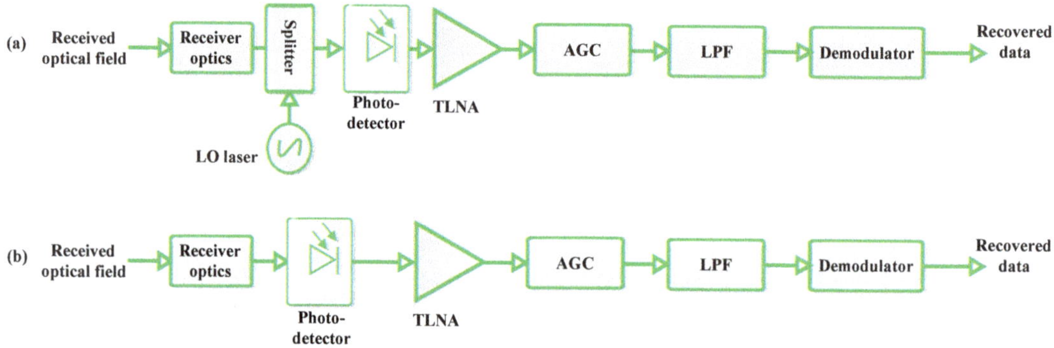

Figure 6. Block diagram of (**a**) coherent and (**b**) IM/DD FSO Rxs. TLNA: Trans-Impedance Low-Noise Amplifier, AGC: Automatic Gain Control, LPF: Low-Pass Filter.

Solid-state PDs can either be Positive–Intrinsic–Negative (PIN) diodes or avalanche PDs (APDs), each offering distinct advantages and disadvantages. PIN diodes are known for their simple structure and reliability, while APDs provide internal gain, enhancing their sensitivity but also introducing higher noise and requiring higher operating voltages. The current output from the PD is converted into a voltage using a Trans-Impedance Amplifier (TIA), typically a low-noise Operational Amplifier (Op-Amp) with a load resistor. The selection of this load resistor is crucial and depends on factors such as the transmission rate, the dynamic range of the converted electrical signal, the thermal noise generated by the Rx, and the need to match impedance with other components in the Rx. Finally, the voltage signal from the TIA is passed through a Low-Pass Filter to reduce thermal and background noise, ensuring a cleaner signal for further processing [1].

2.3.2. Coherent FSO Systems

Commercial FSO communication systems traditionally use IM/DD techniques due to their lower implementation complexity and cost effectiveness [24]. However, there is a growing shift towards coherent FSO systems. This trend is driven by recent advancements in the fabrication of integrated coherent Rxs and high-speed digital signal processing (DSP) integrated circuits, which have significantly enhanced the practicality of coherent detection for wired and wireless communications. Fortunately, the suitable bands for FSO transmission coincide with the wavelengths used in fiber optics, specifically the C band, which lowers the cost of optical and optoelectronic components due to mass production [25–27].

In coherent FSO systems, information is encoded using modulation techniques such as amplitude and phase modulation. When the signal reaches the Rx, it is optically combined with a locally generated reference signal (LO, optical oscillator, usually a laser with the same wavelength of the received optical signal). This mixing process is crucial for detecting and decoding the information carried by the signal [25–27].

Coherent detection in FSO systems retrieves the complete electric field of a signal by measuring both of its field quadratures (i.e., its magnitude and phase) in two polarizations.

This approach maximizes the degrees of freedom (DoF) available in the optical channel for carrying information. At the Rx, the incoming signal beam is combined with a LO beam, which provides a phase reference for downconversion as illustrated in Figure 6a. This mixing process with the LO amplifies the received signal, making the detection process primarily limited by shot noise. As a result, coherent detection is often viewed as a method for enhancing Rx sensitivity in FSO systems. Furthermore, coherent detection is effective in rejecting background noise, thermal noise of the TIA and intentional interferences [28].

An intriguing aspect of coherent systems is the ability to send information via the phase, amplitude, or polarization of the optical field, significantly increasing the system's spectral efficiency (SE). Typical modulation schemes in coherent systems include multilevel Phase-Shift Keying (PSK), Quadrature Amplitude Modulation (QAM), and multilevel Polarization Shift Keying (POLSK). These advanced modulation techniques contribute to the robust and efficient performance of coherent FSO communication systems [1,27].

In the realm of coherent Rxs, there exist two primary methodologies for signal detection: homodyne detection and heterodyne detection. Homodyne detection is known for its superior detection sensitivity, making it highly effective in accurately capturing weak signals. However, this approach necessitates the use of an exceptionally precise optical phase-locked loop (PLL). The implementation of such a highly accurate PLL is technologically demanding and incurs significant costs, rendering it economically challenging to realize on a large scale [27]. As a result of these constraints, heterodyne detection has garnered more widespread attention and consideration in the academic and research literature. While heterodyne detection may not match the sensitivity levels of homodyne detection, it offers a more practical and cost-effective solution. This is because heterodyne systems do not require the same degree of precision in their PLLs, thereby reducing both complexity and expense. Consequently, researchers and engineers have favored heterodyne detection in many applications where the balance of performance, feasibility, and cost is a critical consideration [1,27].

Moreover, the application of DSP techniques to coherent receivers has facilitated the transmission of spectrally efficient advanced modulation formats and effective compensation for impairments. A significant challenge in coherent detection is the synchronization of the optical carrier. To circumvent the need for phase-locked-loop systems, homodyne phase-diversity architectures can operate with free-running (unsynchronized) LO lasers. This approach addresses synchronization issues and additional impairments through the further processing of in-phase (I) and quadrature (Q) baseband signals. This receiver concept, known as the "Intradyne Receiver", presents a viable alternative. In this phase-diversity scheme, it is crucial for the optical frequencies of the Tx and LO lasers to remain as close as possible to avoid detection penalties [29–31]. Frequency offset and phase noise compensation are managed by DSP after the optical front-end in this scheme [31]. Recently, this concept has garnered renewed interest, as advancements in digital signal processors and analog-to-digital converters (ADCs) are achieving speeds that make the hardware implementation of such receivers feasible for higher data rates, such as 10 GBaud [32].

2.3.3. Self-Coherent FSO Systems

Coherent optical transmission systems facilitate efficient information encoding, resulting in improved SE and extended network reach for both digital and analog channels. These systems are especially well suited for medium-to-long-reach links. However, the practical application of full-coherent FSO systems may be limited due to their high power consumption, large physical footprint, and elevated system costs, primarily driven by optical components such as 90° optical hybrids, LO laser, and high-speed DSP [24,26,28,33]. To tackle these challenges, several quasi-coherent approaches with reduced complexity have been proposed to strike an optimal balance between achievable bit-rate and hardware complexity. In this context, various high-capacity DD transmission techniques, which offer notable advantages such as a a smaller footprint, lower power consumption, and significantly reduced costs, have been extensively explored. Despite these benefits, DD systems are susceptible to linear propagation impairments like chromatic dispersion. These include single-polarization intensity modulation (IM) using high-order Pulse-Amplitude Modulation (PAM), such as PAM-4, and discrete multi-tone (DMT) [34–39]. Other techniques under investigation include Stokes vector transceivers [40–42], and Single-Sideband (SSB) DD. Notably, the SSB DD scheme employs self-coherent detection, which can be realized through Signal–Signal Beat Interference (SSBI) cancellation or Kramers–Kronig (KK) field reconstruction. Among the various DD-based transmission systems, the SSB DD scheme stands out for its superior performance in terms of network reach [25,26,28,43–45].

Moreover, achieving optimal performance in DD systems necessitates intricate manipulation of the optical field at both the Tx and Rx ends. For example, SSB techniques have been investigated to capture amplitude and phase information using DD. However, SSB transmission schemes often suffer from inherent signal-to-signal beating noise (SSBN) in DD systems. Although a Stokes vector Rx could potentially address this issue, its practical use is hindered by sensitivity degradation and increased hardware complexity. A more promising and practical solution for SSB DD transmission is the self-coherent, minimum phase signal-based KK field reconstruction algorithm, which significantly reduces SSBN and enhances system performance [26].

The KK algorithm allows the employment of a single PD and ADC per polarization, while maintaining compatibility with phase-preserving Rx-side post-processing and supporting advanced modulation formats. Also, it enables the reconstruction of the complex-valued electric field at the Rx using only intensity measurements, allowing for digital compensation of propagation effects similar to those achieved with more complex coherent Rxs [26,43]. Despite the significant complexity reduction offered by the self-coherent-based KK FSO (KK-FSO) system, several challenges remain. For example, the nonlinear operations (logarithmic and exponential) in the KK method require the DSP to operate at much higher speeds than the Nyquist sampling rate to handle spectral broadening. A potential solution to reduce the sampling rate is to bypass these nonlinear operations. For instance, an upsampling-free KK method employs a Taylor expansion approximation for these operations. However, these approximations generally require a higher carrier-to-signal power ratio (CSPR), leading to a sensitivity penalty. To meet the demands for an increased tone power and high sampling rate of both the traditional KK and the upsampling-free KK methods, a DC-Value (DC-V) method can be used [26,27,33]. This method leverages the SSB and DC-V properties of the minimum phase signal, offering an upsampling-free phase reconstruction process with low tone power operation.

Additionally, a significant challenge in the KK implementation is the necessity for a mandatory carrier tone at the edge of the information spectrum to mitigate the effects of SSBN [27]. This carrier tone can be introduced in the optical domain, RF domain, or digital domain. Both optical and RF methods, however, increase the hardware requirements and complexity. Thus, inserting the virtual carrier in the digital domain is a simpler alternative that avoids these additional resource demands. However, this approach increases the signal's dynamic range, which leads to a quantization noise penalty owing to the finite resolution of the Digital-to-Analog Converter (DAC) [26,27,43].

A digital resolution enhancer (DRE) method has been introduced to mitigate the impacts of quantization noise in low-resolution DACs. Unlike traditional rounding-off quantizers, the DRE employs a dynamic quantization method that efficiently minimizes in-band quantization noise and enhances transmission performance. Building on this DRE technique, a novel self-coherent FSO scheme has been proposed, further simplifying the KK-FSO system. In this new system, the optical carrier is substituted with a virtual carrier, which is directly incorporated into the signal within the digital domain. This approach eliminates the need for additional hardware to generate and synchronize the optical carrier. Additionally, the DRE method allows the use of extremely low-resolution DACs with just three physical numbers of bits, leading to significant reductions in both power consumption and cost. On the Rx side, the KK transform is replaced with the DC-V method, enhancing power efficiency by lowering the requirements for oversampling and CSPR [33]. Key distinctions among non-coherent (IM/DD), coherent, and self-coherent FSO systems are presented in Table 1. Also, a thorough analysis of optical transceivers is available in [27].

Table 1. Fundamental differences between non-coherent (IM/DD), coherent, and self-coherent FSO systems.

Feature	Non-Coherent	Coherent	Self-Coherent
Modulation Techniques	Intensity modulation	Amplitude, frequency, or phase modulation	Phase modulation
DSP Complexity	Low	High	Moderate
Receiver Complexity	Low	High	Moderate
Detection Mechanism	Direct detection of light intensity	Optical mixing with a locally generated optical field	Optical mixing with a delayed version of the received signal
Need for polarization control or diversity	No	Yes	Yes
Need for Local Oscillator	No	Yes	No
Noise Rejection Performance	Moderate	Superior	High
Sensitivity to Atmospheric Turbulence	Higher	Lower	Lower
Receiver Sensitivity	Moderate	High	High
Spectral efficiency	Lower	Higher	Moderate
Reach	Lower	Higher	Moderate
Power consumption	Lower	Higher	Moderate
Footprint	Lower	Higher	Moderate
Cost	Lower	Higher	Moderate
Common Usage	Terrestrial FSO links	Advanced applications requiring high performance	Applications needing a balance of performance and complexity

3. Challenges in FSO Communication

Following the introduction of fundamental concepts in FSO communication systems, this section explores the challenges primarily arising from atmospheric channel effects. Subsequently, we discuss various FSO systems and channel models that can be utilized to assess performance.

3.1. Atmospheric Channel Effects Phenomena

FSO communication is a promising technology with notable benefits, yet it encounters several hurdles that impede its broad implementation. This section examines the primary challenges in FSO communication, such as atmospheric interference, alignment and pointing precision, range restrictions, and security issues. Addressing and overcoming these obstacles is essential for creating robust FSO communication systems that deliver reliable and efficient wireless communication across diverse applications.

3.1.1. Atmospheric Interference

Atmospheric interference poses a major challenge to FSO communication. Various atmospheric conditions, such as weather, aerosols, and particles, can impact the propagation of optical signals. For instance, FSO communication is heavily influenced by weather conditions such as snow, rain, sleet, fog, and haze. These atmospheric phenomena can attenuate, scatter, or absorb optical signals, resulting in significant signal attenuation and weakened signal strength, which degrade performance. Additionally, aerosols and particles in the atmosphere greatly affect FSO signal propagation. These tiny solid or liquid particles can scatter and absorb the optical beams used in FSO communication, causing signal degradation [1,7,20]. The transmit power can be increased to compensate for these losses, but the maximum permissible power is constrained by eye-safety standards [2,20]. To counteract the negative impacts of weather on FSO communication, several strategies such as AO, wavelength selection, hybrid FSO systems, air quality monitoring, redundancy, and diversity techniques can be employed. Further details on atmospheric attenuation models are presented in Section 3.2.

3.1.2. Atmospheric Turbulence

AT arises from irregularities in air temperature, humidity, and pressure, which cause random variations in the atmosphere's refractive index. This results in turbulent eddies of different sizes and strengths, presenting significant challenges for the propagation of FSO signals. Such turbulence leads to detrimental effects like beam wander, beam spreading, scintillation, and phase distortion, all of which can degrade FSO communication system performance [2,15,20]. To develop reliable and efficient FSO systems, it is essential to understand and mitigate these effects. Promising solutions include advances in AO, diversity techniques, ECC, and beam steering technologies. Section 4 provides further details on these mitigation strategies.

Several mathematical models are developed and analyzed to accurately characterize the probability density function (PDF) of random irradiance fluctuations under various turbulent conditions, such as weak, moderate, strong, and saturation. Examples of these models include the log-normal (LN), Gamma–Gamma (ΓΓ), *K*-distribution, and negative exponential (NE) models [15,46,47]. Section 3.2 discusses several AT models for the FSO system.

3.1.3. Alignment and Pointing Accuracy

In FSO communication, maintaining accurate alignment and pointing between the Tx and Rx is crucial. Misalignments, which can occur along both vertical and horizontal axes, are often modeled using Hoyt and Rayleigh distributions [8]. Such misalignments can cause PE that diminish the performance of the FSO system. These errors are quantified as the Euclidean distance between the centers of the photodetector and the beam footprint at the Rx. In addition to causing signal degradation, PE can lead to intermittent or complete signal loss. These errors may arise from various factors, including the movement of mobile stations, platform vibrations, building sway, transceiver sway, seismic activity, inaccuracies in the tracking system, or stress in the system's mechanical or electronic components. An alternative form of PE, known as beam wandering, is due to AT, where large-scale atmospheric eddies cause the transmitted beam to deviate from its intended trajectory [2,7].

Dynamic environments pose considerable challenges for the propagation and stability of optical signals. These environments are marked by ongoing changes in position, orientation, or atmospheric conditions, which can affect the alignment, signal integrity, and overall performance of FSO systems. Therefore, in mobile applications, continuous realignment is essential to sustain a stable connection, necessitating advanced tracking and control systems. A thorough review of pointing, acquisition, and tracking (PAT) mechanisms employed in FSO communication systems is provided in [7].

3.1.4. Reliability and Security Concerns

FSO communication inherently benefits from enhanced security due to its highly directional and narrow optical beams, making interception challenging without direct physical access to the LoS path. However, security concerns in FSO communication, such as eavesdropping, jamming, and physical obstruction, can significantly impact the propagation and reliability of optical signals [2]. These challenges must be addressed through a combination of advanced encryption, AO, dynamic beam management, and robust system design. By mitigating these security risks, FSO communication can offer a secure and efficient alternative to traditional RF-based systems, particularly in applications where high data rates and secure transmission are paramount. This section provides a detailed discussion of these FSO security concerns.

Eavesdropping

Similar to wired links, wireless optical transmission is inherently more secure than RF transmission due to the high directionality of optical beams as opposed to the almost broadcast nature of RF signals, making them significantly harder to intercept, though not impossible [48]. However, there are scenarios where the threat of eavesdropping from one or more malicious actors may arise. Eavesdropping in FSO communication involves intercepting the optical beam to gain unauthorized access to the transmitted data. This can happen because of AT, optical beam divergence, channel scattering, and PE. Consequently, adversaries might access information through scattering over a non-LoS (NLoS) channel and scattering over a LoS channel caused by aerosol particles. Eavesdropping may be feasible if both the legitimate and the adversary Rxs are positioned close to each other within the FSO links [48–50]. Additionally, eavesdropping is possible when part of the beam radiation is reflected by small particles and detected by an external observer not in the LoS of either communication peer. However, the power received by the eavesdropper in such cases will be significantly less than in an equivalent RF scenario. An alternative scenario involves the eavesdropper blocking the laser beam to collect more power. In this case, the legitimate Rx would notice a significant decrease in the average received power, prompting the communication to be stopped for security reasons [48].

Traditionally, upper-layer cryptographic algorithms, which assume a limited computational capability of eavesdroppers, have been sufficient to protect confidential messages. However, due to the complexity of key generation and allocation, along with the current advanced hardware available to eavesdroppers, these cryptographic algorithms may not meet future wireless network security requirements. In addition to these upper-layer algorithms, physical (PHY) layer key generation and encryption methods, which exploit channel reciprocity to generate key packets [51], and PHY layer wiretap coding, which leverages the difference in channel quality between the Tx-to-Rx and Tx-to-eavesdropper channels to protect confidential messages [52,53], can be employed to secure confidential messages in the FSO-RF relay network [50].

The challenge of ensuring secure communications in the presence of external eavesdroppers is a well-known issue in digital communications. There is increasing interest in the PHY layer secrecy performance over fading channels, leading the research community to focus on PHY layer security (PLS) as a means to avoid eavesdropping without relying on complex encryption techniques in the upper layers [54]. Consequently, the PLS of FSO links has been studied over $\mathcal{M}$ [54–56] and $\Gamma\Gamma$ turbulence channels [48]. Additionally, ref. [57] explores the PLS of a LoS FSO link employing Orbital Angular Momentum (OAM) multiplexing. The information security risks of an FSO communication system with an eavesdropper outside the laser beam, who could obtain information through a non-LoS scattering channel, are examined in [58]. Furthermore, key PLS metrics such as secrecy outage probability (SOP), average secrecy capacity (ASC), and the probability of strictly positive secrecy capacity (SPSC) are unified in the context of FSO communication in [59–61].

In [61–63], the secrecy of a mixed RF-FSO system is analyzed, where an adversary intercepts secure information via the FSO link by capturing a fraction of the optical power due to beam scattering. The findings in [62], supported by simulations, demonstrate that factors such as photo-aperture size, fading parameters, and the average eavesdropper signal-to-noise ratio (SNR) are crucial for securing RF-FSO communications. Specifically, the results indicate that under most weather conditions, and with a photo-aperture size of approximately 200 mm, mixed RF-FSO relaying is advantageous for designing a secure RF-FSO relay network. Furthermore, refs. [59,64] propose a hybrid RF/FSO system utilizing RIS reflection to interfere with eavesdroppers in FSO eavesdropping scenarios. The system's PLS performance is also analyzed and optimized.

The performance of PLS in mixed FSO-RF systems has been the focus of several studies. In [50], a PLS secrecy analysis for a mixed FSO-RF system under a ΓΓ and Rayleigh fading scenario is conducted. The study in [65] analyzes the secrecy performance of a mixed FSO-RF simultaneous wireless information and power transfer (SWIPT) system using a decode-and-forward (DF) relaying scheme. These studies consider eavesdropping on the FSO side with full channel state information (CSI) of the eavesdropper's channel. However, the studies reported in [54,63,65–70] are limited to PLS secrecy analysis under the assumption of ideal power amplifier (PA) deployment at the RF front-end. In practice, the non-negligible power consumption by PAs significantly impacts the secrecy performance, making the effects of PA efficiency crucial in PLS secrecy analysis. Although a few studies have examined the impact of hardware imperfections on mixed RF-FSO systems' performance, such as [71,72], only a limited number have done so.

To model these hardware imperfections, [72] considers soft envelope limiter (SEL) PA nonlinearities, while [71] uses a degradation PA model to characterize the hardware impairments. The study in [73] analyzes the secrecy performance of a hybrid RF-FSO system under hardware impairments using the degradation PA model, focusing on scenarios where only the RF eavesdropper intercepts secure information. Additionally, ref. [49] propose a different system structure for a mixed FSO-RF system under hardware imperfections, where all RF Rxs use the SWIPT technique to collect energy from the relay's wireless signals. This study also assumes that both FSO- and RF-side eavesdroppers can overhear the intended information separately and simultaneously.

Jamming

The wireless medium's inherent openness renders it vulnerable to both intentional and unintentional interference. A significant source of unintentional interference is interference from neighboring cells. Conversely, intentional interference, like jamming attacks, represents adversarial actions targeting a victim Rx [74]. Therefore, as wireless links, the reliability and security of the entire FSO system are susceptible to intentional jamming. This threat is particularly significant in military platforms or other security-sensitive communication protocols controlled by FSO networks, which can be targeted and disrupted by adversaries [75]. The impact of jamming on RF systems, including wireless multihop networks, has been extensively studied from various perspectives [74,76–79]. FSO systems are also vulnerable to jamming because the spectral width of the laser light in FSO communications is typically confined to specific wavelengths to ensure compatibility with optical fiber communication. Usually, FSO communication operates within very narrow wavelength ranges, such as 1300 nm or 1550 nm. Consequently, the operating frequency of a legitimate user link is readily accessible in FSO systems, making them highly susceptible to jamming attacks [75].

Additionally, in an FSO system, it is crucial to optimize the field-of-view (FoV) based on the specific application. The FoV is defined as the solid angle through which a detector can receive an incoming optical signal. As illustrated in Figure 7a, the optical signal emitted by a jammer is captured when it falls within the FoV; otherwise, it passes through unobstructed. Typically, the FoV of the receiving aperture is kept wide to maximize signal reception, mitigate the effects of fading due to AT and PE, address aperture misalignment issues, and prevent communication link failures caused by obstacles in the FSO system [75].

In tracking-based FSO communication systems, a wide FoV of up to 60° is employed to increase the tracking range and reduce optical aberrations [80]. Additionally, coarse PAT (CPAT) systems consider FoVs up to 70° [7,81]. Due to the inherent fluctuations of UAVs, the optimal FoVs for UAV-based FSO links are significantly larger than those for ground-based FSO links [82].

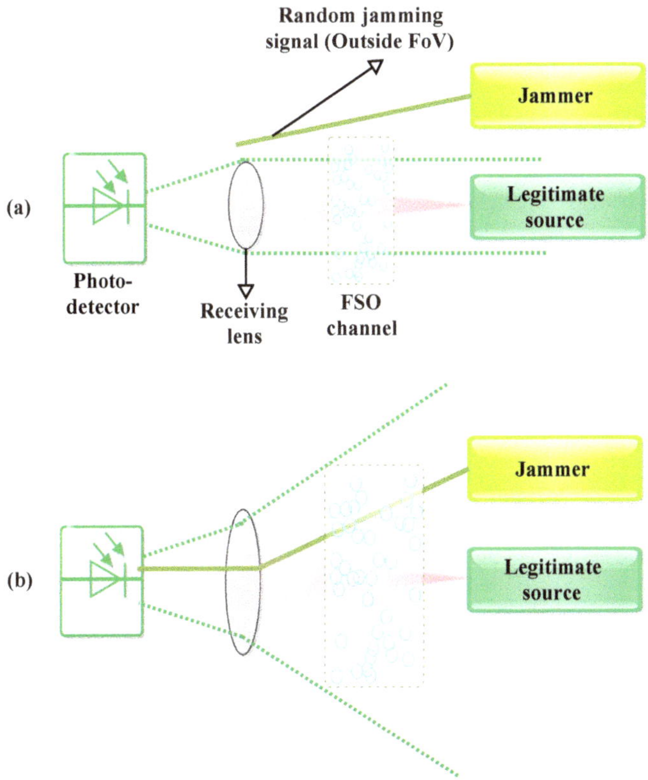

Figure 7. FSO detector with (**a**) a narrow FoV and (**b**) a wide FoV.

Moreover, aperture averaging is an essential technique in FSO systems, as the unpredictable nature of the received signal caused by fading imposes severe constraints on system performance. Aperture averaging is a technique used to control and define the size of the detector or receiving aperture needed to capture the traversing optical field as effectively as possible. This process is particularly crucial for FSO systems operating in strong and saturated regimes. Aperture averaging helps to mitigate signal scintillation or irradiance flux variations (reducing fade probability), minimize the impacts of angle of arrival (AoA) fluctuations and boresight angle, and optimize the average signal power at the detector plane. Additionally, it serves as a primary method to ensure spatial diversity [46,83]. A comprehensive discussion on aperture averaging is provided in Section 4.9. Therefore, to maintain effective FSO communication through the continuous establishment of transmission links, it is crucial to have a larger aperture diameter, which consequently results in a wide FoV [46,82].

A narrow FoV restricts the operational coverage area, making it less ideal for FSO systems. Conversely, a wider FoV increases ambient light background noise, which can be mitigated through various methods to capture a larger portion of the incoming optical beam at the FSO Rx [82,84]. Also, a higher FoV offers a broader coverage area and improved signal reception. For instance, a larger aperture diameter (8 inches) significantly reduces error probability in comparison with a smaller aperture diameter (2 inches) [46].

Moreover, implementing a higher FoV does not necessitate additional circuitry for pointing and tracking mechanisms, thereby guaranteeing reliable wireless connectivity between the transmitting and receiving apertures. As a result, FSO systems with a higher FoV are less susceptible to misalignment errors, which helps to mitigate the blocking probability caused by obstacles between the source and destination points. This reduces the need for stringent alignment control mechanisms. However, the extensive FoV essential for effective signal reception, coupled with the multiple receiving nodes required in cooperative networks, renders FSO links susceptible to jamming attacks at any node. The detector can easily capture the jamming signal, as illustrated in Figure 7b, creating a significant challenge in differentiating the transmitted signal from unwanted jamming signals [46,82]. Moreover, in a dual-hop link, an attack that blocks either node will result in the failure of the entire source–relay–destination link.

Jamming in point-to-point (p2p) FSO communication is studied in [75,85], while a similar scenario for cooperative relaying is explored in [86,87]. In [75], the impact of jamming, where the jammer aims to congest the Rx with unwanted noise, is investigated for FSO communication systems to analyze the outage probability (OP) and bit-error rate (BER) expressions of the system. However, none of these studies address the issue of jamming in the context of aperture averaging and phase error impairment in Exponentiated Weibull (EW) statistics. In [75], it is demonstrated that jamming can severely degrade the performance of single-input single-output (SISO) FSO systems. In [46], the threat to security-constrained reliable FSO communications due to jamming over an exponentially weighted fading channel is examined, taking into account various AT conditions, aperture averaging, and PE. Analytical expressions for the average BER (ABER) and OP are derived for the FSO system under the influence of a random jammer. In [88], jamming for cooperative relaying is further investigated. To address the impact of jamming in FSO communication networks, a buffer-aided relaying approach is proposed. This method employs a relay node with a finite-size buffer and assumes that both relay and destination nodes are susceptible to jamming. Comparisons with a non-buffer-aided (NBA) cooperative FSO system impacted by jamming show that the proposed buffer-aided FSO system considerably outperforms its NBA counterparts.

3.2. *FSO System and Channel Models*

FSO channel models are essential for understanding and predicting system performance. These models account for various factors that affect the propagation of light, such as AT, absorption, scattering, and weather conditions like fog or rain. Accurate channel modeling helps in designing robust FSO systems, optimizing link availability, and ensuring reliable data transmission. By analyzing different channel models, researchers and engineers can develop strategies to mitigate adverse effects, enhancing the efficiency and reliability of FSO communication systems.

3.2.1. FSO System Model

In this section, we analyze a practical FSO link that employs IM/DD with On–Off Keying (OOK) modulation. The data are encoded onto the instantaneous intensity of an optical beam at the Tx. It is assumed that the optical power emitted from the transmit aperture into free space is affected by factors such as misalignment fading (PE), fading induced by AT, and background or ambient noise before it reaches the receive aperture. These factors lead to variations in signal intensity. As a result, the received electrical signal r at the receive aperture can be modeled as follows [3,89–91]:

$$r = \eta_e h x + n, \tag{1}$$

where η_e denotes the effective photoelectric conversion ratio of the Rx, $x \in \{0, 1\}$ represents the transmitted information bit, n represents the additive white Gaussian noise (AWGN) with zero mean and variance $\sigma_n^2 = N_0/2$, N_0 denotes the one-sided noise power spectral density in watts/Hz, and the term $h = h_\ell h_\mathrm{a} h_\mathrm{p}$ signifies the irradiance affecting the channel state, which is a product of the deterministic path loss h_ℓ, the random attenuation due to AT-induced fading h_a, and the random attenuation caused by geometric spread and PE h_p. Both h_a and h_p are random variables characterized by their respective PDFs, $f_{h_\mathrm{a}}(h_\mathrm{a})$ and $f_{h_\mathrm{p}}(h_\mathrm{p})$.

Noise Model

The noise performance of a Rx is influenced by the type of photodetector employed, specifically whether a PIN photodetector or an APD is used. It is important to recognize that each of these photodetectors contributes to the overall noise in distinct ways. In this subsection, we delve into the different noise models associated with both PIN and APD photodetectors, providing a comprehensive review of their respective noise characteristics and how they impact the Rx performance.

If an APD Rx is employed, the overall noise contribution is understood to consist of three primary components: thermal noise, shot noise, and dark current. However, assuming the dark current is insignificant, the noise generated by the APD Rx, denoted as n, can be represented as the sum of the thermal noise, i_{Th}, and the shot noise, i_{Sh}. Therefore, the equation representing the APD Rx noise is [92,93]

$$n = i_{Th} + i_{Sh}, \tag{2}$$

where i_{Sh} and i_{Th} represent the shot noise and thermal noise, respectively. The thermal noise i_{Th} is characterized as a zero-mean stationary Gaussian random process, with its variance defined by [92,93]

$$\sigma_{Th}^2 = 4k_B \frac{T}{R_L} F_n \Delta f, \tag{3}$$

where k_B represents the Boltzmann constant, T denotes the absolute temperature in degrees Kelvin, R_L represents the APD load resistance, F_n represents the amplifier noise figure, and $\Delta f = R_b/2$ denotes the symbol effective noise bandwidth that is associated with the bit rate R_b.

Shot noise is influenced by the APD Rx and can be modeled similarly to thermal noise using a zero-mean stationary Gaussian random process. The variance in shot noise can be described as

$$\sigma_{Sh}^2 = 2q_e \bar{g}^2 F_A \Re h P_t m \Delta f, \tag{4}$$

where q_e is the electron charge, m is the modulation index, P_t represents the transmitted optical power, $\Re$ and $\bar{g}$ are the APD responsivity and average gain, respectively, and F_A represents the excess noise factor of the APD, which is defined by

$$F_A = k_A \bar{g} + (1 - k_A)\left(2 - \frac{1}{\bar{g}}\right), \tag{5}$$

where k_A represents the ionization factor. Utilizing the characteristics of Gaussian random variables, the total APD noise can be modeled as a stationary random process with a mean of zero. The aggregate variance can be defined as [92,93]

$$\sigma_n^2 = \sigma_{th}^2 + \sigma_{Sh}^2 = 2\Delta f \left\{ 2k_B \frac{T}{R_L} F_n + q_e \bar{g}^2 F_A \Re h P_t m \right\}. \tag{6}$$

The instantaneous electrical SNR per symbol, represented by γ_i at the output of the APD, is thus given by

$$\gamma_i = \frac{(\bar{g} \Re h P_t)^2}{2\Delta f \left\{ 2k_B \frac{T}{R_L} F_n + q_e \bar{g}^2 F_A \Re h P_t m \right\}}. \tag{7}$$

3.2.2. FSO Channel Models

This section provides an overview of various channel models that can be utilized to predict the performance of FSO systems. These models are essential for understanding how different environmental factors and conditions affect the effectiveness and reliability of FSO communications. The models incorporate PDFs derived from turbulence intensity to characterize the impact of turbulence on FSO signals. Analyzing these models allows us to gain insights into how factors like AT, weather conditions, and signal attenuation impact system performance, enabling us to optimize and better anticipate the behavior of FSO systems in real-world scenarios.

Atmospheric Attenuation

As an optical beam travels through the atmosphere, it experiences atmospheric loss. The signal power attenuation follows the exponential Beer–Lambert law and is expressed as [3]

$$h_\ell(\lambda, z) = \frac{P(\lambda, z)}{P(\lambda, 0)} = \exp(-\sigma(\lambda)z) \tag{8}$$

where $h_\ell(\lambda, z)$ represents the loss as a function of the propagation path length z at wavelength λ. The signal power at a distance z and the emitted power at $z = 0$ are expressed as $P(\lambda, z)$ and $P(\lambda, 0)$, respectively. The total extinction coefficient or the attenuation coefficient $\sigma(\lambda)$ per unit length is defined by [3]

$$\sigma(\lambda) = \alpha_m(\lambda) + \alpha_a(\lambda) + \beta_m(\lambda) + \beta_a(\lambda), \tag{9}$$

where $\alpha_{m,a}$ are the absorption coefficients for molecules and aerosols, respectively, and $\beta_{m,a}$ represent the scattering coefficients for molecules and aerosols, respectively.

The attenuation h_ℓ is considered a constant scaling factor over an extended period, ensuring consistent behavior. Additionally, it depends on the size and distribution of scattering particles, along with the wavelength used. This attenuation can be described regarding visibility, which is directly measurable from the atmosphere. Empirically, attenuation is characterized in relation to visibility as [3]

$$\sigma(\lambda) = \frac{3.912}{V}\left(\frac{\lambda}{550}\right)^{-q}, \tag{10}$$

where V represents visibility (in kilometers), and q is a parameter influenced by the particle size distribution in the atmosphere as described by the Kruse model [90]:

$$q = \begin{cases} 1.6 & V > 50 \text{ km} \\ 1.3 & 6 \text{ km} < \text{V} < 50 \text{ km} \\ 0.585V^{1/3} & V < 6 \text{ km} \end{cases} \tag{11}$$

Additionally, Kim introduces an extended model designed to improve accuracy in low-visibility situations. The Kim model is described as [90]

$$q = \begin{cases} 1.6 & V > 50 \text{ km} \\ 1.3 & 6 \text{ km} < \text{V} < 50 \text{ km} \\ 0.16V + 0.34 & 1 \text{ km} < \text{V} < 6 \text{ km} \\ V - 0.5 & 0.5 \text{ km} < \text{V} < 1 \text{ km} \\ 0 & V < 0.5 \text{ km} \end{cases} \tag{12}$$

Pointing Error or Misalignment Fading

An FSO link is a LoS communication system that uses a narrow optical beam, which requires precise pointing accuracy to ensure the effective performance and reliability of optical systems [16]. PE and signal degradation at the Rx are typically caused by wind loads and thermal expansions, leading to random movements of the building. Assuming a Gaussian spatial intensity profile characterized by a beam waist w_z at the Rx plane—located a distance z from the Tx—and a circular aperture with a radius r_a, the fraction of collected power resulting from geometric spreading with a radial displacement α_d from the detector's origin can be approximated using a Gaussian distribution as [3,89]

$$h_{\mathrm{p}}(\alpha_d) \approx A_0 \exp\left(-\frac{2\alpha_d^2}{w_{z_{eq}}^2}\right), \tag{13}$$

where $w_{z_{eq}}^2 = w_z^2 \sqrt{\pi}\mathrm{erf}(v)/2v\exp(-v^2)$, $v = \sqrt{\pi} r_a/\sqrt{2} w_z$, $w_{z_{eq}}$ is the equivalent beamwidth, $A_0 = [\mathrm{erf}(v)]^2$, and $\mathrm{erf}(\cdot)$ is the error function defined as [94]

$$\mathrm{erf}(x) = (2/\sqrt{\pi}) \int_0^x e^{-u^2} du. \tag{14}$$

Assuming that the horizontal and elevation sway are independent and follow identical Gaussian distributions, the radial displacement α_d will follow a Rayleigh distribution. Consequently, $f_{h_{\mathrm{p}}}(h_{\mathrm{p}})$ can be expressed as

$$f_{h_{\mathrm{p}}}(h_{\mathrm{p}}) = \frac{\gamma_{\mathrm{p}}^2}{A_0^{\gamma_{\mathrm{p}}^2}} h_{\mathrm{p}}^{\gamma_{\mathrm{p}}^2 - 1}, \quad 0 \le h_{\mathrm{p}} \le A_0, \tag{15}$$

where $\gamma_{\mathrm{p}} = w_{z_{eq}}/2\sigma_s$ denotes the ratio of the equivalent beam radius to twice the standard deviation (jitter) of the PE displacement, and σ_s^2 represents the jitter variance at the Rx.

Atmospheric Turbulence

The intensity fluctuations over the FSO channel have been described using various statistical models in the literature, tailored for different turbulence regimes. The commonly used models encompass the LN distribution, ideal for weak turbulence and frequently applied due to its strong correlation with experimental data, the ΓΓ distribution for moderate to strong turbulence, and the K-distribution for strong turbulence. These models incorporate PDFs based on turbulence intensity to characterize the impact of turbulence on FSO signals. Other widely used models include the I-K distribution, NE, and $\mathcal{M}$ distribution models. However, this work focuses on the LN, ΓΓ, and generic $\mathcal{M}$ distribution models.

Log-normal Distribution: Generally, the LN model is only appropriate for weak turbulence conditions and for link ranges shorter than 100 m [95]. Therefore, the intensity fluctuation PDF for weak turbulence, modeled using the LN distribution, is presented by [96]

$$f_{h_{\mathrm{a}}}(h_{\mathrm{a}}) = \frac{1}{2h_{\mathrm{a}}\sqrt{2\pi\sigma_x^2}} \exp\left(-\frac{(\ln(h_{\mathrm{a}}) + 2\sigma_x)^2}{8\sigma_x^2}\right), \tag{16}$$

where σ_x^2 denotes the log-amplitude variance specified for plane waves and spherical waves, respectively, and is expressed as [96]

$$\sigma_x^2|_{\mathrm{plane}} = 0.307 C_n^2 k^{7/6} L^{11/6}, \tag{17a}$$

$$\sigma_x^2|_{\mathrm{spherical}} = 0.124 C_n^2 k^{7/6} L^{11/6}, \tag{17b}$$

where L denotes the distance, $k = \frac{2\pi}{\lambda}$ denotes the optical wave number, and C_n^2 represents the altitude-dependent index of the refraction structure parameter. Also, with respect to the log-irradiance variance, $\sigma_x^2 = \sigma_l^2/4$, where σ_l^2 is the log-irradiance variance expressed as [96]

$$\sigma_l^2|_{\mathrm{plane}} = 1.23 C_n^2 k^{7/6} L^{11/6}, \tag{18a}$$

$$\sigma_l^2|_{\mathrm{spherical}} = 0.50 C_n^2 k^{7/6} L^{11/6}. \tag{18b}$$

The C_n^2 parameter is crucial for quantifying the refractive index fluctuations in AT. It varies with the wavelength, atmospheric altitude, and temperature. Various C_n^2 profile models have been presented in the literature, with the Hufnagel valley model being the most widely used for altitude and is defined as [6,91,94,96,97]

$$C_n^2(h_n) = 0.00594(v_w/27)^2(10^{-5}h_n)^{10}\exp(-h_n/1000) \\ + 2.7\times 10^{-16}\exp(-h_n/1500) + \hat{A}\exp(-h_n/100), \tag{19}$$

where h_n represents the altitude in meters (m) and $\hat{A}$ denotes the nominal value of $C_n^2(0)$ at ground level in $m^{-2/3}$. The value of C_n^2 for FSO links at ground level is approximately 1.7×10^{-14} $m^{-2/3}$ during the daytime and 8.4×10^{-15} $m^{-2/3}$ at night. Generally, C_n^2 ranges from 10^{-13} $m^{-2/3}$ in strong turbulence to 10^{-17} $m^{-2/3}$ in weak turbulence, with an average typical value of 10^{-15} $m^{-2/3}$ [94,97]. The variable v_w is the root mean square wind speed (pseudo wind) in meters per second (m/s), typically valued at 21 m/s, is given as [94]

$$w = \left[\frac{1}{15\times10^3}\int_{5\times10^3}^{20\times10^3} V^2(h_n)dh_n\right]^{1/2}, \tag{20}$$

where $V(h_n)$ is generally characterized by the Bufton wind model as described by [94]

$$V(h_n) = \omega_s h_n + V_g + 30\exp\left[-\left(\frac{h_n - 9400}{4800}\right)^2\right], \tag{21}$$

where V_g denotes the ground wind speed, and ω_s represents the *slew rate*, which is related to the satellite's movement in relation to the observer on the ground.

Gamma–Gamma Distribution: In many cases, when the LN distribution characterization is invalid due to strong turbulence regimes, the ΓΓ distribution is typically used to model scintillation effects. Furthermore, the ΓΓ model can characterize fading gains across both weak and strong turbulence scenarios. The PDF of h_a based on the ΓΓ distribution is defined as [6,91,94,96,97]

$$f_{h_a}(h_a) = \frac{2(\alpha\beta)^{(\alpha+\beta)/2}}{\Gamma(\alpha)\Gamma(\beta)}(h_a)^{\frac{(\alpha+\beta)}{2}-1}K_{\alpha-\beta}(2\sqrt{\alpha\beta h_a}), \tag{22}$$

where $\Gamma(\cdot)$ denotes the gamma function, while $K_\nu(\cdot)$ represents the modified Bessel function of the second kind of order ν. The parameters α and β refer to the effective number of large-scale and small-scale eddies involved in the scattering process, respectively. These parameters are defined for the plane wave as [91,94,96,97]

$$\alpha = \left[\exp\left(\frac{0.49\sigma_R^2}{(1+1.11\sigma_R^{12/5})^{7/6}}\right) - 1\right]^{-1}, \tag{23a}$$

$$\beta = \left[\exp\left(\frac{0.51\sigma_R^2}{(1+0.69\sigma_R^{12/5})^{5/6}}\right) - 1\right]^{-1}, \tag{23b}$$

and for the spherical wave, they can be expressed as [6,91,98]

$$\alpha = \left[\exp\left(\frac{0.49\sigma_R^2}{(1+0.18d^2+0.56\sigma_R^{12/5})^{7/6}}\right) - 1\right]^{-1}, \tag{24a}$$

$$\beta = \left[\exp\left(\frac{0.51\sigma_R^2(1+0.69\sigma_R^{12/5})^{-5/6}}{(1+0.9d^2+0.62d^2\sigma_R^{12/5})^{5/6}}\right) - 1\right]^{-1}, \tag{24b}$$

where $d \triangleq \left(\frac{kD^2}{4L}\right)^{1/2}$, with D denoting the diameter of the Rx aperture, and σ_R^2 denoting the Rytov variance, a metric for the strength of turbulence fluctuations. The Rytov variance σ_R^2 is expressed for both plane and spherical waves, respectively, as [91,94,97]

$$\sigma_R^2\big|_{\text{plane}} = 1.23\, C_n^2\, k^{7/6}\, L^{11/6}, \tag{25a}$$

$$\sigma_R^2\big|_{\text{spherical}} = 0.492\, C_n^2\, k^{7/6}\, L^{11/6}, \tag{25b}$$

Additionally, the normalized variance in the irradiance, commonly known as the scintillation index (σ_N^2), can be expressed in relation to σ_x^2 and the eddies involved in the scattering process (α and β), respectively, as [6,91,94,98]

$$\sigma_N^2 \triangleq \frac{\langle h_a^2 \rangle - \langle h_a \rangle^2}{\langle h_a \rangle^2} \tag{26a}$$

$$= \frac{\langle h_a^2 \rangle}{\langle h_a \rangle^2} - 1 \tag{26b}$$

$$= exp\left(4\sigma_x^2\right) - 1 \tag{26c}$$

$$= 1/\alpha + 1/\beta + 1/(\alpha\beta). \tag{26d}$$

Málaga ($\mathcal{M}$)-Distribution: The $\mathcal{M}$-distribution serves as a versatile model suitable for characterizing the entire range of turbulent regimes. Table 2 outlines the methods for deriving current AT models from the $\mathcal{M}$-distribution model [99]. The $\mathcal{M}$-distributed fading model incorporates components like U_L, U_S^C, and U_S^G. These components denote the LoS contribution; the scattered component by eddies along the propagation axis coupled to the LoS; and the scattered component by off-axis eddies received by the Rx, respectively. The average power of the LoS component and the total scatter components are given by $\Omega = E[|U_L|^2]$ and $2b_0 = E[|U_S^C|^2 + |U_S^G|^2]$, respectively.

Furthermore, the average power of the scattering component coupled to the LoS and the scattering component received from off-axis eddies are $E[|U_S^C|^2] = 2\rho b_0$ and $E[|U_S^G|^2] = (1-\rho)2b_0$, respectively. The parameter $0 \leq \rho \leq 1$ denotes the proportion of scattering power coupled to the LoS component. The PDF of the optical channel gain h_a for the $\mathcal{M}$-distribution is expressed as [99,100]

$$f_{h_a}(h_a) = A \sum_{k=1}^{\beta} a_k h_a^{\frac{\alpha+k}{2}-1} K_{\alpha-k}\left(2\sqrt{\frac{\alpha\beta h_a}{\mu\beta + \Omega'}}\right) \tag{27}$$

where $\mu = E[|U_S^G|]^2 = (1-\rho)2b_0$, $\Omega' = \Omega + 2\rho b_0 + 2\sqrt{2\rho b_0 \Omega}\cos(\varphi_A - \varphi_B)$, and φ_A and φ_B denote the deterministic phases of the LoS component and the coupled-to-LoS component, respectively. The parameter α is a positive value dependent on the effective number of large-scale cells in the scattering process, while β is a natural number representing the fading parameter. The terms A and a_k can be defined respectively as [99,100]

$$A \triangleq \frac{2\alpha^{\frac{\alpha}{2}}}{\mu^{1+\frac{\alpha}{2}}\Gamma(\alpha)}\left(\frac{\mu\beta}{\mu\beta + \Omega'}\right)^{\beta+\frac{\alpha}{2}} \tag{28a}$$

$$a_k \triangleq \begin{pmatrix} \beta & -1 \\ k & -1 \end{pmatrix} \frac{(\mu\beta + \Omega')^{1-\frac{k}{2}}}{(k-1)!}\left(\frac{\Omega'}{\mu}\right)^{k-1}\left(\frac{\alpha}{\beta}\right)^{\frac{k}{2}} \tag{28b}$$

Table 2. Unified $\mathcal{M}$-distribution framework for various FSO models (adapted from [99]).

	Conditions									
Distribution Model	ρ	**Var[|U$_L$|]**	**Var[G]**	X	Ω'	**Var[X]**	γ	Ω	α	β
Rice-Nakagami	0	0								
Gamma	0						0			
Homodyned K	0		0	γ						
ΓΓ	1				1		0			
Shadowed-Rician						0				
Log-normal	0	0					$\to 0$			
K	0							0		
Exponential	0	0						0	$\to\infty$	
Gamma-Rician										$\to\infty$

Aggregate Attenuation Data: The PDF of $h = h_\ell h_a h_p$, representing the characteristics of the propagation channel, can be represented as [89,101]

$$f_h(h;w_z) = \int f_{h|h_a}(h|h_a) f_{h_a}(h_a) dh_a \tag{29}$$

where $f_{h|h_a}(h|h_a)$ represents the conditional probability given a turbulence state h_a, with its distribution expressed as [89,101]

$$f_{h|h_a}(h|h_a) = \frac{1}{h_a h_\ell} f_{h_p}\left(\frac{h}{h_a h_\ell}\right) = \frac{\gamma_p^2}{A_0^{\gamma_p^2} h_a h_\ell}\left(\frac{h}{h_a h_\ell}\right)^{\gamma_p^2 - 1}, \quad 0 \le h \le A_0 h_a h_\ell. \tag{30}$$

Hence, $f_h(h;w_z)$ can be expressed as [89]

$$f_h(h;w_z) = \frac{\gamma_p^2}{(A_0 h_\ell)^{\gamma_p^2}} h^{\gamma_p^2 - 1} \int\limits_{h/A_0 h_\ell}^{\infty} h_a^{-\gamma_p^2} f_{h_a}(h_a) dh_a. \tag{31}$$

3.2.3. FSO System Model Considering Jammer Interference

The FSO system model described in Equation (1) can be expanded to include a jamming source. In this scenario, a jammer, which is uncorrelated and synchronized with the user's transmission time interval, emits random pulses with a certain level of jamming. Both the jamming and user signals travel through the FSO channel and reach the receiving photodetector. At the Rx, after signal conversion, the relationship between the input and output can be mathematically expressed as [46,86]

$$r = \eta_e h x + \sqrt{\Upsilon_j} h_j \Phi \tag{32}$$

where h_j represents the irradiance of the jamming source at the destination, Υ_j denotes the electrical power of the jammer, and Φ indicates the jamming state defined as

$$\Phi = \begin{cases} 1 & \text{In the presence of a jammer} \\ 0 & \text{otherwise.} \end{cases} \tag{33}$$

4. Enabling Technologies for FSO Communication

In the rapidly evolving field of FSO communication, various enabling technologies are pivotal in enhancing system performance and reliability. This section delves into key advancement such as AO, which corrects wavefront distortions; advanced modulation schemes that improve data throughput; and aperture averaging to reduce signal fading. Additionally, spatial diversity and cooperative relaying techniques are explored for their roles in mitigating adverse atmospheric effects. Adaptive transmission methods and robust ECC are also discussed, highlighting their contributions to maintaining signal integrity and optimizing communication efficiency in FSO systems. Methods for mitigating atmospheric turbulence are shown in Figure 8.

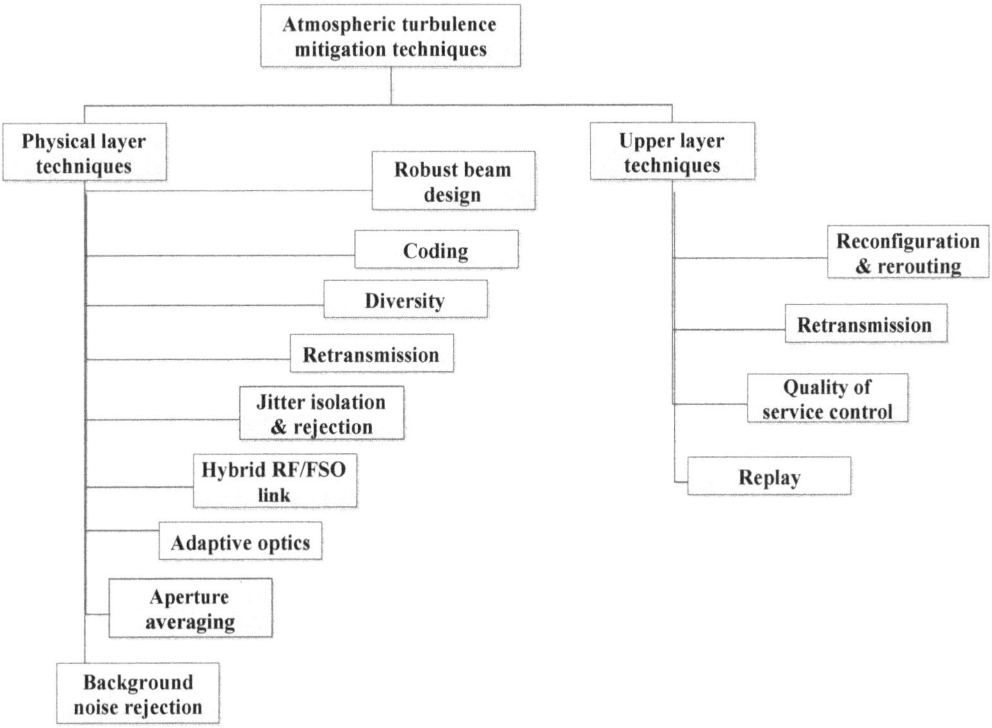

Figure 8. Techniques for mitigating atmospheric turbulence.

4.1. Robust Beam and Multi-Modulation Compatible System Design

The robust beam and multi-modulation compatible laser design focuses on enhancing optical communication systems by integrating a laser capable of supporting multiple modulation formats. This design improves signal integrity and system flexibility while reducing the complexity and size of the communication hardware. By enabling seamless switching between various modulation techniques, it optimizes performance for diverse communication scenarios, ensuring robust and efficient data transmission.

4.1.1. Robust Beam Design

The propagation of laser beams through the atmosphere has been extensively studied both theoretically and experimentally. Recently, significant attention has been focused on laser beams with phase singularities moving through ideal atmospheric turbulence and non-Kolmogorov turbulence. Research indicates that partially coherent vortex beams are less impacted by atmospheric turbulence compared to partially coherent non-vortex beams, and that the topological charge of these beams can serve as an information carrier in optical communications [102].

To mitigate the disturbances caused by turbulence, extensive research has focused on the effects of non-diffracting and pseudo-non-diffracting vortex beams in resisting turbulent disturbances. The findings indicate that these beams' favorable characteristics can significantly reduce the impact of turbulence on the transmission of OAM modes [103–106]. Specifically, experimentalists are seeking laser beams that are either non-diffracting or possess a definite value of photon OAM along the propagation direction. These specialized laser beams have proven useful in various applications, including optical tweezers, optical trapping, laser radar, metrology, remote sensing, microlithography, image processing, astronomical imaging, medical imaging and surgery, as well as wireless, optical, and FSO communications [107].

In OAM-based underwater optical wireless communication (OWC), the OAM mode is influenced by the wavefront spatial structure of the carrier. As a result, absorption, ocean turbulence, and scattering inevitably distort the OAM mode and cause crosstalk between its energy states. To mitigate the perturbations caused by seawater turbulence, employing various beam structures has become a new trend in this field. These structures include the Gaussian Schell-model (GSM) vortex beam [102], random frozen photons (RFP) beam [108], Hermite–Gauss (HG) vortex beam [109], perfect Laguerre–Gauss (PLG) beam [110], and several quasi-non-diffractive beams such as the Bessel–Gauss (BG) vortex beam [111,112], Lommel–Gauss (LMG) beam [113], Airy circle (AiC) vortex beams, Airy ellipse (AiE) vortex beams [114–116], and Hypergeometric-Gaussian (HyGG) vortex beam [103].

Meanwhile, most previous studies on OAM-based systems have concentrated on Laguerre–Gaussian (LG) beams in turbulent atmospheres. In contrast, non-diffraction BG beams exhibit clear advantages over traditional LG beams in reducing turbulence effects and enhancing the performance of OAM-based FSO communication links in such conditions. These benefits arise from two notable characteristics of BG beams: their ability to propagate without altering their intensity profile (non-diffraction nature), and their exceptional capacity to reconstruct after encountering an obstacle (self-healing mechanism) [111,117]. The BG beam is essentially a Bessel beam (BB) truncated by a Gaussian profile, exhibiting nearly diffraction-free properties up to a specific distance known as the non-diffractive range (NDR). The primary advantage of BG beams over BBs is that the radial modulation tapers the aperture fields, thereby reducing the on-axis ripples caused by edge diffraction [118]. These characteristics are crucial for optical communications based on LoS operations. Consequently, for OAM-based FSO communication systems, the non-diffraction and self-recovery properties of BG beams make them highly valuable. The transmission characteristics of BG beams in free-space have been explored, and their non-obstruction properties in FSO communication systems are examined in [117,119].

Moreover, structured light beams, particularly pin beams (PBs), exhibit reduced scintillations in free space compared to Gaussian beams. A novel class of optical beams has been introduced, characterized by a main lobe with a Bessel-like transverse profile and a width that gradually increases during propagation. These are termed inverted pin beams (IPBs). Numerical simulations demonstrate that IPBs have a significantly lower scintillation index than other beam classes such as PBs, BBs, collimated Gaussian beams (CGBs), and focused Gaussian beams (FGBs) in atmospheric turbulence regimes [120,121]. Also, IPBs carrying OAM can further mitigate intensity scintillations under moderate to strong irradiance fluctuation conditions [122].

As a member of the pseudo-non-diffracting OAM-carrying beam family, HyGG vortex beams exhibit self-reconstruction and abrupt autofocusing properties. Notably, HyGG vortex beams can be converted into Gaussian mode, modified BG (MBG) mode, modified exponential Gauss (MEG) mode, and modified LG (MLG) mode by setting specific parameters of the beam source. Due to these characteristics, HyGG beams hold significant potential for use in OWC [103–106,123].

4.1.2. Multi-Modulation Compatible Miniaturization System

Free-space optical (FSO) communication has been extensively used to establish satellite-to-earth, inter-satellite, and deep-space links for purposes such as environmental monitoring, information transmission, and celestial exploration [124]. The electro-optical conversion in FSO links typically relies on various modulation techniques, including intensity modulation (IM) and phase modulation (PM). IM is widely utilized due to its simple architecture and ease of operation [125,126]. In contrast, PM is considered a more suitable solution for long-haul links, offering higher data rates and improved receiving sensitivity [127,128]. Therefore, achieving compatibility between IM and PM within the same communication system is of practical significance.

Traditional IM and PM both rely on the Mach–Zehnder modulator (MZM) and its associated supporting devices [129,130]. As a result, an IM/PM-compatible system requires two independent sets of modulation modules along with numerous auxiliary devices, significantly increasing system complexity. However, the limited resources on satellites impose strict conditions on the size, weight, and power (SWaP) of laser communication terminals in practical applications. Therefore, solutions that enhance compatibility and allow for seamless IM/PM switching while maintaining optimal SWaP are highly desirable.

Multi-modulation-format compatible laser communication refers to a laser communication system designed to support and efficiently manage multiple modulation formats. This capability allows for seamless communication between different devices. Such a system can dynamically switch between or simultaneously support various modulation formats to optimize performance based on changing conditions, including channel conditions, data rate requirements, power efficiency, and compatibility with various systems and standards. By being adaptable to multiple modulation formats, a laser communication system enhances its flexibility, efficiency, and reliability across diverse applications and environments.

Since the intensity or phase change of an optical signal can be controlled by the input electrical level of a chirp-managed laser (CML) [131,132], the CML has emerged as a convenient solution for achieving IM/PM compatibility. This CML-based modulation technique eliminates the need for traditional Mach–Zehnder modulators (MZMs) and their supporting devices, thus reducing the size, weight, and power (SWaP) requirements and aiding in the miniaturization of optical communication terminals. In [133], a CML-based IM/PM compatible transmitter is designed, demonstrating the ability to switch seamlessly between On–Off Keying (OOK) and return-to-zero differential phase-shift keying (RZ-DPSK) using a single set of devices.

4.2. Mode Diversity Reception

Using lasers as information carriers, FSO communication with fiber-based optical receivers presents an excellent solution for achieving high data rates, ultra-wide bandwidth, and low power consumption [134]. However, atmospheric turbulence during free-space propagation induces refractive index disturbances. This results in laser signals experiencing beam spread, beam drift, intensity scintillation, and angle-of-arrival fluctuations. These effects reduce the coupling efficiency of the spatial optical beam to the fiber, leading to significant power fading and scintillation, which consequently degrades the performance of practical laser transmission systems [135,136].

Several methods have been investigated to mitigate the effects of atmospheric turbulence and enhance the performance of high-speed FSO communication systems. One of the most commonly used and effective technologies is the AO system, which improves the coupling efficiency of free-space light into a single-mode fiber (SMF) by correcting wavefront aberrations within the aperture [137,138]. However, AO FSO systems still face challenges in terms of reliability under strong turbulence. Due to the limited spatial resolution and response time of the optics, AO does not achieve perfect wavefront correction in these conditions [139]. Additionally, high-speed FSO communication with bandwidths on the order of GHz demands higher accuracy and faster response times from AO systems, which can result in slight deficiencies in communication system reliability under strong turbulence [140].

Modes diversity offers a method to counteract wavefront distortion within the aperture, thereby enhancing the performance of FSO communication systems under atmospheric turbulence [139]. To this end, mode diversity reception (MDR) has been proposed and examined as an effective strategy for improving the reception performance of high-speed FSO communication systems. This approach involves receiving optical signals using few-mode fibers (FMFs) or multimode fibers (MMFs) that support multiple orthogonal modes [140–142]. Furthermore, the distorted wavefront caused by atmospheric turbulence can be considered a combination of a fundamental mode and multiple high-order modes. MDR captures as many high-order mode optical signals as possible. Using mode demultiplexers, the turbulent light is effectively split into an array of SMFs. Different methods are then used to combine the outputs from the single-mode ends of the demultiplexers to enhance signal quality. The probability of all copies, which carry the same information in the SMF array, simultaneous experiencing deep fading is low. Thus, mode diversity technology effectively improves the quality of the received signal and can be applied to satellite-to-ground optical communication systems [143].

The atmospheric turbulence compensation scheme using MDR involves two key technologies: mode reception and signal combination. In fiber-based MDR systems, mode demultiplexers are essential for decomposing spatial light affected by turbulence into SMFs. The two common types of mode demultiplexers used in these systems are multi-plane light converters (MPLCs) and photonic lanterns (PLs). Additionally, there are three primary methods for combining multichannel signals: equal gain combining (EGC), selection combining (SC), and maximum ratio combining (MRC) [140–142,144]. For different FSO communication scenarios, numerous studies have explored MDR using various mode demultiplexers and combining technologies to enhance system performance. For instance, ref. [145] proposes and verifies two simplified MDM schemes for FSO communication systems under moderate-to-strong turbulence conditions. These MDM schemes are designed and implemented for coherent detection in FSO communication systems using FMF coupling. The first scheme replaces the coherent receiver with a KK receiver (field reconstruction), significantly reducing the number of photodetectors and high-speed analog-to-digital converters by 75%. The second scheme achieves few-mode KK detection by employing a few-mode local oscillator and a single PD, thereby simplifying the MDM system to an extreme level. Experimental results indicate that an FSO communication system utilizing six-mode fiber demonstrates an average received power gain of approximately 6 dB over a system based on SMF under moderate-to-strong turbulence conditions. The two proposed MDRs exhibit similar performance, with an improvement of 5 dB to 6 dB compared to SMF-based heterodyne detection in the same turbulence conditions. Moreover, the simplified receiver schemes reduce both the volume and cost of the optical receiver, offering practical benefits for FSO communication systems.

4.3. Modulation Schemes

Modulation schemes are fundamental to the robustness and efficiency of FSO communication systems. The choice of modulation technique directly impacts the data rate, power efficiency, and resilience of the communication link, particularly under varying atmospheric conditions. For instance, conventional modulation techniques such as OOK and Pulse-Position Modulation (PPM) are commonly used, but they often suffer from performance degradation under adverse atmospheric conditions. Also, advanced modulation schemes like QAM and Orthogonal Frequency-Division Multiplexing (OFDM) can be employed to improve data rates and resilience in FSO systems. In addition, the emerging optical spatial modulation (OSM) technique can address the previously mentioned problems associated with conventional modulation schemes and Multiple-Input Multiple-Output (MIMO) setups. This section delves into both conventional and advanced modulation schemes, examining their characteristics, advantages, and limitations in the context of FSO communication. A summary of the literature on various modulation formats and their associated metrics for FSO is presented in Table 3.

Table 3. Literature summary on modulation formats and associated metrics for FSO.

Modulation Scheme	Metrics	Comments	Ref.
Numerical simulation			
PPM-MSK-SIM	BER	The BER performance of PPM-MSK-SIM surpasses that of both PPM and BPSK-SIM.	[146]
DOSM-PAM	ASER, APEP	A DOSM scheme that eliminates the need for CSIR results in a slight performance degradation compared to the conventional scheme with perfect CSIR.	[147]
Q-ary PPM	BER	The suggested Q-ary PPM FEC-based approach offers outstanding coding gain.	[148]
OOK, DPSK, DQPSK	BER	DPSK offers modulation gains of 3.2 dB over the OOK format in strong turbulence and 4.5 dB in weak turbulence. In moderate and strong turbulence, DPSK and DQPSK formats exhibit nearly identical BER performance.	[149]
MPAM	BER	The adaptive LDPC-coded modulation scheme can tolerate deep fades of 30 dB and higher in strong turbulence conditions.	[150]
MPPM	BER	Simulative evidence indicates that RL-MLC MPPM, when used with standard LDPC codes, can surpass the performance of any PPM scheme under identical transmission constraints.	[151]
OSM	ABEP	OSM can provide performance on par with conventional coherent FSO systems using spatial diversity, while also surpassing them in spectral efficiency and hardware simplicity.	[152]
SIM/SM	BER	The SIM/SM scheme can surpass the performance of conventional SIM in both AWGN and outdoor channels across different spectral efficiencies.	[153]
SCPPM/SCMPPM	BER	In comparison to the SCPPM, the SCMPPM offers improved BER performance under a peak power constraint and is capable of achieving higher data rates.	[154]
CPolSK	BER	The CPolSK system requires $\approx$ 3 dB less SNR compared to the OOK system.	[155]
OOK, M-QAM-OFDM	BER	System utilizing hybrid WDM-PON-FSO technology has the potential to achieve high data rates.	[156]
OSSK	BER, EC	AT and PE have an insignificant impact on the performance of the OSSK-based system.	[157]
COIQSM	BER	COIQSM provides an efficient way to improve the reliability and resilience of FSO systems against atmospheric turbulence.	[158]
Experimental demonstration			
CAP	BER	The proposed CAP and STBC scheme is minimally affected by air turbulence. It can be scaled up for higher speeds and extended for a greater interconnection range.	[159]
PAM-4-PolM-DD	BER	The PolM-DD system offers greater reliability compared to the conventional IM-DD FSO system.	[160]
OOK, BPSK, PPM	SER, Q-factor	PPM delivers superior performance in weak turbulence environments compared to OOK and BPSK.	[161]
SCOM, PM	BER	SCOM reduces RF power fading and interference. PM scheme with remotely injection-locked DFB LD has high resilience to noise and distortion.	[162]
M-QAM-OFDM, WDM	BE, BER	HH, PTS, and AC algorithms provide more than a 21% improvement in capacity. WDM can further enhance this capacity.	[163]
M-QAM-DMT	Q^2 factor, BER	In under weak turbulence conditions, the Q^2 factor of the M-QAM DMT signals with FEC encoding schemes maintains its maximum value and is improved compared to the regular scheme.	[164]

OOK: On–Off Keying, PM: Phase Modulation, PSK: Phase-Shift Keying, DPSK: Differential PSK, DQPSK: Differential Quadrature PSK, PAM: Pulse Amplitude Modulation, PolM: Polarization Modulation, PPM: Pulse-Position Modulation, MPPM: Multipulse PPM, SCPPM: Serially Concatenated PPM, SCMPPM: Serially Concatenated MPPM, CPolSK: Circle Polarization Shift-Keying, M-QAM: M-ary Quadrature Amplitude Modulation, OSSK: Optical Space Shift Keying, COIQSM: Coded-Optical Improved Quadrature Spatial Modulation, CAP: Carrierless-Amplitude-Phase Modulated, DMT: Discrete Multi-Tone, SCOM: Single-Carrier Optical Modulation, SIM: Subcarrier Intensity-Modulation, SM: Spatial Modulation, DD: Direct detection, IM-DD: Intensity Modulation and Direct Detection, OSM: Optical Spatial Modulation, DOSM: Differential OSM, FEC: Forward Error Correction, LDPC: Low-Density-Parity-Check, MLC: Multilevel Coding, RL-MLC: Reduced-Layer MLC, CSIR: Channel State Information at Receiver, BE: Bit Efficiency, BER: Bit-Error Rate, ASER: Average Symbol Error Rate, APEP: Average Pairwise Error Probability, ABEP: Average Bit Error Probability, EC: Ergodic Capacity, DFB LD: Distributed Feedback Laser Diode, AWGN: Additive White Gaussian Noise, AT: Atmospheric Turbulence PE: Pointing Errors, HH: Hughes-Hartogs, PTS: Partial Transmit Sequence, AC: Arithmetic Coding.

4.3.1. Traditional Modulation Techniques

The performance impairments caused by scintillation can be alleviated through various methods, including enhanced error control coding [10,165], advanced modulation schemes [166], and diversity techniques [10,167,168]. In modulation schemes, the three main optical signal parameters that can be adjusted to achieve high data transmission rates are phase, polarization, and intensity [169]. Choosing the appropriate modulation scheme is crucial for mitigating scintillation in FSO communications. Various FSO modulation techniques, such as PSK, OOK, PAM, and Frequency Shift Keying (FSK), have been extensively studied and analyzed over the years [93,147].

Traditionally, OOK using IM/DD is the most commonly used modulation scheme due to its simplicity in design and implementation, as well as its cost effectiveness [170]. However, OOK-based FSO systems are suboptimal in AT channels and provide a lower SE compared to PAM [171]. The high SE of the PAM scheme comes with a trade-off of considerable performance degradation, as higher-order PAM is highly sensitive to Rx noise and AT [147]. Consequently, PPM has been extensively employed in FSO systems, despite its poor bandwidth efficiency and requirement for symbol synchronization between transceivers, adding significant complexity to the system [146]. Phase modulation schemes, including standard and differential PSK (DPSK), are prone to phase noise [172]. For instance, DPSK is frequently employed but experiences fluctuations in the phase of the encoded signal as a result of scintillation. Differential Quadrature PSK (DQPSK) offers double the SE of binary modulations like DPSK and OOK, as it transmits two bits per symbol with four potential optical phase fluctuations in successive symbol periods [169]. Further advancements have utilized coherent digital modulation schemes such as QAM. While coherent modulations have shown improved data rates, their power efficiency limitations and complexity become more significant at higher data rates [158].

4.3.2. Advanced Modulation Techniques

Spectrally efficient QAM is widely used and preferred for various applications because it modulates both amplitude and phase. In fact, QAM encompasses a family of modulation techniques with different constellation types, each corresponding to distinct arrangements of the signal points [173]. So, for spectrally efficient high-data rate transmission, higher-order modulation techniques like the QAM family—which includes rectangular QAM (RQAM), squared QAM (SQAM), hexagonal QAM (HQAM), and cross QAM (XQAM)—are gaining increased attention in wireless communication systems due to their high power and bandwidth efficiency. In SQAM, constellation points are positioned at the vertices of a square lattice, suitable only for constellations with even powers of 2. To better adapt to varying channel conditions and improve spectral efficiency, incorporating constellations with odd powers of 2 offers greater flexibility [174–176]. RQAM is typically preferred for these odd-power constellations, as it is a versatile modulation scheme, encompassing orthogonal binary FSK (OBFSK), QPSK, multilevel Amplitude Shift Keying (ASK), and SQAM as special cases. However, RQAM is not ideal for odd-power constellations; instead, a modified XQAM constellation is favored for its reduced peak-to-average-power ratio (PAPR) and enhanced power efficiency. XQAM constellations are formed by removing the outer corner points of RQAM constellations and arranging them in a cross shape, minimizing the average power of the constellation [174,177].

Furthermore, the growing demand for high-data rates has led to the development of an optimal two-dimensional (2D) hexagonal-shaped constellation known as HQAM. HQAM features the densest 2D packing for a given optimal Euclidean distance between constellation points, resulting in improved power efficiency with lower peak and average powers as the constellation size (M) increases. Consequently, HQAM offers better BER/SER performance compared to other QAM schemes [174,175,177]. Depending on the trade-off between detection complexity and power efficiency, HQAM constellations are classified as regular and irregular HQAM structures. While regular HQAM has simpler detection, its power efficiency and BER performance can be further enhanced for larger M values.

Irregular HQAM provides superior power efficiency and optimal performance, though it comes with increased detection complexity [174].

Additionally, carrier recovery and automatic gain control (AGC) are necessary when using SQAM with differential coding, but these are challenging to implement in practice. Carrier recovery faces significant issues, including false locks. To eliminate the need for AGC and address the false lock problem, the star QAM constellation is introduced. Star QAM, a special case of circular amplitude PSK (APSK), outperforms SQAM in peak power-limited systems. It features multiple concentric PSK circles with equal constellation points and identical phase angles. The amplitude and phase of the constellation points are mutually independent, allowing for successful differential detection rather than coherent detection, thus removing the need for precise phase tracking and channel estimation at the Rx. Due to these advantageous characteristics, star QAM has been adopted in various satellite communication standards.

Conversely, the state of polarization is significantly less affected by turbulence-induced fading compared to phase and amplitude. Consequently, POLSK presents a viable alternative to optical intensity and phase modulation methods, owing to its resistance to phase noise and efficiency for long-distance operation [172]. Based on this advantage, a non-coherent binary POLSK (BPOLSK) technique has been proposed, utilizing the state of polarization of a fully polarized beam with two orthogonal channels as the information-bearing parameter [167,178]. In this context, POLSK-based schemes have been demonstrated to offer 3 dB better sensitivity than OOK. However, the effectiveness of POLSK in standard fibers is limited due to fiber birefringence. The polarization state changes unpredictably, making it difficult to track at the Rx. Conversely, POLSK does not encounter these issues in FSO communication systems. Theoretically, AT causes minimal depolarization, and there is negligible crosstalk between orthogonal polarization states during transmission. The polarization state of an optical wave is well preserved over distances of several kilometers, allowing reliable detection at the Rx [179]. Therefore, since turbulence only causes intensity distortion, while the polarization state remains stable in FSO channels [160,179], the use of polarization modulation (PolM) is both feasible and attractive for enhancing system performance and reducing turbulence-induced fading in FSO systems [167]. Additionally, subtracting the two signals reduces common mode noise. PolM-DD FSO communication systems for OOK signals have been extensively studied and shown to achieve an approximate 3 dB sensitivity gain compared to IM [155,160,180].

Turbulence-induced scintillation can be significantly mitigated by utilizing redundancy offered by multiple Txs and/or Rxs. Consequently, POLSK is also employed in MIMO FSO systems, as demonstrated in [181–183]. These systems typically investigate coherent BPOLSK or coherent multilevel POLSK (MPOLSK). However, the non-coherent BPOLSK system described in [178] has not been taken into account in a MIMO architecture. To address this, an innovative non-coherent MIMO optical modulation technique, termed PPM-BPOLSK, has been introduced. This approach integrates the implementation simplicity and performance benefits of non-coherent BPOLSK with the power efficiency of PPM. It has been shown that the proposed method offers enhanced error performance in AT channels compared to systems that rely solely on BPOLSK or PPM, establishing it as an effective alternative to other IM/DD techniques regarding power efficiency [167]. Also, assuming bandwidth is virtually unlimited, the main design emphasis in FSO systems has been on power efficiency instead of SE [167,184]. Consequently, MIMO FSO systems are typically designed using Space–Time Coding (STC) or repetition coding (RC) on the transmission side (see Section 4.6), along with diversity combining or aperture averaging techniques on the reception side to address scintillation effects. Additionally, utilizing multiple Txs and Rxs helps prevent beam blockage and enables coverage over greater distances [184].

OFDM is an extensively utilized method for broadband wireless communications, renowned for its inherent advantages such as enhanced resilience to frequency-selective fading, low cost, efficient implementation, narrow-band interference, and high spectral and power efficiency. As a result, OFDM has been incorporated into various high-speed digital communication standards and is being regarded as a potential solution for future mobile wireless systems [98,185–187]. Despite its disadvantages, such as sensitivity to phase noise and a high PAPR, OFDM can be optimized to maximize its benefits and mitigate its drawbacks for FSO channels through careful design [186,188]. Coherent detection is another promising technique for FSO channels, significantly enhancing Rx sensitivity [189]. Therefore, combining OFDM with coherent detection presents considerable potential for future FSO communication systems [186,188].

OFDM is a type of multicarrier transmission where high data rate streams are divided into lower rate streams and transmitted simultaneously over multiple narrow-band subcarriers. These subcarriers are modulated using PSK or QAM and are carried on a high-frequency carrier [98,186,187]. For example, in OFDM-based coherent FSO systems using multilevel PSK (MPSK) or M-ary (multilevel) QAM (M-QAM), it is necessary to employ additional pilot tones or training sequences for channel estimation and equalization. This is to counteract the amplitude and phase variations resulting from turbulence in the channel and laser phase noise. Consequently, this increases system complexity and reduces SE. Conversely, multilevel DPSK (MDPSK) shows great potential for OFDM-based coherent FSO systems due to its strong resilience to amplitude fading and phase distortion. MDPSK can mitigate the disadvantages of OFDM's sensitivity to phase noise and ensure good system performance without the need for equalization [188,190].

In OFDM, the spacing between adjacent subcarriers is meticulously chosen so that each subcarrier aligns with the spectral nulls of all the others, allowing them to be packed as closely as possible. This careful arrangement ensures spectral orthogonality, meaning each subcarrier is orthogonal to the others. This orthogonality enables the subcarriers to be separated at the Rx using correlation techniques, thereby eliminating intersymbol interference between channels. The orthogonal carrier set is generated by applying the Inverse Fast Fourier Transform (IFFT) at the Tx and the Fast Fourier Transform (FFT) at the Rx, thus negating the need for equalization. Figure 9 illustrates the fundamental setup of OFDM signal transmission over an FSO link [98,186,187].

The transmission of OFDM through an optical link is a specialized form of multiple-subcarrier modulation (MSM). In this method, multiple independent bit streams are modulated onto subcarriers at different frequencies, combined in the RF domain, and transmitted via high-capacity optical links employing IM/DD. The MSM signal undergoes minimal distortion and does not require equalization at the Rx. Each subcarrier functions as an individual narrow-band signal with a reduced symbol rate [98].

The baseband OFDM signal is inherently complex and bipolar, but an IM/DD optical link requires a real and positive RF signal to drive the LD. To create a real OFDM signal, Hermitian symmetry is applied to the input vector of the Tx's IFFT block. To convert the OFDM signal to unipolar, a DC bias is added (DC-OFDM), making the signal positive. This DC bias must be sufficiently large to avert clipping and distortion in the optical domain. However, the primary drawbacks of IM/DD MSM systems are the inefficiency in average optical power owing to the large DC bias and the distortions caused by the nonlinearities of the LD and optical channel. This issue holds particular significance for IM/DD DC-OFDM systems, where a significant number of subcarriers result in high PAPR [188]. Additionally, the LD's nonlinearity induces interference between subcarriers and broadens the overall signal spectrum. These factors create a complex scenario and impose strict demands on the linearity of optical devices to minimize distortion and ensure satisfactory performance [98].

Furthermore, one way to enhance SE is by using amplitude modulation techniques, such as PAM and QAM, with higher-order constellations [167,184]. However, these multilevel modulation methods come with increased hardware costs, as higher modulation levels lead to greater PAPR and increased sensitivity to noise and device nonlinearities [83]. High PAPR values can push Txs beyond their linear operating range, making large PAM constellations less desirable [184].

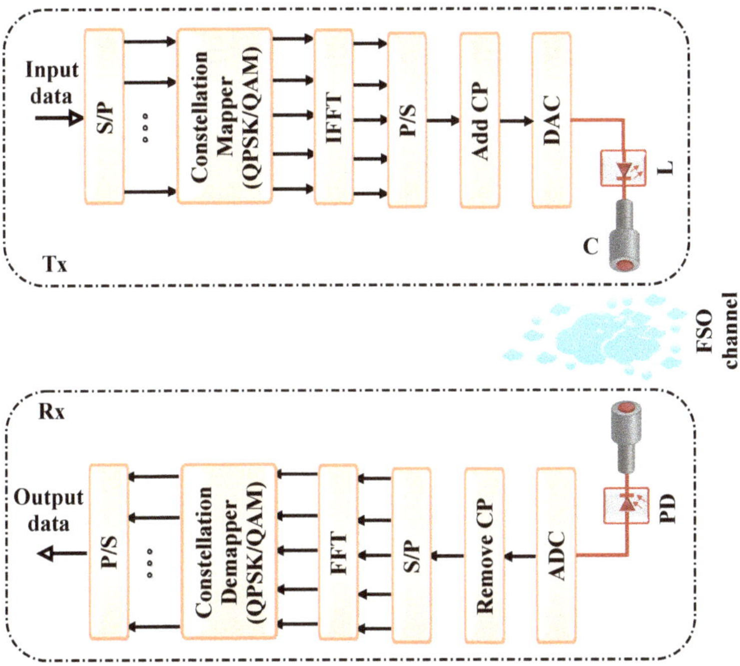

Figure 9. Basic setup for OFDM signal transmission. C: Collimator, PD: Photodetector, ADC: analog-to-digital converter, DAC: Digital-to-Analog Converter, FFT: Fast Fourier Transform, IFFT: Inverse Fast Fourier Transform, CP: Cyclic Prefix, S/P: Serial-to-Parallel, P/S: Parallel-to-Serial, L: Laser, Tx: Transmitter, Rx: Receiver.

Moreover, systems utilizing subcarrier intensity modulation (SIM) can enhance SE [83,167,184,191]. Consequently, the performance of subcarrier binary PSK (BPSK) IM in FSO systems has been thoroughly examined [92,191–195]. When compared to conventional SIM in FSO communication, BPSK-SIM demonstrates superior BER performance relative to other modulation schemes such as QPSK, MPSK, DPSK, and QAM [83,146,192,196]. To further improve the BER performance of FSO systems utilizing BPSK-SIM, a hybrid modulation scheme called PPM-MSK-SIM, which integrates PPM and MSK SIM, has been proposed [146]. However, the complexities of both the Tx and Rx in SIM systems are considerably higher than those found in pulse-based optical communications. Additionally, SIM systems face the challenge of requiring a time-varying DC bias to restrain the transmitted signal from going negative [83,167,184,191].

MIMO spatial multiplexing (SM) techniques are also taken into account when designing spectrally efficient FSO systems. However, these systems face several challenges, including inter-channel interference (ICI) from simultaneous transmissions, increased complexity in transceiver design to mitigate ICI, high computational complexity, the need for inter-antenna synchronization, and the need for numerous components to implement MIMO transceivers [152,167,184]. SM, initially developed for antenna-based systems, marked a shift towards free inter-antenna synchronization (IAS) [197,198]. In this approach, only one antenna is activated during each time slot, keeping the other transmission antennas inactive. This means SM requires just a single RF chain, making it a more cost-effective and less complex MIMO transmission structure that is resilient to IAS and ICI [158].

4.3.3. Advanced Optical Spatial Modulation Techniques

In light-based communication systems, LEDs have a relatively low modulation bandwidth, which greatly restricts their transmission rates. Additionally, conventional SISO techniques do not provide spatial diversity gains. To tackle these challenges, MIMO technology, similar to its use in RF communication, has been integrated into various OWC systems. A MIMO setup using SM can also enhance SE. However, as previously mentioned, the development of multiple antenna transmission schemes encounters several challenges.

The challenges associated with conventional modulation schemes and MIMO setups can be addressed by a novel technique known as OSM, where only a single transmit aperture (LED/laser) is active at any moment [158,199,200]. In these configurations, light sources replace antennas, and optical chains substitute for RF chains. During transmission, information is transmitted through both the signal and antenna spaces simultaneously. Specifically, a portion of the incoming bits is utilized to select a LED/laser, while the remaining bits choose the symbols to be transmitted from the selected LED/laser. By incorporating the LED/laser index as part of the symbol, additional bits can be transmitted, enhancing the system's SE without resorting to higher-order PAM or PPM. A particular case of OSM is optical space shift keying (OSSK), where transmit aperture indices alone are used to transfer information, simplifying the system compared to OSM. Since only one laser is active during each signaling period, issues related to ICI and antenna synchronization are completely avoided in OSM/OSSK, thereby reducing the design complexity of MIMO transceiver chains [152,167,184]. These inherent advantages make OSM/OSSK an ideal modulation technique for light-based communication systems such as underwater OWC (UOWC), visible-light communication (VLC), and FSO communication [201].

Furthermore, in OSM, the constellation size dictates the number of bits transmitted per symbol; larger constellation sizes can enhance SE but may also increase Rx complexity. Additionally, research indicates that the error performance of OSM surpasses that of alternative MIMO transmission techniques with equivalent SE, making it a viable option for FSO communication [152,167,184]. OSM can reduce the number of required constellation levels for a given SE without compromising power efficiency and while simplifying Rx complexity in comparison with MIMO SM techniques. This requirement can be minimized even further by incorporating the polarization dimension. Moreover, utilizing polarization alongside the Tx index and symbol dimensions could enable a corresponding MIMO system with half the number of Txs and Rxs, enhancing space efficiency [167]. Consequently, researchers have begun exploring FSO-OSM systems in combination with coherent detection, capitalizing on the seamless integration of their advantages. This exploration has led to significant improvements in both SE and transmission capacity [158]. An innovative coherent MIMO OSM architecture, which integrates the spatial modulation (SM) principle with MPOLSK, termed SM-MPOLSK, has been proposed. By allocating one bit to the additional polarization dimension, this new OSM approach necessitates smaller constellations or fewer transmit apertures than previous methods. Additionally, due to POLSK's inherent resilience to channel scintillation, this OSM method, which employs the polarization dimension along with multiple transmit/receive apertures, demonstrates enhanced error performance in AT conditions [167].

Researchers have recently created new OSM structures to enhance SE. This innovation enables the use of multiple Txs, leading to increased data rates and a flexible optical chain. This category of OSM is known as an *advanced OSM* system, which employs multiple light sources simultaneously for data transmission, resulting in higher data rates and more effective utilization of available spectrum bandwidths [158]. Advanced techniques for OSM include Optically Enhanced SM (OESM), Optical Generalized SM (OGSM), and Optically Improved Quadrature SM (OIQSM) [202–210].

In OESM, the number of active light sources varies based on the signal constellation. If the size of the primary constellation is m, where $m = \log_2(M)$, then OGSM employs N_{opt} optical chains, where N_{opt} light sources are activated by N_L light sources. The optimal N_{opt} for achieving maximum SE should lie between $\left\lfloor \frac{N_L}{2} \right\rfloor$ and N_L. For enhanced SE, OIQSM separates two symbols from a constellation into their real and imaginary components, which are transmitted independently using two sets of light sources. Table 4 summarizes the SEs and optical chains required for these advanced SM techniques. Overall, OIQSM provides the highest SE with the smallest constellation size, requiring one optical chain for OSM and four for OIQSM.

Although, OSM system improves SE, it also leads to higher hardware costs and increased power consumption. These associated costs vary across different schemes. For instance, the power consumption, $P_{tot}^{(\cdot)}$, number of switches, $N_{sw}^{(\cdot)}$, and total hardware expenses,$C_{(\cdot)}$, for OSM, OESM, OGSM, and OIQSM are expressed as [199]

$$P_{tot}^{OSM} = P_{oc} + \alpha_{tr} P_t + P_{sw} N_{sw}^{OSM}, \tag{34a}$$
$$P_{tot}^{OESM} = 2P_{oc} + \alpha_{tr} P_t + P_{sw} N_{sw}^{OESM}, \tag{34b}$$
$$P_{tot}^{OGSM} = P_{oc} N_{opt} + \alpha_{tr} P_t + P_{sw} N_{sw}^{OGSM}, \tag{34c}$$
$$P_{tot}^{OIQSM} = 4P_{oc} + \alpha_{tr} P_t + P_{sw} N_{sw}^{OIQSM}. \tag{34d}$$

$$N_{sw}^{OSM} = 2^{(\eta_{SE} - log_2(M) - 1)}, \tag{35a}$$
$$N_{sw}^{OESM} = 2^{(\eta_{SE} - log_2(M))/2 - 1}, \tag{35b}$$
$$N_{sw}^{OGSM} = 1/4\left(1 + \sqrt{1 + 4 \times 2^{((\eta_{SE} - 2log_2(M)) + 1)}}\right), \tag{35c}$$
$$N_{sw}^{OIQSM} = 1/4\left(1 + \sqrt{1 + 4 \times 2^{((\eta_{SE} - 2log_2(M))/2 + 1)}}\right). \tag{35d}$$

$$C_{OSM} = C_o + C_{S/P} + C_{sw} N_{sw}^{OSM}, \tag{36a}$$
$$C_{OESM} = 2C_o + C_{S/P} + C_{sw} N_{sw}^{OESM}, \tag{36b}$$
$$C_{OGSM} = C_o N_{opt} + C_{S/P} + C_{sw} N_{sw}^{OGSM}, \tag{36c}$$
$$C_{OIQSM} = 4C_o + C_{S/P} + C_{sw} N_{sw}^{OIQSM}. \tag{36d}$$

where P_t represents the total transmitted optical power, P_{oc} and P_{sw} denote the power consumption of the optical chain and the optical switch, respectively. α_{tr} denotes the slope-dependent load factor, η_{SE} represents the SE, and $C_{S/P}$, C_o, and C_{sw} are the costs associated with the serial-to-parallel converter, the optical chain, and the optical switch, respectively.

Table 4. Spectral efficiency and optical chain for various modulation techniques.

Modulation	Spectral Efficiency (SE) [bpcu]	Optical Chain	References
SM	$\log_2 M + \log_2 N_L$	1	[158,199]
OESM	$\log_2 M + 2\log_2 N_L$	2	[158,199,203,205,206]
QSM OIQSM	$2\left\lfloor \log_2\left(\frac{N_L}{2}\right) \right\rfloor + 2\log_2 M$	4	[158,199,210]
OGSM	$\left\lfloor \log_2\left(\binom{N_L}{N_{\text{opt}}}\right) \right\rfloor + N_{\text{opt}} \log_2 M$	$\left\lfloor \frac{N_L}{2} \right\rfloor \le N_{\text{opt}} \le N_L$	[158,199,210]

bpcu: bits per channel use.

The quantity of switches needed for OGSM, OESM, and OIQSM is significantly lower than that for OSM at a given SE and M-ary modulation scheme. Consequently, the power consumption and switch costs are much lower for higher SEs in OESM, OGSM, and OIQSM. As SE increases, the number of switches in OSM grows exponentially, while in the other schemes, the growth is linear. This results in the total power consumption of OSM exceeding that of the advanced OSM schemes because of the high switch demand. It is important to note that the modest rise in power consumption due to multiple optical chains in advanced OSM schemes is balanced by the significantly higher power consumption of switches in OSM at extremely high SE levels. Additionally, the overall cost of OSM exceeds that of other advanced OSM schemes when achieving such high SE values. Therefore, as SE increases, advanced OSM schemes outperform OSM in both cost and power consumption [199].

In general, modulation schemes have a vital impact on the performance of FSO communication systems. However, the choice of modulation scheme in FSO communication systems involves trade-offs between simplicity, data rate, power efficiency, and resilience to atmospheric conditions. Conventional schemes such as PPM and OOK are easier to implement and more cost-effective but offer lower data rates and are more susceptible to adverse weather conditions. Advanced schemes like QAM and OFDM provide higher data rates and better resilience to AT but come with increased complexity and implementation costs. Also, the issues associated with conventional modulation schemes and MIMO setups can be addressed by OSM and its variants. Ongoing research and development in these advanced techniques are essential to overcoming the inherent challenges of FSO communication and unlocking its full potential in various applications.

4.4. *Adaptive Transmission*

This section presents adaptive transmission methods that dynamically adjust transmission parameters including, code rate, transmit power, and modulation size, to optimize communication performance. By adapting these parameters in real-time, these methods aim to maintain reliable and efficient data transmission even under varying channel conditions and interference levels. Also, by employing these adaptive transmission techniques, wireless communication systems can significantly enhance their performance, achieving a balance between reliability, efficiency, and speed. This dynamic adjustment capability is crucial for modern communication systems, including cellular networks, satellite communications, and emerging technologies like UAV-to-ground FSO systems, where varying conditions are the norm.

4.4.1. Overview

Two important parameters that characterize turbulence-induced fading are d_0, the correlation length of intensity fluctuations, and τ_0, the correlation time of intensity fluctuations. These parameters are defined respectively as [47,211,212].

$$d_0 \approx \sqrt{\lambda L}. \tag{37a}$$

$$\tau_0 = \frac{d_0}{u \perp}. \tag{37b}$$

where $u \perp$ denotes the component of the wind velocity vector perpendicular to the direction of propagation.

When the Rx aperture D_0 exceeds the correlation length d_0, turbulence-induced fading can be significantly mitigated through aperture averaging [211]. However, due to limitations on Rx size, it is not always feasible to ensure that the Rx aperture is much larger than the turbulent correlation length (i.e., $D_0 > d_0$). In scenarios where $D_0 < d_0$, aperture averaging loses its effectiveness, necessitating alternative methods to counteract intensity fluctuations [150,211,212]. These methods are categorized into two main types: spatial-domain techniques, which involve diversity detection using multiple Rxs, and time-domain techniques, which adaptively optimize the decision threshold based on the maximum likelihood criterion.

In temporal-domain techniques, a single Rx is employed. If the Rx knows the marginal fading distribution but lacks information about both the temporal fading correlation and the instantaneous fading state, a maximum-likelihood symbol-by-symbol detection technique can be utilized. Conversely, when the Rx is aware of the joint temporal fading distribution but not the instantaneous fading state, it can apply a maximum-likelihood sequence detection (MLSD) technique to counteract turbulence-induced fading.

Spatial-domain techniques necessitate at least two Rxs to capture signal light from various positions or angles. To optimize diversity reception gain, the Rxs should be positioned as far apart as possible to ensure that turbulence-induced fading remains uncorrelated between them. However, in practice, it may not always be feasible to place the Rxs at such a distance.

Although MLSD offers an optimal solution, it involves complex multidimensional integration and has a computational complexity that increases exponentially with the length of the transmitted bit sequence. To simplify MLSD, suboptimal implementations such as per-survivor processing (PSP) are often preferred in practice [186,212]. However, for BER below 10^{-6}, both MLSD and PSP require an electrical SNR exceeding 20 dB, even in weak turbulence conditions. Such high signal powers are impractical for many applications, necessitating the development of novel modulation techniques for IM/DD FSO systems. In response, various coding approaches have been proposed, including coded OFDM [186], coded MIMO [213], and rateless coding [214,215].

The core idea behind the coded-OFDM method is to reduce the symbol rate through the use of OFDM. When combined with interleaving and low-density parity-check (LDPC) codes, this technique achieves high resilience to the deep fades characteristic of turbulent channels. However, OFDM is more susceptible to phase noise and has a comparatively high PAPR, necessitating meticulous design to maximize its benefits and mitigate its drawbacks. Given that OFDM in IM/DD systems may lack power efficiency, SSB clipped and unclipped OFDM schemes can be employed to enhance power efficiency [186].

In a coded-MIMO approach, multiple optical sources and detectors are utilized, based on either repetition MIMO concepts or Space–Time Coding, along with LDPC codes, to enhance resistance to AT [213]. Temporally correlated stochastic channels can benefit from rateless codes, which are ECC with adjustable rates to accommodate varying channel quality. Examples of such codes include punctured codes and fountain codes, with Raptor codes being a specific instance. Punctured codes vary the rate by dropping parity bits, thereby increasing the effective code rate. This method requires a base code with a minimum rate suitable for the worst-case channel condition. Puncturing schemes are typically defined and tested at each desired rate to ensure optimal performance. Conversely, Raptor codes adjust their rate by extending the codeword length, with the Tx generating coded bits continuously until the Rx decodes the transmitted word without error. Raptor codes have been previously proposed for the network layer of both wired and RF communication systems [215]. Demonstrations of Raptor and punctured codes in the PHY layer of an FSO link, under various turbulence conditions, have shown high tolerance to deep fades caused by AT [214]. However, the achievable information rates (lower bounds on channel capacity) indicate that these approaches are still several decibels below channel capacity, suggesting potential for further improvement [216].

4.4.2. Adaptive Transmission Techniques

Adaptive modulation and coding (AMC), which is already utilized in wireless channels, facilitates reliable and spectrally efficient transmission over channels that vary over time [217]. The core concept involves estimating the channel conditions at the Rx and relaying this estimate back to the Tx via an RF feedback channel, allowing the Tx to adjust according to these conditions. Recently, adaptive transmission techniques have been integrated into FSO systems, showing promise as solutions to counteract AT and enhance FSO link performance [150,214].

A prevalent assumption in the existing literature is that open-loop implementations are used, where the Tx lacks knowledge of the channel conditions. Such open-loop designs are beneficial in time-varying channels, where obtaining feedback on channel estimates can be challenging. However, in the case of quasi-static channels, reliable feedback is attainable, and having available CSI at the Tx can enable the design of adaptive transmission schemes that significantly enhance performance. AT causes slow fading in FSO systems, with the coherence time of the optical channel ranging from 1 to 100 ms. This means that fading can remain constant over hundreds of thousands to millions of sequential bits at standard transmission rates, allowing the channel to be modeled as a slow fading process [20]. Consequently, adaptive transmission presents an attractive option for FSO systems.

Figure 10 illustrates a typical block diagram of an integrated power control and channel coding system in FSO communication. The feedback information needed for adaptive transmission can be relatively easily implemented in FSO systems, as commercially available FSO units typically support bidirectional (full-duplex) communication. A small fraction of the extensive available bandwidth can be assigned for feedback without significantly impacting data rates. For instance, in some hybrid RF/FSO systems, the RF channel can be utilized for CSI feedback. This makes reliable CSI accessible in FSO systems, which is crucial for developing adaptive transmission schemes. As a result, to improve communication reliability and throughput, the Tx can adjust its transmission parameters—such as code rate, transmit power, and modulation size—based on the estimated current CSI [218–222]. Thus, various varieties of adaptive transmission systems have been suggested in the existing research on FSO communication, demonstrating their feasibility and advantages. For example, in [223], a straightforward rate-adaptive transmission technique is introduced that utilizes variable silence intervals and OOK formats with memory.

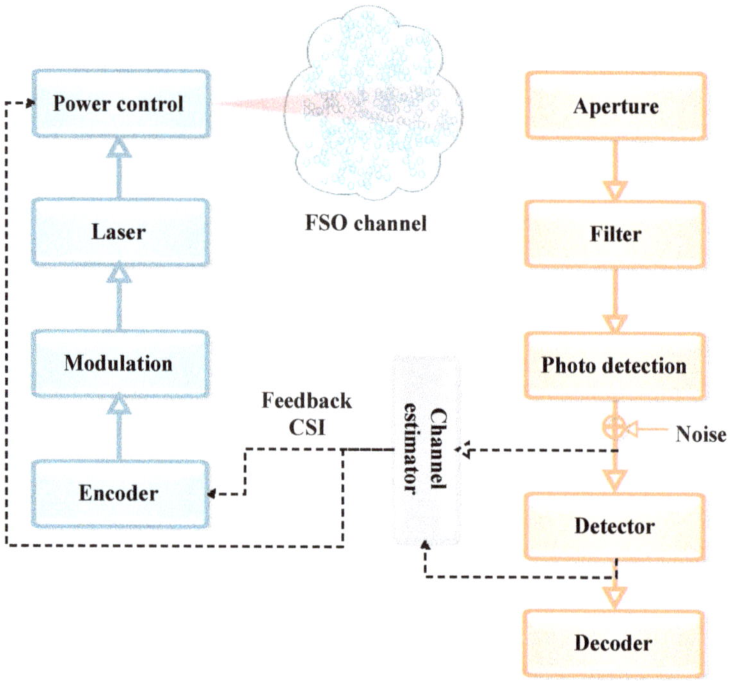

Figure 10. Block diagram of integrated power control and channel coding in an FSO communication system.

The study in [224] presents a statistical-delay quality of service (QoS)-aware adaptive modulation and power allocation method for a dual-channel coherent OWC system facing AT fading. The findings indicate that this adaptive scheme significantly outperforms conventional adaptive transmission methods under strict statistical-delay constraints and highlights the benefits of joint channel optimization in conditions of strong turbulence fading and tight statistical-delay requirements. Additionally, in [225], a joint relaying link selection and power allocation strategy that is statistical delay QoS-aware is proposed for the uplink of a multichannel coherent FSO communication-based fronthaul network. The analysis shows that to maximize the overall end-to-end (e2e) effective capacity, it is essential to account for both the statistical delay QoS needs of the transmitted traffic and the channel gains of the transmission links. The results further demonstrate that this approach enhances the statistical delay QoS-aware throughput despite the challenges posed by AT fading and PE.

An adaptive symbol rate, where the symbol duration adjusts based on channel conditions, along with the simultaneous adaptation of both the symbol rate and power, has been proposed to mitigate the effects of scintillation in FSO links [20]. In [14], the challenges of adaptive FSO transmission over AT channels are revisited, considering practical power constraints and modulation sizes. Two scenarios, (1) power adaptation with a fixed modulation size and (2) joint adaptation of power and modulation size, are explored. Also, an adaptive uncoded transmission algorithm that adjusts both transmit power and modulation size based on channel conditions is introduced. The optimization focuses on these adaptive parameters while adhering to peak power constraints. Simulation results indicate that these adaptive algorithms significantly enhance performance, particularly in conditions of intense turbulence.

In [222], a coded FSO system utilizing a fixed code rate and M-ary PPM under an average power constraint is investigated. The simulation results show that the proposed adaptive scheme offers considerable performance enhancements in relation to non-adaptive alternatives. In [226], the study examines rate-adaptive FSO communication using experimental data collected from a 1.87 km terrestrial FSO link under various weather conditions. To address the fluctuations in channel gain caused by AT and weather changes, a rate-adaptive communication system is implemented that punctures LDPC codes in a rate-compatible manner. In addition to the standard random puncturing method, an optimized intentional puncturing technique is also utilized. The evaluation results indicate that, unlike uncoded OOK, the rate-adaptive method ensures reliable and efficient FSO communication across different weather conditions and scintillation indices. Furthermore, applying intentional puncturing enhances the system's efficiency in terms of throughput.

Moreover, a novel adaptive transmission technique for FSO systems, designed to operate under AT and utilizing subcarrier PSK (S-PSK) intensity modulation, is introduced in [227]. By leveraging the fixed envelope properties of S-PSK, this approach enhances the effective use of the FSO channel capacity by adjusting the modulation order of S-PSK in response to real-time turbulence-induced fading and a specified BER target. Numerical results demonstrate considerable gains in SE without increasing the average transmitted optical power or compromising BER, particularly in conditions of moderate to strong turbulence. Additionally, this variable-rate transmission technique is utilized in MIMO FSO systems, leading to further improvements in SE as the number of transmit and/or receive apertures increases. The performance of an FSO system for ground-to-satellite communication is examined in [228], focusing on the combined impacts of AT and beam wandering using MPSK. To enhance SE, an adaptive transmission strategy is implemented, optimizing channel capacity by adjusting the modulation order of the MPSK scheme based on the current channel conditions and the target BER. Additionally, a single-input multiple-output (SIMO) system employing MRC diversity is utilized to alleviate the impacts of AT and beam wandering.

Optical power allocation strategies for an adaptive Wavelength Division Multiplexing (WDM) radio over FSO (RoFSO) systems, utilizing variable wavelengths to reduce the impact of weather turbulence, are introduced in [220,229]. These strategies optimally distribute transmit power to maximize capacity and enhance the overall effectiveness of the WDM RoFSO system. Thus, within a total power constraint, varying optical powers can be assigned to selected wavelengths, leading to further enhancements in system performance [16]. A practical FSO system featuring an adaptive channel coding and power control scheme is examined in [221]. This approach allows for the simultaneous adjustment of both the coding rate and transmission power to minimize power consumption while adhering to practical constraints, such as the target OP, target BER, and transmit power limits imposed by eye-safety standards.

In [150], an AMC scheme is introduced as an effective method for addressing severe AT. The study explores three distinct adaptive modulation scenarios: (i) channel inversion with a fixed rate, (ii) variable-rate variable-power adaptation, and (iii) truncated channel inversion with a fixed rate. Additionally, the potential enhancements from using adaptive LDPC coded modulation (CM) are examined. The findings indicate that, under strong turbulence conditions, it is possible to withstand significant fades of 30 dB and higher [150].

Additionally, the FSO channel can exhibit a significant level of reciprocity in axially aligned bidirectional systems, meaning that the fading experienced at each end can be closely correlated [230,231]. This characteristic allows for immediate access to CSI at the Tx, eliminating the need for a separate feedback channel. Consequently, it is justifiable to assume the availability of CSI at both the Tx and Rx, which is essential for adaptive transmission [20].

Many proposed adaptive transmission methods based on CSI in FSO communication aim to enhance the reliability and throughput of these systems. However, these methods often assume that turbulence conditions remain constant. In reality, varying turbulence intensities at the same SNR can significantly impact communication quality. This suggests that while adaptive schemes may perform well under stable turbulence, their effectiveness can diminish or even degrade when faced with changing turbulence conditions. For example, an AMC scheme designed for weak turbulence may struggle to maintain the required BER in moderate or strong turbulence, leading to reduced communication quality or interruptions. Conversely, applying an AMC scheme intended for strong turbulence in milder conditions might result in a less efficient coding and modulation choice, ultimately lowering channel utilization and bit rate. To address these challenges, it is essential to develop an enhanced AMC scheme capable of adapting to FSO channels with varying turbulence intensity. Accordingly, an enhanced AMC scheme is proposed, incorporating a hybrid switching standard for UAV-to-ground FSO communication to mitigate the adverse effects of turbulence variation. This approach incorporates a channel estimator based on Gated Recurrent Units (GRU) and utilizes a hybrid switching standard linked to the frame error rate (FER) of LDPC decoding, aiming to optimize bit efficiency across different turbulence conditions.

4.5. Aperture Averaging

Aperture averaging is a crucial technique in FSO communication systems, employed to mitigate the adverse effects of AT. By averaging the optical signal over a larger aperture, this method effectively reduces signal fluctuations and enhances communication reliability. This section will explore the principles of aperture averaging, its advantages in mitigating the impacts of turbulence, and its implications for the future of FSO communication systems.

4.5.1. Overview

A straightforward method to mitigate the fading effect is to employ a relatively large lens at the Rx to smooth out intensity fluctuations. This technique, commonly referred to as *aperture averaging*, serves as an inherent form of receive diversity. It involves adjusting the size of the detector or receive aperture to capture as much of the traversing optical field as possible. In FSO systems operating under strong and saturated conditions, aperture averaging is essential to address (1) signal scintillation or irradiance flux variations [232,233]; (2) address AoA fluctuations and boresight angle effects; and (3) to optimize the average signal power at the detector plane. Additionally, aperture averaging is a fundamental scheme that ensures spatial diversity [46]. In this context, aperture averaging can be considered a basic form of spatial diversity when the Rx lens aperture exceeds the fading correlation length [47,233,234].

Aperture averaging calculations are crucial in Rx design. The Rx needs to be large enough to gather adequate power and minimize scintillation effects at a specific distance, yet small enough to remain practical. Typically, the Rx does not require a significantly large diameter due to the nonlinear decrease in scintillation with increasing diameter. It is important to note that while aperture averaging becomes increasingly effective at reducing intensity variance as turbulence intensifies, this improvement is limited to a certain extent [233,235].

4.5.2. Advantages of Aperture Averaging

To achieve effective FSO communication through continuous transmission link establishment, it is crucial to maintain a larger aperture area, resulting in a wider FoV. A narrow FoV limits the coverage area, which is not ideal for FSO communication systems. Although a larger collection aperture increases ambient light background noise, this can be mitigated by various methods that allow the FSO Rx to capture a larger portion of the incoming optical beam. Conversely, a higher FoV provides a larger coverage area and eliminates the need for additional circuitry for pointing and tracking mechanisms, ensuring reliable wireless connectivity between transmission and reception apertures. Moreover, FSO systems with a larger FoV encounter reduced misalignment errors, reducing the impact of obstacles between the source and destination points and easing the alignment control mechanism [46].

4.5.3. Impact of Aperture Averaging

In FSO systems, the impact of aperture averaging is examined by taking into account various facets of these systems. For aperture-averaged FSO Rxs, it is widely recognized that the $\Gamma\Gamma$ distribution is valid for a point Rx under a bounded Gaussian wave model; however, it does not support aperture averaging phenomena [236,237]. Also, experimental studies indicate that the LN model is valid for a point Rx in the weak turbulence regime and performs well in all turbulence regimes for aperture-averaged data [234,238]. Conversely, while the $\Gamma\Gamma$ model is considered valid for a point Rx across all turbulence regimes, this validity does not extend to scenarios involving aperture averaging [234,238,239]. Additionally, it has been demonstrated that the LN and $\Gamma\Gamma$ models become virtually identical when $D > 6\rho_0$, where ρ_0 is the spatial coherence radius [240]. To analyze the impact, consider the PDF of the irradiance, h_a, which follows the $\Gamma\Gamma$ distribution as defined in Equation (22) [91,94,96,97]:

$$f_{h_a}(h_a) = \frac{2(\alpha\beta)^{(\alpha+\beta)/2}}{\Gamma(\alpha)\Gamma(\beta)}(h_a)^{\frac{(\alpha+\beta)}{2}-1}K_{\alpha-\beta}(2\sqrt{\alpha\beta h_a}), \tag{38}$$

Utilizing the power series expansion of the modified Bessel function around the origin, Equation (38) can be estimated as [241]

$$f_{h_a}(h_a) \approx a_0 I^{\beta-1} + a_1 I^{\beta}, \tag{39}$$

In Equation (39), only the initial two terms of the power series expansion are taken into account, with a_0 and a_1 defined accordingly as [241]

$$a_0 = \frac{\Gamma(\alpha-\beta)}{\Gamma(\alpha)\Gamma(\beta)}(\alpha\beta)^{\beta}, \tag{40a}$$

$$a_1 = \frac{\Gamma(\alpha-\beta)}{\Gamma(\alpha)\Gamma(\beta)}\frac{(\alpha\beta)^{\beta+1}}{(1-\alpha+\beta)} \tag{40b}$$

All the expressions provided for $\Gamma\Gamma$ are valid for point Rxs, where the diameter of the Rx lens (D) is zero. Specifically, the channel parameters α and β in Equation (24a) and (24b), as well as the power series expansions in (39), (40a) and (40b), apply in the case of $D = 0$. For Rxs that utilize aperture averaging, where $D > 0$, there is a reduction in the variance in intensity fluctuations. In such cases, the $\Gamma\Gamma$ statistical model remains applicable but with modified parameters α and β [233,241].

The effect of aperture averaging is that Equations (39), (40a), and (40b) are no longer valid. For $D > 0$, the parameter β may exceed the parameter α, which is different from the situation with point Rxs. Nonetheless, this issue can be readily addressed, as when $\beta > \alpha$, Equation (39) can be reformulated as

$$f_{\gamma\gamma}(I) \approx a_0 I^{\alpha-1} + a_1 I^{\alpha} \tag{41}$$

where (40a) and (40b) can be expressed as

$$a_0 = \frac{\Gamma(\beta-\alpha)}{\Gamma(\alpha)\Gamma(\beta)}(\alpha\beta)^{\alpha}, \tag{42a}$$

$$a_1 = \frac{\Gamma(\beta-\alpha)}{\Gamma(\alpha)\Gamma(\beta)}\frac{(\alpha\beta)^{\alpha+1}}{(1-\beta+\alpha)} \tag{42b}$$

It is important to mention that these expressions are akin to (39), (40a), and (40b), where the roles of α and β should be reversed when $\beta > \alpha$.

Furthermore, examine the scintillation index, $\sigma_N^2(D)$, associated with a Rx lens of diameter D. The impact of aperture averaging is to reduce this variance when $D > 0$. Thus, the effectiveness of aperture averaging is influenced by the propagation conditions [233]. The degree of intensity fluctuation is dependent on the size of the Rx aperture [235]. The extent to which aperture averaging diminishes the variance in intensity fluctuations is measured by a parameter known as the *aperture averaging factor* [242]. The aperture averaging factor, denoted as A_{av}, is described as the ratio of the normalized intensity variance in fluctuations for a Rx with a diameter D compared to that of a point Rx, and is given by [233,235]

$$A_{av} = \frac{\sigma_N^2(D)}{\sigma_N^2(0)}, \tag{43}$$

where $\sigma_N^2(0)$ represents the scintillation index for a point Rx with $D = 0$. In the regime of weak fluctuations, A can be estimated as [233,235]

$$A_{av} \approx \left[1 + 1.062\left(\frac{D^2}{4}\frac{k}{L}\right)\right]^{-7/6}. \tag{44}$$

The concept of aperture averaging to mitigate scintillation effects has been extensively studied both in theoretical research and practical applications [232,234,243]. The reduction in fading is commonly measured by the aperture averaging factor (refer to Equation (43)). Research indicates that a significant reduction in scintillation can be achieved, particularly under conditions of strong turbulence. Enhanced scintillation reduction might be possible by employing multiple beams at the Tx and/or multiple lenses at the Rx [233,244,245].

Additionally, the quest for a distribution that can precisely model the PDF of irradiance data across all AT conditions while considering aperture averaging continues to inspire significant research efforts [238]. To tackle this challenge, an alternative family of PDFs is presented to represent the PDF of received optical power in FSO communications, specifically the Weibull and EW distributions [46,246]. Notably, the EW distribution demonstrates a remarkable fit to both simulation and experimental data across all aperture averaging scenarios, including weak and moderate turbulence regimes, and point-like apertures as well. Another appealing aspect of these distributions is the straightforward closed-form expressions for their corresponding PDF and cumulative distribution function (CDF) [238,247–251].

4.6. Spatial Diversity

Spatial diversity is a key technique employed in FSO systems to enhance communication reliability and performance. By utilizing multiple transmission and/or reception points, spatial diversity effectively mitigates the effects of AT, scattering, and signal fading, which are prevalent challenges in FSO communications. This section explores the principles of spatial diversity, its implementation strategies, and the associated benefits, including improved signal quality and increased system robustness. Through a detailed examination of various spatial diversity configurations, we aim to highlight their significance in achieving more resilient FSO links, particularly in dynamic and unpredictable environmental conditions.

4.6.1. Overview

It has been shown that the performance of SISO FSO links is significantly impacted by strong turbulence, falling short of the typical BER targets for FSO applications within practical SNR ranges. This highlights the need for the implementation of robust fading-mitigation techniques [252]. Historically, various techniques have been employed in FSO systems to address the detrimental effects of the atmospheric channel. These include error control coding combined with interleaving [165,253] and multiple-symbol detection through MLSD [47,212]. However, both methods present certain practical challenges [252]. For example, relying solely on an ECC is not an effective strategy for mitigating fades, as it necessitates a significantly large interleaver, around 100 Gb for a 10 Gb/s channel [245,254,255]. Moreover, when the link functions as the PHY layer of a data network, the upper layer protocols struggle to perform optimally in the presence of these fades and the prolonged delays associated with interleaving and deinterleaving, resulting in reduced throughput and poor efficiency [233,254]. It has also been noted that channel coding is effective primarily in weak turbulence conditions or when a large aperture size is utilized. In reality, coding is not efficient in countering fading. Its advantages become more pronounced only in low fading channels or when the effects of turbulence are significantly mitigated, for instance, through techniques like aperture averaging, AO, or spatial diversity [233]. Conversely, MLSD is hindered by high computational complexity [47,252].

Another promising approach is the implementation of spatial diversity, a well-established technique in RF systems. By utilizing multiple apertures at either the Tx or Rx, spatial diversity can significantly enhance performance by introducing additional DoF in the spatial dimension. This method is particularly appealing for reducing the frequency and duration of signal fades [256]. It is especially critical in strong turbulence channels, where SISO links demonstrate extremely poor performance (i.e., exceeding 10^{-3} in the SNR range of 30–50 dB) [252]. Furthermore, spatial diversity minimizes the likelihood of temporal blockage of laser beams caused by obstructions, enabling communication over longer distances even in challenging weather conditions [252].

Furthermore, unlike spatial diversity in wireless systems, spatial diversity in atmospheric optical systems can be implemented compactly, as the coherence length is typically on the order of centimeters. This means that multiple Txs or Rxs can be positioned just centimeters apart to experience nearly independent channel fades [245,254,255]. Consequently, effective fading reduction can be achieved through spatial diversity techniques, such as employing multiple beams at the Tx (MISO) [257,258], using multiple apertures at the Rx (SIMO) [233,245,259], or utilizing a combination of both (MIMO) [244,260,261]. Among these approaches, using multiple lenses at the Rx is the simplest, offering a balance between gain and system complexity in achieving spatial diversity [233,262].

The overall spatial diversity system configuration, illustrated in Figure 11, consists of M Txs and N Rxs, where $M,\ N \geq 1$ and at least one of M or N is greater than one. The outputs from the Rxs can be combined in various ways to generate the final observations, which are used to determine whether a 0 bit or a 1 bit is transmitted [91,255]. In various application scenarios, it is reasonable to assume that the Tx subapertures are spaced apart by more than a coherence length, which is particularly significant for ground-to-space links [254,255].

FSO links utilizing transmit and receive diversity can be effectively modeled as equivalent SISO systems by appropriately scaling the channel variance. In essence, spatial diversity leads to a reduction in channel variance. Additionally, the performance degradation resulting from spatial correlation can be significant, highlighting the importance of efficient separation between apertures and precise co-alignment to realize the expected diversity benefits from multiple Txs and Rxs [263]. A key observation is that as the degree of diversity—whether spatial, temporal, or frequency—increases, performance improves significantly and rapidly [47,254].

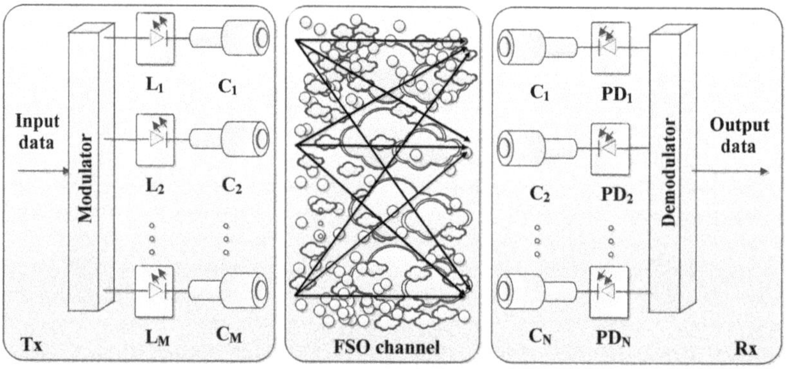

Figure 11. Generic spatial diversity setup.

4.6.2. MISO, SIMO, and MIMO Free-Space Optical Systems

To improve the reliability and performance of FSO communication systems under varying atmospheric conditions, advanced configurations such as Multiple-Input Single-Output (MISO), SIMO, and MIMO are employed. This section delves into these configurations, exploring their architectures, benefits, and the unique challenges they address in the realm of FSO communications. By leveraging multiple Txs and/or Rxs, these systems aim to alleviate the impacts of AT and enhance link robustness and capacity.

Transmit Diversity FSO Systems

Utilizing optical arrays, akin to antenna-array technology in microwave systems, is viewed as a strategy for addressing the fading effects of FSO channels. In particular, MISO (transmit diversity) FSO systems are based on spatial diversity methods with multiple lasers at the Tx end, all of which are pointed towards a distant photodetector. In these systems, the sources and the detector are positioned such that the Rx can simultaneously observe all Txs [264]. Also, this diversity technique has been used to improve PLS in FSO systems [60].

In FSO systems utilizing IM/DD, heuristic STC designs, including RC and orthogonal space–time block codes (OSTBCs), have been suggested as simple alternatives to the MISO FSO diversity scheme [265–267]. In the RC system, the same information signal is simultaneously transmitted from all the transmit apertures (laser sources). This is highly effective for reducing fading at the Rx [1,263], and the superiority in RC performance has been shown in [268]. For instance, research indicates that the performance of Alamouti code is inferior to that of RC over FSO links [266,269,270]. Additionally, it has been demonstrated that OSTBC yields worse performance than RC in FSO systems utilizing IM/DD and OOK. Furthermore, studies have shown that RC outperforms OSTBC in a MIMO FSO system using SIM [268,269,271].

Additionally, when CSI is accessible at the Tx, a transmit diversity technique can be employed that selects the optical path with the highest scintillation value. Under IM/DD, this selection transmit diversity (STD) method, also referred to as transmit laser selection (TLS) can achieve full diversity while outperforming other alternatives, such as STCs and RCs. These alternatives implement transmit diversity without relying on CSI, meaning they do not require any feedback from the Rx to the Tx [264,272].

In scenarios where the Tx has imperfect CSI, various transmission strategies are explored by proposing an extension of the STD scheme to situations where a feedback with a limited number of bits is available [272]. In this scheme, the transmit apertures are divided into various partition sizes, and one partition is selected for signal transmission based on feedback bits, utilizing RC. However, optimizing the partition selection is quite complex and necessitates the numerical solution of an infinite integral [273]. Additionally, more intricate signaling schemes can be employed to enhance coding gain alongside the benefits of diversity. For example, a combination of STD and space–time trellis coding (STTC) for MISO FSO communication systems using IM/DD over strong AT channels is analyzed in [274]. Additionally, since a feedback-based beamforming scheme enhances channel-adaptive signaling in wireless communication systems and significantly improves performance, an imperfect quantized feedback-based beamforming approach for a generalized MISO FSO system operating over ΓΓ fading channels with PE is proposed in [275]. Similarly, a simplified version of the STD scheme, which utilizes only one-bit feedback regarding the instantaneous CSI available at the Tx in a MIMO FSO system using SIM BPSK and functioning under ΓΓ fading, is introduced to mitigate the high sensitivity to feedback errors associated with the STD method [273].

The influence of MISO relay-assisted FSO systems on diversity order is being examined using the STD scheme, specifically in the context of ΓΓ fading channels that include PE. The findings indicate that the gain in diversity order is significantly influenced by the position of the relay. Furthermore, it is noted that this gain can significantly diminish when PEs affect the diversity order [276].

Receive Diversity FSO Systems

A straightforward approach to mitigate the fading effect is to employ a larger lens at the Rx to average out intensity fluctuations. This method, commonly referred to as aperture averaging, can be regarded as a form of inherent receive diversity. Efficient fading reduction can also be achieved by utilizing multiple apertures at the Rx (SIMO). Specifically, instead of employing a single large aperture, several smaller apertures can be deployed at the Rx. This configuration allows each Rx aperture to experience some degree of aperture averaging, which is less than that of a single large aperture. Moreover, this approach benefits from spatial diversity by combining the signals from the different apertures. Assuming that the signals from the various apertures are subject to uncorrelated fading, the multiple aperture solution outperforms the single large aperture solution, given that both configurations have the same total effective aperture area. It is important to note that using multiple apertures is particularly advantageous in strong turbulence conditions [1,233]. Also, it is important to recognize that, from a practical perspective, utilizing an excessively large lens requires a photodetector with a larger active area to effectively capture the received photons at the lens's focal plane. This, in turn, significantly constrains the system's data rate due to the relatively high parasitic capacitance associated with such a component [1].

In SIMO (Receive Diversity) FSO systems, diversity gain can be achieved by averaging multiple independent laser beams. At the receiving end, three techniques can be utilized: MRC, SC, and EGC. Among these, the MRC scheme provides a higher diversity gain, resulting in an optimal signal-to-interference-plus-noise ratio (SINR). In contrast, EGC is often favored over the other two methods due to its simplicity and lower installation costs [245,263,277–280].

MIMO FSO Systems

A highly effective approach to address the dispersive nature of FSO channels involves utilizing multiple lasers and photodetectors, similar to the RF MIMO wireless communication systems. This configuration can enhance channel capacity linearly, scaling with the number of antennas [47,148,281]. MIMO FSO systems are linked to either RC [265,268] or TLS [264,272]. In the case of RC, the same information symbol is sent from all apertures, allowing for implementation without CSI or feedback. Conversely, the TLS approach involves activating the laser that provides the highest irradiance at the Rx, necessitating both a feedback link and the acquisition of CSI [241].

MIMO FSO systems are particularly adept at mitigating AT fading by averaging its effects, thereby increasing the reliability of FSO links [91,282]. Furthermore, the arrangement of multiple sources and detectors can indirectly assist with pointing issues. For instance, a horizontal rooftop transmit/receive array is likely to experience cyclostationary fading rather than complete pointing loss in the event of significant sway [265,283].

In a MIMO FSO system, the laser sources and photodetectors are arranged such that the fading encountered between source–detector pairs is statistically independent, allowing for the exploitation of diversity benefits from the MIMO channel [148,244,265]. Additionally, because of the incoherence of the lasers across the array, their powers combine effectively at the Rx array. In practice, however, MIMO FSO systems may not consistently deliver the theoretical gains they promise. This limitation arises in situations where the assumption of independent channel fading across all links in the MIMO FSO system fails, often due to the proximity of the devices [244,265]. In this context, the limited distance between adjacent optical sources at the Tx or neighboring photodetectors at the Rx can easily lead to spatially correlated fading, which reduces the efficiency of the MIMO FSO system [284,285]. Also, because both the path loss and fading statistics of the channel depend on distance, a significant number of transmit and/or receive apertures is necessary for long-range links to attain the desired performance improvements. This requirement increases the complexity of MIMO FSO systems [244].

The performance of MIMO FSO systems has been explored in various previous studies. For instance, the BER performance of FSO links under LN AT with spatial diversity is examined in [263]. In [286], the study investigates MIMO FSO systems operating in strong turbulence channels, utilizing a single ΓΓ approximation for the distribution of the sum of independent ΓΓ variates and employing EGC at the Rx. Additionally, ref. [287] presents infinite series representations for the error performance of MIMO FSO systems over independent and identically distributed (i.i.d.) ΓΓ fading channels. Furthermore, in [288], the $\alpha - \mu$ distribution is employed to approximate the sum of arbitrary i.i.d. ΓΓ variates. This approach leads to simple yet accurate approximations for the OP and average bit error probability (ABEP) of MIMO FSO systems that operate over i.i.d. ΓΓ channels with EGC at the Rx.

For MIMO-FSO systems operating over correlated ΓΓ channels, numerous studies have utilized a simplified channel correlation model, such as the exponential correlation model [263,289]. However, this exponential correlation model, originally developed for RF communications, is inadequate for accurately describing the correlated AT channels. Consequently, recent research has focused on exploring the characteristics of fading correlation induced by AT [258,290,291]. Specifically, studies [258,290] employ wave optics and Monte Carlo simulations to elucidate the effects of turbulence strength, link span, aperture diameter, and separation on channel correlation for both receive and transmit diversity systems. However, since simulation methods can only focus on specific configurations, extending the results to general cases can be challenging. To address this issue, the correlation coefficient as a function of the aforementioned parameters is presented in [291]. In [292], the average achievable rate (AAR) of spatial diversity MIMO-FSO systems over correlated ΓΓ fading channels with arbitrary correlation is examined. The AAR over LN fading channels is considered in [293,294], both with and without bandwidth constraints, respectively. Additionally, ref. [295] investigates the AAR for SM and spatial diversity MIMO FSO links with PE over LN channels.

4.7. Error Correction Codes and Modulated Signal Optimization

ECC are crucial in reducing the impact of signal degradation in FSO communication. By detecting and correcting errors in transmitted data, ECC improves the reliability and robustness of FSO links, ensuring consistent data transmission even in challenging atmospheric conditions. However, Forward Error Correction (FEC) alone cannot meet the capacity demands of optical wireless systems. Incorporating high-order modulations, such as PAM and QAM, can enhance channel capacity. Despite this, a gap still exists between the Shannon limit and the capacity of uniform symbols. This section presents advanced schemes designed to enhance communication reliability and efficiency. It also outlines the essential requirements that transceivers must meet to approach channel capacity. We explore ECC, which improves data integrity by detecting and correcting errors. CM techniques are introduced to increase SE by combining coding and modulation processes. Lastly, we delve into capacity-approaching Probabilistic Amplitude Shaping (PAS), a cutting-edge method that optimizes data transmission rates close to the channel capacity limit. Each scheme contributes uniquely to advancing communication technologies, ensuring robust and high-performance data transfer.

4.7.1. Requirements for Channel Capacity-Approaching Transceivers

To develop transceivers that approach the channel capacity in terms of transmission rate, three key requirements must be met. First, the symbol distribution should align with the channel's capacity-achieving distribution. Second, optimal, adequately long channel codes are necessary. Third, the transmission rate must be finely adjusted according to the channel conditions, meaning the encoder should support a wide range of transmission modes across various SNRs. However, designing efficient CM systems that meet these criteria is challenging. Typically, the input distribution does not match the capacity-achieving distribution, resulting in a shaping gap. Additionally, practical implementations use finite-length Forward Error Correction (FEC) codes, leading to a coding gap. Lastly, the number of modulation orders and coding rates available is restricted by system complexity, resulting in only partially adaptive systems [296–298].

Concerning the shaping gap, optimizing the shape of the modulated signal constellation can reduce the discrepancy between the Shannon's limit and transmission rate. The primary methods of constellation shaping are geometric shaping and probabilistic shaping (PS). In geometric shaping, the constellation symbols are equiprobable but unevenly spaced. Conversely, a probabilistically shaped constellation features uniformly spaced symbols with differing probabilities assigned to each symbol. The latter has garnered significant attention in recent years owing to its rate adaptability, higher shaping gain, and the feasibility of employing Gray code for symbol labeling [296,297,299].

4.7.2. Coded Modulation and Error Correction Codes

As previously mentioned, the performance of an FSO system regarding availability and reliability is greatly influenced by unpredictable variations resulting from AT, which is the primary factor hindering transmission performance [300]. Higher laser powers are essential to extend transmission distances. However, the free-space, optical IM/DD channel presents modem designers with unique challenges. Most practical FSO systems utilize LEDs or LDs as Txs, along with PIN PDs or APDs as Rxs, creating a relatively straightforward communication system. These devices are designed to modulate and detect only the intensity of the carrier wave, rather than its phase, meaning that all transmitted signal intensities must be non-negative. Additionally, considerations for biological safety (specifically eye safety) restrict the average optical power emitted, which in turn restricts the average signal amplitude. The effects of multipath distortion in signal propagation, along with the limited response times of optoelectronic components, impose significant restrictions on the channel bandwidth. Standard signal-space models and CM techniques for electrical channels cannot be directly applied to this scenario because they do not account for the constraints on signal amplitude [301,302]. Consequently, there is a need for power-efficient modulation schemes [148].

As highlighted in Section 4.3, modulation schemes like PAM and QAM, often employed in wireless communications, offer high SE but lack power efficiency, making them less suitable for certain applications. In contrast, Q-ary PPM presents a compelling power-efficient alternative for FSO links [244,265,303,304]. The significant bandwidth of FSO links, when contrasted with RF links, mitigates concerns regarding the low SE of PPM [302]. Additionally, as the number of slots in Q-ary PPM increases, the impact of background radiation diminishes, largely due to the shorter PD integration interval [244,265]. To enhance the performance of FSO systems utilizing PPM, various FEC strategies have been introduced, such as convolutional codes, Reed–Solomon (RS) codes, turbo codes, and trellis-CM (TCM) [302,305,306]. While RS and convolutional codes provide adequate coding gains in conditions of weak turbulence and without fog, their effectiveness declines under strong turbulence or in deep fog scenarios. In such cases, more sophisticated FEC solutions, including turbo and LDPC codes, become necessary [302,307]. A power-efficient CM scheme employing multilevel coding with LDPC codes as component codes has been proposed for implementation in repetition MIMO FSO systems with Q-ary PPM [148].

Moreover, FEC techniques, such as polar codes and LDPC codes, are extensively utilized in FSO to counteract turbulence, achieving substantial coding gains under varying turbulence conditions [164,186,307,308]. For instance, LDPC-coded optical OFDM has been shown to vastly outperform LDPC-coded OOK in AT channels in terms of both SE and coding gain. In strong turbulence conditions at a BER of 10^{-5}, the enhancement in coding gain of the LDPC-coded single-side band unclipped-OFDM system with 64 subcarriers exceeds that of the LDPC-coded OOK system by 20.2 dB for QPSK and by 23.4 dB for BPSK [186]. Additionally, the combination of FEC and spatial diversity has been demonstrated to mitigate fading [186,213,285]. The LDPC-coded MIMO FSO communication system, which employs LDPC codes and spatial diversity reception to combat turbulence fading when sub-channels are independent, has been examined [186]. In practice, deterioration due to spatially correlated fading is considerable, especially in moderate to strong turbulence channels [290]. Polar codes, a novel FEC technique, are the only ones proven to achieve the Shannon limit with low encoding and decoding complexity and have been selected for the 5G channel coding standard because of their distinctive benefits [309,310]. Polar-coded MIMO-FSO systems with spatially correlated fading using OOK modulations are introduced in [285], where polar codes outperform LDPC codes under the same conditions. However, OOK modulations cannot satisfy the capacity demands of future optical communication. Unipolar M-ary PAM (M-PAM) is a candidate for achieving near capacity in FSO systems [311], and the non-uniform distributions of M-PAM can significantly increase the capacity of FSO systems [297,312].

Furthermore, efficient FSO communications can be realized by integrating higher-order modulation schemes with channel coding, a method referred to in the literature as CM [217,313]. This approach improves system reliability by combining high-order modulations with FEC codes and offers flexible data rates through adjustments in modulation schemes or code rates [164,308].

4.7.3. Probabilistic Shaping

PS with non-uniform probabilities has been shown to reduce capacity gaps and facilitate fine-grained rate adaptation [296]. This approach has become a significant topic in optical communications. By adjusting the signal probability distribution, PS can modify the entropy and improve system performance [314]. Additionally, it can be used in bit-interleaved CM (BICM) systems utilizing PAS, whose advantages have been experimentally confirmed [308,314].

PAS integrates CM with PS [296,315]. This technique has been utilized in coherent communications to enhance SNR in AWGN channels, providing greater flexibility and precise control over bitrate. Recently, time-adaptive PAS has been presented for FSO links, leading to significant capacity improvements under time-varying power fading conditions [316,317]. Nevertheless, the PAS cannot be directly applied to FSO communications because it relies on the symmetry of bipolar signals, whereas FSO systems use unipolar (asymmetric) signals. To address this limitation, the modified Maxwell–Boltzmann distribution based on PAS (PAS-MMB) can be utilized; however, its distribution remains pairwise, which is not ideal for unipolar symbols [309,311]. Additionally, unipolar M-PAM are promising candidates for achieving near-capacity in FSO systems [311]. The non-uniform distributions of M-PAM can further enhance the capacity of these systems. Research in [297,300] has introduced numerical simulations of LDPC and polar-coded PS PAM schemes to optimize the capacity of FSO channels with unipolar signals. In this context, pairwise distributions have proven effective for PAM8, resulting in only minor performance losses [308].

Hybrid Constellation Shaping

In addition to PAS, another approach utilizing PS is Sparse–Dense Transmission (SDT). Research indicates that SDT-based OOK modulation can achieve approximately 1 dB of shaping gain in AWGN channels [309]. Furthermore, the capacity can be enhanced through entropy-loading DMT systems based on PS. Entropy-loading is a robust and adaptable method for capacity enhancement, allowing continuous adjustments to the source entropy for each subcarrier. However, implementing entropy-loading DMT presents a trade-off between optimization complexity and achievable capacity [318,319]. In comparison, Hybrid Constellation Shaping (HCS) offers a lower-complexity solution that yields greater shaping gains by combining PS with geometric shaping (GS), particularly for non-Gray mapped QAM [315,320]. Nevertheless, determining appropriate probabilities and geometric shapes to align with existing FEC schemes poses challenges. To address some of the implementation issues associated with HCS designs, a rate-flexible HCS scheme has been presented for DMT-FSO systems [164,321].

Bitrate Adaptation Using PCS

PCS is recognized as an effective method to boost system capacity by reducing the energy of the transmitted signal. In AWGN channels, the lowest transmitted signal energy for a specified bitrate is obtained by adjusting the probabilities of QAM symbols to match the Maxwell–Boltzmann distribution [316,322] given as [316,317]

$$P_{x_n} = \frac{\exp(-\lambda_{sp}|x_n|)^2}{\sum_{n=1}^{M} \exp(-\lambda_{sp}|x_n|)^2}, \tag{45}$$

where $\lambda_{sp} \geq 0$ denotes the shaping parameter, M represents the constellation size of the chosen QAM template for PCS, and x_n represents the n-th symbol in the square M-QAM constellation alphabet, expressed as [317]

$$x_n = 2\Big((n-1) \bmod \sqrt{M}\Big) - \sqrt{M} + 1 + j\left(2\left\lceil \frac{n}{\sqrt{M}} \right\rceil - \sqrt{M} - 1\right). \tag{46}$$

In addition to reducing the average signal energy, PCS can also be utilized to adapt the entropy of the transmitted signal, defined as [316]

$$H_{\mathrm{PCS}} = -\sum_{n=1}^{M} P_{x_n} \log_2(P_{x_n}). \tag{47}$$

Thus, by utilizing Equations (45) and (47), the net bitrate can be adjusted by modifying the shaping parameter λ_{sp}. Also, in a system utilizing PCS modulation and FEC, the relationship between the system bitrate and the constellation entropy is defined as

$$R_b = N_{\mathrm{pol}} R_S [H_{\mathrm{PCS}} - (1 - R_{\mathrm{FEC}}) \log_2 M], \tag{48}$$

where R_b denotes the net bitrate, N_{pol} represents the number of polarizations, R_S denotes the operating symbol rate, and R_{FEC} denotes the FEC rate.

To accurately set the PCS shaping factor for maximizing channel capacity, it is necessary to first evaluate the achievable information rate for a given SNR. For this purpose, the normalized generalized mutual information (NGMI) is often utilized as a pre-FEC performance metric in systems employing soft-decision FEC and is given by [317]

$$\mathrm{NGMI} = 1 - \frac{G(M, P_x, \sigma^2)}{\log_2(M)}, \tag{49}$$

where σ^2 denotes the noise variance in the AWGN channel.

It can be demonstrated that the minimum NGMI required for error-free transmission matches the FEC rate. However, because of the imperfect performance of FEC, it is necessary to consider a slightly higher NGMI threshold, denoted as $\mathrm{NGMI}_{\mathrm{th}}$, than the theoretical minimum. To control the transmitted net bitrate, adjust the shaping parameter λ_{sp} according to Equation (45).

4.8. *Cooperative Relaying*

In this section, we present relay-assisted transmission as a method to enhance communication reliability and coverage. By utilizing intermediate relay nodes, this technique helps overcome obstacles and extend the range of transmission, ensuring more robust and efficient communication links. We will explore the principles, advantages, and implementation strategies of relay-assisted systems in various communication scenarios.

4.8.1. Relay-Assisted Transmission

In contrast to the localized diversity techniques used in MIMO systems, relay-assisted solutions provide distributed spatial diversity methods, where neighboring nodes (or relays) support communication between a source node and a destination node. Moreover, cooperative communication is a dynamic area of research because of its potential to improve reliability and broaden the reach of wireless networks without the need for a new infrastructure [323]. It has also become a viable method for achieving spatial diversity benefits [324–326]. The core concept of cooperative diversity (parallel relaying) is founded on the observation that, in a wireless RF channel, the signal sent by the source node is received by other nodes known as partners or relays. These partners, along with the source, can collaboratively manage and convey their information, forming a virtual antenna array despite each node having only one antenna. Another relay-assisted transmission scheme is multihop transmission (serial relaying), which employs relays arranged in a sequential configuration [327–331]. This approach is typically employed to extend the coverage area for Txs with limited power. It does not enhance performance against fading influences in wireless RF environments, which means it does not improve the diversity order [332]. Also, cooperative techniques are extensively researched in FSO communication systems to counteract the adverse effects of turbulence-induced atmospheric scintillation [323,332–336]. Table 5 presents some of the related work in this area.

Current research in cooperative FSO communications primarily focuses on two approaches: all-active relaying, which is used when CSI is unavailable, and selective relaying, which is applied when CSI is present. All-active relaying is a straightforward and efficient two-slot scheme. In the first slot, the source (S) sends the information packet to the relays (Rs). The relays then decode (or amplify) the received symbols and, in the second slot, forward the processed signals to the destination (D). In this setup, no priority is given to any relay, irrespective of the strengths of the source-relay and relay-destination links [323,332,334,335]. In selective relaying, either direct transmissions occur, or one relay is selected from the pool of available relays to forward the information symbols to the destination based on the channel state [336–338]. This approach prioritizes the transmission along the strongest e2e link. Instead of distributing power across multiple relays, the power is concentrated and directed to the *best relay* to improve signal reconstruction quality at this relay. The criteria for selecting the optimal path involve minimizing the conditional error probability for quantum-limited systems without background noise [337], and minimizing the OP for systems with Gaussian noise [336]. While all-active relaying is simple and does not require any CSI, selective relaying provides better performance but comes with increased system complexity. This complexity arises from the need to acquire complete CSI for all intermediate channels and to establish feedback links between the communicating nodes [241,323,339]. Furthermore, it has been observed that the strength of the two-hop source–relay–destination (S-R-D) link depends on its weakest hop. This protocol shows that both all-active and selective relaying methods can achieve full diversity gains, with the selective method offering an improved coding gain. Recently, inter-relay cooperation has been introduced to enhance the reliability of FSO networks, especially in scenarios where FSO links are established between the relays [340].

Table 5. Literature on cooperative hybrid FSO.

Link		Channel Model					
1st Hop	**2nd Hop**	**1st Hop**	**2nd Hop**	**Relaying Mode**	**Detection Technique/Modulation**	**Metrics**	**Reference**
FSO	RF	DGG	EGK	AF, DF	IM/DD, HD	ABER, OP, EC	[13]
RF	FSO	SR	GG	AF	QPSK	SER	[341]
FSO	FSO	NE	NE	AF	IM/DD OOK	ABER, OP	[88]
RF	FSO	Rayleigh	GG	DF	IM/DD, HD	ABER, OP	[342]
RF	FSO	N-m	GG	DF	IM/DD OOK	OP	[343]
RF	FSO	Rayleigh	GG	DF	SIM	IC	[344]
RF	FSO	SR	GG	DF, AF	IM/DD OOK	SOP, ASC	[345]
FSO	FSO	GG	$\alpha-\mu$	DF	IM/DD	EC	[346]
FSO	FSO	GG	GG	DF	SIM DBPSK	BER	[347]
RF	FSO	Rayleigh	GG	DF	IM/DD, HD	ABER, OP, EC	[348]
FSO	RF	GG	GN-m	AF	IM/DD, HD	ABER, OP, EC	[67]
FSO	RF	GG	GK	AF	SIM	OP	[349]
FSO	RF	EW	N-m	DF	IM/DD	OP, ABER	[350]
FSO	RF	$\mathcal{M}$	$\kappa-\mu$	AF	IM/DD	OP, EC	[351]
RF	FSO	Rayleigh	GG	AF	IM/DD	BER, EC	[352]
FSO	RF	GG	FTR	AF, DF	SIM, IM/DD, HD	OP, BER, EC	[353]
RF	FSO	N-m	GG	AF	IM/DD, HD	OP, BER, EC	[354]
RF	FSO	Rayleigh	GG	AF	SIM	OP, BER	[355]
RF	FSO	BX	$\mathcal{M}$	DF	IM/DD, HD	OP, BER, EC	[356]
RF	FSO	Rayleigh	GG	AF	IM/DD	OP, BER, EC	[6]
RF	FSO	$\alpha-\mu$	GG	AF	IM/DD	OP, BER	[357]
RF	FSO	$\alpha-\mu$	DGG	AF	IM/DD	OP, BER, EC	[358]
FSO	RF	$\alpha-\mu$	GG	DF	IM/DD	OP, BER, EC	[359]

SER: Symbol error rate, AF: Amplify-and-forward, DF: Decode-and-forward, QPSK Quadrature Phase-Shift Keying, OOK: on–off keying, DBPSK: differential binary phase-shift keying, NE: negative exponential, GG: Gamma–Gamma Distribution, $\alpha-\mu$: Alpha–Mu Fading, SR: shadowed-Rician fading distribution, N-m: Nakagami-m fading, FTR: Fluctuating two-ray fading, GN-m: Generalized Nakagami-m fading, GK: Generalized-$\mathcal{K}$ fading, Málaga-$\mathcal{M}$ distribution, EW: Exponentiated Weibull fading, BX: Beaulieu–Xie fading, DGG: Double Generalized Gamma Distribution, EGK: Extended Generalized-K fading, OP: Outage probability, SOP: secrecy outage probability, ASC/EC/IC: Average Secrecy/Ergodic/Intercept capacity, BER: bit-error rate, ABER: average bit error rate, IM/DD: intensity-modulation direct detection, HD: heterodyne detection, SIM: subcarrier intensity modulation scheme.

In multi-node scenarios, cooperative communication acts as a unique variant of MIMO by utilizing relay nodes to improve the link quality between the source and destination nodes. This approach not only offers a novel method to boost transmission reliability in FSO systems but also provides a viable solution for FSO networking [332,338,360]. One benefit of this virtual multiple-aperture system, as opposed to MIMO FSO systems, is that the multiple paths are spatially separated, which guarantees the independence of the associated fading channels [336]. Also, compared to the conventional MIMO FSO system, the cooperative FSO system allows for the flexible selection of relay nodes, which helps to diversify transmission paths and effectively eliminate channel correlation. Thus, in cooperative FSO communication systems, data are transmitted through a direct link from a source node to a destination node, as well as through one or more relay nodes. These relay nodes aid in enhancing signal strength and reliability, particularly in adverse atmospheric conditions [284].

Additionally, several cooperative protocols have been suggested for use in relay-assisted FSO links. The traditional methods include amplify-and-forward (AF) [331,332,334, 361–365] and DF [331,363] protocols. In AF relaying, optical-to-electrical (OE) conversion is performed at each relay, followed by electrical amplification and modulation of a laser source. The electrical amplifier gain at each relay is established based on the CSI of the previous hop, necessitating continuous estimation of the channel fading amplitude. As with all AF systems, the noise introduced at each relay accumulates along the transmission path, limiting the maximum transmission distance [331,361,363,364,366,367]. DF FSO systems, on the other hand, involve additional electrical processing steps after OE conversion. To mitigate the propagation of background noise, the received electrical signal at each relay is decoded and re-encoded before retransmission in DF systems. While DF significantly enhances performance, it introduces considerable encoding/decoding delays and requires more complex hardware [88,332,366–369]. The need for more processors results in higher component error rates and increased costs [370,371].

The modified versions of the DF strategy, such as Bit Detect and Forward, Adaptive Bit Detect and Forward, and Adaptive Decode and Forward, have been considered [372]. Additionally, FSO DF relaying techniques have progressed from buffer-free (BF) relaying to buffer-aided (BA) relaying [324,333]. These FSO BA relaying methods are inspired by their RF counterparts [373–375]. In mixed RF/FSO systems, buffers play a significant role by queuing data collected from mobile RF users before they are multiplexed along the backhaul FSO link [376]. Implementing BA relay selection strategies requires the acquisition of CSI and/or buffer state information (BSI). CSI knowledge can be either full or partial [373,375,377]. BSI involves acquiring the number of packets stored in the relays' buffers, which enhances performance but increases system complexity. Recently, deep reinforcement learning (DRL) has been applied to maximize the throughput of RF BA networks, and a relay clustering approach has been suggested to improve performance in asymmetrical RF networks [373–375].

4.8.2. Cooperative FSO Transmission Schemes

In this section, we explore the synchronous and asynchronous cooperative FSO transmission schemes. These schemes represent critical strategies for enhancing the performance and reliability of FSO systems. Synchronous cooperative transmission involves the precise coordination of multiple transmitting sources, ensuring that their signals reach the Rx in a well-aligned manner. This alignment can significantly improve signal quality and reduce error rates. Conversely, asynchronous cooperative transmission schemes do not require such strict timing coordination, offering greater flexibility and robustness in environments where perfect synchronization is challenging. By comparing these two approaches, we aim to highlight their respective advantages, limitations, and potential applications in various FSO communication scenarios.

Synchronous Transmission Scheme

Synchronous transmission, a fundamental technique, has been utilized since the introduction of cooperative communication in radio systems. In this method, the source node transmits identical data on both the relay links and the direct link, enabling the destination node to combine the received signals synchronously, resulting in a stronger SNR. In this regard, all nodes involved in the cooperative FSO system transmit their signals in a coordinated manner, ensuring that the signals arrive at the destination simultaneously. This coordination requires precise timing synchronization among the source and relay nodes. This approach aims to enhance the reliability of the communication by providing multiple copies of the same signal. The destination node combines these received signals to improve signal quality and mitigate errors caused by AT and path loss. While cooperative communication has been extensively investigated in FSO systems, the redundancy inherent in synchronous transmission persists. Consequently, the FSO system must address the challenge of channel underutilization to achieve spatial diversity through cooperative communication [284].

Asynchronous Transmission Scheme

An asynchronous transmission scheme is designed to minimize the substantial link redundancy in cooperative FSO systems (typically seen in synchronous transmission). In contrast to synchronous transmission, the asynchronous scheme enables the source node to send different data over both the direct and relay links, facilitating more flexible and independent data transmission. To leverage the diversity gain from these links, which carry distinct data, a squared signal combination method can be utilized to merge the signals received from multiple links. By jointly processing the signals before and after the squaring operation, the data on each link can be accurately recovered. As a result, asynchronous transmission provides diversity gain for cooperative FSO systems, improves channel utilization, and ultimately increases the effective data rate compared to synchronous transmission [284].

4.8.3. Cooperative FSO System Model

The system model for parallel cooperative FSO communication comprises a source node S, multiple relay nodes R_i, and a destination node D. This system features two primary communication links: a direct link from S to D, enabling S to send data directly to D, and a relay-assisted link where data are transmitted from S to D via relay nodes R_i. The relay node index i ranges from 1 to M. The relay node employs DF or AF protocols to assist in the transmission. As illustrated in Figure 12, X_0 denotes the modulated optical signal transmitted directly from S to D, while X_i represents the signal transmitted from S to D via R_i.

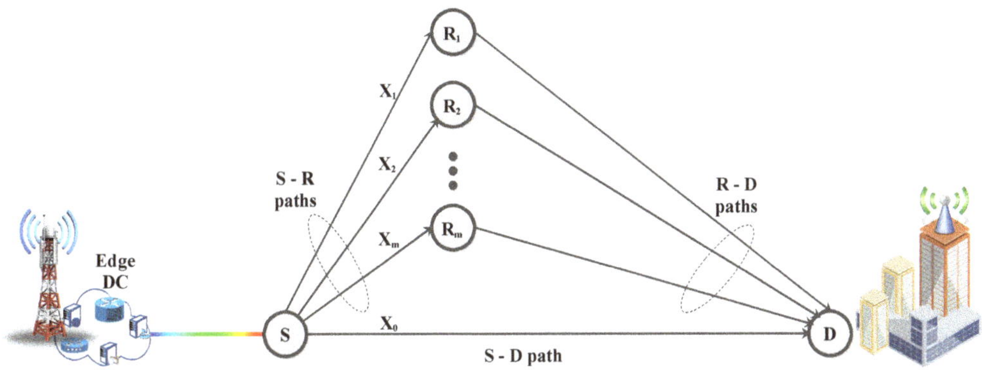

Figure 12. A typical model of a cooperative FSO communication system.

Synchronous Transmission Scheme

In the conventional system model, the optical signal transmitted over each link remains identical, a concept referred to as synchronous transmission. This implies that $X_0 = X_1 = X_2 = \ldots = X_M$. Consequently, the received signal after electrical combination in a synchronous transmission system is expressed as [284,360]

$$Y_{sc} = \sum_{i=0}^{M} (Y_i + N_i), \tag{50}$$

where $Y_i = \bar{g}|X_i|^2 H_i G_i$ represents the electrical signal received from either the direct link (for $i = 0$) or the relay links (for $i = 1, \ldots, M$); G_i represents the gain for optical-to-electrical conversion; N_i denotes the AWGN, encompassing thermal noise and background light noise; $|X_i|^2$ is the average optical power of X_i; and H_i denotes the combined channel effect, expressed as

$$H_i \triangleq h_\ell h_a. \tag{51}$$

The destination node combines the received signals using techniques such as MRC to enhance the signal quality.

Asynchronous Transmission Scheme

Unlike the FSO system that uses synchronous transmission, asynchronous scheme allows the source node to transmit different data to each relay node in an uncoordinated manner. In this transmission method, $X_0 \neq X_1 \neq \ldots \neq X_M$. To recover the data and attain diversity gain from the different links transmitting various data, an electronic signal combiner and a squared signal combination method can be employed at the Rx. This method involves squaring the received signals before combining them. During this process, the j-th combined and squared signal (for $j = 1, \ldots, H$) can be mathematically expressed as

$$Z_j = \left[\sum_{i=0}^{M} w_{ji}(Y_i + N_i)\right]^2, \tag{52}$$

where the electrical signal Y_i is anticipated to be bipolar with fewer amplitudes after applying the square operator. The coefficient $w_{j,i}$ functions as a relay selection indicator, taking the value of 1 if the relay node R_i is chosen to produce Z_j, and 0 otherwise.

The data on each link can be recovered by jointly processing the signals before and after the square operation, allowing for improved data recovery and reduced redundancy. In this context, the complexity of relay selection can be simplified by employing the pairwise combination method, where each link is supported by another link in pairs for signal combination. The corresponding coefficient matrix can then be denoted as [284]

$$w = \begin{bmatrix} w_{00} & \cdots & w_{0M} \\ \vdots & \cdots & \vdots \\ w_{M0} & \cdots & w_{MM} \end{bmatrix} = \begin{bmatrix} 1 & 1 & 0 & 0 & \cdots & 0 \\ 0 & 0 & 1 & 1 & 0 & \vdots \\ \vdots & \ddots & \ddots & \ddots & \ddots & 0 \\ 0 & 0 & \cdots & 0 & 1 & 1 \end{bmatrix}. \tag{53}$$

Additionally, a Least Mean Squares algorithm can be used to eliminate the product term of signal and noise from Z_j. The resulting signal, Y^j_{ac1}, is given by

$$Y^j_{ac1} = \left(\sum_{i=0}^{M} w_{j,i} \cdot Y_i\right)^2 + \left(\sum_{i=0}^{M} w_{j,i} \cdot N_i\right)^2 \tag{54}$$

The square operator improves the SNR of the received signal without increasing its amplitudes. However, it does not allow for the direct determination of the bipolar amplitude (i.e., whether the amplitude is positive or negative) from the squared signal Y^j_{ac1}. It is essential to recognize that the combined signal prior to squaring retains the bipolarity, which aids in determining the amplitude. Therefore, the merged signal before the square operator can be defined as

$$Y^j_{ac2} = \sum_{i=0}^{M} w_{j,i} \cdot (Y_i + N_i). \tag{55}$$

In principle, by jointly sampling the signals Y^j_{ac1} and Y^j_{ac2}, the transmitted data can be recovered with a lower BER.

Both synchronous and asynchronous transmission schemes have their advantages and challenges. Synchronous transmission ensures coordinated signal arrival, which simplifies signal processing at the destination node. However, it requires precise timing synchronization, which can be challenging in practical implementations. On the other hand, asynchronous transmission offers more flexibility and reduces link redundancy by allowing independent data transmission on different links. The squared signal combination method in asynchronous transmission provides diversity gain, enhancing data recovery. Consequently, the choice between synchronous and asynchronous transmission schemes in cooperative FSO systems is contingent upon the specific requirements and limitations of the application. Asynchronous transmission, with its reduced redundancy and flexibility, presents a promising approach for enhancing the performance and reliability of FSO communication systems.

4.8.4. Hybrid Satellite–Terrestrial FSO Cooperative System

In [378], it is demonstrated that portable cooperative relaying ground stations can be utilized to offer broadband access in disaster-stricken areas. These ground stations receive signals from a satellite and relay them to a hidden destination node via an FSO link. These hybrid satellite–terrestrial FSO cooperative schemes are not only cost effective but also can be deployed rapidly. In [341], an AF protocol-based hybrid satellite–terrestrial FSO cooperative system is analyzed. In this system, the satellite-to-relay link follows the shadowed-Rician land mobile satellite model, while the relay-to-destination FSO link follows the ΓΓ fading model. This hybrid system is distinct from a typical tandem RF/FSO link, where the RF link follows simpler fading statistics. The moment generating function (m.g.f.) for the hybrid link is derived, and this m.g.f. is used to approximate the average symbol error rate (SER) of the cooperative system. Additionally, the analytical diversity order of the AF-based hybrid cooperative system is determined.

4.9. *Adaptive Optics*

AO is a crucial technology in FSO communication systems, designed to counteract the negative impacts of AT [8,379]. Turbulence can cause phase distortions in the optical wavefront, leading to beam spreading and signal degradation. AO systems utilize real-time feedback to adjust the shape of the optical wavefront, compensating for these distortions and thereby enhancing signal quality [277]. Consequently, through advanced wavefront sensing, real-time correction, and continuous innovation, the reliability and quality of FSO communication links can be improved significantly by AO systems, enabling more widespread and efficient deployment of this promising communication technology. This section explores the principles of AO, its implementation in FSO systems, and the current advancements in AO technology, highlighting its significance in enhancing signal quality and reliability.

4.9.1. Principles of Adaptive Optics

AO systems are designed to correct phase distortions in the optical wavefront caused by AT. These distortions, resulting from variations in air temperature and pressure, can lead to beam spreading, scintillation, and other forms of signal degradation. The application of AO in turbulent conditions mitigates optical wavefront distortion, resulting in notable performance enhancements. Initially developed for astronomy and medical science, AO have gained prominence in imaging, FSO communications, and various military uses. In a turbulent medium, the AO correction process acts as a multiplicative spatial filter on the power spectrum. As defined in [380], AT causes specific aberrations and wavefront distortions, which are statistically characterized. The use of AO for correction is further analyzed through the Zernike polynomials representation in [8,381]. Studies on AO correction with a deformable mirror (DM) as a spatial filter and the associated fitting errors are documented in [382,383]. Research on the outer scale effect on wavefront distortion reveals significant impacts on AO systems [384]. Regarding communication systems, investigations have explored the decrease in the scintillation index for Gaussian beams in non-Kolmogorov turbulent atmospheres and the BER of IM systems with AO [385,386]. These findings collectively demonstrate the efficacy of AO in mitigating turbulence effects. Figure 13 illustrates a simplified AO system from the perspective of FSO. The key principles of AO include the following [277,387]:

(i) Wavefront Sensing:

- Shack–Hartmann Wavefront Sensor: One of the most common devices used in AO systems is the Shack–Hartmann wavefront sensor. It comprises a series of lenslets that focus portions of the incoming wavefront onto a detector. The displacements of the focal spots on the detector are used to reconstruct the wavefront distortions [387–389].
- Other Sensors: There are various other wavefront sensors, including curvature sensors and pyramid sensors, each with specific advantages depending on the application [390].

(ii) Wavefront Correction:

- Deformable Mirrors: These mirrors have surfaces that can be precisely shaped using actuators to counteract the detected wavefront distortions. The most commonly used DMs are based on piezoelectric or micro-electro-mechanical systems (MEMS) technology [391–393].
- Spatial Light Modulators (SLMs): These devices modulate the phase of the light passing through them and can also be used for wavefront correction [394–399].

(iii) Real-Time Feedback:

- The AO system continuously monitors the wavefront distortions and adjusts the DM or SLM in real time to compensate for these distortions. This real-time feedback loop is crucial for maintaining high signal quality [277,395,397].

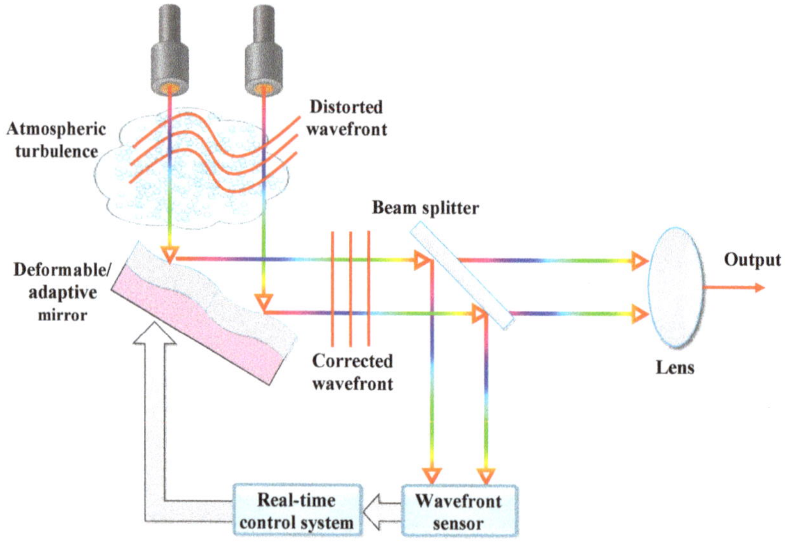

Figure 13. Depiction of an adaptive optics system.

4.9.2. Implementation in FSO Systems

FSO technology has been explored for applications involving both stationary and mobile nodes. In 6G mobile networks, it is anticipated that many antennas utilizing high-frequency bands (such as THz) will be installed on a variety of stationary nodes like traffic lights and signboards, as well as mobile nodes like trains and drones [400]. FSO presents a promising solution to deliver high-capacity mobile fronthaul connections for these antennas [9,12].

In scenarios where FSO is used with moving nodes, an FSO node must adjust its output beam to ensure the FSO signal power surpasses the minimum Rx sensitivity at the receiving node's aperture [9]. In these FSO communication applications, a significant technical challenge is the PAT. Optical beam steering is a crucial function in PAT [401]. This involves tracking the position of the paired FSO node and adjusting the beam direction accordingly as the paired node moves. The position is monitored using a position detection method, such as a position-sensitive detector. The beam direction is then altered using a rotating mechanism [9]. Traditionally, this has been achieved through mechanical mechanisms such as fast steering mirrors (FSMs), two-axis gimbals, and MEMS mirrors. While effective, these methods are often heavy, bulky, slow to adjust, and prone to mechanical failure [402–405]. Alternative non-mechanical approaches, such as those using optical phased arrays or liquid crystals, have also been explored. However, optical phased arrays are complicated in terms of control and calibration, and liquid crystals offer a limited steering range. Furthermore, these non-mechanical methods are not yet sufficiently advanced for use in FSO communication systems [401].

Furthermore, to maintain allowable PE, one approach is to use a beam Rx with a larger radius. However, this is challenging to mount on a moving FSO node due to severe constraints on size, weight, and power consumption. Another approach is to utilize a beam with a larger divergence angle at the Tx. Nevertheless, fixing the divergence angle at a larger value reduces the maximum FSO link distance [9]. Recent proposals and demonstrations of adaptive beam divergence control have shown its effectiveness in improving PAT despite PE. By adjusting the beam divergence angle according to channel conditions, this method counteracts the negative impacts of PE. Studies have demonstrated that adaptive beam divergence control can enhance the range of an FSO link and reduce OP in comparison to systems utilizing fixed beam control [9,401].

4.9.3. Pointing, Acquisition, and Tracking

In FSO communication systems, PAT mechanisms can reduce or minimize PE through the continuous monitoring of performance metrics like the Strehl ratio and received signal power. The Strehl ratio is defined as the ratio of the intensity along the optical axis of an aberrated or corrected point spread function (PSF) to the intensity of the diffraction-limited PSF on the optical axis. These metrics help in adjusting correction elements, including mirrors, gimbals, or AO [7].

The PAT mechanism has several key tasks: orienting the Tx towards the Rx, capturing the incoming light signal, and ensuring the FSO link remains stable by monitoring the position of the remote FSO terminal [81]. Pointing involves adjusting the Tx to align with the FoV of the Rx. Signal acquisition refers to aligning the Rx with the direction from which the beam arrives. Tracking entails continuously maintaining both pointing and signal acquisition during the optical communication between terminals [7].

Alongside PAT mechanisms, choosing the right beam divergence can help reduce PE. Specifically, as the beam footprint at the Rx increases, the PE decreases. However, increasing the divergence angle also leads to higher geometric path loss, which is the spread of transmitted power over the communication distance. This loss is influenced by the distance between stations, the divergence angle of the beam, and the size of the Rx aperture. Various optimization models have been proposed to determine the optimal beam width to minimize the BER and OP of an optical wireless link, taking into account PE and SNR [7,23,406,407].

Moreover, due to the increased geometric path loss associated with greater beam divergence, FSO communications typically use a narrow beam with a divergence angle no larger than 1 milliradian (mrad). This narrow beam enhances received power density and boosts link margin under various weather conditions. However, it also necessitates precise alignment between the transceivers due to the beam's small footprint at the Rx. Achieving this alignment becomes more challenging when at least one of the parties involved in communication is mobile [7,408,409].

PAT mechanisms are typically utilized in stationary and mobile FSO communication systems when employing a narrow beam. The primary motivations for using such beams include the high data rate associated with concentrated light intensity and the extended reach they provide. However, utilizing a wide beam can potentially reduce or even eliminate the performance demands on the PAT mechanism. Certain FSO applications have been proposed to take advantage of a wide beam to ease the requirements of PAT mechanisms. For example, utilizing the divergence angle of a wide beam can reduce the required steering speed of the FSM, a crucial component of the PAT mechanism in an FSO transceiver. It has been shown that a wide beam's divergence angle offers a greater link range and more effective coverage length compared to a narrow beam, making it advantageous for applications such as high-speed train communications [410].

PAT mechanisms in FSO communication systems can be categorized based on their principles of operation, including mirror-based, gimbal-based, gimbal–mirror hybrid, liquid crystal, AO, and RF-FSO hybrid methods [411,412]. Additionally, their operating principles and mechanics allow for classification based on dimensionality (i.e., 2D or 3D) as illustrated in Figure 14 [7].

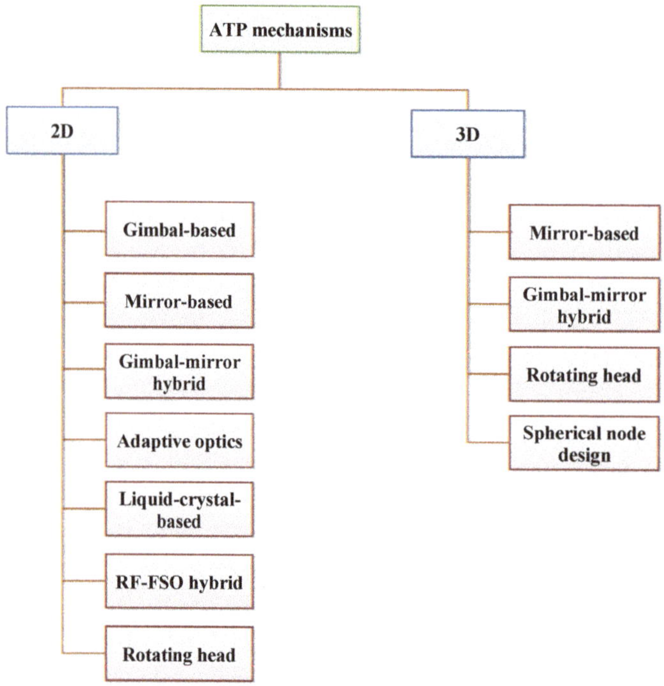

Figure 14. Categorization of PAT mechanisms in FSO communications based on their dimensionality.

4.10. Adaptive Detection Thresholds

This section introduces Adaptive Detection Threshold (ADT) schemes, which are essential for optimizing communication system performance in varying conditions. ADT schemes dynamically adjust detection thresholds based on real-time environmental and signal quality metrics, thereby enhancing reliability and efficiency. By continuously adapting to changing conditions, these schemes mitigate the impact of noise, interference, and signal fading, ensuring robust and efficient data transmission.

4.10.1. Background

In the absence of scintillation, the BER of an FSO system can be improved significantly through the use of an optimal detection threshold based on the maximum likelihood principle, assuming the bit level means and variances are known. However, in an atmospheric channel, turbulence-induced scintillation causes considerable fluctuations in the bit means and variances, which in turn affects the optimal threshold. To address this, a new thresholding method has been developed that replaces the constant threshold with an adaptive one. This method utilizes a Kalman filter algorithm to continuously track and adjust the bit means and variances. In this context, conventional fixed detection thresholds (FDTs) are insufficient for the dynamic environments of FSO systems, as they fail to consider real-time fluctuations in signal quality. Conversely, ADTs improve system performance by dynamically adjusting threshold levels based on the existing channel conditions, thereby optimizing the detection process [413].

Typically, the standard approach for adaptive detection assumes perfect knowledge of the instantaneous CSI, utilizing the instantaneous SNR to detect each data symbol [414–417]. To ensure reliable communication, adaptive detection techniques can be used to dynamically adjust the Rx sensitivity in response to these fluctuations [418–422]. Consequently, ADT methods have garnered significant interest as effective strategies for alleviating the impacts of AT in FSO. ADT enables systems to attain improved average BER with low complexity [418]. For instance, it is well known that an OOK IM/DD system with a fixed decision techniques leads to an irreducible error floor and is power inefficient. Optimal OOK demodulation necessitates time-varying ADTs based on perfect CSI for detecting each data symbol [194,415,416,421]. However, this approach presents clear practical challenges for OOK IM/DD operation with nanosecond data symbol durations (i.e., Gb/s rates) and millisecond turbulence coherence times. Specifically, it necessitates rapid adjustments to the detection threshold on the millisecond timescale of turbulence coherence times [415,416,423]. In this context, practical FSO systems will not be able to accurately track changes in the means and variances at the bit level. Therefore, optimizing the threshold to a constant value based on slowly varying statistical quantities, such as average signal power and turbulence strength, can also be adopted as a suboptimal method [413,422].

Although the performance of the Rx heavily relies on the chosen threshold level [420], determining the appropriate ADT remains a challenge [418]. Previous studies often assume detection with instantaneous CSI [10,293]. To address these practical issues, an electrical-SNR-optimized detection system has been proposed [211,424]. This system requires changing the detection thresholds only over extended timescales, such as seconds or minutes, during which a stationary turbulence channel assumption is valid [425]. However, this method necessitates knowledge of the turbulence channel's PDF, thereby increasing computational complexity [414]. For instance, research has indicated that estimating the fading intensity requires using redundant overhead, such as pilot symbols or training sequences [426]. In [424], pilot-symbol assisted modulation (PSAM) and the maximum likelihood method are introduced as methods to reduce the effects of turbulence and improve system performance. However, PSAM can cause delays at the Rx because the entire frame must be stored before decoding. In practice, to simplify commercial implementation and ensure infrastructure transparency, detection without CSI and the PDF of the turbulence channel is the favored approach [194,424].

In [189], the performance enhancements of OOK FSO communication systems under AT conditions are investigated, focusing on the dynamic decision threshold and coherent detection schemes. The findings indicate that the dynamic decision threshold scheme achieves a 5.7 dB reduction in the power needed to reach a BER of 10^{-3} in comparison to the conventional FDT approach. Similarly, in [427], it is shown that an OOK IM/DD system can utilize two laser wavelengths at the Tx and two photodetectors at the Rx to operate in differential mode, leading to outstanding BER performance with a detection threshold set to zero. This study is also extended to include PPM. However, this scheme is limited by low throughput, as it requires two lasers to transmit the same information during each symbol duration.

Furthermore, to enhance performance beyond what is achieved in [427], a source information transformation system has been introduced in [414]. This system utilizes multiple optical wavelengths to determine ADTs. It has been demonstrated that this approach can achieve a satisfactory BER performance without needing instantaneous CSI or the PDF of the turbulence model. Numerical simulations indicate that the proposed system delivers performance similar to a theoretical adaptive detection scheme, while significantly reducing implementation complexity. Additionally, it incurs only a 1.8 dB loss in SNR at a BER of 1×10^{-9} for a lognormal turbulence channel. Also, to enhance threshold accuracy and overall system performance, an FSO communication system using OOK with coherent detection-double ADT has been developed [418]. The Tx employs 3-bit encoding to relay information, while the Rx features two adaptive thresholds to identify the in-phase and quadrature components of the signal, operating without instantaneous CSI or the PDF of the channel model. Additionally, a decision-aided scheme is proposed to establish the second detection threshold, further refining precision. Numerical analyses indicate that this system achieves a BER performance comparable to that of an idealized adaptive detection system.

4.10.2. Mechanism of ADTs

ADTs work by continuously monitoring the received signal's parameters, such as BER, SNR, and scintillation index. Algorithms, often based on ML or statistical analysis, process these real-time data to determine the optimal threshold. The primary mechanisms include the following:

(i) Feedback Loops: Real-time feedback from the Rx to the Tx helps adjust the power levels and detection thresholds. This feedback loop ensures that the system adapts to varying conditions promptly [194,424].
(ii) Threshold Adjustment Algorithms: Algorithms such as the Kalman filter, maximum likelihood estimation, and adaptive filtering techniques dynamically adjust the detection thresholds. These algorithms analyze the statistical properties of the received signal and make predictive adjustments [413].
(iii) Diversity Techniques: Implementing spatial or temporal diversity techniques can also aid in adaptive threshold adjustment. By combining multiple received signals, the system can better estimate the optimal threshold [423].

4.10.3. Benefits of ADTs

ADTs offer several advantages in FSO communication systems:

(i) Improved Signal Detection: By continuously adjusting to the optimal threshold, the Rx can better distinguish between the signal and noise, leading to lower BER and enhanced data integrity [418,427].
(ii) Enhanced System Reliability: Adaptive thresholds help maintain consistent communication quality even under adverse weather conditions and AT, thereby increasing the system's reliability [413].
(iii) Efficient Use of Power: Adaptive thresholds can lead to the more efficient use of power by ensuring that the system only uses the necessary amount of power to maintain communication, which is particularly beneficial in power-constrained applications [189,413,422].

4.10.4. Implementation Challenges

Despite their advantages, implementing ADTs in FSO systems comes with challenges:

(i) Complexity: The algorithms required for real-time threshold adjustment can be complex and computationally intensive, necessitating advanced processing capabilities.
(ii) Latency: Real-time adjustments must be swift to be effective. Any latency in the feedback loop or threshold adjustment process can degrade performance.
(iii) Calibration: Adaptive systems require initial calibration and ongoing tuning to ensure they respond appropriately to changing conditions.

4.11. Hybrid FSO Systems

The integration of FSO communication with RF, mmWave, and THz technologies has emerged as a promising approach to overcoming the limitations of standalone FSO systems. Hybrid FSO systems leverage the strengths of these technologies to provide enhanced reliability and coverage. This section examines the architecture of hybrid FSO systems, their operational principles, and the benefits they offer in terms of performance and reliability.

4.11.1. Overview

FSO, mmWave, and THz communications have proven to be effective solutions for high-data-rate wireless transmission over short distances. They offer a promising response to the increasing demand for higher data rates amid the growing scarcity of RF spectrum, positioning them as strong candidates for joint deployment to ensure reliable, high-data-rate wireless backhaul [428]. Nevertheless, these technologies fall short of achieving the desired carrier-grade availability of 99.999% due to their related transmission issues. A possible solution to enhance the FSO link's availability and reliability is to integrate it with the RF (or mmWave/THz) link (hereinafter referred to as an alternative link), which can reach data rates similar to those of an FSO link, to create a hybrid FSO/RF (or mmWave/THz) communication system (hereinafter referred to as a hybrid system) [2,362,365,429]. As a result, a hybrid system is better equipped to maintain connectivity compared to a purely FSO system. By integrating the advantages of various technologies and addressing the limitations of each, the hybrid system greatly improves network performance. It enhances network resilience and offers a backup solution in difficult atmospheric conditions, thereby contributing to overall network reliability [2,365].

4.11.2. Architecture of Hybrid FSO Systems

Hybrid FSO systems typically combine FSO links with RF, mmWave, or THz links to create a robust communication network. In this system, the efficiency and reliability of signal transmission are significantly influenced by the switching configurations employed. This section delves into two primary switching architectures: soft-switching and hard-switching. Soft-switching aims to minimize losses and electromagnetic interference by ensuring smoother transitions between states, thereby enhancing system performance and longevity. Conversely, hard-switching, characterized by abrupt transitions, can lead to increased efficiency under certain conditions but may also introduce greater stress on system components. Understanding the nuances of these configurations is crucial for optimizing FSO systems for various operational scenarios. Figure 15 illustrates two potential setups for a hybrid FSO/RF link.

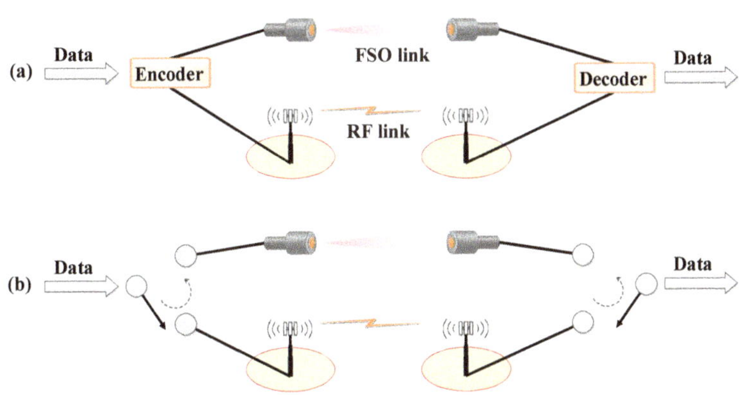

Figure 15. Hybrid FSO switching configurations: (**a**) soft-switching and (**b**) hard-switching.

Hard-Switching Configurations

One switching approach is a switch-over hybrid technique that employs hard-switching between the FSO and alternative links to make use of their complementary natures [430,431]. In this approach, as shown in Figure 15a, the Tx and Rx collaboratively decide between the FSO and an alternative channel for data transmission. The FSO link is chosen only when the channel conditions allow for reliable communication; otherwise, all data are transmitted via the alternative channel [432,433]. In this regard, commercially available hybrid solutions simply use the RF link as a hot-standby backup for the FSO link, which is to be used only when the FSO channel is inoperative [434]. It is important to note that the Tx and Rx need to be coordinated through feedback to choose the appropriate channel for transmission. One major drawback of hard-switching is that one link remains idle at any given time. In reality, scintillation or loss in the sensitive FSO channel leads to infrequent selection of that link. Consequently, once the RF link is chosen, the channel capacity of the FSO link goes unused [432]. Also, this method results in frequent switching between the FSO and alternative links [429,430].

Soft-Switching Configurations

One solution to address the challenges of hard-switching is to synchronize data transmission across both links using channel coding, as illustrated in Figure 15b. In this method, both the FSO and alternative links are utilized simultaneously for data transmission. One approach involves sending identical data concurrently over both links and employing diversity combining techniques for the received signals from each link [435]. This method, however, limits the system's data rate to that of the slower link. Alternatively, the data stream can be jointly coded and divided between the two links, potentially leading to a significant enhancement in overall system capacity [431]. In this context, data are encoded using a single LDPC code, with part of the codeword allocated to both FSO and alternative links, and the rate is modified through puncturing based on the current channel conditions [429,432]. While this approach offers improvements over hard-switching, it necessitates knowledge of channel conditions at both the Tx and Rx, as well as complex soft decoding, which poses challenges at FSO data rates. Recently, Raptor codes have also been explored for hybrid FSO/RF links [214,434,436,437]. These codes operate without the need for channel knowledge at the Tx and automatically adjust the rate between RF and FSO links based on channel conditions, utilizing just a single bit of feedback per message [432].

Generally, hard-switching schemes tend to result in some rate loss compared to soft-switching schemes. Both approaches require that the FSO and alternative links remain continuously active, even during periods of poor quality. Consequently, both links must be operational even when the FSO link has good quality and can independently meet the required BER. In such situations, these solutions are inefficient regarding bandwidth utilization and resource usage across both channels [429]. For instance, the power used for alternative link transmission is wasted, leading to unnecessary interference with the environment [428,430]. To tackle these issues of frequent switching, unnecessary RF interference, and RF power wastage, a new hybrid FSO/RF transmission scheme with adaptive combining is proposed in [430]. In this approach, the FSO link is utilized for data transmission as long as its quality remains acceptable, while the RF link is placed in standby mode. When the quality of the FSO link deteriorates, the system activates the RF link and employs MRC for signals received from both the FSO and RF links.

The complementary characteristics of FSO and RF channels have prompted several proposals for hybrid RF/FSO systems. A combination of high-rate FSO with low-rate RF systems operating in the 2.4 GHz frequency band is suggested in [438] and experimentally validated in [433,439]. However, the maximum data rate achievable in such hybrid systems is constrained by the limited RF bandwidth available in the 2.4 GHz band. This limitation can be addressed by utilizing RF systems in the mmWave frequency band, which supports data rates comparable to those of FSO [440–442]. When instantaneous CSI is available at the Tx, a straightforward CSI-dependent switching mechanism between the FSO and RF systems can be implemented [438].

Additionally, much of the prior research on hybrid FSO/RF systems has concentrated on the joint design of coding schemes for both FSO and RF links, facilitating soft-switching between the two links [429,434,443]. For instance, to enhance the performance of hybrid FSO/RF systems, ref. [443] employs hybrid channel codes that not only optimally achieve the combined capacity of the FSO and RF channels but also ensure carrier-grade reliability (99.999%) for the FSO link. The coding scheme for the hybrid channel is based on non-uniform punctured LDPC codes. This scheme relies on knowledge of the instantaneous channel conditions at the Tx and the appropriate adjustment of code rates for RF and FSO transmission [428].

An alternative approach to joint FSO/RF channel coding using a practical implementation of rateless Fountain codes is presented in [434]. The scheme belongs to the category of hybrid automatic repeat request (HARQ) that utilizes incremental redundancy coding, and as such does not require rate adjustment prior to transmission. This is a clear distinction relative to rate-adaptive coding techniques like the ones proposed in [429,443]. A bit-interleaved CM scheme for hybrid systems is introduced in [431]. Additionally, ref. [432] proposes the use of short-length Raptor codes. The link availability of hybrid FSO/RF systems is explored from an information theory perspective in [444]. Meanwhile, ref. [435] presents diversity combining for parallel FSO and RF channels, assuming both links transmit identical information simultaneously. The performance of a similar hybrid FSO/RF system subject to non-Gaussian noise is analyzed in [445].

5. Emerging Trends in FSO Communication

FSO communication is evolving rapidly, integrating cutting-edge technologies to overcome conventional limitations and enhance performance. This section delves into the latest advancements shaping the future of FSO systems. Key trends include the incorporation of emerging technologies, which bolster system capabilities and adaptability. Enhanced FSO transceiver design aims to improve reliability and efficiency. The integration of quantum cryptography offers unprecedented security for data transmission. OAM-aided FSO systems introduce a novel method for multiplexing, significantly increasing data throughput. Additionally, the application of AI in FSO systems promises to optimize performance through intelligent decision-making and adaptive strategies. Together, these innovations are poised to revolutionize FSO communication, making it a more viable and robust solution for modern communication needs [2].

5.1. Incorporation of Emerging Technologies

The landscape of wireless communication is rapidly evolving, fueled by the increasing need for higher data rates, improved reliability, and seamless connectivity. As mentioned in the preceding section, FSO communication, known for its high bandwidth and security, is being integrated with other advanced technologies such as RF, mmWave, and THz systems [2,93]. This section discusses these emerging trends in detail, highlighting how these integrations can enhance communication systems' overall performance and capabilities.

5.1.1. Hybrid FSO/RF System

FSO and RF technologies have complementary strengths that can be harnessed to overcome the limitations of each. While FSO provides high bandwidth and secure communication, RF is less affected by weather conditions and can penetrate obstacles. This integration enhances reliability and resilience by providing alternative communication paths in adverse conditions. Also, in the hybrid system, RF links can serve as a backup for FSO links during adverse weather conditions (e.g., fog, rain) that degrade optical signals. This redundancy ensures continuous and reliable communication. Likewise, advanced algorithms can switch dynamically between RF and FSO links based on real-time environmental conditions and link quality, optimizing the communication system's performance.

Because of the offered reliability and redundancy, the hybrid FSO/RF system is ideal for providing last-mile connectivity in urban areas where physical obstructions and varying weather conditions pose challenges. Also, integration with RF enables the deployment of FSO links in mobile networks, supporting high-speed data transfer for moving platforms like vehicles and ships. Also, the scheme's high-speed data transmission capabilities make it a suitable solution for backhaul and fronthaul applications, improving the overall performance of the 5G infrastructure [2,93,365].

5.1.2. Hybrid FSO/mmWave System

mmWave frequencies (30–300 GHz) offer higher bandwidth compared to traditional RF frequencies, supporting gigabit per second (Gbps) data rates. Similar to FSO, mmWave communication utilizes highly directional beams, which can reduce interference and enhance security. So, combining FSO and mmWave technologies can significantly increase the available bandwidth, enabling ultra-high-speed communication. Also, both FSO and mmWave systems benefit from advanced beam steering and alignment techniques, often driven by AI and ML algorithms, to maintain optimal connectivity. Due to the high frequency, mmWave signals are best suited for short-range applications. Integrating them with FSO can extend the range and improve the reliability of these short-range links. Based on this, the scheme is crucial for the deployment of 5G networks and the development of 6G, providing the high data rates and low latency required for these advanced networks [12]. Also, the hybrid system can be used for high-speed data transfer within data centers and for backhaul connections between network nodes [2].

5.1.3. Hybrid FSO/THz System

THz frequencies (0.1–10 THz) offer even higher data rates than mmWave, potentially supporting tens to hundreds of Gbps. The THz spectrum provides ample bandwidth, which is crucial for future high-capacity communication systems. FSO can complement THz communication by providing high-capacity links in scenarios where THz propagation is limited due to atmospheric absorption and scattering. Integrating FSO with THz technologies can leverage advanced modulation techniques to maximize data throughput and SE. AI-driven adaptive systems can dynamically allocate FSO and THz resources based on real-time performance metrics, optimizing the communication link's quality and capacity. So, the scheme is a key enabler for wireless terabit networks, which are essential to enable future data-intensive applications, including holographic communications and immersive virtual reality. Hybrid FSO/THz systems can enhance space communication links, providing high-speed data transfer between satellites and ground stations or inter-satellite links [2].

In general, the integration of FSO communication with RF, mmWave, and THz technologies represents a notable trend in the evolution of wireless communication systems. These hybrid systems leverage the strengths of each technology to overcome their individual limitations, resulting in enhanced performance, reliability, and capacity. As research and development continue, these integrated systems are expected to be instrumental in the deployment of next-generation networks, supporting a broad spectrum of applications from urban connectivity to space communications. This convergence of technologies represents a significant advancement in realizing the vision of ubiquitous, high-speed wireless communication [2,365].

5.2. Enhanced FSO Transceiver Design

A roadmap for future wireless communications is expected to exploit all transmission-suitable spectrum bands, from the microwave to the optical frequencies, to support orders of magnitude faster data transfer with much lower latency than the deployed solutions nowadays. The currently under-exploited mid-infrared (mid-IR) spectrum is an essential building block for such an envisioned all-spectra wireless communication paradigm. FSO communications in the mid-IR region have recently attracted great interest due to their intrinsic merits of low propagation loss and high tolerance of atmospheric perturbations [446].

Meanwhile, FSO technologies have been proposed, studied, and even developed for decades by reusing the fiber-optic components in the near-IR band (1–2 μm). Unfortunately, these have not been considered reliable building blocks of the modern ICT infrastructure. The main concern of these near-IR FSO technologies is that they are susceptible to the dynamic atmospheric environment. For instance, with the development of passive and active fiber components and devices, the 1.55 μm band, also referred to as the telecom band, has been applied for commercial FSO communication systems. However, the 1.55 μm band is sensitive to atmospheric channel perturbations caused by fog, dust, and turbulence [447].

Recently, the under-exploited mid-IR spectral region, particularly the two atmospheric transmission windows at the mid-wavelength IR (MWIR, 3–5 μm) and the long-wavelength IR (LWIR, 9–12 μm) [448], are attracting increasing interest for FSO communications. The interest in a MWIR platform stems from the intrinsic PHY layer advantage of low absorption when propagating in the atmosphere even under adverse conditions (for instance, scattering by aerosols, rain, and snow, as well as beam broadening and scintillation by turbulence effects), robustness of the wavefront during long-distance propagation compared with the near-IR FSO, and the absence of regulations and restrictions for this range of wavelengths [446,448–450]

The future development of viable mid-IR FSO transceivers requires a semiconductor source to fulfill the high-bandwidth, low-energy-consumption, and small-footprint requirements. In this context, quantum cascade devices, including quantum cascade lasers (QCLs) and quantum cascade detectors (QCDs) appear as promising technological candidates to fulfill this role [446,447,451,452]. Quantum cascade devices are based on inter-sub-band transitions to cover a wide spectral range from the mid-IR to THz. Particularly, directly modulated-QCLs have ultra-short carrier relaxation lifetime, leading to a high intrinsic modulation bandwidth and making the laser response over-damped, thus repressing the resonance frequency [449,451,452]. It is a promising candidate for solid-state FSO Txs at the MWIR and LWIR bands. Combined with advanced modulation formats (e.g., PAM and DMT) and DSP techniques, capacity-approaching transmission performance in both optical and wireless communication systems can be achieved [447].

5.3. *Quantum Cryptography*

Quantum cryptography represents a groundbreaking advancement in secure communication, utilizing the concepts of quantum mechanics to ensure the confidentiality and integrity of data. When integrated with FSO communication systems, quantum cryptography, particularly Quantum Key Distribution (QKD), offers unparalleled security, providing ultra-secure communication channels, particularly in scenarios requiring high levels of data protection. Also, integrating QKD into FSO involves transmitting qubits via the same optical channel, requiring precise alignment and stable atmospheric conditions to minimize decoherence and loss. So, AT, scattering, and absorption can impact the fidelity of quantum states. Advanced error correction and AO schemes can be used to alleviate these effects. Also, maintaining accurate alignment between the Tx and Rx is crucial. Automated tracking systems and fine-tuning mechanisms help maintain the integrity of the quantum channel [2,453–455].

In general, the integration of quantum cryptography with FSO communication systems holds immense potential for revolutionizing secure communications, particularly within the framework of 5G and future networks. By leveraging the principles of quantum mechanics, QKD provides a level of security that is fundamentally unattainable with classical cryptographic methods. As we advance towards more connected and data-driven environments, the role of QKD in ensuring ultra-secure communication channels will become increasingly vital, offering robust solutions across various sectors, from finance and healthcare to government and smart cities [453,454,456].

5.4. Orbital Angular Momentum-Aided FSO Systems

Several multiplexing methods, such as WDM, space division multiplexing (SDM), and polarization division multiplexing (PDM), have been employed to enhance the transmission capacity of FSO communication systems. However, the numerous limitations and the complex architecture of FSO Rxs make it challenging to significantly boost the transmission capacity further. Recently, a new multiplexing technique known as OAM has gained prominence among researchers for its potential to further increase the capacity of FSO systems [457]. So, in future ultra-fast optical communication, besides the widely applied characteristics of light, including amplitude, phase, polarization, and wavelength, the OAM will offer a new degree of freedom of light beams. In principle, light beams carrying different OAM states are mutually orthogonal, and the number of OAM eigenstates is infinite. Therefore, the application of OAM provides alternative technologies to meet the ever-increasing bandwidth demands in wireless communication systems [2,458].

Moreover, to enhance the data transmission capacity of FSO communication, techniques utilizing OAM—a fundamental property of Laguerre–Gaussian (LG) beams—have been proposed and garnered significant attention [459]. Leveraging the theoretically unlimited range of available OAM states [460], OAM division multiplexing [461,462] and OAM-shift keying (OAM-SK) [50,463,464] can significantly improve the capacity of FSO communication systems. Specifically, in OAM division multiplexing, LG beams carrying various OAM states act as channels to multiplex information streams [458]. Traditionally, these beams are multiplexed at the Tx and then transmitted through the atmosphere to the Rx, where they are demultiplexed due to their orthogonal properties [465]. Meanwhile, in OAM-SK, information is modulated into LG beams with different OAM modes, representing a new modulation format. The OAM-SK-FSO communication system offers distinct advantages, including reduced equipment costs, increased transmission distances, enhanced photon efficiency, and improved information security. This makes it particularly suitable for cost-effective emergency communication and medium-distance FSO applications, such as high-dimensional QKD and deep-space or near-Earth communications. Additionally, advancements in SLMs can further enhance the capacity of OAM-SK communication systems.

5.5. Artificial Intelligence-Driven FSO Systems

The integration of AI and ML into FSO communication systems represents a significant advancement and is a rapidly growing trend, offering new possibilities for optimizing system performance and reliability. AI and ML technologies offer powerful tools to address the inherent challenges of FSO communication, such as atmospheric interference, alignment issues, and error correction. In this regard, AI-driven algorithms can be used to predict atmospheric conditions, estimate channels, modulate and demodulate signals, optimize beam alignment, monitor performance, and enhance error correction processes. For example, ML models can analyze vast amounts of meteorological data to predict weather patterns with high accuracy. By integrating these predictions into FSO systems, operators can anticipate adverse conditions and take preemptive measures [2,404,466–474].

Furthermore, AI-driven systems can continuously monitor atmospheric conditions using sensors and adjust system parameters like the transmission power, modulation schemes, and other parameters accordingly in real-time to maintain optimal performance. In this regard, recent advancements in deep learning (DL) have enabled the development of sophisticated models for processing FSO signals. These models can filter out noise, enhance signal clarity, and improve data rates [467]. For example, FSO communication systems utilizing DL for various roles—including channel estimation, detection, joint constellation shaping, and as a combined channel estimator and detector—are discussed in [475]. In [476], a design of an optical feedback network employing ML techniques is presented, demonstrating its effectiveness in correcting the effects of turbulent propagation on optical modes. This artificial neural network (ANN) approach relies solely on measuring the intensity profile of the distorted modes, making it both simple and robust. Additionally, a novel joint AT detection and adaptive demodulation technique utilizing convolutional neural networks (CNNs) is proposed to enhance the reliability and flexibility of OAM-FSO communication [469]. An m-ary adaptive demodulator based on the maximum likelihood for light beams carrying OAM over FSO turbulence channels has also been proposed and demonstrated [458]. Also, in [477], a new turbo-coded 16-ary OAM-SK-FSO communication system is introduced, which incorporates a CNN-based adaptive demodulator designed for operation under strong AT. In [465], a method is proposed for demultiplexing OAM-carrying beams by capturing an image of their unique multiplexing intensity pattern and training a CNN to serve as a classifier. This CNN-based approach simplifies the demultiplexing process, as it eliminates the need for alignment, relaxes orthogonality constraints, and reduces the reliance on expensive optical hardware.

AI is being used to manage and optimize entire FSO networks. This includes tasks such as traffic routing, load balancing, and resource allocation, ensuring efficient and reliable network performance. The research presented in [474,478] utilizes the actor–critic method from reinforcement learning for power adaptation in SM FSO systems. Additionally, the study in [466] addresses the broader issue of resource allocation aimed at reducing channel fading effects in FSO communications. This resource allocation problem is framed within a constrained stochastic optimization framework, encompassing a range of FSO scenarios that involve power adaptation, relay selection, and their combined allocation. In a similar vein, ref. [479] explores optimal resource allocation in FSO fronthaul networks. This optimal allocation aims to maximize the average weighted sum-capacity while adhering to constraints related to power limitations and data congestion. The study takes into account both adaptive power assignment and node selection, leveraging the instantaneous CSI of the links. By parameterizing the resource allocation policy, the problem is framed as an unsupervised statistical learning challenge. Furthermore, the use of graph neural networks (GNN) for policy parameterization is proposed to effectively utilize the FSO network structure with a reduced number of training parameters.

Maintaining precise alignment between the Tx and Rx is crucial for FSO communication. AI algorithms can process data from various sensors to automatically adjust the alignment of the optical beam [480]. This includes compensating for environmental factors such as wind-induced vibrations or the movement of mobile platforms. ML models can predict the best alignment strategies based on historical data and real-time feedback, ensuring the beam remains accurately directed towards the Rx even in dynamic conditions [467,481]. For instance, sensor-less AO is among the most promising techniques for compensating significant wavefront disturbances in FSO. A backpropagation ANN can be utilized in sensor-less AO systems to create a distortion correction scheme. This approach requires only one or a few online measurements to correct wavefront distortion, unlike other model-based methods, thereby enhancing the system's real-time capabilities [473].

One of the mostly used DL algorithms in optical communication applications is a deep neural network (DNN). The DNN is an extension of the ANN [482]. It is flexible and could be used for various applications, such as modulation format identification [483,484], mitigating the fiber effects [485], optical amplifier control [486], optical performance monitoring [50,487], along with various optical network applications such as quality of transmission (lightpaths) prediction [488] and an AI-driven routing framework for multi-domain optical networks [489]. For instance, a novel block-wise detection technique utilizing support vector machines (SVMs) to address scintillation noise in FSO links is proposed in [471]. Given that turbulence fading occurs on a millisecond scale, while signaling rates typically reach hundreds of Mbps, the SVM approach offers an exceptionally large data block proportional to the bit rates. Additionally, the SVM calculation is independent of the channel's turbulence conditions and the received signal models, requiring no prior information or heuristic assumptions. Furthermore, ref. [490] introduces a new detection method for OOK-modulated signals using various DL models across different strengths of FSO turbulent channels, without needing prior knowledge of the channel parameters. The DL decoders enhance the performance of the FSO turbulent channel and reduce power consumption. Notably, these DL models also operate faster than maximum likelihood methods with perfect channel estimation, achieving slightly better performance under turbulent conditions, thus facilitating FSO communications over turbulent atmospheric channels.

In general, AI-driven FSO systems could enable autonomous vehicles, drones, and other robotic platforms to communicate seamlessly, even in challenging environments. This includes applications in disaster response, remote sensing, and smart cities. AI and ML could play a critical impact in the development of quantum FSO systems, optimizing the transmission and reception of quantum signals for ultra-secure communication. Also, AI-aided FSO systems can lead to the development of smart infrastructure, where buildings, streetlights, and other urban elements are equipped with FSO transceivers managed by AI, providing ubiquitous connectivity [2,458,469,491]. In this context, AI-driven FSO communication systems are transforming the field, offering innovative solutions to long-standing challenges and paving the way for new applications. By leveraging AI-driven algorithms for predicting atmospheric conditions, optimizing beam alignment, and enhancing error correction, FSO systems can achieve higher performance and reliability. As research and development in this area continue to advance, the potential applications of AI and ML in FSO communication are vast, promising a future where seamless, high-speed optical communication is a reality [466,479,481,492].

6. Open Challenges, Future Prospects and Research Directions

FSO communication systems have developed into an innovative technology for high-speed wireless communication. Despite their potential, various challenges need to be overcome to fully harness their capabilities. This section explores the key open challenges, discusses future prospects, and outlines potential research directions in FSO communication. By identifying critical areas, such as AT mitigation, aperture averaging effects, and advanced modulation techniques, we aim to offer an in-depth overview of the current landscape, fostering a deeper understanding of where focused efforts and innovative solutions are needed. This discussion is crucial for guiding ongoing and future research, ensuring continued advancement and addressing the critical needs of the field.

6.1. *FSO Scalability and Reliability for 5G and Beyond Networks*

The scalability and reliability of FSO systems are critical factors for their adoption in next-generation networks. As the world transitions to 5G and beyond, the demand for high-capacity, low-latency, and resilient communication networks is skyrocketing. This section assesses the potential of FSO to scale with the increasing demands of 5G and beyond, addressing key issues such as network architecture, deployment strategies, resource management, and system integration.

6.1.1. Network Architecture

The architecture of a network determines its ability to scale and adapt to new demands. In the realm of FSO systems, several architectural considerations are crucial. FSO systems can complement existing RF networks by providing high-capacity backhaul links. Hybrid architectures that seamlessly integrate FSO with RF can offer improved redundancy, resilience, and bandwidth. Also, implementing mesh network topologies with multiple FSO nodes can enhance coverage and provide alternative pathways for data transmission, improving reliability and scalability.

Distributed control mechanisms can enhance the scalability of FSO networks by enabling more efficient resource management and reducing the load on central controllers. Also, integrating edge computing capabilities with FSO nodes can reduce latency and improve the performance of latency-sensitive applications, which is critical for 5G and beyond [478,493].

6.1.2. Deployment Strategies

Effective deployment strategies are essential for the successful scaling of FSO systems. In this regard, site selection and planning are very important. For example, thorough LoS analysis is required to identify optimal locations for FSO transceivers, minimizing the impact of physical obstructions. Also, installing FSO equipment on rooftops, towers, and other elevated structures can enhance coverage and reduce the likelihood of obstructions.

Additionally, FSO can be effectively used for backhauling small cell networks in dense urban areas, where laying fiber may be impractical or cost prohibitive. High-density deployment of FSO nodes can support the massive device connectivity required for 5G and beyond, providing high-capacity links in congested areas. Also, deploying mobile FSO units on vehicles or drones can provide temporary high-capacity links in emergency situations or during large events, offering flexibility and rapid deployment capabilities.

6.1.3. Resource Management

Effective resource management is essential for the scalability and reliability of FSO systems. Dynamic resource allocation is an important aspect of resource management. In this regard, implementing AMC schemes can optimize the use of available bandwidth and maintain link quality under varying atmospheric conditions. Likewise, dynamic load balancing across multiple FSO links can enhance network performance and prevent congestion.

Utilizing WDM technology can significantly increase the capacity of FSO links by transmitting multiple data streams on different wavelengths over a single optical path. Efficient channel allocation strategies can minimize interference and maximize the utilization of available optical spectrum.

Furthermore, priority-based traffic management can facilitate network performance. For example, implementing QoS mechanisms that prioritize critical traffic can ensure reliable performance for latency-sensitive applications, such as real-time video and Internet of Things (IoT) services. Similarly, advanced error correction techniques can alleviate the impact of signal deterioration caused by atmospheric conditions, enhancing the reliability of FSO links.

6.1.4. System Integration

Integrating FSO systems with existing network infrastructure and emerging technologies is vital for scalability. Key integration consideration include compatibility with existing networks. So, ensuring interoperability between FSO systems and existing network equipment, such as routers and switches, can facilitate seamless integration and operation. Also, adopting standardized protocols and interfaces can simplify the integration process and enhance compatibility across different vendors and technologies.

In addition, for effective integration with 5G and beyond technologies, FSO systems must be designed to support the requirements of 5G and future networks, including high data rates, ultra-low latency, and massive connectivity. Also, integrating FSO with IoT infrastructure and smart city applications can provide high-capacity, low-latency links for different use cases, such as smart lighting, traffic management, and environmental monitoring. Implementing centralized network management systems can provide real-time monitoring and control of FSO links, facilitating proactive maintenance and optimization. Also, leveraging AI and ML techniques can enhance network management by predicting and mitigating issues related to atmospheric conditions, traffic patterns, and potential obstructions.

In general, the potential of FSO communication systems to scale with the increasing demands of 5G and beyond is promising, provided that key challenges related to network architecture, deployment strategies, resource management, and system integration are addressed. By leveraging hybrid network architectures, adopting efficient deployment strategies, implementing dynamic resource management techniques, and ensuring seamless integration with existing and emerging technologies, FSO can be crucial in the next generation of high-capacity, reliable communication networks.

6.2. Standardization and Regulatory Challenges

For FSO communication to achieve widespread adoption, it is essential to establish standardized protocols and address regulatory challenges. This section discusses the current state of FSO standardization efforts and the regulatory issues that need to be resolved to facilitate the deployment of FSO systems on a global scale. By analyzing the current state of FSO standardization, this section highlights the progress made, the challenges remaining, and the significance of these efforts in facilitating the growth and integration of FSO systems into global communication networks.

6.2.1. Current State of FSO Standardization Efforts

The standardization of FSO systems is crucial for ensuring interoperability, reliability, and widespread adoption. This section explores the ongoing efforts to develop and implement standards for FSO technology. It examines the roles of various international bodies and industry groups in establishing guidelines and protocols that address key aspects such as performance metrics, safety regulations, and compatibility with the existing communication infrastructure.

IEEE Standards

The Institute of Electrical and Electronics Engineers (IEEE) has been instrumental in developing standards for various communication technologies. For FSO, the IEEE 802.15.7-2018 standard focuses on OWCs, including FSO. This standard outlines the PHY layer and medium access control (MAC) layer specifications for short-range OWC, which can be adapted for FSO applications [494–497]. IEEE is also working on extending these standards to accommodate long-range FSO communication, which is crucial for bridging the gap between current capabilities and future needs.

ITU Recommendations

It is important to note that the International Telecommunication Union (ITU) has recognized the potential of FSO systems, and through the ITU—Telecommunications Standardization Sector (ITU-T), it has approved Recommendation G.640 [498], providing a procedure for enabling the interference-free coexistence of more than one p2p FSO communication system at a location. The Radio Communication Sector of ITU (ITU-R) has also issued a report on fixed service use cases using FSO links, defining device characteristics, possible applications, as well as technical and operational aspects of FSO systems [499]. Also, ITU has developed recommendations that cover aspects of optical communication, including FSO. ITU-T G.9991, for instance, deals with high-speed indoor OWC transceivers, which can inform FSO developments [500–502]. ITU-R P.1817 provides guidelines for the performance and availability of optical free-space links, addressing atmospheric effects and link budgets. These recommendations are critical for ensuring reliable FSO communication across different environments [503,504]. The recommendation in [505] provides propagation prediction methods for planning terrestrial FSO systems. It includes methods to estimate attenuation in clear air, fog, and rain and snow precipitation. It also covers scintillation and impairments by sunlight.

6.2.2. Regulatory Issues

The deployment of FSO communication systems faces several regulatory barriers that must be resolved to guarantee widespread adoption. Key issues include spectrum allocation, compliance with international standards, and the establishment of safety guidelines to prevent interference with other communication systems and ensure eye safety. This section explores these regulatory hurdles, discusses current policies and their implications, and outlines the necessary steps for creating a supportive regulatory framework that can facilitate the efficient and safe deployment of FSO systems.

Spectrum Allocation

Unlike RF communication, which operates within regulated frequency bands, FSO communication utilizes unlicensed optical spectrum. While this provides a significant advantage in terms of deployment flexibility, it also raises concerns about potential interference and the need for global harmonization. Also, regulatory bodies need to establish guidelines for the use of optical spectrum, ensuring that FSO systems can operate without interference from other optical devices and across national borders.

Safety Regulations

FSO communication involves high-powered lasers, which can pose safety risks to humans and animals if not properly regulated. The International Electrotechnical Commission (IEC) provides standards such as IEC 60825-1, refs. [3,506] which outlines safety requirements for laser products. Compliance with laser safety standards is critical for the safe deployment of FSO systems. Regulatory bodies must enforce these standards and conduct regular inspections to prevent accidents and ensure public safety.

Deployment Permits and Zoning Laws

Installing FSO equipment, especially in urban areas, requires navigating complex zoning laws and obtaining necessary permits. Local regulations can vary significantly, impacting the speed and cost of FSO deployment. Streamlining the permitting process and creating uniform regulations can facilitate faster and more efficient deployment of FSO systems, particularly in densely populated regions.

Cross-Border Coordination

For international FSO links, especially those connecting different countries, coordination between national regulatory bodies is essential. Issues such as signal interference, spectrum management, and compliance with safety standards must be harmonized to ensure seamless cross-border communication. International cooperation and agreements are necessary to address these challenges, fostering a regulatory environment conducive to global FSO deployment.

Environmental Regulations

The deployment of FSO systems must also consider environmental regulations, particularly in sensitive areas such as wildlife reserves or historical sites. Ensuring that FSO infrastructure does not negatively impact the environment is a key regulatory concern. Regulatory frameworks should include provisions for environmental impact assessments and the implementation of mitigation measures to protect ecological and cultural heritage.

Moreover, to realize the full potential of FSO communication, stakeholders must collaborate to establish comprehensive standards and regulatory frameworks. By addressing these standardization and regulatory challenges, the FSO communication industry can pave the way for widespread adoption, enabling high-speed, secure, and reliable wireless communication on a global scale.

6.3. Open Challenges

Despite the significant advancements in FSO technology, several open challenges remain that require further research. These include the development of more robust mitigation techniques for AT, the integration of FSO with emerging technologies like Light Fidelity (Li-Fi), and the exploration of new materials and devices for FSO systems. This section identifies the key areas where further future research is needed to fully realize the potential of FSO communication in next-generation networks.

6.3.1. Mitigation Techniques for Atmospheric Turbulence

AT remains one of the primary obstacles in FSO communication, causing signal degradation and fluctuations in signal strength. Future research should focus on developing advanced AO systems that can dynamically adjust the optical beam in real-time to compensate for turbulence-induced distortions. Also, enhancing error correction techniques to effectively mitigate the impact of turbulence on data integrity is important. Likewise, extensive research on the utilization of ML algorithms to predict atmospheric conditions and adjust FSO parameters proactively to maintain optimal performance is required. Also, investigating hybrid systems that combine FSO with other technologies, such as RF, to provide redundancy and maintain connectivity in turbulent conditions is important to realize the full potential of FSO communication.

6.3.2. Integration with Emerging Technologies

The integration of FSO communication with emerging technologies like Li-Fi and 5G/6G networks presents exciting opportunities and challenges. This involves exploring the seamless integration of FSO with Li-Fi to enable high-speed, high-capacity indoor and outdoor communication networks. This includes addressing handover issues between indoor Li-Fi and outdoor FSO links [201]. Similarly, it entails investigating the role of FSO in supporting ultra-high-speed data rates and low-latency communication in 5G and future 6G networks. This involves the development of protocols and standards for FSO integration with existing and future wireless networks. This will facilitate the development of FSO-based solutions to support the growing demand for high-capacity, low-latency communication in IoT applications and smart city infrastructures [378,493].

6.3.3. New Materials and Devices

Advancements in materials science and photonic devices are crucial for the evolution of FSO systems. Researching new materials and designs for high-performance photodetectors that offer higher sensitivity, faster response times, and better performance in diverse environmental conditions is essential. Also, developing low-loss optical components, such as lenses, mirrors, and modulators, to improve the efficiency and extend the range of FSO systems is highly imperative, as is investigating the use of quantum dots and other nanomaterials to create novel light sources and detectors that can operate efficiently under varying atmospheric conditions. Another important aspect is creating flexible and wearable FSO devices to expand the application scope of FSO communication in areas like personal area networks and medical devices [446,448–450,493,507].

6.3.4. Security Enhancements

Ensuring the security of FSO communication is vital, especially for sensitive applications. Developing advanced encryption techniques tailored for the high data rates and unique characteristics of FSO links is vital to ensure robust FSO systems. Also, researching methods to enhance PLS, such as beamforming and spatial diversity, to protect against eavesdropping and jamming is important, especially in applications for finance and healthcare. In this regard, exploring the use of QKD over FSO links to provide unbreakable encryption keys for secure communication require further research [46,86,453–455,493,508,509].

6.3.5. Network Architecture and Protocols

The development of new network architectures and protocols is essential to support the deployment and scalability of FSO systems. In this regard, certain areas such as designing network topologies that can adjust dynamically to varying conditions and optimize FSO link performance require further research: creating efficient routing and handover protocols to manage the transition between FSO and other communication technologies, ensuring seamless connectivity, and leveraging software-defined networking to enable flexible and programmable FSO networks that can be easily managed and optimized for different applications and environments [75,340,478,508,509].

Addressing the previously mentioned key areas of research is essential for realizing the complete capabilities of FSO communication in next-generation networks. By developing robust mitigation techniques for AT, integrating FSO with emerging technologies, exploring new materials and devices, enhancing security, and innovating network architectures and protocols, researchers can pave the way for more reliable, efficient, and versatile FSO communication systems. These advancements will enable FSO to play a pivotal role in meeting the growing need for high-speed, high-capacity wireless communication in the future.

6.4. Future Directions

This section explores future directions in optical communication technologies, focusing on several key advancements. It delves into novel modulation schemes, adaptive transmission techniques, and aperture averaging methods to enhance signal quality. The discussion extends to spatial diversity and cooperative relaying strategies aimed at improving communication reliability. Additionally, the integration of AO and ADTs is examined for their potential to mitigate atmospheric distortions. The potential of hybrid FSO systems, along with advanced ECC and modulated signal optimization, is also highlighted, underscoring the continuous evolution of this field. The future directions are highlighted in Figure 16 and explored in detail in the following subsections.

6.4.1. Modulation Schemes

The future of FSO communications is poised for considerable advancements through the continuous enhancement and integration of various modulation schemes. These advancements will enable FSO systems to achieve higher capacities, enhanced reliability, and greater resilience against atmospheric challenges. Several key directions in the enhancement and integration of traditional, advanced, and SM techniques promise to significantly improve the performance of FSO communication networks.

Adaptive Modulation

Traditional modulation techniques, while foundational, are being complemented and surpassed by more advanced methods. Among these, adaptive modulation stands out. Consequently, adaptive modulation techniques will be essential in future FSO systems. These techniques adjust the modulation parameters in real-time based on the current channel conditions. For instance, during periods of favorable atmospheric conditions, higher-order modulation schemes such as 64-QAM can be used to enhance data rates. Conversely, during periods of high turbulence or fog, the system can switch to more robust modulation schemes like BPSK to maintain link reliability. The dynamic nature of adaptive modulation ensures that FSO systems can optimize their performance under a diverse array of environmental conditions [93].

Advanced Modulation Schemes in Conjunction with Sophisticated Error Correction

Advanced error correction techniques are also playing a crucial role. By correcting errors that occur during data transmission, these methods enhance the reliability and efficiency of the systems, even in the presence of atmospheric disturbances like fog, rain, and turbulence. Thus, future modulation schemes will also incorporate advanced error correction methods to alleviate the impact of channel impairments. Techniques such as LDPC codes and Turbo codes will be integrated into the modulation process to detect and correct errors in the transmitted data. These error correction methods will enhance the reliability of FSO communications, ensuring that data are accurately received even amid interference and noise [93].

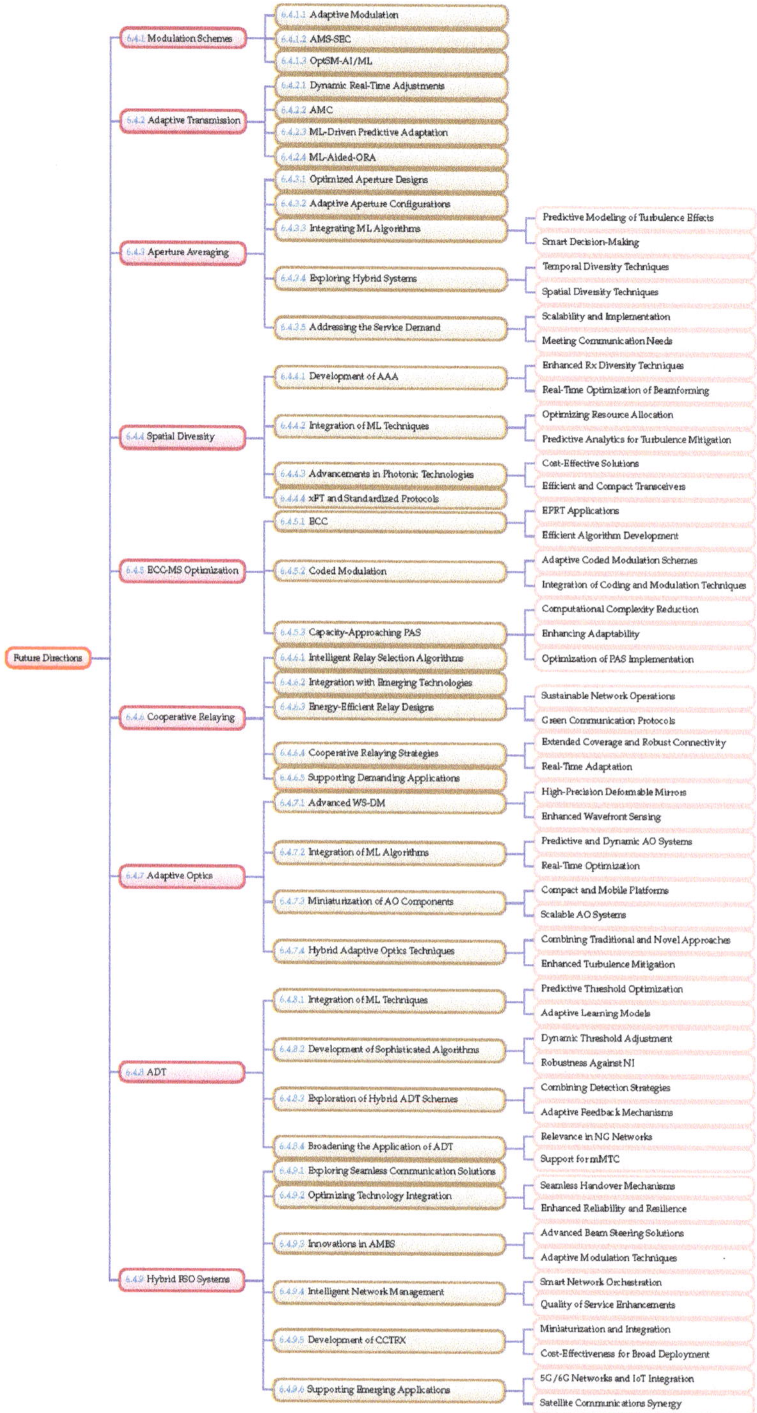

Figure 16. Schematic of the tutorial organization with related content. AMS-SEC: Advanced Modulation Schemes in conjunction with Sophisticated Error Correction, OptSM-AI/ML: Optimized Spatial Modulation with AI and

ML, ML-Aided-ORA: ML-Aided Optimization and Resource Allocation, AMC: Adaptive modulation and coding, AAA: Advanced Adaptive Algorithms, ECC-MS: Error Correction Codes and Modulated Signal, EPRT: Enhanced Performance in Real-Time, PAS: Probabilistic Amplitude Shaping, xFT: Extensive Field Trials, WS-DM: Wavefront Sensors and Deformable Mirrors, NI: Noise and Interference, NG: Next-Generation, mMTC: Massive Machine-Type Communications, ADT: Adaptive Detection Thresholds, AMBS: Adaptive Modulation and Beam Steering, CCTRX: Compact and Cost-Effective Transceivers.

Optimizing Spatial Modulation with AI and ML

SM techniques, which exploit multiple spatial channels, are proving essential in increasing data rates and system capacity without requiring additional bandwidth. Also, AI and ML are emerging as transformative technologies in this domain. The application of AI and ML algorithms can further optimize SM schemes. AI and ML can analyze large volumes of data to predict channel conditions and dynamically adjust modulation parameters in real-time. For instance, ML models can predict periods of high turbulence and preemptively switch to more robust modulation schemes, reducing the risk of data loss. These predictive capabilities will ensure that FSO systems maintain optimal performance in real-time, enhancing their overall efficiency and reliability.

6.4.2. Adaptive Transmission

Future research on adaptive transmission schemes for FSO communication systems promises to bring substantial advancements to the field. In this context, adaptive transmission schemes in FSO communication systems are crucial for overcoming the inherent challenges posed by atmospheric conditions and ensuring robust, high-quality data transmission. So, as FSO technology advances, several promising directions for future research and development can significantly advance adaptive transmission techniques, enhancing their reliability, efficiency, and overall performance.

Dynamic Real-Time Adjustments

One primary focus is the development of more sophisticated algorithms capable of dynamically adjusting to real-time channel conditions. These algorithms will leverage advanced signal processing techniques to continuously monitor and respond to changes in the atmospheric environment, such as fog, rain, or turbulence. By doing so, they will ensure that the communication link remains robust and efficient, regardless of external conditions. In this regard, future research will likely focus on the creation of more sophisticated algorithms capable of making dynamic adjustments to transmission parameters in real-time. These algorithms will analyze the present condition of the communication channel, including factors such as AT, alignment errors, and environmental changes, and adjust modulation schemes, power levels, beam steering, and/or other transmission parameters accordingly, all performed in real-time. By doing so, they will maintain optimal communication performance and minimize errors and data loss.

Adaptive Modulation and Coding

Advances in AMC techniques will be pivotal. Algorithms that can switch between different modulation formats and ECC based on real-time channel assessments will enhance the resilience of FSO systems. For instance, during periods of high turbulence, the system could switch to a more robust modulation scheme to ensure data integrity, while reverting to higher data rate schemes when conditions are favorable.

ML-Driven Predictive Adaptation

The integration of ML and AI techniques represents another promising direction. These technologies can be used to develop predictive adaptation methods that preemptively mitigate the effects of adverse conditions. In this context, ML techniques hold significant promise for enhancing adaptive transmission in FSO systems. By leveraging vast amounts of historical and real-time data, ML models can predict atmospheric conditions and channel states, allowing FSO systems to preemptively adjust their transmission parameters. Predictive adaptation allows the system to prepare for potential disruptions before they occur and can mitigate adverse effects such as signal fading and scattering, ensuring a more stable and reliable communication link.

ML-Aided Optimization and Resource Allocation

ML algorithms can also optimize resource allocation within FSO networks. By analyzing usage patterns and channel conditions, these algorithms can allocate bandwidth and power resources more efficiently, enhancing overall network performance. For example, during periods of peak demand, the system could dynamically allocate more resources to critical communication links, ensuring consistent service quality.

6.4.3. Aperture Averaging

As FSO communication systems advance, exploring new developments in aperture averaging offers several promising directions that can significantly enhance system performance. Aperture averaging, which mitigates the impact of AT by averaging the received optical signal over a larger aperture, is critical for improving signal quality and reliability in FSO systems. Future research in this area can explore several innovative approaches to optimize and advance aperture averaging techniques.

Optimized Aperture Designs

One promising direction is the optimization of aperture designs. By refining the size, shape, and configuration of optical apertures, researchers can further reduce the impacts of turbulence and enhance the quality of the received signal. In this regard, they can identify designs that minimize the scattering and diffraction of optical signals as they travel through the atmosphere. Optimized aperture designs can effectively average out the scintillations caused by atmospheric variations, leading to more stable and reliable communication links. This can involve the use of novel materials and engineering approaches to create apertures that are more efficient and effective in mitigating turbulence.

Adaptive Aperture Configurations

Innovations in AO and dynamic aperture configurations also hold significant potential for real-time performance enhancement. AO systems, which use DMs and other technologies to adjust the optical path in real-time, can be integrated with aperture averaging to dynamically counteract turbulence effects. This real-time adjustment capability allows the FSO system to maintain high performance under varying environmental conditions. Thus, dynamic aperture configurations, where the aperture size, shape and/or arrangement can be altered on-the-fly, can also adapt to changing turbulence conditions. This adaptability will help maintain signal integrity and reduce errors in data transmission.

Integrating Machine Learning Algorithms

The integration of ML algorithms into aperture averaging techniques is another exciting avenue for future research. ML can be employed to develop predictive models that anticipate the effects of AT. Also, ML can also facilitate more intelligent decision-making within FSO systems [400].

Predictive Modeling of Turbulence Effects: The integration of ML algorithms into aperture averaging techniques can revolutionize how FSO systems handle atmospheric disturbances. ML models can analyze historical and real-time data to predict turbulence patterns and their effects on optical signals. By employing predictive modeling, FSO systems can proactively adjust their aperture configurations or signal processing parameters, enhancing resilience against turbulence-related challenges.

Smart Decision-Making: ML can also enable smarter decision-making within FSO systems. For example, algorithms could assess the current atmospheric conditions and determine the optimal aperture configuration or modulation technique to employ, ensuring that the system adapts to real-time changes. This capability will improve link reliability and enhance overall system performance, particularly in dynamic environments where conditions fluctuate rapidly.

Exploring Hybrid Systems

Investigating hybrid systems that integrate aperture averaging with other diversity techniques—such as spatial or temporal diversity—holds promise for further improving FSO communication. By integrating these diversity techniques with aperture averaging, researchers can create hybrid systems that offer enhanced resilience against turbulence and other atmospheric disturbances.

Spatial Diversity Techniques: By leveraging the strengths of various approaches, hybrid systems can provide enhanced link reliability and greater overall communication efficiency. For example, spatial diversity techniques can be combined with aperture averaging to exploit multiple paths for signal transmission, mitigating the effect of turbulence and improving the robustness of the communication link.

Temporal Diversity Techniques: Temporal diversity, which involves transmitting signals over different time slots, can also be effectively integrated with aperture averaging. By employing adaptive transmission schemes that adjust the timing of signal transmission based on atmospheric conditions, FSO systems can minimize the risk of signal degradation. This combination could lead to a more resilient system designed to sustain high data rates even under challenging weather conditions.

Addressing the Service Demand

Advancements in aperture averaging will be pivotal in meeting the high-capacity demands of 5G and future networks.

Meeting Communication Needs: As the demand for high-capacity optical communication solutions grows—driven by applications such as 5G, remote sensing, and satellite communications—the advancements in aperture averaging will play a crucial role. Enhanced signal quality and link reliability achieved through optimized aperture designs and innovative adaptive techniques will ensure that FSO systems can support the increasing data traffic and diverse application requirements in both urban and remote settings.

Scalability and Implementation: Future research should also focus on ensuring that these advancements in aperture averaging are scalable and practical for deployment in various environments. Developing cost-effective solutions and robust implementation strategies will be vital for the widespread adoption of FSO systems. This includes creating standardized protocols for the design and deployment of aperture averaging techniques, ensuring interoperability across different platforms and applications.

6.4.4. Spatial Diversity

Spatial diversity techniques in FSO communication, specifically MISO, SIMO, and MIMO systems, are becoming increasingly essential in addressing the difficulties presented by AT and signal degradation. Future research in these areas is set to explore several key avenues aimed at enhancing performance, reliability, and applicability of FSO systems.

Development of Advanced Adaptive Algorithms

Adaptive approaches will greatly improve link reliability and data throughput, especially in dynamic environments, where the atmospheric conditions can vary rapidly.

Real-Time Optimization of Beamforming: One of the promising research directions is the development of advanced adaptive algorithms that optimize beamforming and Rx diversity in real-time. By flexibly adjusting the transmission and reception parameters depending on current environmental conditions, these algorithms can significantly improve link reliability and quality. For instance, algorithms that can assess atmospheric conditions—such as turbulence and scatter—will allow for intelligent adjustments in beam alignment and power levels, ensuring that the optical signal remains robust even in challenging situations. Also, adaptive beamforming techniques can direct the optical beams in real-time to the most favorable paths.

Enhanced Rx Diversity Techniques: In addition to beamforming, enhancing Rx diversity techniques is critical for mitigating the effects of multipath propagation caused by atmospheric disturbances. Rx diversity algorithms can combine signals from multiple Rxs to maximize signal quality. Future algorithms may incorporate feedback mechanisms that evaluate signal quality across multiple Rxs, allowing systems to select the optimal Rx in real-time. Such adaptive diversity strategies will ensure a more stable communication link and minimize the potential for signal dropouts.

Integration of Machine Learning Techniques

Integrating ML techniques into FSO communication offers another exciting research avenue for spatial diversity systems.

Predictive Analytics for Turbulence Mitigation: Integrating ML techniques into spatial diversity systems presents a significant opportunity for improved performance. ML models can analyze vast amounts of historical and real-time data to predict turbulence patterns and their potential impact on signal quality. By leveraging predictive analytics, FSO systems can preemptively adjust transmission parameters, such as modulation schemes and power levels, thereby enhancing turbulence mitigation strategies and resource allocation.

Optimizing Resource Allocation: ML can assist in resource allocation by determining the most efficient use of available communication channels and transceivers based on real-time and historical data. This predictive capability can greatly improve the resilience and efficiency of FSO systems. For instance, ML can play a pivotal role in optimizing resource allocation in MIMO systems. By intelligently managing the distribution of available bandwidth and power across multiple transmission paths, these algorithms can enhance throughput and efficiency, especially in dense urban environments where interference is prevalent. The ability to dynamically allocate resources based on real-time conditions will be crucial for maintaining high-quality communication links.

Advancements in Photonic Technologies

Advancements in photonic technologies and materials will be essential for the development of more efficient, compact, and cost-effective transceivers in spatial diversity systems. This can lead to the development of FSO systems that are easier to deploy and maintain.

Efficient and Compact Transceivers: Research should focus on miniaturizing components, integrating multiple functionalities into single devices, and improving energy efficiency. Innovations in photonic integration, such as the use of photonic integrated circuits (PICs), have the potential to decrease the size and power consumption of FSO devices while enhancing their performance. These innovative designs will facilitate the creation of lightweight transceivers that can be deployed in a diverse array of applications, from small-scale IoT devices to large-scale urban infrastructure [510].

Cost-Effective Solutions: Reducing the cost of these technologies will be crucial for enabling the widespread adoption of advanced spatial diversity systems. By streamlining manufacturing processes and exploring new materials that offer improved performance at lower costs, researchers can ensure that these technologies are accessible for various applications, including broadband access in underserved areas. Additionally, new materials with superior optical properties can improve the efficiency and reliability of FSO transceivers.

Extensive Field Trials and Standardized Protocols

Extensive field trials are essential not only for the widespread adoption of advanced FSO systems but also for validating the performance of advanced spatial diversity systems and their associated new technologies and algorithms in real-world environments. This process will provide essential data for further refinement to enhance system performance. In this regard, the trials will provide valuable insights into system behavior under various atmospheric conditions and help refine algorithms and protocols for optimal performance. By conducting tests in diverse settings—rural, urban, and coastal—researchers can assess the adaptability and reliability of FSO systems across different scenarios.

6.4.5. Error Correction Codes and Modulated Signal Optimization

The future directions for ECC and modulated signal optimization in FSO communication are promising and diverse, encompassing advancements in algorithm development, integrated coding and modulation techniques, and innovative approaches like capacity-approaching PAS.

Error Correction Codes

Research in ECC will likely focus on creating more efficient algorithms capable of handling higher data rates and more complex error patterns, especially in real-time use cases.

Efficient Algorithm Development: The development of these algorithms will be driven by the need to correct errors introduced by AT, scattering, and absorption, which can significantly degrade the quality of FSO links. Future ECC will need to be adaptive, adjusting to the varying error characteristics of the FSO channel in real-time. This could involve ML techniques to predict and correct errors dynamically, thereby improving the efficiency and reliability of data transmission.

Enhanced Performance in Real-Time Applications: As real-time applications like video streaming and online gaming become more prevalent, ECC must evolve to provide low-latency error correction. Research will focus on reducing the computational complexity and latency of these codes while maintaining high error correction performance. This could involve the use of parallel processing techniques and hardware accelerators to achieve faster error correction without compromising the integrity of the data.

Coded Modulation

Future work in CM will aim to further integrate coding and modulation techniques. Additionally, efforts will focus on developing adaptive CM schemes.

Integration of Coding and Modulation Techniques: Coding and modulation techniques will be combined to optimize SE and enhance resilience against interference. This integration is crucial for FSO systems, where SE is paramount due to the limited available bandwidth. By combining coding and modulation, these systems can transmit more information over the same bandwidth while enhancing resistance to errors caused by atmospheric disturbances.

Adaptive Coded Modulation Schemes: Future research will likely focus on developing adaptive CM schemes that can adjust to changing channel conditions in real-time. These schemes will use feedback from the Rx to adapt the modulation order and coding rate, optimizing the trade-off between the error resilience and data rate based on current channel conditions. This adaptability will be essential for maintaining high performance in the variable and unpredictable FSO environment.

Capacity-Approaching Probabilistic Amplitude Shaping

Future research in capacity-approaching PAS will aim to further optimize PAS implementation, enhance adaptability, and reduce computational complexity. The work will focus on ensuring efficient use of the communication channel by aligning the signal's amplitude distribution with the channel's capacity.

Optimization of PAS Implementation: Future research will focus on optimizing PAS implementation in practical systems, ensuring that it can be deployed effectively in real-world FSO networks. This will involve developing efficient algorithms for shaping and decoding that minimize computational complexity while maximizing performance.

Enhancing Adaptability: Efforts will also be directed at enhancing the adaptability of PAS to different channel conditions. This adaptability is crucial for FSO systems, where channel conditions can change rapidly due to weather and other environmental factors. Researchers will work on creating PAS schemes that can dynamically adjust the signal shaping based on real-time feedback from the channel, ensuring optimal performance under all conditions.

Reducing Computational Complexity: Reducing the computational complexity of PAS is another critical area of focus. While PAS offers significant benefits in terms of SE and robustness, its implementation can be computationally intensive. Future research will aim to streamline the algorithms used for PAS, making them more efficient and practical for deployment in FSO systems. This could involve leveraging advanced hardware, such as Field-Programmable Gate Arrays and Graphics Processing Units, to accelerate the necessary computations.

6.4.6. Cooperative Relaying

The future directions of cooperative relaying in FSO communication are centered on enhancing network performance, reliability, and efficiency to address the requirements of increasingly demanding use cases. This involves advancements in relay selection algorithms, integration with emerging technologies, energy-efficient designs, and innovative cooperative relaying strategies.

Intelligent Relay Selection Algorithms

One of the key areas of future research is the development of intelligent relay selection algorithms that utilize ML to dynamically optimize the selection of relay nodes based on real-time network conditions. These algorithms can analyze data such as signal quality, node locations, and atmospheric conditions to make informed decisions on the best relay nodes to use at any given time. This dynamic optimization helps to maintain robust communication links, reduce latency, and improve overall network performance. ML techniques like reinforcement learning and neural networks can be particularly effective in predicting and adapting to changing conditions in real-time, ensuring optimal relay performance.

Integration with Emerging Technologies

Future research also focuses on integrating relay-assisted transmission with innovative technologies like 5G and future advancements. This integration will enable ultra-reliable low-latency communication (uRLLC) and massive machine-type communications (mMTC), which are crucial for applications like autonomous vehicles, smart cities, and industrial automation. By combining the low-latency and high data rates of 5G with the robust, high-capacity links provided by FSO, these hybrid systems can deliver seamless and reliable connectivity. This approach also supports the dense deployment of IoT devices and ensures that critical applications maintain connectivity even in challenging environments.

Energy-Efficient Relay Designs

Energy efficiency is gaining increasing importance in communication networks. Consequently, future research is expected to concentrate on exploring sustainable network operations and the development of green communication protocols.

Sustainable Network Operations: Future research will explore energy-efficient relay designs to extend network coverage and improve energy sustainability. This includes the development of low-power relays and the utilization of renewable energy sources like solar power for relay nodes. Energy-efficient algorithms that minimize power consumption while maintaining high performance will be critical for ensuring the sustainability of large-scale FSO networks. These advancements are crucial for lowering the carbon footprint of communication networks and for deploying FSO systems in remote or off-grid locations [511,512].

Green Communication Protocols: The development of green communication protocols that prioritize energy efficiency without compromising performance will be another focus area. These protocols can manage the power usage of relay nodes and optimize data transmission to reduce energy consumption, contributing to the overall sustainability of the network [511,512].

Cooperative Relaying Strategies

Innovative cooperative relaying strategies will be investigated to extend network coverage and enhance connectivity. This will encompass the real-time adaptation of relay nodes.

Extended Coverage and Robust Connectivity: Cooperative relaying involves multiple relay nodes working together to transmit data, which can significantly improve the robustness and reliability of communication links. By leveraging spatial diversity, these strategies can alleviate the impacts of AT and ensure consistent performance across the network. Future research will investigate new cooperative techniques, such as network coding and relay selection diversity, to maximize the benefits of cooperative relaying.

Real-Time Adaptation: The ability of relay nodes to adapt in real-time to changing network conditions will be a key focus. This includes developing protocols that allow relay nodes to switch roles dynamically, share resources, and balance the load efficiently. Real-time adaptation will ensure that the network can handle varying traffic demands and maintain high performance even under challenging conditions.

Supporting Demanding Applications

The advancements in cooperative relaying are aimed at supporting increasingly demanding applications such as autonomous vehicles and IoT networks. Autonomous vehicles require ultra-reliable, low-latency communication to ensure safety and efficient operation. Similarly, IoT networks involve a vast number of connected devices that need robust and seamless connectivity. The future research directions in cooperative relaying will ensure that FSO systems can meet these demands, providing high-capacity, reliable, and efficient communication links.

6.4.7. Adaptive Optics

Future research in AO for FSO communication is poised to significantly enhance system reliability and performance by addressing the challenges posed by AT and environmental variability. Key areas of focus include the development of advanced wavefront sensors, DMs, the integration of ML algorithms, miniaturization of AO components, and the exploration of hybrid AO techniques.

Advanced Wavefront Sensors and Deformable Mirrors

The development of more advanced wavefront sensors is critical for improving the ability of AO systems to detect and correct for atmospheric distortions in real-time.

Enhanced Wavefront Sensing: Future research will likely explore new sensor technologies that offer higher sensitivity, faster response times, and greater accuracy. Innovations in wavefront sensing could include the use of novel materials and designs that improve performance while reducing power consumption and cost. These sensors will be essential for detecting subtle changes in the optical wavefront caused by turbulence, allowing for more precise corrections.

High-Precision Deformable Mirrors: DMs are central to the operation of AO systems, as they physically adjust to correct distortions in the optical wavefront. Future advancements will focus on developing high-precision DMs that can respond more quickly and accurately to detected distortions. This includes the exploration of new actuator technologies and mirror materials that offer improved performance characteristics. Additionally, research will aim to increase the number of actuators on DMs, providing finer control over the mirror surface and enabling better correction of complex wavefront distortions.

Integration of Machine Learning Algorithms

ML algorithms are anticipated to be pivotal in developing predictive and dynamic AO systems with real-time optimization.

Predictive and Dynamic AO Systems: ML algorithms are anticipated to play a significant role in optimizing AO systems by predicting and compensating for dynamic environmental changes. ML can be used to analyze historical and real-time data to predict atmospheric conditions and preemptively adjust the AO system. For example, neural networks can be trained to recognize patterns in AT and optimize the response of wavefront sensors and DMs accordingly. This predictive capability will enhance the robustness and efficiency of FSO communication by reducing latency and improving correction accuracy [389].

Real-Time Optimization: ML algorithms can also be employed for the real-time optimization of AO systems. By continuously learning from the incoming data and adjusting system parameters on-the-fly, these algorithms can ensure optimal performance under varying conditions. This dynamic adaptation will be particularly valuable for mobile FSO platforms, such as UAVs, where environmental conditions can change rapidly.

Miniaturization of AO Components

The miniaturization of AO components is essential for integrating these systems into a variety of use cases.

Compact and Mobile Platforms: Miniaturizing AO components is crucial for integrating these systems into compact and mobile platforms, such as UAVs, satellites, and portable ground stations. Future research will focus on developing smaller, lighter, and more energy-efficient wavefront sensors and DMs without compromising performance. This involves leveraging progress in MEMS technology and nanotechnology to create miniaturized components that can be easily integrated into various platforms.

Scalable AO Systems: The miniaturization of AO components will also enable the development of scalable AO systems that can be deployed across a broad spectrum of applications, from small handheld devices to large-scale ground stations. Scalable systems will offer greater flexibility and adaptability, making AO technology accessible for diverse FSO communication scenarios.

Hybrid Adaptive Optics Techniques

Investigating hybrid AO techniques can offer more comprehensive solutions for FSO communication systems.

Combining Traditional and Novel Approaches: Exploring hybrid AO techniques that combine traditional methods with novel approaches, such as digital holography, can further enhance the robustness and efficiency of FSO communication. Digital holography involves capturing and reconstructing the wavefront information using digital sensors and processing algorithms. Integrating this technique with conventional AO methods can enhance the system's ability to correct for complex distortions and improve overall performance.

Enhanced Turbulence Mitigation: Hybrid AO techniques can provide more comprehensive solutions for mitigating turbulence effects by leveraging the strengths of different approaches. For instance, digital holography can offer high-resolution wavefront reconstruction, while traditional AO methods can provide real-time correction. Combining these techniques can result in more effective turbulence mitigation and better communication link stability.

6.4.8. Adaptive Detection Thresholds

The future of ADT schemes in FSO communication presents a wealth of opportunities aimed at improving system performance, reliability, and adaptability. As the demand for high-capacity and resilient communication networks continues to grow, the development of advanced ADT strategies will be critical. The following sections elaborate on the key avenues for future research and development in ADT schemes.

Integration of Machine Learning Techniques

One of the most promising directions for ADT schemes is the integration of ML techniques to enhance the adaptability and precision of detection thresholds.

Predictive Threshold Optimization: By leveraging real-time data, ML algorithms can predict the optimal thresholds required for different conditions, including varying noise levels and interference patterns. This predictive capability allows for proactive adjustments to the detection thresholds, ensuring that the system can maintain optimal performance even in dynamic environments.

Adaptive Learning Models: Implementing adaptive learning models that continuously refine their predictions based on incoming data will be essential. These models can be trained using historical data to recognize patterns in environmental changes and system performance, allowing them to adjust thresholds dynamically. Over time, the algorithms will improve their accuracy and responsiveness, making the communication link more reliable and efficient.

Development of Sophisticated Algorithms

Future research should concentrate on creating more sophisticated algorithms that can dynamically adjust detection thresholds in response to environmental changes.

Dynamic Threshold Adjustment: This involves creating algorithms that monitor various parameters, such as ambient light levels, atmospheric conditions, and interference sources, in real-time. By correlating these parameters with the performance metrics of the communication system, the algorithms can adapt the thresholds to optimize detection sensitivity and minimize false positives or negatives.

Robustness Against Noise and Interference: The ability to adjust detection thresholds based on the specific characteristics of noise and interference is crucial for improving system robustness. Advanced statistical methods, including Bayesian inference and stochastic modeling, can be employed to develop thresholds that are not only reactive but also proactive, enabling the system to anticipate potential performance degradation before it occurs. Such algorithms would be particularly beneficial in urban environments where interference from multiple sources is common.

Exploration of Hybrid ADT Schemes

Exploring hybrid ADT schemes that combine multiple detection strategies will be another key area for future research.

Combining Detection Strategies: By integrating different detection approaches, such as energy detection, matched filtering, and decision feedback techniques, hybrid schemes can achieve improved performance across a broader range of operating conditions. This approach can leverage the strengths of each method while mitigating their weaknesses, leading to a more robust overall detection mechanism.

Adaptive Feedback Mechanisms: Hybrid ADT schemes can also include adaptive feedback mechanisms that allow for the real-time adjustment of detection strategies based on performance outcomes. For example, if one detection method is underperforming due to a sudden increase in noise, the system can switch to another method that may be more suitable under the current conditions. This flexibility will enhance the system's overall reliability and efficiency.

Broadening the Application of ADT Schemes

Expanding the application of ADT schemes to emerging communication technologies, such as 5G and beyond, will ensure their relevance and effectiveness in next-generation networks.

Relevance in Next-Generation Networks: As FSO communication is increasingly integrated into 5G architectures for high-speed data transmission, ADTs will play a crucial role in managing the complexities of multi-user environments, varying signal quality, and high data rates.

Support for Massive Machine-Type Communications: The rise in mMTC and the IoT presents unique challenges for detection schemes. ADT strategies will need to adapt to the high density of devices and the potential for increased interference. The research will focus on developing ADT mechanisms that can efficiently manage these challenges, ensuring that each device can maintain reliable communication despite the presence of numerous concurrent transmissions.

6.4.9. Hybrid FSO Systems

Hybrid FSO systems are poised to revolutionize optical communication by leveraging the unique strengths of FSO alongside other complementary technologies. These systems address individual limitations, offering enhanced performance in various environmental conditions. The future of hybrid FSO systems will revolve around several key research areas aimed at optimizing integration, enhancing reliability, and supporting the next generation of communication technologies.

Exploring Seamless Communication Solutions

The exploration of hybrid FSO-RF systems offers a promising avenue for ensuring seamless communication across diverse environments. By combining the high-capacity and low-latency characteristics of FSO with the robustness and reliability of RF technologies, these systems can maintain connectivity even in the face of adverse atmospheric conditions. Future research should prioritize the development of intelligent algorithms that can determine the most suitable transmission medium based on environmental assessments, user demands, and application requirements.

Optimizing Technology Integration

The synergy between FSO and alternative technologies will allow for redundancy and resilience in communication networks. This flexibility will be vital in applications requiring uninterrupted service, such as critical communications for emergency services and real-time data transmission for autonomous vehicles. One way to achieve this is by implementing advanced handover protocols and algorithms to enhance the reliability of hybrid systems.

Seamless Handover Mechanisms: Future research will focus on developing seamless handover mechanisms between hybrid FSO systems and complementary communication technologies, such as RF and microwave links. Achieving smooth transitions between different transmission mediums is critical for maintaining continuous connectivity, particularly in mobile scenarios. By implementing advanced handover protocols and algorithms, hybrid systems can automatically switch between FSO and RF links based on real-time assessments of signal quality, distance, and atmospheric conditions. This ensures uninterrupted service, even in challenging environments.

Enhanced Reliability and Resilience: Improving the reliability of hybrid systems is essential, especially in the face of AT and link blockages. Future studies will explore the integration of robust error correction methods and adaptive modulation techniques that can respond dynamically to environmental fluctuations. The use of intelligent algorithms to assess real-time conditions and adjust transmission parameters accordingly will be crucial. For instance, adaptive modulation schemes can vary the modulation format and coding rate based on current channel conditions, enhancing data throughput while minimizing errors.

Innovations in Adaptive Modulation and Beam Steering

Innovations in adaptive modulation are crucial for maximizing the performance of hybrid FSO systems. Beam steering technology is another critical area of development.

Adaptive Modulation Techniques: By employing techniques that can dynamically adjust modulation schemes depending on the quality of the communication link, these systems can optimize data rates and reliability. Future research may focus on integrating

ML algorithms to predict optimal modulation schemes based on historical and real-time data, enabling systems to adapt more effectively to varying conditions.

Advanced Beam Steering Solutions: Future hybrid FSO systems will benefit from advanced beam steering mechanisms that enhance the alignment of the optical beam, compensating for the effects of turbulence and movement. Techniques such as phased array systems and smart optics can be employed to achieve high-precision steering, ensuring that the optical signal remains focused on the intended Rx. The integration of intelligent beam steering algorithms will allow these systems to respond to environmental changes dynamically, further improving performance.

Intelligent Network Management

As hybrid FSO systems become more prevalent, intelligent network management will be essential to optimize resource allocation and enhance overall network performance. Also, ensuring QoS is paramount in hybrid FSO systems, especially as user demands for high-speed, low-latency communication increase.

Smart Network Orchestration: Future research will likely focus on developing advanced orchestration algorithms that can analyze network traffic, device connectivity, and environmental conditions to optimize the distribution of data across different transmission mediums. This will involve the employment of AI and ML techniques to create predictive models that facilitate efficient network management and minimize latency.

Quality-of-Service Enhancements: Future research will explore QoS management techniques that prioritize traffic based on application requirements, ensuring that critical applications such as real-time video conferencing and remote surgery receive the necessary bandwidth and reliability. These enhancements will be vital for supporting a diverse range of applications across hybrid communication networks.

Development of Compact and Cost-Effective Transceivers

The development of compact, low-power, and cost-effective transceivers will be pivotal in making hybrid FSO systems more accessible and practical for widespread deployment. Also, addressing the cost barriers associated with deploying hybrid FSO systems will also be a significant area of focus.

Miniaturization and Integration: Future research will focus on miniaturizing optical components and integrating them into single packages, reducing the overall footprint and power requirements. Advances in photonics and microfabrication technologies will facilitate the creation of lightweight and energy-efficient transceivers that can be easily installed in various environments, from urban infrastructures to remote rural areas.

Cost Effectiveness for Broad Deployment: Research will aim to streamline manufacturing processes and explore innovative funding models to make these technologies affordable for various applications, including municipal networks, broadband access, and IoT infrastructures. The availability of cost-effective solutions will accelerate the adoption of hybrid FSO systems across different market segments.

Supporting Emerging Applications

With the advancement of hybrid FSO systems, they will possess significant potential to support emerging applications.

5G/6G Networks and IoT Integration: As hybrid FSO systems evolve, their potential to support emerging applications like 5G and 6G networks, as well as the IoT, will be significant. The integration of hybrid systems into these next-generation networks will enable uRLLC, and facilitate advanced applications such as autonomous vehicles, smart cities, and augmented reality. By providing high-capacity links that can dynamically adapt to varying conditions, hybrid FSO systems will be integral in meeting the demands of an increasingly connected world.

Satellite Communications Synergy: Furthermore, hybrid FSO systems will play a crucial role in enhancing satellite communications by offering high-speed, low-latency links between satellites and ground stations. By combining FSO with traditional RF communication, satellite systems can achieve greater resilience and capacity, addressing the challenges of high-speed data transfer in diverse atmospheric conditions.

Moreover, hybrid free-space optical communication systems hold significant promise for the future of Space–Air–Ground–Sea Integrated Networks (SAGSINs). These systems offer high bandwidth, low latency, and secure data transmission, making them ideal for enhancing connectivity in SAGSINs. Future advancements are expected to focus on improving link reliability, atmospheric compensation techniques, and seamless integration with existing RF systems to ensure robust, high-speed communication across diverse environments.

7. Conclusions

In conclusion, FSO communication represents a transformative technology poised to address the escalating demands for high-speed, low-latency communication in the 5G and beyond era. This survey has highlighted the critical enabling technologies, such as adaptive optics, advanced modulation schemes, and robust error correction codes, that are essential for overcoming the inherent challenges associated with FSO systems, including atmospheric disturbances and alignment issues. The exploration of hybrid solutions integrating FSO with radio frequency, millimeter-wave, and Terahertz technologies demonstrates the potential for enhancing network reliability and coverage. Moreover, the incorporation of artificial intelligence and machine learning techniques marks a significant trend toward optimizing FSO performance in dynamic environments, paving the way for smarter and more resilient communication systems. By systematically analyzing current trends and challenges, this paper underscores the capacity of FSO to not only meet but exceed the communication requirements of future networks. Looking ahead, several open challenges remain, particularly concerning scalability, deployment strategies, and real-time adaptive capabilities in diverse atmospheric conditions. Addressing these challenges will be pivotal in realizing the full potential of FSO technology. Future research should focus on developing innovative solutions to enhance the robustness and efficiency of FSO systems, ensuring their successful integration into next-generation communication infrastructures. Ultimately, the insights presented in this paper contribute to a deeper understanding of the role of FSO in revolutionizing wireless communication and set the stage for ongoing advancements in this promising field.

Author Contributions: Conceptualization, I.A.A. and P.P.M.; methodology, I.A.A. and P.P.M.; software, I.A.A.; validation, I.A.A. and P.P.M.; formal analysis, P.P.M.; investigation, I.A.A. and P.P.M.; resources, I.A.A. and P.P.M.; data curation, I.A.A. and P.P.M.; writing—original draft preparation, I.A.A.; writing—review and editing, I.A.A. and P.P.M.; visualization, I.A.A. and P.P.M.; project administration, I.A.A. and P.P.M.; funding acquisition, I.A.A. and P.P.M. All authors have read and agreed to the published version of the manuscript.

Funding: This work is supported by the Fundação para a Ciência e a Tecnologia (FCT). The author gratefully acknowledges the support of the FCT under the grant agreement "Ultra-High-Capacity Optical Wireless Communication with Low-Complexity Transceiver for Effective Spectrum Utilization in Beyond-5G Networks," DOI 10.54499/2021.03815.CEECIND/CP1653/CT0005 (https://doi.org/10.54499/2021.03815.CEECIND/CP1653/CT0005). Acknowledgment is also extended to the European Regional Development Fund (FEDER) through the Competitiveness and Internationalization Operational Programme (CENTRO 2020) for the ORCIP project, under Grant CENTRO-01-0145-FEDER-022141. Additionally, the technical support provided by the Marie Skłodowska-Curie Actions (MSCA) through the Horizon 2020 Research and Innovation Staff Exchange (RISE) programme, specifically the Deep Intelligent Optical and Radio Communication Networks (DIOR) project under Grant Agreement ID: 101008280, is deeply appreciated.

Conflicts of Interest: The authors declare no conflicts of interest.

Abbreviations

The following abbreviations are used in this manuscript:

2D	Two-Dimensional
3D	Three-Dimensional
5G	Fifth Generation
6G	Sixth Generation
ABER	Average BER
ADC	Analog-to-Digital Converter
ADT	Adaptive Detection Threshold
AF	Amplify-and-Forward
AGC	Automatic Gain Control
AI	Artificial Intelligence
AMC	Adaptive Modulation and Coding
ANN	Artificial Neural Network
AO	Adaptive Optics
APDs	Avalanche PDs
APSK	Amplitude PSK
ASC	Average Secrecy Capacity
ASK	Amplitude Shift Keying
AT	Atmospheric Turbulence
AWGN	Additive White Gaussian Noise
BA	Buffer-Aided
BER	Bit-Error Rate
BF	Buffer-Free
BICM	Bit-Interleaved CM
BPOLSK	Binary POLSK
BPSK	Binary PSK
BSI	Buffer State Information
CNNs	Convolutional Neural Networks
CPAT	Coarse PAT
CSI	Channel State Information
CSPR	Carrier-to-Signal Power Ratio
DAC	Digital-to-Analog Converter
DD	Direct Detection
DF	Decode-and-Forward
DFB	Distributed Feedback
DL	Deep Learning
DM	Deformable Mirror
DMT	Discrete Multi-Tone
DNN	Deep Neural Network
DoF	Degrees of Freedom
DPSK	Differential PSK
DQPSK	Differential Quadrature PSK
DRE	Digital Resolution Enhancer
DRL	Deep Reinforcement Learning
DSP	Digital Signal Processing
e2e	End-to-End
ECC	Error Correction Code
EGC	Equal Gain Combining
EW	Exponentiated Weibull
FDT	Fixed Detection Threshold
FEC	Forward Error Correction
FER	Frame Error Rate
FFT	Fast Fourier Transform
FoV	Field-of-View
FP	Fabry–Pérot

FSK	Frequency Shift Keying
FSM	Fast Steering Mirror
FSO	Free-Space Optical
Ge	Germanium
GNN	Graph Neural Network
GRU	Gated Recurrent Unit
GS	Geometric Shaping
HARQ	Hybrid Automatic Repeat Request
HCS	Hybrid Constellation Shaping
HQAM	Hexagonal QAM
i.i.d.	Independent and Identically Distributed
IAS	Inter-Antenna Synchronization
ICI	Inter-Channel Interference
IEC	International Electrotechnical Commission
IEEE	Institute of Electrical and Electronics Engineers
IFFT	Inverse Fast Fourier Transform
IM	Intensity Modulation
IM/DD	Intensity Modulation with Direct Detection
InGaAs	Indium Gallium Arsenide
IR	Infrared
ITU	International Telecommunication Union
KK	Kramers–Kronig
LD	Laser Diode
LDPC	Low-Density Parity Check
LED	Light-Emitting Diode
LG	Laguerre–Gaussian
Li-Fi	Light Fidelity
LN	Log-Normal
LO	Local Oscillator
LoS	Line-of-Sight
m.g.f.	Moment Generating Function
MAC	Medium Access Control
MDPSK	Multilevel DPSK
MEMS	Micro-Electro-Mechanical Systems
MIMO	Multiple-Input Multiple-Output
MISO	Multiple-Input Single-Output
ML	Machine Learning
MLSD	Maximum-Likelihood Sequence Detection
mMTC	Massive Machine-Type Communications
mmWave	Millimeter-Wave
M-PAM	M-ary PAM
MPOLSK	Multilevel POLSK
MPSK	Multilevel PSK
M-QAM	M-ary QAM
MRC	Maximum Ratio Combining
MSM	Multiple-Subcarrier Modulation
MTBF	Mean Time Between Failures
NBA	Non-Buffer-Aided
NE	Negative Exponential
NGMI	Normalized Generalized Mutual Information
NLoS	Non-LoS
OAM	Orbital Angular Momentum
OBFSK	Orthogonal Binary FSK
OE	Optical-to-Electrical
OESM	Optically Enhanced SM
OFDM	Orthogonal Frequency-Division Multiplexing
OGSM	Optical Generalized SM

OIQSM	Optically Improved Quadrature SM
OOK	On–Off Keying
OP	Outage Probability
Op-Amp	Operational Amplifier
OSM	Optical Spatial Modulation
OSSK	Optical Space Shift Keying
OSTBCs	Orthogonal Space–Time Block Codes
OWC	Optical Wireless Communication
p2p	Point-to-Point
PAM	Pulse-Amplitude Modulation
PAPR	Peak-to-Average-Power Ratio
PAS	Probabilistic Amplitude Shaping
PAT	Pointing, Acquisition, and Tracking
PD	Photodiode
PDF	Probability Density Function
PDM	Polarization Division Multiplexing
PE	Pointing Errors
PHY	Physical
PICs	Photonic Integrated Circuits
PIN	Positive–Intrinsic–Negative
PLL	Phase-Locked Loop
PLS	PHY Layer Security
PolM	Polarization Modulation
POLSK	Polarization Shift Keying
PPM	Pulse-Position Modulation
PS	Probabilistic Shaping
PSAM	Pilot-Symbol Assisted Modulation
PSF	Point Spread Function
PSK	Phase-Shift Keying
PSP	Per-Survivor Processing
QAM	Quadrature Amplitude Modulation
QCDs	Quantum Cascade Detectors
QCLs	Quantum Cascade Lasers
QKD	Quantum Key Distribution
QoS	Quality of Service
RC	Repetition Coding
RF	Radio Frequency
RoFSO	Radio over FSO
RQAM	Rectangular QAM
RS	Reed–Solomon
Rx	Receiver
SC	Selection Combining
SDM	Space Division Multiplexing
SDT	Sparse–Dense Transmission
SE	Spectral Efficiency
SER	Symbol Error Rate
Si	Silicon
SIM	Subcarrier Intensity Modulation
SIMO	Single-Input Multiple-Output
SINR	Signal-To-Interference-Plus-Noise Ratio
SISO	Single-Input Single-Output
SLMs	Spatial Light Modulators
SM	Spatial Modulation
SNR	Signal-to-Noise Ratio

SOP	Secrecy Outage Probability
SPSC	Strictly Positive Secrecy Capacity
S-PSK	Subcarrier PSK
SQAM	Squared QAM
SSB	Single-Sideband
SSBI	Signal–Signal Beat Interference
SSBN	Signal-to-Signal Beating Noise
STC	Space–Time Coding
STD	Selection Transmit Diversity
STTC	Space–Time Trellis Coding
SVM	Support Vector Machines
SWIPT	Simultaneous Wireless Information and Power Transfer
TCM	Trellis-Coded Modulation
THz	Terahertz
TIA	Trans-Impedance Amplifier
TLS	Transmit Laser Selection
Tx	Transmitter
UAVs	Unmanned Aerial Vehicles
UOWC	Underwater OWC
uRLLC	Ultra-Reliable Low-Latency Communication
UV	Ultraviolet
VCSELs	Vertical-Cavity Surface-Emitting Lasers
VLC	Visible-Light Communication
WDM	Wavelength Division Multiplexing
XQAM	Cross QAM

References

1. Khalighi, M.A.; Uysal, M. Survey on Free Space Optical Communication: A Communication Theory Perspective. *IEEE Commun. Surv. Tutor.* **2014**, *16*, 2231–2258. [CrossRef]
2. Alimi, I.A.; Monteiro, P.P., Free-Space Optical Communication for Future Broadband Access Networks. In *Handbook of Radio and Optical Networks Convergence*; Kawanishi, T., Ed.; Springer Nature: Singapore, 2023; pp. 1–28. [CrossRef]
3. Alimi, I.; Shahpari, A.; Sousa, A.; Ferreira, R.; Monteiro, P.; Teixeira, A. Challenges and Opportunities of Optical Wireless Communication Technologies. In *Optical Communication Technology*; Pinho, P., Ed.; IntechOpen: Rijeka, Croatia, 2017; Chapter 2. [CrossRef]
4. Fernandes, M.A.; Fernandes, G.M.; Monteiro, P.P.; Guiomar, F.P. High-Capacity Coherent FSO. In *Handbook of Radio and Optical Networks Convergence*; Kawanishi, T., Ed.; Springer Nature: Singapore, 2023; pp. 1–31. [CrossRef]
5. Fernandes, M.A.; Fernandes, G.M.; Brandão, B.T.; Freitas, M.M.; Kaai, N.; van Der Wielen, B.; Reid, J.; Raiteri, D.; Monteiro, P.P.; Guiomar, F.P. Unraveling "Fiber in the Sky": Terabit Capacity Enabled by Coherent Optical Wireless. *IEEE Commun. Mag.* **2024**, *62*, 40–46. [CrossRef]
6. Alimi, I.A.; Monteiro, P.P.; Teixeira, A.L. Analysis of multiuser mixed RF/FSO relay networks for performance improvements in Cloud Computing-Based Radio Access Networks (CC-RANs). *Opt. Commun.* **2017**, *402*, 653–661. [CrossRef]
7. Kaymak, Y.; Rojas-Cessa, R.; Feng, J.; Ansari, N.; Zhou, M.; Zhang, T. A Survey on Acquisition, Tracking, and Pointing Mechanisms for Mobile Free-Space Optical Communications. *IEEE Commun. Surv. Tutor.* **2018**, *20*, 1104–1123. [CrossRef]
8. Ata, Y.; Alouini, M.S. HAPS Based FSO Links Performance Analysis and Improvement with Adaptive Optics Correction. *IEEE Trans. Wirel. Commun.* **2023**, *22*, 4916–4929. [CrossRef]
9. Harada, R.; Shibata, N.; Kaneko, S.; Imai, T.; Kani, J.I.; Yoshida, T. Adaptive Beam Divergence for Expanding Range of Link Distance in FSO with Moving Nodes Toward 6G. *IEEE Photonics Technol. Lett.* **2022**, *34*, 1061–1064. [CrossRef]
10. Uysal, M.; Li, J.; Yu, M. Error rate performance analysis of coded free-space optical links over gamma-gamma atmospheric turbulence channels. *IEEE Trans. Wirel. Commun.* **2006**, *5*, 1229–1233. [CrossRef]
11. Alsaedi, W.K.; Ahmadi, H.; Khan, Z.; Grace, D. Spectrum Options and Allocations for 6G: A Regulatory and Standardization Review. *IEEE Open J. Commun. Soc.* **2023**, *4*, 1787–1812. [CrossRef]
12. Alimi, I.A.; Patel, R.K.; Zaouga, A.; Muga, N.J.; Pinto, A.N.; Teixeira, A.L.; Monteiro, P.P. 6G CloudNet: Towards a Distributed, Autonomous, and Federated AI-Enabled Cloud and Edge Computing. In *6G Mobile Wireless Networks*; Wu, Y., Singh, S., Taleb, T., Roy, A., Dhillon, H.S., Kanagarathinam, M.R., De, A., Eds.; Springer International Publishing: Cham, Switzerland, 2021; pp. 251–283. [CrossRef]
13. Ashrafzadeh, B.; Soleimani-Nasab, E.; Kamandar, M.; Uysal, M. A Framework on the Performance Analysis of Dual-Hop Mixed FSO-RF Cooperative Systems. *IEEE Trans. Commun.* **2019**, *67*, 4939–4954. [CrossRef]
14. Karimi, M.; Uysal, M. Novel Adaptive Transmission Algorithms for Free-Space Optical Links. *IEEE Trans. Commun.* **2012**, *60*, 3808–3815. [CrossRef]

15. Alimi, I.A.; Muga, N.J. Simple and robust transmit diversity based free-space optical communications for 5G and beyond networks. *Opt. Commun.* **2020**, *476*, 126306. [CrossRef]
16. Alimi, I.A.; Patel, R.K.; Muga, N.J.; Pinto, A.N.; Teixeira, A.L.; Monteiro, P.P. Towards Enhanced Mobile Broadband Communications: A Tutorial on Enabling Technologies, Design Considerations, and Prospects of 5G and beyond Fixed Wireless Access Networks. *Appl. Sci.* **2021**, *11*, 427. [CrossRef]
17. Korevaar, C.W. Optical Satellite Communications: Design Considerations and Research Outlook. In *Handbook of Radio and Optical Networks Convergence*; Kawanishi, T., Ed.; Springer Nature: Singapore, 2024; pp. 945–961. [CrossRef]
18. Bloom, S.; Korevaar, E.; Schuster, J.; Willebrand, H. Understanding the performance of free-space optics [Invited]. *J. Opt. Netw.* **2003**, *2*, 178–200. [CrossRef]
19. Naimullah, B.; Othman, M.; Rahman, A.K.; Sulaiman, S.; Ishak, S.; Hitam, S.; Aljunid, S. Comparison of wavelength propagation for Free Space Optical Communications. In Proceedings of the 2008 International Conference on Electronic Design, Penang, Malaysia, 1–3 December 2008; pp. 1–5. [CrossRef]
20. Salah, M.S.; Cowley, W.G.; Nguyen, K.D. Adaptive transmission schemes for free-space optical channel. In Proceedings of the 2014 8th International Conference on Signal Processing and Communication Systems (ICSPCS), Gold Coast, QLD, Australia, 15–17 December 2014; pp. 1–7. [CrossRef]
21. Manor, H.; Arnon, S. Performance of an optical wireless communication system as a function of wavelength. *Appl. Opt.* **2003**, *42*, 4285–4294. [CrossRef]
22. Xu, Z.; Sadler, B.M. Ultraviolet Communications: Potential and State-Of-The-Art. *IEEE Commun. Mag.* **2008**, *46*, 67–73. [CrossRef]
23. Alkholidi, A.G.; Altowij, K.S. Free Space Optical Communications—Theory and Practices. In *Contemporary Issues in Wireless Communications*; Khatib, M., Ed.; IntechOpen: Rijeka, Croatia, 2014; Chapter 5. [CrossRef]
24. Deep Verma, G.; Mathur, A.; Bhatnagar, M.R. Differential Chaos Shift Keying for FSO Systems: A Novel Approach Under Turbulence and Boresight Pointing Errors. *IEEE Open J. Commun. Soc.* **2024**, *5*, 3263–3276. [CrossRef]
25. Mecozzi, A.; Antonelli, C.; Shtaif, M. Kramers–Kronig coherent receiver. *Optica* **2016**, *3*, 1220–1227. [CrossRef]
26. Muga, N.J.; Patel, R.K.; Alimi, I.A.; Silva, N.A.; Pinto, A.N. Self-coherent optical detection for access and metro networks. In Proceedings of the 2019 21st International Conference on Transparent Optical Networks (ICTON), Angers, France, 9–13 July 2019; pp. 1–4. [CrossRef]
27. Alimi, I.; Patel, R.; Silva, N.; Sun, C.; Ji, H.; Shieh, W.; Pinto, A.; Muga, N. A Review of Self-Coherent Optical Transceivers: Fundamental Issues, Recent Advances, and Research Directions. *Appl. Sci.* **2021**, *11*, 7554. [CrossRef]
28. Chou, E.S.; Srinivas, H.; Kahn, J.M. Phase Retrieval-Based Coherent Receivers: Signal Design and Degrees of Freedom. *J. Light. Technol.* **2022**, *40*, 1296–1307. [CrossRef]
29. Tabares, J.; Polo, V.; Cano, I.; Prat, J. Automatic λ -Control with Offset Compensation in DFB Intradyne Receiver for udWDM-PON. *IEEE Photonics Technol. Lett.* **2015**, *27*, 443–446. [CrossRef]
30. Nelson, L.E.; Woodward, S.L.; Foo, S.; Moyer, M.; Beckett, D.J.S.; O'Sullivan, M.; Magill, P.D. Detection of a Single 40 Gb/s Polarization-Multiplexed QPSK Channel with a Real-Time Intradyne Receiver in the Presence of Multiple Coincident WDM Channels. *J. Light. Technol.* **2010**, *28*, 2933–2943. [CrossRef]
31. Schaefer, S.; Rosenkranz, W.; Gregory, M. Comparison of homodyne and intradyne detection for high-order modulation schemes in optical intersatellite communication systems. In Proceedings of the 2015 IEEE International Conference on Space Optical Systems and Applications (ICSOS), New Orleans, LA, USA, 26–28 October 2015; pp. 1–5. [CrossRef]
32. Leven, A.; Kaneda, N.; Koc, U.V.; Chen, Y.K. Frequency Estimation in Intradyne Reception. *IEEE Photonics Technol. Lett.* **2007**, *19*, 366–368. [CrossRef]
33. Patel, R.K.; Domingues, G.A.; Muga, N.J.; Pinto, A.N.; Fernandes, M.A.; Fernandes, G.M.; Monteiro, P.P.; Guiomar, F.P. Simplified Self-Coherent FSO Transmission Boosted by Digital Resolution Enhancer. *J. Light. Technol.* **2023**, *41*, 5958–5965. [CrossRef]
34. Eiselt, N.; Griesser, H.; Wei, J.; Dochhan, A.; Eiselt, M.; Elbers, J.P.; Olmos, J.J.V.; Monroy, I.T. Real-time evaluation of 26-GBaud PAM-4 intensity modulation and direct detection systems for data-center interconnects. In Proceedings of the 2016 Optical Fiber Communications Conference and Exhibition (OFC), Anaheim, CA, USA, 20–24 March 2016; pp. 1–3.
35. Li, Q.; Yang, H.; Xiang, M.; Cheng, W.; Wu, R.; Zhang, L.; Fu, S.; Qin, Y. Optical Amplifier-Free 100 Gbaud PAM-4 Transmission Over 10 Km SSMF Based on O-Band SiP Transceiver. *J. Light. Technol.* **2024**, *42*, 5175–5181. [CrossRef]
36. Lee, B.Y.; Cheng, C.H.; Tsai, C.T.; Lin, H.S.; Kao, S.C.; Chiang, P.; Lee, H.C.; Shih, T.T.; Kuo, H.C.; Lin, G.R. Si Mach-Zehnder Modulator for PAM-4, QAM-OFDM, and DMT Transmission at C-Band. *IEEE J. Sel. Top. Quantum Electron.* **2023**, *29*, 1–12. [CrossRef]
37. Chen, Z.; Liang, Z.; Wu, J.; Lu, W.; Chang, X.; Dai, X.; Yang, Q.; Liu, C.; Cheng, M.; Deng, L.; et al. Cost-Effective 40-Gbaud THP PS-PAM-4 C-Band IM/DD Transmission for 50-km Inter-DCI Utilizing 2-bit DAC. *IEEE Photonics J.* **2024**, *16*, 1–7. [CrossRef]
38. Maneekut, R.; Elson, D.J.; Wakayama, Y.; Yoshikane, N.; Kaewplung, P. Reach Extension of a 53.12 Gbps PAM-4 IM/DD Signals with Optimized Extinction Ratio in Amplified Multi-Span O-Band Transmission. *IEEE Access* **2024**, *12*, 34656–34667. [CrossRef]
39. Karar, A.S.; Falou, A.R.E.; Barakat, J.M.H.; Gürkan, Z.N.; Zhong, K. Recent Advances in Coherent Optical Communications for Short-Reach: Phase Retrieval Methods. *Photonics* **2023**, *10*, 308. [CrossRef]
40. Morsy-Osman, M.; Chagnon, M.; Plant, D.V. Four-Dimensional Modulation and Stokes Direct Detection of Polarization Division Multiplexed Intensities, Inter Polarization Phase and Inter Polarization Differential Phase. *J. Light. Technol.* **2016**, *34*, 1585–1592. [CrossRef]

41. Wu, Q.; Zhu, Y.; Jiang, H.; Zhuge, Q.; Hu, W. Optical Field Recovery in Jones Space. *J. Light. Technol.* **2023**, *41*, 66–74. [CrossRef]
42. Kwapisz, J.; Roudas, I.; Fink, E. Mode Vector Modulation: Extending Stokes Vector Modulation to Higher Dimensions. *J. Light. Technol.* **2024**, *42*, 1966–1990. [CrossRef]
43. Le, S.T.; Schuh, K.; Chagnon, M.; Buchali, F.; Dischler, R.; Aref, V.; Buelow, H.; Engenhardt, K.M. 1.72-Tb/s Virtual-Carrier-Assisted Direct-Detection Transmission Over 200 km. *J. Light. Technol.* **2018**, *36*, 1347–1353. [CrossRef]
44. Tavakkolnia, I.; Chen, C.; Tan, Y.; Haas, H. Terabit Optical Wireless-Fiber Communication with Kramer-Kronig Receiver—Part II. *IEEE Commun. Lett.* **2022**, *26*, 1969–1973. [CrossRef]
45. Tavakkolnia, I.; Chen, C.; Tan, Y.; Haas, H. Terabit Optical Wireless-Fiber Communication with Kramer-Kronig Receiver—Part I. *IEEE Commun. Lett.* **2022**, *26*, 1964–1968. [CrossRef]
46. Paul, P.; Bhatnagar, M.R.; Nebhen, J. Pulse Jamming in Aperture-Averaged FSO Receiver Over Exponentiated Weibull Fading Channel. *IEEE Trans. Wirel. Commun.* **2022**, *21*, 4242–4254. [CrossRef]
47. Alimi, I.A.; Shahpari, A.; Monteiro, P.P.; Teixeira, A.L. Effects of diversity schemes and correlated channels on OWC systems performance. *J. Mod. Opt.* **2017**, *64*, 2298–2305. [CrossRef]
48. Lopez-Martinez, F.J.; Gomez, G.; Garrido-Balsells, J.M. Physical-Layer Security in Free-Space Optical Communications. *IEEE Photonics J.* **2015**, *7*, 1–14. [CrossRef]
49. Singh, R.; Rawat, M.; Jaiswal, A. On the Physical Layer Security of Mixed FSO-RF SWIPT System with Non-Ideal Power Amplifier. *IEEE Photonics J.* **2021**, *13*, 1–17. [CrossRef]
50. Wang, X.; Song, Y. Encoding and decoding by the states of vector modes for vortex beams propagating in air-core fiber. *Opt. Express* **2017**, *25*, 29342–29355. [CrossRef]
51. Lai, L.; Liang, Y.; Du, W. Cooperative Key Generation in Wireless Networks. *IEEE J. Sel. Areas Commun.* **2012**, *30*, 1578–1588. [CrossRef]
52. Shiu, Y.S.; Chang, S.Y.; Wu, H.C.; Huang, S.C.H.; Chen, H.H. Physical layer security in wireless networks: A tutorial. *IEEE Wirel. Commun.* **2011**, *18*, 66–74. [CrossRef]
53. Poor, H.V. Information and inference in the wireless physical layer. *IEEE Wirel. Commun.* **2012**, *19*, 40–47. [CrossRef]
54. Saber, M.J.; Sadough, S.M.S. On Secure Free-Space Optical Communications Over Málaga Turbulence Channels. *IEEE Wirel. Commun. Lett.* **2017**, *6*, 274–277. [CrossRef]
55. Ai, Y.; Mathur, A.; Verma, G.D.; Kong, L.; Cheffena, M. Comprehensive Physical Layer Security Analysis of FSO Communications Over Málaga Channels. *IEEE Photonics J.* **2020**, *12*, 1–17. [CrossRef]
56. Pattanayak, D.R.; Dwivedi, V.K.; Karwal, V.; Yadav, P.K.; Singh, G. Physical Layer Security Analysis of Multi-Hop Hybrid RF/FSO System in Presence of Multiple Eavesdroppers. *IEEE Photonics J.* **2022**, *14*, 1–12. [CrossRef]
57. Sun, X.; Djordjevic, I.B. Physical-Layer Security in Orbital Angular Momentum Multiplexing Free-Space Optical Communications. *IEEE Photonics J.* **2016**, *8*, 1–10. [CrossRef]
58. Zou, D.; Xu, Z. Information Security Risks Outside the Laser Beam in Terrestrial Free-Space Optical Communication. *IEEE Photonics J.* **2016**, *8*, 1–9. [CrossRef]
59. Sun, Q.; Yan, D.; Wang, Y. Research on physical layer security performance of hybrid RF/FSO system based on RIS assistance under FSO eavesdropping environment. *IEEE Photonics J.* **2024**, *16*, 1–9. [CrossRef]
60. Boluda-Ruiz, R.; Tokgoz, S.C.; García-Zambrana, A.; Qaraqe, K. Enhancing Secrecy Capacity in FSO Links via MISO Systems Through Turbulence-Induced Fading Channels with Misalignment Errors. *IEEE Photonics J.* **2020**, *12*, 1–13. [CrossRef]
61. Pan, X.; Ran, H.; Pan, G.; Xie, Y.; Zhang, J. On Secrecy Analysis of DF Based Dual Hop Mixed RF-FSO Systems. *IEEE Access* **2019**, *7*, 66725–66730. [CrossRef]
62. Erdogan, E.; Altunbas, I.; Kurt, G.K.; Yanikomeroglu, H. The Secrecy Comparison of RF and FSO Eavesdropping Attacks in Mixed RF-FSO Relay Networks. *IEEE Photonics J.* **2022**, *14*, 1–8. [CrossRef]
63. Abd El-Malek, A.H.; Salhab, A.M.; Zummo, S.A.; Alouini, M.S. Security-Reliability Trade-Off Analysis for Multiuser SIMO Mixed RF/FSO Relay Networks with Opportunistic User Scheduling. *IEEE Trans. Wirel. Commun.* **2016**, *15*, 5904–5918. [CrossRef]
64. Rahman, M.M.; Badrudduza, A.S.M.; Sarker, N.A.; Ibrahim, M.; Ansari, I.S.; Yu, H. RIS-Aided Mixed RF-FSO Wireless Networks: Secrecy Performance Analysis with Simultaneous Eavesdropping. *IEEE Access* **2023**, *11*, 126507–126523. [CrossRef]
65. Lei, H.; Dai, Z.; Park, K.H.; Lei, W.; Pan, G.; Alouini, M.S. Secrecy Outage Analysis of Mixed RF-FSO Downlink SWIPT Systems. *IEEE Trans. Commun.* **2018**, *66*, 6384–6395. [CrossRef]
66. Saber, M.J.; Keshavarz, A.; Mazloum, J.; Sazdar, A.M.; Piran, M.J. Physical-Layer Security Analysis of Mixed SIMO SWIPT RF and FSO Fixed-Gain Relaying Systems. *IEEE Syst. J.* **2019**, *13*, 2851–2858. [CrossRef]
67. Zedini, E.; Soury, H.; Alouini, M.S. On the Performance Analysis of Dual-Hop Mixed FSO/RF Systems. *IEEE Trans. Wirel. Commun.* **2016**, *15*, 3679–3689. [CrossRef]
68. Saber, M.J.; Mazloum, J.; Sazdar, A.M.; Keshavarz, A.; Piran, M.J. On Secure Mixed RF-FSO Decode-and-Forward Relaying Systems with Energy Harvesting. *IEEE Syst. J.* **2020**, *14*, 4402–4405. [CrossRef]
69. Lei, H.; Dai, Z.; Ansari, I.S.; Park, K.H.; Pan, G.; Alouini, M.S. On Secrecy Performance of Mixed RF-FSO Systems. *IEEE Photonics J.* **2017**, *9*, 1–14. [CrossRef]
70. Yang, L.; Liu, T.; Chen, J.; Alouini, M.S. Physical-Layer Security for Mixed η–μ and $\mathcal{M}$-Distribution Dual-Hop RF/FSO Systems. *IEEE Trans. Veh. Technol.* **2018**, *67*, 12427–12431. [CrossRef]

71. Makki, B.; Svensson, T.; Brandt-Pearce, M.; Alouini, M.S. On the Performance of Millimeter Wave-Based RF-FSO Multi-Hop and Mesh Networks. *IEEE Trans. Wirel. Commun.* **2017**, *16*, 7746–7759. [CrossRef]
72. Balti, E.; Guizani, M.; Hamdaoui, B.; Khalfi, B. Mixed RF/FSO Relaying Systems with Hardware Impairments. In Proceedings of the GLOBECOM 2017—2017 IEEE Global Communications Conference, Singapore, 4–8 December 2017; pp. 1–6. [CrossRef]
73. Ai, Y.; Mathur, A.; Lei, H.; Cheffena, M.; Ansari, I.S. Secrecy enhancement of RF backhaul system with parallel FSO communication link. *Opt. Commun.* **2020**, *475*, 126193. [CrossRef]
74. Amuru, S.; Dhillon, H.S.; Buehrer, R.M. On Jamming Against Wireless Networks. *IEEE Trans. Wirel. Commun.* **2017**, *16*, 412–428. [CrossRef]
75. Paul, P.; Bhatnagar, M.R.; Jaiswal, A. Jamming in Free Space Optical Systems: Mitigation and Performance Evaluation. *IEEE Trans. Commun.* **2020**, *68*, 1631–1647. [CrossRef]
76. Arkin, E.; Cassuto, Y.; Efrat, A.; Grebla, G.; Mitchell, J.S.B.; Sankararaman, S.; Segal, M. Optimal placement of protective jammers for securing wireless transmissions in a geographic domain. In Proceedings of the 14th International Conference on Information Processing in Sensor Networks (IPSN '15), New York, NY, USA, 13–16 April 2015; pp. 37–46. [CrossRef]
77. Grover, K.; Lim, A.; Yang, Q. Jamming and anti-jamming techniques in wireless networks: A survey. *Int. J. Ad Hoc Ubiquitous Comput.* **2014**, *17*, 197–215. [CrossRef]
78. Tague, P.; Slater, D.; Poovendran, R.; Noubir, G. Linear programming models for jamming attacks on network traffic flows. In Proceedings of the 2008 6th International Symposium on Modeling and Optimization in Mobile, Ad Hoc, and Wireless Networks and Workshops, Berlin, Germany, 1–3 April 2008; pp. 207–216. [CrossRef]
79. Yan, Q.; Zeng, H.; Jiang, T.; Li, M.; Lou, W.; Hou, Y.T. Jamming Resilient Communication Using MIMO Interference Cancellation. *IEEE Trans. Inf. Forensics Secur.* **2016**, *11*, 1486–1499. [CrossRef]
80. Deng, P.; Yuan, X.; Zhao, M.; Zeng, Y.; Kavehrad, M. Off-axis catadioptric fisheye wide field-of-view optical receiver for free space optical communications. *Opt. Eng.* **2012**, *51*, 063002. [CrossRef]
81. Ho, T.H. *Pointing, Acquisition, and Tracking Systems for Free-Space Optical Communication Links*; University of Maryland: College Park, MD, USA, 2007.
82. Dabiri, M.T.; Sadough, S.M.S.; Khalighi, M.A. Channel Modeling and Parameter Optimization for Hovering UAV-Based Free-Space Optical Links. *IEEE J. Sel. Areas Commun.* **2018**, *36*, 2104–2113. [CrossRef]
83. Ghassemlooy, Z.; Popoola, W.; Rajbhandari, S. *Optical Wireless Communications: System and Channel Modelling with MATLAB®*, 2nd ed.; CRC Press: Boca Raton FL, USA, 2019.
84. Kaushal, H.; Kaddoum, G. Optical Communication in Space: Challenges and Mitigation Techniques. *IEEE Commun. Surv. Tutor.* **2017**, *19*, 57–96. [CrossRef]
85. Paul, P.; Bhatnagar, M.R. Jamming Threats in Free-Space Optics. *IEEE Commun. Mag.* **2022**, *60*, 104–108. [CrossRef]
86. Paul, P.; Bhatnagar, M.R. Random Relay Jamming in Cooperative Free Space Optical Systems. *IEEE Syst. J.* **2022**, *16*, 2413–2424. [CrossRef]
87. Paul, P.; Bhatnagar, M.R. DF Relaying in Cooperative Free Space Optical Communication System in Presence of Jammer. In Proceedings of the 2020 12th International Symposium on Communication Systems, Networks and Digital Signal Processing (CSNDSP), Porto, Portugal, 20–22 July 2020; pp. 1–6. [CrossRef]
88. Shukla, A.K.; Bhatnagar, M.R. Mitigating Jamming in FSO Cooperative Networks: A Buffer-Aided Relaying Approach. *IEEE Commun. Lett.* **2023**, *27*, 2078–2082. [CrossRef]
89. Farid, A.A.; Hranilovic, S. Outage Capacity Optimization for Free-Space Optical Links with Pointing Errors. *J. Light. Technol.* **2007**, *25*, 1702–1710. [CrossRef]
90. Al Naboulsi, M.; Sizun, H.; de Fornel, F. Fog attenuation prediction for optical and infrared waves. *Opt. Eng.* **2004**, *43*, 319–329. [CrossRef]
91. Alimi, I.; Shahpari, A.; Ribeiro, V.; Sousa, A.; Monteiro, P.; Teixeira, A. Channel characterization and empirical model for ergodic capacity of free-space optical communication link. *Opt. Commun.* **2017**, *390*, 123–129. [CrossRef]
92. Luong, D.A.; Thang, T.C.; Pham, A.T. Effect of avalanche photodiode and thermal noises on the performance of binary phase-shift keying-subcarrier-intensity modulation/free-space optical systems over turbulence channels. *IET Commun.* **2013**, *7*, 738–744. [CrossRef]
93. Magidi, S.; Jabeena, A. Free Space Optics, Channel Models and Hybrid Modulation Schemes: A Review. *Wirel. Pers. Commun.* **2021**, *119*, 2951–2974. [CrossRef]
94. Andrews, L.; Phillips, R. *Laser Beam Propagation Through Random Media*; Press Monographs; SPIE Press: Bellingham, DC, USA, 2005.
95. Kiasaleh, K. Performance of APD-based, PPM free-space optical communication systems in atmospheric turbulence. *IEEE Trans. Commun.* **2005**, *53*, 1455–1461. [CrossRef]
96. Alimi, I.; Shahpari, A.; Ribeiro, V.; Kumar, N.; Monteiro, P.; Teixeira, A. Optical wireless communication for future broadband access networks. In Proceedings of the 2016 21st European Conference on Networks and Optical Communications (NOC), Lisbon, Portugal, 1–3 June 2016; pp. 124–128. [CrossRef]
97. Ghassemlooy, Z.; Popoola, W.; Rajbhandari, S. *Optical Wireless Communications: System and Channel Modelling with MATLAB®*; CRC Press: Boca Raton FL, USA, 2012.

98. Bekkali, A.; Naila, C.B.; Kazaura, K.; Wakamori, K.; Matsumoto, M. Transmission Analysis of OFDM-Based Wireless Services Over Turbulent Radio-on-FSO Links Modeled by Gamma–Gamma Distribution. *IEEE Photonics J.* **2010**, *2*, 510–520. [CrossRef]
99. Jurado-Navas, A.; Garrido-Balsells, J.M.; Paris, J.F.; Puerta-Notario, A. *A Unifying Statistical Model for Atmospheric Optical Scintillation*; IntechOpen: Rijeka, Croatia; 2011; pp. 181–206.
100. Yang, L.; Hasna, M.O.; Gao, X. Performance of Mixed RF/FSO with Variable Gain over Generalized Atmospheric Channels. *IEEE J. Sel. Areas Comm.* **2015**, *33*, 1913–1924. [CrossRef]
101. Sandalidis, H.G.; Tsiftsis, T.A.; Karagiannidis, G.K. Optical Wireless Communications with Heterodyne Detection Over Turbulence Channels with Pointing Errors. *J. Light. Technol.* **2009**, *27*, 4440–4445. [CrossRef]
102. Huang, Y.; Zhang, B.; Gao, Z.; Zhao, G.; Duan, Z. Evolution behavior of Gaussian Schell-model vortex beams propagating through oceanic turbulence. *Opt. Express* **2014**, *22*, 17723–17734. [CrossRef] [PubMed]
103. Wang, X.; Wang, L.; Zhao, S. Research on Hypergeometric-Gaussian Vortex Beam Propagating under Oceanic Turbulence by Theoretical Derivation and Numerical Simulation. *J. Mar. Sci. Eng.* **2021**, *9*, 442. [CrossRef]
104. Wang, X.; Wang, L.; Zheng, B.; Yang, Z.; Zhao, S. Effects of Oceanic Turbulence on the Propagation of Hypergeometric-Gaussian Beam Carrying Orbital Angular Momentum. In Proceedings of the 2020 IEEE International Conference on Communications Workshops (ICC Workshops), Dublin, Ireland, 7–11 June 2020; pp. 1–5. [CrossRef]
105. Kotlyar, V.V.; Kovalev, A.A. Family of hypergeometric laser beams. *J. Opt. Soc. Am. A* **2008**, *25*, 262–270. [CrossRef] [PubMed]
106. Karimi, E.; Zito, G.; Piccirillo, B.; Marrucci, L.; Santamato, E. Hypergeometric-Gaussian modes. *Opt. Lett.* **2007**, *32*, 3053–3055. [CrossRef]
107. Suresh, P.; Rajesh, K.; Pillai, T. Tight focusing of amplitude modulated HyGG type—II beam using high NA lens axicon. In Proceedings of the 2012 International Conference on Fiber Optics and Photonics (PHOTONICS), Chennai India, 9–12 December 2012; pp. 1–3.
108. Zhang, Y.; Deng, S.; Yang, H. Transmittance freezing of a random frozen photons beam in a turbulent ocean. *OSA Contin.* **2020**, *3*, 1084–1095. [CrossRef]
109. Li, Y.; Yu, L.; Zhang, Y. Influence of anisotropic turbulence on the orbital angular momentum modes of Hermite-Gaussian vortex beam in the ocean. *Opt. Express* **2017**, *25*, 12203–12215. [CrossRef]
110. Yang, H.; Yan, Q.; Wang, P.; Hu, L.; Zhang, Y. Bit-error rate and average capacity of an absorbent and turbulent underwater wireless communication link with perfect Laguerre-Gauss beam. *Opt. Express* **2022**, *30*, 9053–9064. [CrossRef]
111. Cheng, M.; Guo, L.; Li, J.; Zhang, Y. Channel Capacity of the OAM-Based Free-Space Optical Communication Links with Bessel-Gauss Beams in Turbulent Ocean. *IEEE Photonics J.* **2016**, *8*, 1–11. [CrossRef]
112. Xu, Y.; Shi, H.; Zhang, Y. Effects of anisotropic oceanic turbulence on the power of the bandwidth-limited OAM mode of partially coherent modified Bessel correlated vortex beams. *J. Opt. Soc. Am. A* **2018**, *35*, 1839–1845. [CrossRef]
113. Yu, L.; Zhang, Y. Analysis of modal crosstalk for communication in turbulent ocean using Lommel-Gaussian beam. *Opt. Express* **2017**, *25*, 22565–22574. [CrossRef]
114. Li, H.; Liu, H.; Chen, X. Nonlinear generation of Airy vortex beam. *Opt. Express* **2018**, *26*, 21204–21209. [CrossRef] [PubMed]
115. Wang, X.; Yang, Z.; Zhao, S. Influence of oceanic turbulence on propagation of Airy vortex beam carrying orbital angular momentum. *Optik* **2019**, *176*, 49–55. [CrossRef]
116. Zhang, Q.; Liu, Z.; Wang, X. Generation and characteristics of an Airy vortex beam from the anomalous vortex beam. *Results Phys.* **2022**, *35*, 105389. [CrossRef]
117. Khonina, S.N.; Kazanskiy, N.L.; Karpeev, S.V.; Butt, M.A. Bessel Beam: Significance and Applications—A Progressive Review. *Micromachines* **2020**, *11*, 997. [CrossRef]
118. Paković, S.; Zhou, S.; González-Ovejero, D.; Pavone, S.C.; Grbic, A.; Ettorre, M. Bessel–Gauss Beam Launchers for Wireless Power Transfer. *IEEE Open J. Antennas Propag.* **2021**, *2*, 654–663. [CrossRef]
119. Zhu, K.; Zhou, G.; Li, X.; Zheng, X.; Tang, H. Propagation of Bessel-Gaussian beams with optical vortices in turbulent atmosphere. *Opt. Express* **2008**, *16*, 21315–21320. [CrossRef]
120. Droulias, S.; Loulakis, M.; Papazoglou, D.G.; Tzortzakis, S.; Chen, Z.; Efremidis, N.K. Inverted pin beams for robust long-range propagation through atmospheric turbulence. *Opt. Lett.* **2023**, *48*, 5467–5470. [CrossRef]
121. Droulias, S.; Loulakis, M.; Papazoglou, D.G.; Tzortzakis, S.; Chen, Z.; Efremidis, N.K. Inverted Pin Beams for Robust Propagation Through Atmospheric Turbulence. In Proceedings of the 2024 Conference on Lasers and Electro-Optics (CLEO), Charlotte, NC, USA, 5–10 May 2024; pp. 1–2.
122. Droulias, S.; Loulakis, M.; Papazoglou, D.G.; Tzortzakis, S.; Chen, Z.; Efremidis, N.K. Vortex inverted pin beams: Mitigation of scintillations in strong atmospheric turbulence. *Opt. Lett.* **2024**, *49*, 4811–4814. [CrossRef]
123. Li, J.; Chen, Y.; Cao, Q. Analytical vectorial structure of truncated modified Laguerre–Gaussian beam in the far field. *Opt. Laser Technol.* **2012**, *44*, 1247–1255. [CrossRef]
124. Willner, A.E.; Liu, C. Perspective on using multiple orbital-angular-momentum beams for enhanced capacity in free-space optical communication links. *Nanophotonics* **2021**, *10*, 225–233. [CrossRef]
125. Wang, L.; Wang, J.; Tang, X.; Chen, H.; Chen, X. Performance analysis of a spatial diversity coherent free-space optical communication system based on optimal branch block phase correction. *Opt. Express* **2022**, *30*, 7854–7869. [CrossRef] [PubMed]
126. Geng, C.; Li, F.; Zuo, J.; Liu, J.; Yang, X.; Yu, T.; Jiang, J.; Li, X. Fiber laser transceiving and wavefront aberration mitigation with adaptive distributed aperture array for free-space optical communications. *Opt. Lett.* **2020**, *45*, 1906–1909. [CrossRef]

127. Hong, Y.Q.; Han, S.K. Polarization-dependent SOA-based PolSK modulation for turbulence-robust FSO communication. *Opt. Express* **2021**, *29*, 15587–15594. [CrossRef]
128. Toyoshima, M. Recent Trends in Space Laser Communications for Small Satellites and Constellations. *J. Light. Technol.* **2021**, *39*, 693–699. [CrossRef]
129. Amin, R.; Maiti, R.; Gui, Y.; Suer, C.; Miscuglio, M.; Heidari, E.K.; Khurgin, J.B.; Chen, R.T.; Dalir, H.; Sorger, V.J. Heterogeneously integrated ITO plasmonic Mach–Zehnder interferometric modulator on SOI. *Sci. Rep.* **2021**, *11, 1287*. [CrossRef]
130. Daulay, O.; Liu, G.; Marpaung, D. Microwave photonic notch filter with integrated phase-to-intensity modulation transformation and optical carrier suppression. *Opt. Lett.* **2021**, *46*, 488–491. [CrossRef]
131. Alam, M.S.; Maram, R.; Shahriar, K.A.; Ricciardi, P.; Plant, D.V. Chirped Managed Laser for Multilevel Modulation Formats: A Semi-Analytical Approach for Efficient Filter Design. *J. Light. Technol.* **2024**, *42*, 2351–2361. [CrossRef]
132. Franklin, J.; Kil, L.; Mooney, D.; Mahgerefteh, D.; Zheng, X.; Matsui, Y.; McCallion, K.; Fan, F.; Tayebati, P. Generation of RZ-DPSK using a Chirp-Managed Laser (CML). In Proceedings of the OFC/NFOEC 2008—2008 Conference on Optical Fiber Communication/National Fiber Optic Engineers Conference, San Diego, CA, USA, 24–28 February 2008; pp. 1–3. [CrossRef]
133. Gao, D.; Li, T.; Bai, Z.; Ma, R.; Xie, Z.; Jia, S.; Wang, W.; Xie, X. Multi-modulation compatible miniaturization system for FSO communication assisted by chirp-managed laser. *Opt. Express* **2022**, *30*, 32306–32316. [CrossRef]
134. Hu, Z.; Chen, Z.; Li, Y.; Benton, D.M.; Ali, A.A.I.; Patel, M.; Lavery, M.P.; Ellis, A.D. Adaptive Transceiver Design for High-Capacity Multi-Modal Free-Space Optical Communications with Commercial Devices and Atmospheric Turbulence. *J. Light. Technol.* **2023**, *41*, 3397–3406. [CrossRef]
135. Kaur, S.; Sachdeva, S.; Sindhwani, M. 400 Gb/s free space optical communication (FSOC) system using OAM multiplexing and PDM-QPSK with DSP. *J. Opt. Commun.* **2024**, *45*, 917–923. [CrossRef]
136. Khan, A.N.; Saeed, S.; Naeem, Y.; Zubair, M.; Massoud, Y.; Younis, U. Atmospheric Turbulence and Fog Attenuation Effects in Controlled Environment FSO Communication Links. *IEEE Photonics Technol. Lett.* **2022**, *34*, 1341–1344. [CrossRef]
137. Chen, M.; Liu, C.; Xian, H. Experimental demonstration of single-mode fiber coupling over relatively strong turbulence with adaptive optics. *Appl. Opt.* **2015**, *54*, 8722–8726. [CrossRef] [PubMed]
138. Li, S.; Wang, J. Compensation of a distorted N-fold orbital angular momentum multicasting link using adaptive optics. *Opt. Lett.* **2016**, *41*, 1482–1485. [CrossRef] [PubMed]
139. Zheng, D.; Li, Y.; Zhou, H.; Bian, Y.; Yang, C.; Li, W.; Qiu, J.; Guo, H.; Hong, X.; Zuo, Y.; et al. Performance enhancement of free-space optical communications under atmospheric turbulence using modes diversity coherent receipt. *Opt. Express* **2018**, *26*, 28879–28890. [CrossRef]
140. Wang, F.; Qiu, C.; Zhang, M.; Hu, G. Free-Space Optical Communication Based on Mode Diversity Reception Using a Nonmode Selective Photonic Lantern and Equal Gain Combining. *IEEE Photonics J.* **2023**, *15*, 1–7. [CrossRef]
141. Huang, Y.; Huang, H.; Chen, H.; Alvarado, J.C.; Fontaine, N.K.; Mazur, M.; Zhang, Q.; Ryf, R.; Amezcua-Correa, R.; Song, Y.; et al. Free-Space Optics Communications Employing Elliptical-Aperture Multimode Diversity Reception Under Anisotropic Turbulence. *J. Light. Technol.* **2022**, *40*, 1502–1508. [CrossRef]
142. Ozdur, I.; Toliver, P.; Woodward, T.K. Photonic-lantern-based coherent LIDAR system. *Opt. Express* **2015**, *23*, 5312–5316. [CrossRef]
143. Guo, W.; Li, Y.; Chen, J.; Jin, T.; Jiao, S.; Wu, J.; Qiu, J.; Guo, H. Satellite-to-ground optical downlink model using mode mismatching multi-mode photonic lanterns. *Opt. Express* **2023**, *31*, 35041–35053. [CrossRef]
144. Fontaine, N.K.; Carpenter, J.; Gross, S.; Leon-Saval, S.; Jung, Y.; Richardson, D.J.; Amezcua-Correa, R. Photonic Lanterns, 3-D Waveguides, Multiplane Light Conversion, and Other Components That Enable Space-Division Multiplexing. *Proc. IEEE* **2022**, *110*, 1821–1834. [CrossRef]
145. Wang, F.; Qiu, C.; Chen, Y.; Hu, G. Performance of improved mode diversity reception for free-space optical communication under atmospheric turbulence. *J. Opt. Commun. Netw.* **2022**, *14*, 725–732. [CrossRef]
146. Liu, H.; Liao, R.; Wei, Z.; Hou, Z.; Qiao, Y. BER Analysis of a Hybrid Modulation Scheme Based on PPM and MSK Subcarrier Intensity Modulation. *IEEE Photonics J.* **2015**, *7*, 1–10. [CrossRef]
147. Jaiswal, A.; Bhatnagar, M.R.; Soni, P.; Jain, V.K. Differential Optical Spatial Modulation Over Atmospheric Turbulence. *IEEE J. Sel. Top. Signal Process.* **2019**, *13*, 1417–1432. [CrossRef]
148. Djordjevic, I.; Vasic, B.; Neifeld, M. Multilevel coding in free-space optical MIMO transmission with Q-ary PPM over the atmospheric turbulence channel. *IEEE Photonics Technol. Lett.* **2006**, *18*, 1491–1493. [CrossRef]
149. Wang, Z.; Zhong, W.D.; Fu, S.; Lin, C. Performance comparison of different modulation formats over free-space optical (FSO) turbulence links with space diversity reception technique. *IEEE Photonics J.* **2009**, *1*, 277–285. [CrossRef]
150. Djordjevic, I.B. Adaptive Modulation and Coding for Free-Space Optical Channels. *J. Opt. Commun. Netw.* **2010**, *2*, 221–229. [CrossRef]
151. Nguyen, T.T.; Lampe, L. Coded multipulse pulse-position modulation for free-space optical communications. *IEEE Trans. Commun.* **2010**, *58*, 1036–1041. [CrossRef]
152. Peppas, K.P.; Mathiopoulos, P.T. Free-Space Optical Communication with Spatial Modulation and Coherent Detection Over H-K Atmospheric Turbulence Channels. *J. Light. Technol.* **2015**, *33*, 4221–4232. [CrossRef]
153. Hwang, S.H.; Cheng, Y. SIM/SM-Aided Free-Space Optical Communication with Receiver Diversity. *J. Light. Technol.* **2014**, *32*, 2443–2450. [CrossRef]

154. Tan, Y.; Liu, W.; Guo, J.; Ai, Y. Iterative Coded Multipulse Pulse Position Modulation for High Data Rate Optical Space Communication. *Chin. J. Electron.* **2011**, *20*, 121–124.
155. Zhao, X.; Yao, Y.; Sun, Y.; Liu, C. Circle Polarization Shift Keying with Direct Detection for Free-Space Optical Communication. *J. Opt. Commun. Netw.* **2009**, *1*, 307–312. [CrossRef]
156. El-Nahal, F.; Xu, T.; AlQahtani, D.; Leeson, M. A Bidirectional WDM-PON Free Space Optical (FSO) System for Fronthaul 5 G C-RAN Networks. *IEEE Photonics J.* **2023**, *15*, 1–10. [CrossRef]
157. Vishwakarma, N.; Swaminathan, R.; Diamantoulakis, P.D.; Karagiannidis, G.K. Performance Analysis of Optical Reflecting Surface-Assisted Optical Space Shift Keying-Based MIMO-FSO System. *IEEE Trans. Commun.* **2023**, *71*, 4751–4763. [CrossRef]
158. Matarneh, A.M. Coded-Optical Improved Quadrature Spatial Modulation Over Gamma–Gamma Atmospheric Strong Turbulence Channels with Pointing Errors. *IEEE Access* **2024**, *12*, 83513–83524. [CrossRef]
159. Wang, K.; Lim, C.; Wong, E.; Alameh, K.; Kandeepan, S.; Skafidas, E. High-Speed Reconfigurable Free-Space Optical Interconnects with Carrierless-Amplitude-Phase Modulation and Space-Time-Block Code. *J. Light. Technol.* **2019**, *37*, 627–633. [CrossRef]
160. Zhang, R.; Peng, P.C.; Li, X.; Liu, S.; Zhou, Q.; He, J.; Chen, Y.W.; Shen, S.; Yao, S.; Chang, G.K. 4×100 -Gb/s PAM-4 FSO Transmission Based on Polarization Modulation and Direct Detection. *IEEE Photonics Technol. Lett.* **2019**, *31*, 755–758. [CrossRef]
161. Rajbhandari, S.; Ghassemlooy, Z.; Haigh, P.A.; Kanesan, T.; Tang, X. Experimental Error Performance of Modulation Schemes Under a Controlled Laboratory Turbulence FSO Channel. *J. Light. Technol.* **2015**, *33*, 244–250. [CrossRef]
162. Lu, H.H.; Li, C.Y.; Tsai, W.S.; Peng, Y.E.; Fan, W.C.; Lin, C.J.; Tang, Y.S.; Chen, Y.X.; Lin, Y.S. Two-Way 5G NR Fiber-Wireless Systems Using Single-Carrier Optical Modulation for Downstream and Phase Modulation Scheme for Upstream. *J. Light. Technol.* **2023**, *41*, 1749–1758. [CrossRef]
163. Jian, Y.H.; Chen, J.W.; Wang, C.C.; Wei, T.C.; Chow, C.W. High Spectral Efficient Free-Space Optical Communication Using Multi-Stage Noise Shaping Delta-Sigma Modulation with Partial Transmit Sequence Algorithm and Arithmetic Coding. *J. Light. Technol.* **2024**, *42*, 6767–6773. [CrossRef]
164. Liu, X.; Yang, W.; Liu, Z.; Xiao, S. Performance Enhancement of FEC-Coded 32QAM DMT Signals with Rate-Flexible HCS in FSO Systems. *IEEE Photonics Technol. Lett.* **2024**, *36*, 1081–1084. [CrossRef]
165. Zhu, X.; Kahn, J. Performance bounds for coded free-space optical communications through atmospheric turbulence channels. *IEEE Trans. Commun.* **2003**, *51*, 1233–1239. [CrossRef]
166. Kahn, J.M. Modulation and Detection Techniques for Optical Communication Systems. In Proceedings of the Optical Amplifiers and Their Applications/Coherent Optical Technologies and Applications, Whistler, BC, Canada, 25–30 June 2006; p. CThC1. [CrossRef]
167. Özbilgin, T.; Koca, M. Polarization Diversity in MIMO Communication Over Atmospheric Turbulence Channels. *J. Light. Technol.* **2016**, *34*, 3981–3992. [CrossRef]
168. Yang, L.; Gao, X.; Alouini, M.S. Performance Analysis of Free-Space Optical Communication Systems with Multiuser Diversity Over Atmospheric Turbulence Channels. *IEEE Photonics J.* **2014**, *6*, 1–17. [CrossRef]
169. Sangeetha, R.; Hemanth, C.; Jaiswal, I. Performance of different modulation scheme in free space optical transmission—A review. *Optik* **2022**, *254*, 168675. [CrossRef]
170. Liao, R.; Liu, H.; Qiao, Y. New hybrid reverse differential pulse position width modulation scheme for wireless optical communication. *Opt. Eng.* **2014**, *53*, 056112. [CrossRef]
171. Ma, J.; Jiang, Y.; Yu, S.; Tan, L.; Du, W. Packet error rate analysis of OOK, DPIM and PPM modulation schemes for ground-to-satellite optical communications. *Opt. Commun.* **2010**, *283*, 237–242. [CrossRef]
172. Karbassian, M.M.; Ghafouri-Shiraz, H. Transceiver Architecture for Incoherent Optical CDMA Networks Based on Polarization Modulation. *J. Light. Technol.* **2008**, *26*, 3820–3828. [CrossRef]
173. Nguyen, T.V.; Pham, T.V.; Dang, N.T.; Pham, A.T. Performance of Generalized QAM/FSO Systems with Pointing Misalignment and Phase Error Over Atmospheric Turbulence Channels. *IEEE Access* **2020**, *8*, 203631–203644. [CrossRef]
174. Singya, P.K.; Kumar, N.; Bhatia, V.; Alouini, M.S. On the Performance Analysis of Higher Order QAM Schemes Over Mixed RF/FSO Systems. *IEEE Trans. Veh. Technol.* **2020**, *69*, 7366–7378. [CrossRef]
175. Singya, P.K.; Kumar, N.; Bhatia, V. Impact of Imperfect CSI on ASER of Hexagonal and Rectangular QAM for AF Relaying Network. *IEEE Commun. Lett.* **2018**, *22*, 428–431. [CrossRef]
176. Singya, P.K.; Kumar, N.; Bhatia, V.; Alouini, M.S. On Performance of Hexagonal, Cross, and Rectangular QAM for Multi-Relay Systems. *IEEE Access* **2019**, *7*, 60602–60616. [CrossRef]
177. Kumar, N.; Singya, P.K.; Bhatia, V. ASER Analysis of Hexagonal and Rectangular QAM Schemes in Multiple-Relay Networks. *IEEE Trans. Veh. Technol.* **2018**, *67*, 1815–1819. [CrossRef]
178. Ghassemlooy, Z. Experimental investigation of polarisation modulated free space optical communication with direct detection in a turbulence channel. *IET Commun.* **2012**, *6*, 1489–1494. [CrossRef]
179. Trisno, S.; Davis, C.C. Performance of free space optical communication systems using polarization shift keying modulation. In *Proceedings of the Free-Space Laser Communications VI*; Majumdar, A.K., Davis, C.C., Eds.; International Society for Optics and Photonics, SPIE: San Diego, CA, USA, 2006; Volume 6304, p. 63040V. [CrossRef]
180. Su, Y.; Sato, T. Analysis of CPolSK-based FSO system working in space-to-ground channel. *Opt. Commun.* **2018**, *410*, 660–667. [CrossRef]

181. Prabu, K.; Kumar, D.S. MIMO free-space optical communication employing coherent BPOLSK modulation in atmospheric optical turbulence channel with pointing errors. *Opt. Commun.* **2015**, *343*, 188–194. [CrossRef]
182. Tang, X.; Xu, Z.; Ghassemlooy, Z. Coherent Polarization Modulated Transmission through MIMO Atmospheric Optical Turbulence Channel. *J. Light. Technol.* **2013**, *31*, 3221–3228. [CrossRef]
183. Tang, X.; Ghassemlooy, Z.; Rajbhandari, S.; Popoola, W.O.; Lee, C.G. Coherent Heterodyne Multilevel Polarization Shift Keying with Spatial Diversity in a Free-Space Optical Turbulence Channel. *J. Light. Technol.* **2012**, *30*, 2689–2695. [CrossRef]
184. Özbilgin, T.; Koca, M. Optical Spatial Modulation Over Atmospheric Turbulence Channels. *J. Light. Technol.* **2015**, *33*, 2313–2323. [CrossRef]
185. Alimi, I.A.; Popoola, J.J.; Akingbade, K.F.; Kolawole, M.O. Interference Management in MIMO-OFDM-Based Emerging Wireless Systems. *Am. J. Inf. Sci. Comput. Eng.* **2015**, *1*, 9193027.
186. Djordjevic, I.B. LDPC-coded MIMO optical communication over the atmospheric turbulence channel using Q-ary pulse-position modulation. *Opt. Express* **2007**, *15*, 10026–10032. [CrossRef]
187. Alimi, I.A. Demonstration of Effectiveness of OFDM for Mobile Communication Systems. *Am. J. Mob. Syst. Appl. Serv.* **2015**, *1*, 35–45.
188. Chen, C.; Zhong, W.D.; Li, X.; Wu, D. MDPSK-Based Nonequalization OFDM for Coherent Free-Space Optical Communication. *IEEE Photonics Technol. Lett.* **2014**, *26*, 1617–1620. [CrossRef]
189. Wang, Z.; Zhong, W.D.; Yu, C. Performance Improvement of OOK Free-Space Optical Communication Systems by Coherent Detection and Dynamic Decision Threshold in Atmospheric Turbulence Conditions. *IEEE Photonics Technol. Lett.* **2012**, *24*, 2035–2037. [CrossRef]
190. Zhong, K.; Tjhung, T.T.; Adachi, F. A general SER formula for an OFDM system with MDPSK in frequency domain over Rayleigh fading channels. *IEEE Trans. Commun.* **2004**, *52*, 584–594. [CrossRef]
191. Popoola, W. Free-space optical communication employing subcarrier modulation and spatial diversity in atmospheric turbulence channel. *IET Optoelectron.* **2008**, *2*, 16–23. [CrossRef]
192. Prabu, K.; Bose, S.; Sriram Kumar, D. BPSK based subcarrier intensity modulated free space optical system in combined strong atmospheric turbulence. *Opt. Commun.* **2013**, *305*, 185–189. [CrossRef]
193. Song, X.; Cheng, J. Subcarrier Intensity Modulated Optical Wireless Communications Using Noncoherent and Differentially Coherent Modulations. *J. Light. Technol.* **2013**, *31*, 1906–1913. [CrossRef]
194. Li, J.; Liu, J.Q.; Taylor, D.P. Optical Communication Using Subcarrier PSK Intensity Modulation Through Atmospheric Turbulence Channels. *IEEE Trans. Commun.* **2007**, *55*, 1598–1606. [CrossRef]
195. Popoola, W.O.; Ghassemlooy, Z. BPSK Subcarrier Intensity Modulated Free-Space Optical Communications in Atmospheric Turbulence. *J. Light. Technol.* **2009**, *27*, 967–973. [CrossRef]
196. Strickland, B.R.; Lavan, M.J.; Woodbridge, E.; Chan, V. Effects of fog on the bit-error rate of a free-space laser communication system. *Appl. Opt.* **1999**, *38*, 424–431. [CrossRef]
197. Jeganathan, J.; Ghrayeb, A.; Szczecinski, L. Spatial modulation: Optimal detection and performance analysis. *IEEE Commun. Lett.* **2008**, *12*, 545–547. [CrossRef]
198. Mesleh, R.Y.; Haas, H.; Sinanovic, S.; Ahn, C.W.; Yun, S. Spatial Modulation. *IEEE Trans. Veh. Technol.* **2008**, *57*, 2228–2241. [CrossRef]
199. Bhowal, A.; Kshetrimayum, R.S. Advanced Optical Spatial Modulation Techniques for FSO Communication. *IEEE Trans. Commun.* **2021**, *69*, 1163–1174. [CrossRef]
200. Mesleh, R.; Elgala, H.; Haas, H. Optical Spatial Modulation. *J. Opt. Commun. Netw.* **2011**, *3*, 234–244. [CrossRef]
201. Alimi, I.A.; Abdalla, A.M.; Rodriguez, J.; Pereira Monteiro, P.; Teixeira, A.L.; Zvánovec, S.; Ghassemlooy, Z., Enabling VLC and WiFi Network Technologies and Architectures Toward 5G. In *Optical and Wireless Convergence for 5G Networks*; John Wiley & Sons, Ltd.: Hoboken, NJ, USA, 2019; Chapter 5, pp. 101–122. [CrossRef]
202. Cheng, C.C.; Sari, H.; Sezginer, S.; Su, Y.T. Enhanced Spatial Modulation with Multiple Signal Constellations. *IEEE Trans. Commun.* **2015**, *63*, 2237–2248. [CrossRef]
203. Wang, K.; Kandeepan, S. Filter-Enhanced Spatial Modulation and Multiplexing in Indoor Optical Wireless Systems. *IEEE Photonics Technol. Lett.* **2021**, *33*, 1479–1482. [CrossRef]
204. Zhan, Y.; Huang, F. Generalized Spatial Modulation with Multi-Index Modulation. *IEEE Commun. Lett.* **2020**, *24*, 585–588. [CrossRef]
205. Qasem, Z.A.H.; Esmaiel, H.; Sun, H.; Wang, J.; Miao, Y.; Anwar, S. Enhanced Fully Generalized Spatial Modulation for the Internet of Underwater Things. *Sensors* **2019**, *19*, 1519. [CrossRef]
206. Castillo-Soria, F.; Cortez, J.; Gutiérrez, C.; Luna-Rivera, M.; Garcia-Barrientos, A. Extended quadrature spatial modulation for MIMO wireless communications. *Phys. Commun.* **2019**, *32*, 88–95. [CrossRef]
207. Datta, T.; Chockalingam, A. On generalized spatial modulation. In Proceedings of the 2013 IEEE Wireless Communications and Networking Conference (WCNC), Shanghai, China, 7–10 April 2013; pp. 2716–2721. [CrossRef]
208. Celik, Y. Fully Improved Quadrature Spatial Modulation. *Arab. J. Sci. Eng.* **2021**, *46*, 9639–9647. [CrossRef]
209. Celik, Y.; Aldirmaz-Colak, S. Extended fully improved quadrature spatial shift keying modulation. *Signal Image Video Process.* **2022**, *16*, 783–788. [CrossRef]

210. Vo, B.; Nguyen, H.H. Improved Quadrature Spatial Modulation. In Proceedings of the 2017 IEEE 86th Vehicular Technology Conference (VTC-Fall), Toronto, ON, Canada, 24–27 September 2017; pp. 1–5. [CrossRef]
211. Zhu, X.; Kahn, J. Free-space optical communication through atmospheric turbulence channels. *IEEE Trans. Commun.* **2002**, *50*, 1293–1300. [CrossRef]
212. Zhu, X.; Kahn, J. Markov chain model in maximum-likelihood sequence detection for free-space optical communication through atmospheric turbulence channels. *IEEE Trans. Commun.* **2003**, *51*, 509–516. [CrossRef]
213. Djordjevic, I.B.; Denic, S.; Anguita, J.; Vasic, B.; Neifeld, M.A. LDPC-Coded MIMO Optical Communication Over the Atmospheric Turbulence Channel. In Proceedings of the IEEE GLOBECOM 2007—IEEE Global Telecommunications Conference, Washington, DC, USA, 26–30 November 2007; pp. 2220–2225. [CrossRef]
214. Anguita, J.A.; Neifeld, M.A.; Hildner, B.; Vasic, B. Rateless Coding on Experimental Temporally Correlated FSO Channels. *J. Light. Technol.* **2010**, *28*, 990–1002. [CrossRef]
215. Castura, J.; Mao, Y. Rateless coding over fading channels. *IEEE Commun. Lett.* **2006**, *10*, 46–48. [CrossRef]
216. Denic, S.Z.; Djordjevic, I.; Anguita, J.; Vasic, B.; Neifeld, M.A. Information Theoretic Limits for Free-Space Optical Channels with and without Memory. *J. Light. Technol.* **2008**, *26*, 3376–3384. [CrossRef]
217. Goldsmith, A.; Chua, S.G. Adaptive coded modulation for fading channels. *IEEE Trans. Commun.* **1998**, *46*, 595–602. [CrossRef]
218. Zhang, Q.; Liu, B.; Chen, G.; Zhan, S.; Li, Z.; Zhang, J.; Jiang, N.; Cao, B.; Li, Z. An Improved Adaptive Coding and Modulation Scheme with Hybrid Switching Standard for UAV-to-Ground Free Space Optical Communication. *IEEE Photonics J.* **2024**, *16*, 1–8. [CrossRef]
219. Li, R.; Chen, T.; Fan, L.; Dang, A. Performance Analysis of a Multiuser Dual-Hop Amplify-and-Forward Relay System with FSO/RF Links. *J. Opt. Commun. Netw.* **2019**, *11*, 362–370. [CrossRef]
220. Zhou, H.; Mao, S.; Agrawal, P. Optical power allocation for adaptive transmissions in wavelength-division multiplexing free space optical networks. *Digit. Commun. Networks* **2015**, *1*, 171–180. [CrossRef]
221. Safi, H.; Sharifi, A.A.; Dabiri, M.T.; Ansari, I.S.; Cheng, J. Adaptive Channel Coding and Power Control for Practical FSO Communication Systems Under Channel Estimation Error. *IEEE Trans. Veh. Technol.* **2019**, *68*, 7566–7577. [CrossRef]
222. Hariq, S.H.; Odabasioglu, N.; Uysal, M. An adaptive modulation scheme for coded free-space optical systems. In Proceedings of the 2014 19th European Conference on Networks and Optical Communications—(NOC), Milan, Italy, 4–6 June 2014; pp. 132–135. [CrossRef]
223. Jurado-Navas, A.; Garrido-Balsells, J.M.; Castillo-Vázquez, M.; Puerta-Notario, A. An efficient rate-adaptive transmission technique using shortened pulses for atmospheric optical communications. *Opt. Express* **2010**, *18*, 17346–17363. [CrossRef] [PubMed]
224. Hassan, Z.; Hossain, J.; Cheng, J.; Leung, V.C. Delay-QoS-aware adaptive modulation and power allocation for dual-channel coherent OWC. *J. Opt. Commun. Netw.* **2018**, *10*, 138–151. [CrossRef]
225. Hassan, M.Z.; Hossain, M.J.; Cheng, J.; Leung, V.C.M. Statistical Delay-QoS Aware Joint Power Allocation and Relaying Link Selection for Free Space Optics Based Fronthaul Networks. *IEEE Trans. Commun.* **2018**, *66*, 1124–1138. [CrossRef]
226. Liu, L.; Safari, M.; Hranilovic, S. Rate-adaptive FSO communication via rate-compatible punctured LDPC codes. In Proceedings of the 2013 IEEE International Conference on Communications (ICC), Budapest, Hungary, 9–13 June 2013; pp. 3948–3952. [CrossRef]
227. Chatzidiamantis, N.D.; Lioumpas, A.S.; Karagiannidis, G.K.; Arnon, S. Adaptive Subcarrier PSK Intensity Modulation in Free Space Optical Systems. *IEEE Trans. Commun.* **2011**, *59*, 1368–1377. [CrossRef]
228. Alshaer, N.; Ismail, T.; Nasr, M.E. Enhancing earth-to-satellite FSO system spectrum efficiency with adaptive M-ary PSK and SIMO in presence of scintillation and beam wander. *AEU—Int. J. Electron. Commun.* **2020**, *125*, 153366. [CrossRef]
229. Zhou, H.; Mao, S.; Agrawal, P. Optical power allocation for adaptive WDM transmissions in free space optical networks. In Proceedings of the 2014 IEEE Wireless Communications and Networking Conference (WCNC), Istanbul, Turkey, 6–9 April 2014; pp. 2677–2682. [CrossRef]
230. Giggenbach, D.; Cowley, W.; Grant, K.; Perlot, N. Experimental verification of the limits of optical channel intensity reciprocity. *Appl. Opt.* **2012**, *51*, 3145–3152. [CrossRef]
231. Shapiro, J.H. Reciprocity of the Turbulent Atmosphere. *J. Opt. Soc. Am.* **1971**, *61*, 492–495. [CrossRef]
232. Churnside, J.H. Aperture averaging of optical scintillations in the turbulent atmosphere. *Appl. Opt.* **1991**, *30*, 1982–1994. [CrossRef]
233. Khalighi, M.A.; Schwartz, N.; Aitamer, N.; Bourennane, S. Fading Reduction by Aperture Averaging and Spatial Diversity in Optical Wireless Systems. *J. Opt. Commun. Netw.* **2009**, *1*, 580–593. [CrossRef]
234. Vetelino, F.S.; Young, C.; Andrews, L. Fade statistics and aperture averaging for Gaussian beam waves in moderate-to-strong turbulence. *Appl. Opt.* **2007**, *46*, 3780–3789. [CrossRef]
235. Yuksel, H.; Davis, C. Aperture averaging experiment for optimizing receiver design and analyzing turbulence on free space optical communication links. In Proceedings of the Conference on Lasers and Electro-Optics (CLEO), Baltimore, MD, USA, 22–27 May 2005; Volume 1, pp. 743–745. [CrossRef]
236. Wayne, D.T.; Phillips, R.L.; Andrews, L.C.; Leclerc, T.; Sauer, P.; Stryjewski, J. Comparing the Log-Normal and Gamma-Gamma model to experimental probability density functions of aperture averaging data. In *Proceedings of the Free-Space Laser*

Communications X; Majumdar, A.K., Davis, C.C., Eds.; International Society for Optics and Photonics, SPIE: Bellingham, Wash, 2010; Volume 7814, p. 78140K. [CrossRef]

237. Wayne, D.T.; Phillips, R.L.; Andrews, L.C.; Leclerc, T.T.; Sauer, P.; Stryjewski, J.S. Comparing the Log-Normal and Gamma-Gamma model to experimental probability density functions of aperture averaging data. In Proceedings of the Optical Engineering + Applications, San Diego, CA, USA, 1–5 August 2010.
238. Barrios, R.; Dios, F. Exponentiated Weibull distribution family under aperture averaging for Gaussian beam waves. *Opt. Express* **2012**, *20*, 13055–13064. [CrossRef] [PubMed]
239. Epple, B. Simplified Channel Model for Simulation of Free-Space Optical Communications. *J. Opt. Commun. Netw.* **2010**, *2*, 293–304. [CrossRef]
240. Khalighi, M.; Aitamer, N.; Schwartz, N.; Bourennane, S. Turbulence mitigation by aperture averaging in wireless optical systems. In Proceedings of the 2009 10th International Conference on Telecommunications, Zagreb, Croatia, 8–10 June 2009; pp. 59–66. https://ieeexplore.ieee.org/document/5206375.
241. Abou-Rjeily, C. Performance Analysis of FSO Communications with Diversity Methods: Add More Relays or More Apertures? *IEEE J. Sel. Areas Commun.* **2015**, *33*, 1890–1902. [CrossRef]
242. Agarwal, D.; Bansal, A. Unified error performance of a multihop DF-FSO network with aperture averaging. *J. Opt. Commun. Netw.* **2019**, *11*, 95–106. [CrossRef]
243. Andrews, L.C. Aperture-averaging factor for optical scintillations of plane and spherical waves in the atmosphere. *J. Opt. Soc. Am. A* **1992**, *9*, 597–600. [CrossRef]
244. Wilson, S.; Brandt-Pearce, M.; Cao, Q.; Baedke, M. Optical repetition MIMO transmission with multipulse PPM. *IEEE J. Sel. Areas Commun.* **2005**, *23*, 1901–1910. [CrossRef]
245. Lee, E.; Chan, V. Part 1: Optical communication over the clear turbulent atmospheric channel using diversity. *IEEE J. Sel. Areas Commun.* **2004**, *22*, 1896–1906. [CrossRef]
246. Barrios, R. Exponentiated weibull fading channel model in free-space optical communications under atmospheric turbulence. Ph.D. Thesis, Universitat Politècnica de Catalunya, Barcelona, Spain, 2013.
247. Xu, G.; Zhang, Q. Mixed RF/FSO Deep Space Communication System Under Solar Scintillation Effect. *IEEE Trans. Aerosp. Electron. Syst.* **2021**, *57*, 3237–3251. [CrossRef]
248. Bashir, M.S.; Alouini, M.S. Optimal Power Allocation Between Beam Tracking and Symbol Detection Channels in a Free-Space Optical Communications Receiver. *IEEE Trans. Commun.* **2021**, *69*, 7631–7646. [CrossRef]
249. Pang, W.; Wang, P.; Han, M.; Li, S.; Yang, P.; Li, G.; Guo, L. Optical Intelligent Reflecting Surface for Mixed Dual-Hop FSO and Beamforming-Based RF System in C-RAN. *IEEE Trans. Wirel. Commun.* **2022**, *21*, 8489–8506. [CrossRef]
250. Yahia, O.B.; Erdogan, E.; Kurt, G.K.; Altunbas, I.; Yanikomeroglu, H. HAPS Selection for Hybrid RF/FSO Satellite Networks. *IEEE Trans. Aerosp. Electron. Syst.* **2022**, *58*, 2855–2867. [CrossRef]
251. Yang, L.; Zhu, Q.; Yan, X.; Li, S.; Jiang, H. Performance Analysis of Mixed PLC-FSO Dual-Hop Communication Systems. *IEEE Internet Things J.* **2022**, *9*, 19307–19317. [CrossRef]
252. Tsiftsis, T.A.; Sandalidis, H.G.; Karagiannidis, G.K.; Uysal, M. Optical wireless links with spatial diversity over strong atmospheric turbulence channels. *IEEE Trans. Wirel. Commun.* **2009**, *8*, 951–957. [CrossRef]
253. Uysal, M.; Navidpour, S.; Li, J. Error rate performance of coded free-space optical links over strong turbulence channels. *IEEE Commun. Lett.* **2004**, *8*, 635–637. [CrossRef]
254. Chan, V.W.S. Free-Space Optical Communications. *J. Light. Technol.* **2006**, *24*, 4750–4762. [CrossRef]
255. Shin, E.; Chan, V. Optical communication over the turbulent atmospheric channel using spatial diversity. In Proceedings of the Global Telecommunications Conference (GLOBECOM '02), Taipei, Taiwan, 17–21 November 2002; Volume 3, pp. 2055–2060. [CrossRef]
256. Lee, E.J.; Chan, V.W.S. Diversity Coherent Receivers for Optical Communication over the Clear Turbulent Atmosphere. In Proceedings of the 2007 IEEE International Conference on Communications, Glasgow, UK, 24–28 June 2007; pp. 2485–2492. [CrossRef]
257. Polynkin, P.; Peleg, A.; Klein, L.; Rhoadarmer, T.; Moloney, J. Optimized multiemitter beams for free-space optical communications through turbulent atmosphere. *Opt. Lett.* **2007**, *32*, 885–887. [CrossRef]
258. Anguita, J.A.; Neifeld, M.A.; Vasic, B.V. Spatial correlation and irradiance statistics in a multiple-beam terrestrial free-space optical communication link. *Appl. Opt.* **2007**, *46*, 6561–6571. [CrossRef]
259. Razavi, M.; Shapiro, J. Wireless optical communications via diversity reception and optical preamplification. *IEEE Trans. Wirel. Commun.* **2005**, *4*, 975–983. [CrossRef]
260. O'Brien, D.C.; Quasem, S.; Zikic, S.; Faulkner, G.E. Multiple input multiple output systems for optical wireless: challenges and possibilities. In *Proceedings of the Free-Space Laser Communications VI*; Majumdar, A.K., Davis, C.C., Eds.; International Society for Optics and Photonics, SPIE: San Diego, CA, USA, 2006; Volume 6304, p. 630416. [CrossRef]
261. Bushuev, D.; Arnon, S. Analysis of the performance of a wireless optical multi-input to multi-output communication system. *J. Opt. Soc. Am. A* **2006**, *23*, 1722–1730. [CrossRef]
262. Priyadarshani, R.; Bhatnagar, M.R.; Ghassemlooy, Z.; Zvanovec, S. Outage Analysis of a SIMO FSO System Over an Arbitrarily Correlated $\mathcal{M}$ -Distributed Channel. *IEEE Photonics Technol. Lett.* **2018**, *30*, 141–144. [CrossRef]

263. Navidpour, S.M.; Uysal, M.; Kavehrad, M. BER Performance of Free-Space Optical Transmission with Spatial Diversity. *IEEE Trans. Wirel. Commun.* **2007**, *6*, 2813–2819. [CrossRef]
264. Garcia-Zambrana, A.; Castillo-Vazquez, C.; Castillo-Vazquez, B.; Hiniesta-Gomez, A. Selection Transmit Diversity for FSO Links Over Strong Atmospheric Turbulence Channels. *IEEE Photonics Technol. Lett.* **2009**, *21*, 1017–1019. [CrossRef]
265. Wilson, S.; Brandt-Pearce, M.; Cao, Q.; Leveque, J. Free-space optical MIMO transmission with Q-ary PPM. *IEEE Trans. Commun.* **2005**, *53*, 1402–1412. [CrossRef]
266. Bayaki, E.; Schober, R. On space-time coding for free-space optical systems. *IEEE Trans. Commun.* **2010**, *58*, 58–62. [CrossRef]
267. Simon, M.; Vilnrotter, V. Alamouti-type space-time coding for free-space optical communication with direct detection. *IEEE Trans. Wirel. Commun.* **2005**, *4*, 35–39. [CrossRef]
268. Safari, M.; Uysal, M. Do We Really Need OSTBCs for Free-Space Optical Communication with Direct Detection? *IEEE Trans. Wirel. Commun.* **2008**, *7*, 4445–4448. [CrossRef]
269. Song, X.; Cheng, J. Subcarrier intensity modulated MIMO optical communications in atmospheric turbulence. *J. Opt. Commun. Netw.* **2013**, *5*, 1001–1009. [CrossRef]
270. Bhatnagar, M.R.; Anees, S. On the Performance of Alamouti Scheme in Gamma-Gamma Fading FSO Links with Pointing Errors. *IEEE Wirel. Commun. Lett.* **2015**, *4*, 94–97. [CrossRef]
271. Priyadarshani, R.; Bhatnagar, M.R.; Ghassemlooy, Z.; Zvanovec, S. Effect of Correlation on BER Performance of the FSO-MISO System with Repetition Coding Over Gamma–Gamma Turbulence. *IEEE Photonics J.* **2017**, *9*, 1–15. [CrossRef]
272. Abou-Rjeily, C. On the Optimality of the Selection Transmit Diversity for MIMO-FSO Links with Feedback. *IEEE Commun. Lett.* **2011**, *15*, 641–643. [CrossRef]
273. Bhatnagar, M.R. A One Bit Feedback Based Beamforming Scheme for FSO MISO System Over Gamma-Gamma Fading. *IEEE Trans. Commun.* **2015**, *63*, 1306–1318. [CrossRef]
274. García-Zambrana, A.; Castillo-Vázquez, C.; Castillo-Vázquez, B. Space-time trellis coding with transmit laser selection for FSO links over strong atmospheric turbulence channels. *Opt. Express* **2010**, *18*, 5356–5366. [CrossRef] [PubMed]
275. Garg, A.; Bhatnagar, M.R.; Berder, O.; Vrigneau, B. Imperfect-quantized-feedback-based beamforming for an FSO MISO system over Gamma-Gamma fading with pointing errors. *J. Opt. Commun. Netw.* **2017**, *9*, 1005–1018. [CrossRef]
276. Boluda-Ruiz, R.; García-Zambrana, A.; Castillo-Vázquez, B.; Castillo-Vázquez, C. MISO Relay-Assisted FSO Systems Over Gamma–Gamma Fading Channels with Pointing Errors. *IEEE Photonics Technol. Lett.* **2016**, *28*, 229–232. [CrossRef]
277. Jahid, A.; Alsharif, M.H.; Hall, T.J. A contemporary survey on free space optical communication: Potentials, technical challenges, recent advances and research direction. *J. Netw. Comput. Appl.* **2022**, *200*, 103311. [CrossRef]
278. Shakir, W.M.R.; Mahdi, A.S.; Hamdan, H.; Charafeddine, J.; Satai, H.A.; Akrache, R.; Haddad, S.; Sayah, J. Novel Approximate Distribution of the Generalized Turbulence Channels for MIMO FSO Communications. *IEEE Photonics J.* **2024**, *16*, 1–15. [CrossRef]
279. Narang, G.; Aggarwal, M.; Kaushal, H.; Kumar, A.; Ahuja, S. Performance Analysis of Differential Chaos Shift Keying in Free Space Optical Communication with Diversity Techniques. *IEEE Access* **2023**, *11*, 54438–54447. [CrossRef]
280. Anandkumar, D.; Sangeetha, R.G. Performance Evaluation of LDPC-Coded Power Series Based Málaga ($\mathcal{M}$) Distributed MIMO/FSO Link with M-QAM and Pointing Error. *IEEE Access* **2022**, *10*, 62037–62055. [CrossRef]
281. Alimi, I.A.; Abdalla, A.M.; Rodriguez, J.; Monteiro, P.P.; Teixeira, A.L. Spatial Interpolated Lookup Tables (LUTs) Models for Ergodic Capacity of MIMO FSO Systems. *IEEE Photonics Technol. Lett.* **2017**, *29*, 583–586. [CrossRef]
282. Qin, M.; Chen, L.; Wang, W. Generalized Selection Multiuser Scheduling for the MIMO FSO Communication System and Its Performance Analysis. *IEEE Photonics J.* **2016**, *8*, 1–9. [CrossRef]
283. Bhatnagar, M.R. Differential Decoding of SIM DPSK over FSO MIMO Links. *IEEE Commun. Lett.* **2013**, *17*, 79–82. [CrossRef]
284. Liu, Y.; He, Y.; Chen, K.; Guo, L. Asynchronous Transmission for Cooperative Free-Space Optical Communication System. *IEEE Wirel. Commun. Lett.* **2022**, *11*, 766–770. [CrossRef]
285. Fang, J.; Bi, M.; Xiao, S.; Yang, G.; Liu, L.; Zhang, Y.; Hu, W. Polar-coded MIMO FSO communication system over gamma-gamma turbulence channel with spatially correlated fading. *J. Opt. Commun. Netw.* **2018**, *10*, 915–923. [CrossRef]
286. Chatzidiamantis, N.D.; Karagiannidis, G.K.; Michalopoulos, D.S. On the Distribution of the Sum of Gamma-Gamma Variates and Application in MIMO Optical Wireless Systems. In Proceedings of the GLOBECOM 2009—2009 IEEE Global Telecommunications Conference, Honolulu, HI, USA, 30 Novembe–4 December 2009; pp. 1–6. [CrossRef]
287. Bayaki, E.; Schober, R.; Mallik, R.K. Performance analysis of MIMO free-space optical systems in gamma-gamma fading. *IEEE Trans. Commun.* **2009**, *57*, 3415–3424. [CrossRef]
288. Peppas, K.P. A Simple, Accurate Approximation to the Sum of Gamma–Gamma Variates and Applications in MIMO Free-Space Optical Systems. *IEEE Photonics Technol. Lett.* **2011**, *23*, 839–841. [CrossRef]
289. Peppas, K. Multivariate gamma–gamma distribution with exponential correlation and its applications in radio frequency and optical wireless communications. *IET Microwaves, Antennas Propag.* **2011**, *5*, 364–371. [CrossRef]
290. Yang, G.; Khalighi, M.A.; Ghassemlooy, Z.; Bourennane, S. Performance evaluation of receive-diversity free-space optical communications over correlated Gamma–Gamma fading channels. *Appl. Opt.* **2013**, *52*, 5903–5911. [CrossRef]
291. Chen, Z.; Yu, S.; Wang, T.; Wu, G.; Wang, S.; Gu, W. Channel correlation in aperture receiver diversity systems for free-space optical communication. *J. Opt.* **2012**, *14*, 125710. [CrossRef]
292. Pham, T.V.; Thang, T.C.; Pham, A.T. Average achievable rate of spatial diversity MIMO-FSO over correlated Gamma–Gamma fading channels. *J. Opt. Commun. Netw.* **2018**, *10*, 662–674. [CrossRef]

293. Haas, S.; Shapiro, J. Capacity of wireless optical communications. *IEEE J. Sel. Areas Commun.* **2003**, *21*, 1346–1357. [CrossRef]
294. Chakraborty, K. Capacity of the MIMO optical fading channel. In Proceedings of the International Symposium on Information Theory (ISIT 2005), Adelaide, SA, Australia, 4–9 September 2005; pp. 530–534. [CrossRef]
295. Dabiri, M.T.; Saber, M.J.; Sadough, S.M.S. On the performance of multiplexing FSO MIMO links in log-normal fading with pointing errors. *J. Opt. Commun. Netw.* **2017**, *9*, 974–983. [CrossRef]
296. Böcherer, G.; Steiner, F.; Schulte, P. Bandwidth Efficient and Rate-Matched Low-Density Parity-Check Coded Modulation. *IEEE Trans. Commun.* **2015**, *63*, 4651–4665. [CrossRef]
297. Elzanaty, A.; Alouini, M.S. Adaptive Coded Modulation for IM/DD Free-Space Optical Backhauling: A Probabilistic Shaping Approach. *IEEE Trans. Commun.* **2020**, *68*, 6388–6402. [CrossRef]
298. Forney, G.; Gallager, R.; Lang, G.; Longstaff, F.; Qureshi, S. Efficient Modulation for Band-Limited Channels. *IEEE J. Sel. Areas Commun.* **1984**, *2*, 632–647. [CrossRef]
299. Böcherer, G.; Schulte, P.; Steiner, F. Probabilistic Shaping and Forward Error Correction for Fiber-Optic Communication Systems. *J. Light. Technol.* **2019**, *37*, 230–244. [CrossRef]
300. Liu, X.; Fang, J.; Xiao, S.; Zheng, L.; Fu, X.; Hu, W. Probabilistically shaped polar-coded MIMO-FSO communication systems with spatially correlated fading. *Opt. Express* **2022**, *30*, 30980–30990. [CrossRef]
301. Hranilovic, S.; Kschischang, F. Optical intensity-modulated direct detection channels: Signal space and lattice codes. *IEEE Trans. Inf. Theory* **2003**, *49*, 1385–1399. [CrossRef]
302. Muhammad, S.; Leitgeb, E.; Koudelka, O. Multilevel Modulation and Channel Codes for Terrestrial FSO links. In Proceedings of the 2005 2nd International Symposium on Wireless Communication Systems, Siena, Italy, 5 December 2005; pp. 795–799. [CrossRef]
303. Letzepis, N.; Holland, I.; Cowley, W. The Gaussian free space optical MIMO channel with Q-ary pulse position modulation. *IEEE Trans. Wirel. Commun.* **2008**, *7*, 1744–1753. [CrossRef]
304. Park, H.; Barry, J. Trellis-coded multiple-pulse-position modulation for wireless infrared communications. *IEEE Trans. Commun.* **2004**, *52*, 643–651. [CrossRef]
305. McEliece, R. Practical codes for photon communication. *IEEE Trans. Inf. Theory* **1981**, *27*, 393–398. [CrossRef]
306. Fares, D.A. Performance of convolutional codes with multipulse signaling in optical channels. *Microw. Opt. Technol. Lett.* **1990**, *3*, 406–410. [CrossRef]
307. Anguita, J.; Djordjevic, I.; Neifeld, M.; Vasic, B. Shannon capacities and error-correction codes for optical atmospheric turbulentchannels. *J. Opt. Netw.* **2005**, *4*, 586–601. [CrossRef]
308. Liu, X.; Liu, Z.; Xiao, S.; Fu, X.; Yang, W.; Hu, W. Performance of polar coded probabilistic shaped PAM8 with systematic interleaver and identically distributed pilot in weak turbulence. *Opt. Express* **2023**, *31*, 28900–28911. [CrossRef] [PubMed]
309. Liu, X.; Fang, J.; Xiao, S.; Zheng, L.; Hu, W. Adaptive Probabilistic Shaping Using Polar Codes for FSO Communication. *IEEE Photonics J.* **2022**, *14*, 1–6. [CrossRef]
310. Arikan, E. Channel Polarization: A Method for Constructing Capacity-Achieving Codes for Symmetric Binary-Input Memoryless Channels. *IEEE Trans. Inf. Theory* **2009**, *55*, 3051–3073. [CrossRef]
311. Farid, A.A.; Hranilovic, S. Channel capacity and non-uniform signalling for free-space optical intensity channels. *IEEE J. Sel. Areas Commun.* **2009**, *27*, 1553–1563. [CrossRef]
312. Fang, J.; Bi, M.; Xiao, S.; Yang, H.; Chen, Z.; Liu, Z.; Yang, F.; Hu, W. Polar coded probabilistic amplitude shaping for the free space optical atmospheric turbulence channel. *Opt. Express* **2020**, *28*, 33208–33219. [CrossRef]
313. Sandalidis, H.G. Coded Free-Space Optical Links over Strong Turbulence and Misalignment Fading Channels. *IEEE Trans. Commun.* **2011**, *59*, 669–674. [CrossRef]
314. Zhang, R.; Hsu, C.W.; Tang, X.; Zhou, Q.; Chang, G.K. LDPC Coded PAM-4/8 Transmission in Fiber-FSO Link Using Unipolar Probability Distribution and Pre-distortion. In Proceedings of the 2021 Optical Fiber Communications Conference and Exhibition (OFC), San Francisco, CA, USA, 6–10 June 2021; pp. 1–3.
315. Zhang, S.; Qu, Z.; Yaman, F.; Mateo, E.; Inoue, T.; Nakamura, K.; Inada, Y.; Djordjevic, I.B. Flex-Rate Transmission using Hybrid Probabilistic and Geometric Shaped 32QAM. In Proceedings of the 2018 Optical Fiber Communications Conference and Exposition (OFC), San Diego, CA, USA, 11–15 March 2018; pp. 1–3.
316. Brandão, B.T.; Fernandes, M.A.; Loureiro, P.A.; Guiomar, F.P.; Monteiro, P.P. Cooperative FSO and mmWave System for Reliable 200G Wireless Transmission. *IEEE Photonics Technol. Lett.* **2022**, *34*, 1333–1336. [CrossRef]
317. Guiomar, F.P.; Lorences-Riesgo, A.; Ranzal, D.; Rocco, F.; Sousa, A.N.; Fernandes, M.A.; Brandão, B.T.; Carena, A.; Teixeira, A.L.; Medeiros, M.C.R.; et al. Adaptive Probabilistic Shaped Modulation for High-Capacity Free-Space Optical Links. *J. Light. Technol.* **2020**, *38*, 6529–6541. [CrossRef]
318. Sun, L.; Du, J.; Wang, C.; Li, Z.; Xu, K.; He, Z. Frequency-resolved adaptive probabilistic shaping for DMT-modulated IM-DD optical interconnects. *Opt. Express* **2019**, *27*, 12241–12254. [CrossRef] [PubMed]
319. Che, D.; Shieh, W. Approaching the Capacity of Colored-SNR Optical Channels by Multicarrier Entropy Loading. *J. Light. Technol.* **2018**, *36*, 68–78. [CrossRef]
320. Li, Y.; Chang, H.; Gao, R.; Zhang, Q.; Tian, F.; Yao, H.; Tian, Q.; Wang, Y.; Xin, X.; Wang, F.; et al. End-to-End Deep Learning of Joint Geometric Probabilistic Shaping Using a Channel-Sensitive Autoencoder. *Electronics* **2023**, *12*, 4234. [CrossRef]

321. Liu, X.; Liu, Z.; Xiao, S.; Yang, W.; Hu, W. Rate-flexible hybrid constellation shaping for polar-coded 32QAM in FSO systems. In Proceedings of the 2024 Optical Fiber Communications Conference and Exhibition (OFC), San Diego, CA, USA, 24–28 March 2024; pp. 1–3.
322. Kschischang, F.; Pasupathy, S. Optimal nonuniform signaling for Gaussian channels. *IEEE Trans. Inf. Theory* **1993**, *39*, 913–929. [CrossRef]
323. Abou-Rjeily, C.; Fawaz, W. Buffer-Aided Relaying Protocols for Cooperative FSO Communications. *IEEE Trans. Wirel. Commun.* **2017**, *16*, 8205–8219. [CrossRef]
324. Abou-Rjeily, C. A Distance-Aware Buffer-Aided Relaying Protocol for Cooperative FSO Communications. *IEEE Wirel. Commun. Lett.* **2024**, *13*, 2275–2279. [CrossRef]
325. Laneman, J.; Tse, D.; Wornell, G. Cooperative diversity in wireless networks: Efficient protocols and outage behavior. *IEEE Trans. Inf. Theory* **2004**, *50*, 3062–3080. [CrossRef]
326. Sendonaris, A.; Erkip, E.; Aazhang, B. User cooperation diversity. Part I. System description. *IEEE Trans. Commun.* **2003**, *51*, 1927–1938. [CrossRef]
327. Francis, F.; Manivasakan, R. A Performance Limit Estimation Framework for Multihop Repeated/Regenerated Optical Links. *IEEE Access* **2022**, *10*, 70016–70031. [CrossRef]
328. Hasna, M.; Alouini, M.S. Outage probability of multihop transmission over Nakagami fading channels. *IEEE Commun. Lett.* **2003**, *7*, 216–218. [CrossRef]
329. Karagiannidis, G.; Tsiftsis, T.; Mallik, R. Bounds for multihop relayed communications in nakagami-m fading. *IEEE Trans. Commun.* **2006**, *54*, 18–22. [CrossRef]
330. Soleimani-Nasab, E.; Ghassemlooy, Z. Multihop radio and optical wireless relaying systems over EGK, DGG, and CU fading channels. *J. Opt. Commun. Netw.* **2022**, *14*, 426–438. [CrossRef]
331. Peppas, K.P.; Stassinakis, A.N.; Nistazakis, H.E.; Tombras, G.S. Capacity analysis of dual amplify-and-forward relayed free-space optical communication systems over turbulence channels with pointing errors. *J. Opt. Commun. Netw.* **2013**, *5*, 1032–1042. [CrossRef]
332. Safari, M.; Uysal, M. Relay-assisted free-space optical communication. *IEEE Trans. Wirel. Commun.* **2008**, *7*, 5441–5449. [CrossRef]
333. Zhu, B.; Cheng, J.; Alouini, M.S.; Wu, L. Relay Placement for FSO Multihop DF Systems with Link Obstacles and Infeasible Regions. *IEEE Trans. Wirel. Commun.* **2015**, *14*, 5240–5250. [CrossRef]
334. Karimi, M.; Nasiri-Kenari, M. Free Space Optical Communications via Optical Amplify-and-Forward Relaying. *J. Light. Technol.* **2011**, *29*, 242–248. [CrossRef]
335. Yang, L.; Gao, X.; Alouini, M.S. Performance Analysis of Relay-Assisted All-Optical FSO Networks Over Strong Atmospheric Turbulence Channels with Pointing Errors. *J. Light. Technol.* **2014**, *32*, 4613–4620. [CrossRef]
336. Chatzidiamantis, N.D.; Michalopoulos, D.S.; Kriezis, E.E.; Karagiannidis, G.K.; Schober, R. Relay selection protocols for relay-assisted free-space optical systems. *J. Opt. Commun. Netw.* **2013**, *5*, 92–103. [CrossRef]
337. Abou-Rjeily, C.; Haddad, S. Cooperative FSO Systems: Performance Analysis and Optimal Power Allocation. *J. Light. Technol.* **2011**, *29*, 1058–1065. [CrossRef]
338. Varshney, N.; Jagannatham, A.K.; Hanzo, L. Asymptotic SER Analysis and Optimal Power Sharing for Dual-Phase and Multi-Phase Multiple-Relay Cooperative Systems. *IEEE Access* **2018**, *6*, 50404–50423. [CrossRef]
339. Abou-Rjeily, C. Performance Analysis of Selective Relaying in Cooperative Free-Space Optical Systems. *J. Light. Technol.* **2013**, *31*, 2965–2973. [CrossRef]
340. Abou-Rjeily, C.; Noun, Z. Impact of Inter-Relay Co-Operation on the Performance of FSO Systems with Any Number of Relays. *IEEE Trans. Wirel. Commun.* **2016**, *15*, 3796–3809. [CrossRef]
341. Bhatnagar, M.R.; Arti, M.K. Performance Analysis of Hybrid Satellite-Terrestrial FSO Cooperative System. *IEEE Photonics Technol. Lett.* **2013**, *25*, 2197–2200. [CrossRef]
342. Varshney, N.; Jagannatham, A.K.; Varshney, P.K. Cognitive MIMO-RF/FSO Cooperative Relay Communication with Mobile Nodes and Imperfect Channel State Information. *IEEE Trans. Cogn. Commun. Netw.* **2018**, *4*, 544–555. [CrossRef]
343. Varshney, N.; Jagannatham, A.K. Cognitive Decode-and-Forward MIMO-RF/FSO Cooperative Relay Networks. *IEEE Commun. Lett.* **2017**, *21*, 893–896. [CrossRef]
344. Hu, Z.; Chen, C.; Zhang, Z.; Zhang, H. Secure Cooperative Transmission for Mixed RF/FSO Spectrum Sharing Networks. *IEEE Trans. Commun.* **2020**, *68*, 3010–3023. [CrossRef]
345. Ai, Y.; Mathur, A.; Cheffena, M.; Bhatnagar, M.R.; Lei, H. Physical Layer Security of Hybrid Satellite-FSO Cooperative Systems. *IEEE Photonics J.* **2019**, *11*, 1–14. [CrossRef]
346. Boluda-Ruiz, R.; García-Zambrana, A.; Castillo-Vázquez, B.; Castillo-Vázquez, C. Ergodic Capacity Analysis of Decode-and-Forward Relay-Assisted FSO Systems Over Alpha–Mu Fading Channels Considering Pointing Errors. *IEEE Photonics J.* **2016**, *8*, 1–11. [CrossRef]
347. Bhatnagar, M.R. Average BER Analysis of Differential Modulation in DF Cooperative Communication System over Gamma-Gamma Fading FSO Links. *IEEE Commun. Lett.* **2012**, *16*, 1228–1231. [CrossRef]
348. Varshney, N.; Puri, P. Performance Analysis of Decode-and-Forward-Based Mixed MIMO-RF/FSO Cooperative Systems with Source Mobility and Imperfect CSI. *J. Light. Technol.* **2017**, *35*, 2070–2077. [CrossRef]

349. Zdravković, N.; Petkovic, M.I.; Djordjevic, G.T.; Kansanen, K. Outage Analysis of Mixed FSO/WiMAX Link. *IEEE Photonics J.* **2016**, *8*, 1–14. [CrossRef]
350. Jing, Z.; Shang-hong, Z.; Wei-hu, Z.; Ke-fan, C. Performance analysis for mixed FSO/RF Nakagami-m and Exponentiated Weibull dual-hop airborne systems. *Opt. Commun.* **2017**, *392*, 294–299. [CrossRef]
351. Trigui, I.; Cherif, N.; Affes, S. Relay-Assisted Mixed FSO/RF Systems Over Málaga- $\mathcal{M}$ and $\kappa - \mu$ Shadowed Fading Channels. *IEEE Wirel. Commun. Lett.* **2017**, *6*, 682–685. [CrossRef]
352. Ansari, I.S.; Yilmaz, F.; Alouini, M.S. Impact of Pointing Errors on the Performance of Mixed RF/FSO Dual-Hop Transmission Systems. *IEEE Wirel. Commun. Lett.* **2013**, *2*, 351–354. [CrossRef]
353. Zhang, Y.; Zhang, J.; Yang, L.; Ai, B.; Alouini, M.S. On the Performance of Dual-Hop Systems Over Mixed FSO/mmWave Fading Channels. *IEEE Open J. Commun. Soc.* **2020**, *1*, 477–489. [CrossRef]
354. Zedini, E.; Ansari, I.S.; Alouini, M.S. Performance Analysis of Mixed Nakagami- m and Gamma–Gamma Dual-Hop FSO Transmission Systems. *IEEE Photonics J.* **2015**, *7*, 1–20. [CrossRef]
355. Djordjevic, G.T.; Petkovic, M.I.; Cvetkovic, A.M.; Karagiannidis, G.K. Mixed RF/FSO Relaying with Outdated Channel State Information. *IEEE J. Sel. Areas Commun.* **2015**, *33*, 1935–1948. [CrossRef]
356. Hu, J.; Zhang, Z.; Dang, J.; Wu, L.; Zhu, G. Performance of Decode-and-Forward Relaying in Mixed Beaulieu-Xie and $\mathcal{M}$ Dual-Hop Transmission Systems with Digital Coherent Detection. *IEEE Access* **2019**, *7*, 138757–138770. [CrossRef]
357. Li, S.; Yang, L.; Hu, Y. Performance Analysis of Mixed THz/FSO Relaying Systems. In Proceedings of the 2021 13th International Conference on Wireless Communications and Signal Processing (WCSP), Changsha, China, 20–22 October 2021; pp. 1–5. [CrossRef]
358. Li, S.; Yang, L.; Zhang, J.; Bithas, P.S.; Tsiftsis, T.A.; Alouini, M.S. Mixed THz/FSO Relaying Systems: Statistical Analysis and Performance Evaluation. *IEEE Trans. Wirel. Commun.* **2022**, *21*, 10996–11010. [CrossRef]
359. Singya, P.K.; Makki, B.; D'Errico, A.; Alouini, M.S. Hybrid FSO/THz-Based Backhaul Network for mmWave Terrestrial Communication. *IEEE Trans. Wirel. Commun.* **2023**, *22*, 4342–4359. [CrossRef]
360. Abou-Rjeily, C.; Kaddoum, G. Free Space Optical Cooperative Communications via an Energy Harvesting Harvest-Store-Use Relay. *IEEE Trans. Wirel. Commun.* **2020**, *19*, 6564–6577. [CrossRef]
361. Anees, S.; Bhatnagar, M.R. Performance of an amplify-and-forward dual-hop asymmetric RF–FSO communication system. *J. Opt. Commun. Netw.* **2015**, *7*, 124–135. [CrossRef]
362. Alimi, I.A.; Monteiro, P.P. RF/FSO and THz/FSO Relaying Systems. In *Handbook of Radio and Optical Networks Convergence*; Kawanishi, T., Ed.; Springer Nature: Singapore, 2023; pp. 1–26. [CrossRef]
363. Ji, J.; Zhang, J.; Wang, K.; Xu, M.; Song, Y. Design and Experimental Investigation of Decode-and-Forward and Amplify-and-Forward Relaying FSO-CDMA. *IEEE Photonics J.* **2022**, *14*, 1–6. [CrossRef]
364. Datsikas, C.K.; Peppas, K.P.; Sagias, N.C.; Tombras, G.S. Serial Free-Space Optical Relaying Communications Over Gamma-Gamma Atmospheric Turbulence Channels. *J. Opt. Commun. Netw.* **2010**, *2*, 576–586. [CrossRef]
365. Alimi, I.A.; Monteiro, P.P. Performance analysis of 5G and beyond mixed THz/FSO relaying communication systems. *Opt. Laser Technol.* **2024**, *176*, 110917. [CrossRef]
366. Kazemlou, S.; Hranilovic, S.; Kumar, S. All-Optical Multihop Free-Space Optical Communication Systems. *J. Light. Technol.* **2011**, *29*, 2663–2669. [CrossRef]
367. Sun, Q.; Hu, Q.; Wu, Y.; Chen, X.; Zhang, J.; López-Benítez, M. Performance Analysis of Mixed FSO/RF System for Satellite-Terrestrial Relay Network. *IEEE Trans. Veh. Technol.* **2024**, *73*, 11378–11393. [CrossRef]
368. Bashir, M.S.; Alouini, M.S. Energy Optimization of a Laser-Powered Hovering-UAV Relay in Optical Wireless Backhaul. *IEEE Trans. Wirel. Commun.* **2023**, *22*, 3216–3230. [CrossRef]
369. Narang, G.; Aggarwal, M.; Kaushal, H.; Kumar, A.; Ahuja, S.; Shukla, N.K. Performance Evaluation of Dual Hop Mixed FSO RF System Using Differential Chaos Shift Keying with Secrecy Analysis. *IEEE Trans. Veh. Technol.* **2024**, *73*, 17347–17358. [CrossRef]
370. Hasna, M.; Alouini, M.S. A performance study of dual-hop transmissions with fixed gain relays. *IEEE Trans. Wirel. Commun.* **2004**, *3*, 1963–1968. [CrossRef]
371. Park, J.; Lee, E.; Chae, C.B.; Yoon, G. Outage Probability Analysis of a Coherent FSO Amplify-and-Forward Relaying System. *IEEE Photonics Technol. Lett.* **2015**, *27*, 1204–1207. [CrossRef]
372. Karimi, M.; Nasiri-Kenari, M. BER Analysis of Cooperative Systems in Free-Space Optical Networks. *J. Light. Technol.* **2009**, *27*, 5639–5647. [CrossRef]
373. Krikidis, I.; Charalambous, T.; Thompson, J.S. Buffer-Aided Relay Selection for Cooperative Diversity Systems without Delay Constraints. *IEEE Trans. Wirel. Commun.* **2012**, *11*, 1957–1967. [CrossRef]
374. Abou-Rjeily, C. Relay-Grouping: A New Paradigm for Enhancing the Performance of Buffer-Aided Relaying Networks. *IEEE Wirel. Commun. Lett.* **2024**, *13*, 153–157. [CrossRef]
375. Huang, C.; Chen, G.; Gong, Y. Delay-Constrained Buffer-Aided Relay Selection in the Internet of Things with Decision-Assisted Reinforcement Learning. *IEEE Internet Things J.* **2021**, *8*, 10198–10208. [CrossRef]
376. Jamali, V.; Michalopoulos, D.S.; Uysal, M.; Schober, R. Link Allocation for Multiuser Systems with Hybrid RF/FSO Backhaul: Delay-Limited and Delay-Tolerant Designs. *IEEE Trans. Wirel. Commun.* **2016**, *15*, 3281–3295. [CrossRef]
377. Xu, P.; Ding, Z.; Krikidis, I.; Dai, X. Achieving Optimal Diversity Gain in Buffer-Aided Relay Networks with Small Buffer Size. *IEEE Trans. Veh. Technol.* **2016**, *65*, 8788–8794. [CrossRef]

378. Leitgeb, E.; Birnbacher, U.; Muhammad, S.S.; Koudelka, O.; Kandus, G.; Gebhart, M.; Schrotter, P.; Chlestil, C.; Merdonig, A. Hybrid Wireless Networks combining WLAN, FSO and Satellite Technology for Disaster Recovery. In Proceedings of the Proceedings 15th IST Mobile Wireless Commun. Summit, Dresden, Germany, 19 June 2005; pp. 1–5.
379. Ata, Y. Angle-of-Arrival Fluctuations and Adaptive Optics Correction in Non-Kolmogorov Turbulent Atmosphere. *IEEE Trans. Antennas Propag.* **2022**, *70*, 12061–12070. [CrossRef]
380. Fried, D.L. Statistics of a Geometric Representation of Wavefront Distortion. *J. Opt. Soc. Am.* **1965**, *55*, 1427–1435. [CrossRef]
381. Noll, R.J. Zernike polynomials and atmospheric turbulence. *J. Opt. Soc. Am.* **1976**, *66*, 207–211. [CrossRef]
382. Tyson, R.K. Adaptive optics system performance approximations for atmospheric turbulence correction. *Opt. Eng.* **1990**, *29*, 1165–1173. [CrossRef]
383. Tyson, R.K. Using the deformable mirror as a spatial filter: Application to circular beams. *Appl. Opt.* **1982**, *21*, 787–793. [CrossRef] [PubMed]
384. Voitsekhovich, V.V.; Cuevas, S. Adaptive optics and the outer scale of turbulence. *J. Opt. Soc. Am. A* **1995**, *12*, 2523–2531. [CrossRef]
385. Toselli, I.; Gladysz, S. Efficiency of adaptive optics correction for Gaussian beams propagating through non-Kolmogorov turbulence. In Proceedings of the Imaging and Applied Optics, Seattle, DC, USA, 13–17 July 2014; p. PM3E.6. [CrossRef]
386. Tyson, R.K. Bit-error rate for free-space adaptive optics laser communications. *J. Opt. Soc. Am. A* **2002**, *19*, 753–758. [CrossRef]
387. Zhang, M.; Guo, Y. Performance Characterization of Deep-Phase-Retrieval Shack-Hartmann Wavefront Sensors. *IEEE Photonics J.* **2023**, *15*, 1–11. [CrossRef]
388. Li, C.; Hall, G.; Zhu, D.; Li, H.; Eliceiri, K.W.; Jiang, H. Three-Dimensional Surface Profile Measurement of Microlenses Using the Shack–Hartmann Wavefront Sensor. *J. Microelectromechanical Syst.* **2012**, *21*, 530–540. [CrossRef]
389. Zhao, M.; Zhao, W.; Wang, S.; Yang, P.; Yang, K.; Lin, H.; Kong, L. Centroid-Predicted Deep Neural Network in Shack-Hartmann Sensors. *IEEE Photonics J.* **2022**, *14*, 1–10. [CrossRef]
390. Rodríguez-Ramos, L.F.; Montilla, I.; Trujillo, J.M.; Rodríguez-Ramos, J.M. Phase structure function measurement with a plenoptic camera used as wavefront sensor. In Proceedings of the 2013 12th Workshop on Information Optics (WIO), Tenerife, Spain, 15–19 July 2013; pp. 1–3. [CrossRef]
391. Jung, I.W.; Peter, Y.A.; Carr, E.; Wang, J.S.; Solgaard, O. Single-Crystal-Silicon Continuous Membrane Deformable Mirror Array for Adaptive Optics in Space-Based Telescopes. *IEEE J. Sel. Top. Quantum Electron.* **2007**, *13*, 162–167. [CrossRef]
392. Dagel, D.; Cowan, W.; Spahn, O.; Grossetete, G.; Grine, A.; Shaw, M.; Resnick, P.; Jokiel, B. Large-stroke MEMS deformable mirrors for adaptive optics. *J. Microelectromechanical Syst.* **2006**, *15*, 572–583. [CrossRef]
393. Kanno, I.; Kunisawa, T.; Suzuki, T.; Kotera, H. Development of Deformable Mirror Composed of Piezoelectric Thin Films for Adaptive Optics. *IEEE J. Sel. Top. Quantum Electron.* **2007**, *13*, 155–161. [CrossRef]
394. Jian, Y.H.; Yang, K.J.; Wei, T.C.; Wang, C.C.; Chow, C.W. Hardware and Software Improvements of Spatial Light Modulator (SLM)-Based Optical Wireless Communication Supporting Four Users Using Non-Orthogonal Multiple Access (NOMA). *J. Light. Technol.* **2024**, *42*, 5839–5845. [CrossRef]
395. Cox, M.A.; Mphuthi, N.; Nape, I.; Mashaba, N.; Cheng, L.; Forbes, A. Structured Light in Turbulence. *IEEE J. Sel. Top. Quantum Electron.* **2021**, *27*, 1–21. [CrossRef]
396. Pinho, C.; Alimi, I.; Lima, M.; Monteiro, P.; Teixeira, A. Spatial Light Modulation as a Flexible Platform for Optical Systems. In *Telecommunication Systems*; Alimi, I.A., Monteiro, P.P., Teixeira, A.L., Eds.; IntechOpen: Rijeka, Croatia, 2019; Chapter 7. [CrossRef]
397. Quintana, C.; Wang, Q.; Jakonis, D.; Oberg, O.; Erry, G.; Platt, D.; Thueux, Y.; Faulkner, G.; Chun, H.; Gomez, A.; et al. A High Speed Retro-Reflective Free Space Optics Links with UAV. *J. Light. Technol.* **2021**, *39*, 5699–5705. [CrossRef]
398. Li, Y.; Chen, Z.; Hu, Z.; Benton, D.M.; Ali, A.A.I.; Patel, M.; Lavery, M.P.J.; Ellis, A.D. Enhanced Atmospheric Turbulence Resiliency with Successive Interference Cancellation DSP in Mode Division Multiplexing Free-Space Optical Links. *J. Light. Technol.* **2022**, *40*, 7769–7778. [CrossRef]
399. Pinho, C.; Shahpari, A.; Alimi, I.; Lima, M.; Teixeira, A. Optical transforms and CGH for SDM systems. In Proceedings of the 2016 18th International Conference on Transparent Optical Networks (ICTON), Trento, Italy, 10–14 July 2016; pp. 1–4. [CrossRef]
400. Alimi, I.A.; Patel, R.K.; Zaouga, A.; Muga, N.J.; Xin, Q.; Pinto, A.N.; Monteiro, P.P. Trends in Cloud Computing Paradigms: Fundamental Issues, Recent Advances, and Research Directions toward 6G Fog Networks. In *Moving Broadband Mobile Communications Forward*; Haidine, A., Ed.; IntechOpen: Rijeka, Croatia, 2021; Chapter 1. [CrossRef]
401. Mai, V.; Kim, H. Beam Steering and Divergence Control Using Variable Focus Liquid Lenses for WDM FSO Communications. *IEEE Photonics Technol. Lett.* **2022**, *34*, 1226–1229. [CrossRef]
402. Dong, Q.; Liu, Y.; Zhang, Y.; Gao, S.; Chen, T. Improved ADRC with ILC Control of a CCD-Based Tracking Loop for Fast Steering Mirror System. *IEEE Photonics J.* **2018**, *10*, 1–14. [CrossRef]
403. Peng, J.; Qi, B.; Zhang, Y.; Li, Z.; Mao, Y. ESPGD Algorithm to Improve the Convergence Speed for Adaptive Single-Mode Fiber Coupling. *IEEE Photonics J.* **2024**, *16*, 1–7. [CrossRef]
404. Bekkali, A.; Fujita, H.; Hattori, M. New Generation Free-Space Optical Communication Systems with Advanced Optical Beam Stabilizer. *J. Light. Technol.* **2022**, *40*, 1509–1518. [CrossRef]
405. Tsai, J.; Wu, M. Gimbal-less MEMS two-axis optical scanner array with high fill-factor. *J. Microelectromechanical Syst.* **2005**, *14*, 1323–1328. [CrossRef]

406. LoPresti, P.; Refai, H.; Sluss, J.; Varela-Cuadrado, I. Adaptive divergence and power for improving connectivity in free-space optical mobile networks. *Appl. Opt.* **2006**, *45*, 6591–6597. [CrossRef]
407. Arnon, S.; Rotman, S.; Kopeika, N.S. Beam width and transmitter power adaptive to tracking system performance for free-space optical communication. *Appl. Opt.* **1997**, *36*, 6095–6101. [CrossRef] [PubMed]
408. He, J.; Norwood, R.A.; Brandt-Pearce, M.; Djordjevic, I.B.; Cvijetic, M.; Subramaniam, S.; Himmelhuber, R.; Reynolds, C.; Blanche, P.; Lynn, B.; et al. A survey on recent advances in optical communications. *Comput. Electr. Eng.* **2014**, *40*, 216–240. [CrossRef]
409. Killinger, D. Free Space Optics for Laser Communication Through the Air. *Opt. Photon. News* **2002**, *13*, 36–42. [CrossRef]
410. Kaymak, Y.; Rojas-Cessa, R.; Feng, J.; Ansari, N.; Zhou, M. On Divergence-Angle Efficiency of a Laser Beam in Free-Space Optical Communications for High-Speed Trains. *IEEE Trans. Veh. Technol.* **2017**, *66*, 7677–7687. [CrossRef]
411. Tian, J.; Guo, T.; He, N.; Du, J.; Huang, Q.; Hong, X.; Fei, C.; Wang, Y.; Zhang, T.; Zhang, J.; et al. Wide-Field-of-View Modulating Retro-Reflector System Based on a Telecentric Lens for High-Speed Free-Space Optical Communication. *IEEE Photonics J.* **2023**, *15*, 1–8. [CrossRef]
412. He, N.; Guo, T.; Tian, J.; Du, J.; Hu, Z.; Xing, Y.; Jin, Y.; Zhang, J.; He, S. High-Speed Duplex Free Space Optical Communication System Assisted by a Wide-Field-of-View Metalens. *ACS Photonics* **2023**, *10*, 3052–3059. [CrossRef]
413. Burris, H.; Reed, A.; Namazi, N.; Scharpf, W.; Vicheck, M.; Stell, M.; Suite, M. Adaptive thresholding for free-space optical communication receivers with multiplicative noise. In Proceedings of the IEEE Aerospace Conference, Big Sky, MT, USA, 9–16 March 2002; Volume 3, p. 3. [CrossRef]
414. Yang, L.; Zhu, B.; Cheng, J.; Holzman, J.F. Free-Space Optical Communications Using on–off Keying and Source Information Transformation. *J. Light. Technol.* **2016**, *34*, 2601–2609. [CrossRef]
415. Shapiro, J.H.; Harney, R.C. Burst-mode atmospheric optical communication. In Proceedings of the NTC 1980; National Telecommunications Conference, Houston, TX , USA, 30 November–4 December 1980; Volume 2, pp. 27.5.1–27.5.7.
416. Tycz, M.; Fitzmaurice, M.; Premo, D. Optical Communication System Performance with Tracking Error Induced Signal Fading. *IEEE Trans. Commun.* **1973**, *21*, 1069–1072. [CrossRef]
417. Choi, J.Y.; Lee, J.W.; Han, S.K. Stochastic Approximation Aided Adaptive Thresholding for Optical Detection in PAM4 Based FSO Transmission. *IEEE Access* **2021**, *9*, 106451–106458. [CrossRef]
418. Li, L.; Geng, T.; Wang, Y.; Li, X.; Wu, J.; Li, Y.; Ma, S.; Gao, S.; Wu, Z. Free-Space Optical Communication Using Coherent Detection and Double Adaptive Detection Thresholds. *IEEE Photonics J.* **2019**, *11*, 1–17. [CrossRef]
419. Shrivastava, A.K.; Das, D.; Mahapatra, R. Adaptive Threshold Detection and ISI Mitigation in Mobile Molecular Communication. In Proceedings of the 2020 IEEE Wireless Communications and Networking Conference (WCNC), Seoul, Republic of Korea, 25–28 May 2020; pp. 1–6. [CrossRef]
420. Sterckx, K.; Elmirghani, J.; Cryan, R. On the use of adaptive threshold detection in optical wireless communication systems. In Proceedings of the Globecom '00—IEEE. Global Telecommunications Conference. Conference Record (Cat. No.00CH37137), San Francisco, CA, USA, 27 November 2000; Volume 2, pp. 1242–1246. [CrossRef]
421. Yang, L.; Cheng, J.; Holzman, J.F. Electrical-SNR-optimized detection threshold for OOK IM/DD optical wireless communications. In Proceedings of the 2013 13th Canadian Workshop on Information Theory, Toronto, ON, Canada, 18–21 June 2013; pp. 186–189. [CrossRef]
422. Fei, L.; Houbing, L. Optimum detection threshold for free-space optical communication with atmospheric scintillation. In Proceedings of the 2015 International Conference on Optoelectronics and Microelectronics (ICOM), Changchun, China, 16–18 July 2015; pp. 191–196. [CrossRef]
423. Riediger, M.L.; Schober, R.; Lampe, L. Fast multiple-symbol detection for free-space optical communications. *IEEE Trans. Commun.* **2009**, *57*, 1119–1128. [CrossRef]
424. Moradi, H.; Refai, H.H.; LoPresti, P.G. Thresholding-Based Optimal Detection of Wireless Optical Signals. *J. Opt. Commun. Netw.* **2010**, *2*, 689–700. [CrossRef]
425. de Wolf, D.A. *J. Opt. Soc. Am.* **1969**, *59*, 1455–1460. [CrossRef]
426. Zhu, X.; Kahn, J.M. Pilot-symbol assisted modulation for correlated turbulent free-space optical channels. In *Proceedings of the Free-Space Laser Communication and Laser Imaging*; Voelz, D.G., Ricklin, J.C., Eds.; International Society for Optics and Photonics, SPIE: San Diego, CA, USA, 2001; Volume 4489, pp. 138–145. [CrossRef]
427. Khalighi, M.; Xu, F.; Jaafar, Y.; Bourennane, S. Double-Laser Differential Signaling for Reducing the Effect of Background Radiation in Free-Space Optical Systems. *J. Opt. Commun. Netw.* **2011**, *3*, 145–154. [CrossRef]
428. Usman, M.; Yang, H.C.; Alouini, M.S. Practical Switching-Based Hybrid FSO/RF Transmission and Its Performance Analysis. *IEEE Photonics J.* **2014**, *6*, 1–13. [CrossRef]
429. Vangala, S.; Pishro-Nik, H. Optimal Hybrid RF-Wireless Optical Communication for Maximum Efficiency and Reliability. In Proceedings of the 2007 41st Annual Conference on Information Sciences and Systems, Baltimore, MD, USA, 14–16 March 2007; pp. 684–689. [CrossRef]
430. Rakia, T.; Yang, H.C.; Alouini, M.S.; Gebali, F. Outage Analysis of Practical FSO/RF Hybrid System with Adaptive Combining. *IEEE Commun. Lett.* **2015**, *19*, 1366–1369. [CrossRef]
431. He, B.; Schober, R. Bit-interleaved coded modulation for hybrid RF/FSO systems. *IEEE Trans. Commun.* **2009**, *57*, 3753–3763. [CrossRef]

432. Zhang, W.; Hranilovic, S.; Shi, C. Soft-Switching Hybrid FSO/RF Links Using Short-Length Raptor Codes: Design and Implementation. *IEEE J. Sel. Areas Commun.* **2009**, *27*, 1698–1708. [CrossRef]
433. Akbulut, A.; Ilk, H.; Ari, F. Design, availability and reliability analysis on an experimental outdoor FSO/RF communication system. In Proceedings of the 2005 7th International Conference Transparent Optical Networks, Barcelona, Spain, 7 July 2005; Volume 1, pp. 403–406. [CrossRef]
434. Abdulhussein, A.; Oka, A.; Nguyen, T.T.; Lampe, L. Rateless coding for hybrid free-space optical and radio-frequency communication. *IEEE Trans. Wirel. Commun.* **2010**, *9*, 907–913. [CrossRef]
435. Chatzidiamantis, N.D.; Karagiannidis, G.K.; Kriezis, E.E.; Matthaiou, M. Diversity Combining in Hybrid RF/FSO Systems with PSK Modulation. In Proceedings of the 2011 IEEE International Conference on Communications (ICC), Kyoto, Japan, 5–9 June 2011; pp. 1–6. [CrossRef]
436. Lu, M.; Liu, L.; Hranilovic, S. Raptor-coded free-space optical communications experiment. *J. Opt. Commun. Netw.* **2016**, *8*, 398–407. [CrossRef]
437. Nguyen, C.T.; Le, H.D.; Nguyen, C.T.; Pham, A.T. Toward Practical HARQ-Based RC-LDPC Design for Optical Satellite-Assisted Vehicular Networks. *IEEE Trans. Aerosp. Electron. Syst.* **2024**, *60*, 8619–8634. [CrossRef]
438. Kim, I.I.; Korevaar, E.J. Availability of free-space optics (FSO) and hybrid FSO/RF systems. In *Proceedings of the Optical Wireless Communications IV*; Korevaar, E.J., Ed.; International Society for Optics and Photonics, SPIE: Denver, CO, USA, 2001; Volume 4530, pp. 84–95. [CrossRef]
439. Derenick, J.; Thorne, C.; Spletzer, J. On the deployment of a hybrid free-space optic/radio frequency (FSO/RF) mobile ad-hoc network. In Proceedings of the 2005 IEEE/RSJ International Conference on Intelligent Robots and Systems, Edmonton, AB, Canada, 2–6 August 2005; pp. 3990–3996. [CrossRef]
440. Trinh, P.V.; Cong Thang, T.; Pham, A.T. Mixed mmWave RF/FSO Relaying Systems Over Generalized Fading Channels with Pointing Errors. *IEEE Photonics J.* **2017**, *9*, 1–14. [CrossRef]
441. Tachikawa, W.; Suetsuna, A.; Wang, M.; Yoshii, K.; Shimamoto, S. Applying PDMA for Ground-HAPS Uplink Network with Hybrid FSO/RF Communication. In Proceedings of the 2023 IEEE Wireless Communications and Networking Conference (WCNC), Glasgow, United Kingdom, 26–29 March 2023; pp. 1–6. [CrossRef]
442. Nafees, M.; Huang, S.; Thompson, J.; Safari, M. Leveraging Hybrid UAV Relays in Adverse Weather for FSO Link Capacity Maximization. In Proceedings of the 2022 IEEE Wireless Communications and Networking Conference (WCNC), Austin, TX, USA, 10–13 April 2022; pp. 178–183. [CrossRef]
443. Vangala, S.; Pishro-Nik, H. A Highly Reliable FSO/RF Communication System Using Efficient Codes. In Proceedings of the IEEE GLOBECOM 2007—IEEE Global Telecommunications Conference, Washington, DC, USA, 26–30 November 2007; pp. 2232–2236. [CrossRef]
444. Letzepis, N.; Nguyen, K.D.; Guillen i Fabregas, A.; Cowley, W.G. Outage analysis of the hybrid free-space optical and radio-frequency channel. *IEEE J. Sel. Areas Commun.* **2009**, *27*, 1709–1719. [CrossRef]
445. Kamboj, A.; Mallik, R.K.; Agrawal, M.; Schober, R. Diversity combining in FSO systems in presence of non-Gaussian noise. In Proceedings of the 2012 International Conference on Signal Processing and Communications (SPCOM), Bangalore, India, 22–25 July 2012; pp. 1–5. [CrossRef]
446. Pang, X.; Schatz, R.; Joharifar, M.; Udalcovs, A.; Bobrovs, V.; Zhang, L.; Yu, X.; Sun, Y.T.; Maisons, G.; Carras, M.; et al. Direct Modulation and Free-Space Transmissions of up to 6 Gbps Multilevel Signals with a 4.65-μm Quantum Cascade Laser at Room Temperature. *J. Light. Technol.* **2022**, *40*, 2370–2377. [CrossRef]
447. Han, M.; Joharifar, M.; Wang, M.; Fan, Y.; Maisons, G.; Abautret, J.; Sun, Y.T.; Teissier, R.; Zhang, L.; Bobrovs, V.; et al. Long-Wave Infrared Discrete Multitone Free-Space Transmission Using a 9.15-μm Quantum Cascade Laser. *IEEE Photonics Technol. Lett.* **2023**, *35*, 489–492. [CrossRef]
448. Delga, A.; Leviandier, L. Free-space optical communications with quantum cascade lasers. In Proceedings of the *Quantum Sensing and Nano Electronics and Photonics XVI*; Razeghi, M., Lewis, J.S., Tournié, E., Khodaparast, G.A., Eds.; International Society for Optics and Photonics, SPIE: San Francisco, CA, USA, 2019; Volume 10926, p. 1092617. [CrossRef]
449. Didier, P.; Dely, H.; Bonazzi, T.; Spitz, O.; Awwad, E.; Rodriguez, É.; Vasanelli, A.; Sirtori, C.; Grillot, F. High-capacity free-space optical link in the midinfrared thermal atmospheric windows using unipolar quantum devices. *Adv. Photonics* **2022**, *4*, 056004. [CrossRef]
450. Corrigan, P.; Martini, R.; Whittaker, E.A.; Bethea, C. Quantum cascade lasers and the Kruse model in free space optical communication. *Opt. Express* **2009**, *17*, 4355–4359. [CrossRef]
451. Joharifar, M.; Dely, H.; Pang, X.; Schatz, R.; Gacemi, D.; Salgals, T.; Udalcovs, A.; Sun, Y.T.; Fan, Y.; Zhang, L.; et al. High-Speed 9.6-μm Long-Wave Infrared Free-Space Transmission with a Directly-Modulated QCL and a Fully-Passive QCD. *J. Light. Technol.* **2023**, *41*, 1087–1094. [CrossRef]
452. Han, M.; Joharifar, M.; Wang, M.; Schatz, R.; Puerta, R.; Sun, Y.t.; Fan, Y.; Maisons, G.; Abautret, J.; Teissier, R.; et al. High Spectral Efficiency Long-Wave Infrared Free-Space Optical Transmission with Multilevel Signals. *J. Light. Technol.* **2023**, *41*, 6514–6520. [CrossRef]
453. Trinh, P.V.; Pham, A.T.; Carrasco-Casado, A.; Toyoshima, M. Quantum Key Distribution over FSO: Current Development and Future Perspectives. In Proceedings of the 2018 Progress in Electromagnetics Research Symposium (PIERS-Toyama), Toyama, Japan, 1–4 August 2018; pp. 1672–1679. [CrossRef]

454. Roediger, J.; Perlot, N.; Freund, R. Quantum cryptography over the FSO channel with PPM and FSK modulations. In Proceedings of the Broadband Coverage in Germany. 9th ITG Symposium, Berlin, Germany, 20–21 April 2015; pp. 1–5.
455. Alshaer, N.; Moawad, A.; Ismail, T. Reliability and Security Analysis of an Entanglement-Based QKD Protocol in a Dynamic Ground-to-UAV FSO Communications System. *IEEE Access* **2021**, *9*, 168052–168067. [CrossRef]
456. Alshaer, N.; Ismail, T. Performance Evaluation and Security Analysis of UAV-Based FSO/CV-QKD System Employing DP-QPSK/CD. *IEEE Photonics J.* **2022**, *14*, 1–11. [CrossRef]
457. Sinha, S.; Kumar, C.; Armghan, A.; Singh, M.; Alsharari, M.; Aliqab, K. Simulative Investigation of the Hybrid, Spatially Multiplexed MIMO-FSO Transmission System Under Atmospheric Turbulence. *IEEE Access* **2023**, *11*, 55071–55080. [CrossRef]
458. Li, J.; Zhang, M.; Wang, D. Adaptive Demodulator Using Machine Learning for Orbital Angular Momentum Shift Keying. *IEEE Photonics Technol. Lett.* **2017**, *29*, 1455–1458. [CrossRef]
459. Allen, L.; Beijersbergen, M.W.; Spreeuw, R.J.C.; Woerdman, J.P. Orbital angular momentum of light and the transformation of Laguerre-Gaussian laser modes. *Phys. Rev. A* **1992**, *45*, 8185–8189. [CrossRef]
460. Wang, J.; Yang, J.Y.; Fazal, I.; Ahmed, N.; Yan, Y.; Huang, H.; Ren, Y.; Yue, Y.; Dolinar, S.; Tur, M.; et al. Terabit free-space data transmission employing orbital angular momentum multiplexing. *Nat. Photonics* **2012**, *6*, 488 – 496. [CrossRef]
461. Ren, Y.; Wang, Z.; Liao, P.; Li, L.; Xie, G.; Huang, H.; Zhao, Z.; Yan, Y.; Ahmed, N.; Willner, A.; et al. Experimental characterization of a 400 Gbit/s orbital angular momentum multiplexed free-space optical link over 120 m. *Opt. Lett.* **2016**, *41*, 622–625. [CrossRef]
462. Lei, T.; Zhang, M.; Li, Y.; Jia, P.; Liu, G.N.; Xu, X.; Li, Z.; Min, C.; Lin, J.; Chen, Y.; et al. Massive individual orbital angular momentum channels for multiplexing enabled by dammann gratings. *Light Sci. Appl.* **2015**, *4*, e257–e257. [CrossRef]
463. Trichili, A.; Salem, A.B.; Dudley, A.; Zghal, M.; Forbes, A. Encoding information using Laguerre Gaussian modes over free space turbulence media. *Opt. Lett.* **2016**, *41*, 3086–3089. [CrossRef]
464. Zhu, L.; Liu, J.; Mo, Q.; Du, C.; Wang, J. Encoding/decoding using superpositions of spatial modes for image transfer in km-scale few-mode fiber. *Opt. Express* **2016**, *24*, 16934–16944. [CrossRef]
465. Doster, T.; Watnik, A.T. Machine learning approach to OAM beam demultiplexing via convolutional neural networks. *Appl. Opt.* **2017**, *56*, 3386–3396. [CrossRef]
466. Gao, Z.; Eisen, M.; Ribeiro, A. Resource Allocation via Model-Free Deep Learning in Free Space Optical Communications. *IEEE Trans. Commun.* **2022**, *70*, 920–934. [CrossRef]
467. Li, Z.; Cao, J.; Zhao, X.; Liu, W. Swarm intelligence for atmospheric compensation in free space optical communication—Modified shuffled frog leaping algorithm. *Opt. Laser Technol.* **2015**, *66*, 89–97. [CrossRef]
468. Babatunde, K.S.; Ibikunle, F.A.; Arowolo, M.O.; Alabi, A.J.; Jiya, E.A.; Kehinde, O.P. Machine Learning Model for Classifying Free Space Optics Channel Impairments. In Proceedings of the 2022 5th Information Technology for Education and Development (ITED), Abuja, Nigeria, 1–3 November 2022; pp. 1–8. [CrossRef]
469. Li, J.; Zhang, M.; Wang, D.; Wu, S.; Zhan, Y. Joint atmospheric turbulence detection and adaptive demodulation technique using the CNN for the OAM-FSO communication. *Opt. Express* **2018**, *26*, 10494–10508. [CrossRef] [PubMed]
470. Maharjan, N.; Kim, B.W. Reduction of Beam Pointing Error on Free Space Optical Communication links using Tree Based Machine Learning Multi-Output Regression Model. In Proceedings of the 2023 Fourteenth International Conference on Ubiquitous and Future Networks (ICUFN), Paris, France, 4–7 July 2023; pp. 664–668. [CrossRef]
471. Zheng, C.; Yu, S.; Gu, W. A SVM-based processor for free-space optical communication. In Proceedings of the 2015 IEEE 5th International Conference on Electronics Information and Emergency Communication, Beijing, China, 14–16 May 2015; pp. 30–33. [CrossRef]
472. Rahman, K.F.; Afrin, F.; Taip, F.S.b. Optical Network Performance Analysis through Prediction of Q Factor on Simulated and Validated Dataset. In Proceedings of the 2018 IEEE International Conference on Artificial Intelligence in Engineering and Technology (IICAIET), Kota Kinabalu, Malaysia, 8 November 2018; pp. 1–4. [CrossRef]
473. Li, Z.; Zhao, X. BP artificial neural network based wave front correction for sensor-less free space optics communication. *Opt. Commun.* **2017**, *385*, 219–228. [CrossRef]
474. Li, Y.; Geng, T.; Tian, R.; Gao, S. Power allocation in a spatial multiplexing free-space optical system with reinforcement learning. *Opt. Commun.* **2021**, *488*, 126856. [CrossRef]
475. Amirabadi, M.A.; Kahaei, M.H.; Nezamalhosseini, S.A.; Vakili, V.T. Deep Learning for channel estimation in FSO communication system. *Opt. Commun.* **2020**, *459*, 124989. [CrossRef]
476. Lohani, S.; Glasser, R.T. Turbulence correction with artificial neural networks. *Opt. Lett.* **2018**, *43*, 2611–2614. [CrossRef]
477. Tian, Q.; Li, Z.; Hu, K.; Zhu, L.; Pan, X.; Zhang, Q.; Wang, Y.; Tian, F.; Yin, X.; Xin, X. Turbo-coded 16-ary OAM shift keying FSO communication system combining the CNN-based adaptive demodulator. *Opt. Express* **2018**, *26*, 27849–27864. [CrossRef]
478. Li, G.; Guo, S.; Yang, Y.; Yang, Y. Traffic Load Minimization in Software Defined Wireless Sensor Networks. *IEEE Internet Things J.* **2018**, *5*, 1370–1378. [CrossRef]
479. Gao, Z.; Eisen, M.; Ribeiro, A. Resource Allocation via Graph Neural Networks in Free Space Optical Fronthaul Networks. In Proceedings of the GLOBECOM 2020 - 2020 IEEE Global Communications Conference, Taipei, Taiwan, 7–11 December 2020; pp. 1–6. [CrossRef]
480. Dehnaw, A.M.; Manie, Y.C.; Du, L.Y.; Yao, C.K.; Li, Y.L.; Hayle, S.T.; Peng, P.C. Bidirectional Free Space Optics Communication for Long-Distance Sensor System. *J. Light. Technol.* **2023**, *41*, 5870–5878. [CrossRef]

481. Song, S.; Liu, Y.; Xu, T.; Liao, S.; Guo, L. Demonstration of Channel-Predictable Free Space Optical Communication System Using Machine Learning. In Proceedings of the 2021 Optical Fiber Communications Conference and Exhibition (OFC), San Francisco, CA, USA, 6–10 June 2021; pp. 1–3.
482. Amirabadi, M.A.; Kahaei, M.H.; Nezamalhosseini, S.A. A Deep Learning based Detector for FSO System Considering Imperfect CSI Scenario. In Proceedings of the 2020 3rd West Asian Symposium on Optical and Millimeter-wave Wireless Communication (WASOWC), Tehran, Iran, 24–25 November 2020; pp. 1–5. [CrossRef]
483. Wang, D.; Zhang, M.; Li, Z.; Li, J.; Fu, M.; Cui, Y.; Chen, X. Modulation Format Recognition and OSNR Estimation Using CNN-Based Deep Learning. *IEEE Photonics Technol. Lett.* **2017**, *29*, 1667–1670. [CrossRef]
484. Khan, F.N.; Zhong, K.; Zhou, X.; Al-Arashi, W.H.; Yu, C.; Lu, C.; Lau, A.P.T. Joint OSNR monitoring and modulation format identification in digital coherent receivers using deep neural networks. *Opt. Express* **2017**, *25*, 17767–17776. [CrossRef] [PubMed]
485. Giacoumidis, E.; Lin, Y.; Wei, J.; Aldaya, I.; Tsokanos, A.; Barry, L.P. Harnessing machine learning for fiber-induced nonlinearity mitigation in long-haul coherent optical OFDM. *Future Internet* **2019**, *11*, 2. [CrossRef]
486. Huang, Y.; Gutterman, C.L.; Samadi, P.; Cho, P.B.; Samoud, W.; Ware, C.; Lourdiane, M.; Zussman, G.; Bergman, K. Dynamic mitigation of EDFA power excursions with machine learning. *Opt. Express* **2017**, *25*, 2245–2258. [CrossRef]
487. Thrane, J.; Wass, J.; Piels, M.; Diniz, J.C.M.; Jones, R.; Zibar, D. Machine Learning Techniques for Optical Performance Monitoring From Directly Detected PDM-QAM Signals. *J. Light. Technol.* **2017**, *35*, 868–875. [CrossRef]
488. Rottondi, C.; Barletta, L.; Giusti, A.; Tornatore, M. Machine-learning method for quality of transmission prediction of unestablished lightpaths. *J. Opt. Commun. Netw.* **2018**, *10*, A286–A297. [CrossRef]
489. Zhong, Z.; Hua, N.; Yuan, Z.; Li, Y.; Zheng, X. Routing without Routing Algorithms: An AI-Based Routing Paradigm for Multi-Domain Optical Networks. In Proceedings of the Optical Fiber Communication Conference (OFC), San Francisco, CA, USA, 3–7 March 2019; p. Th2A.24. [CrossRef]
490. Darwesh, L.; Kopeika, N.S. Deep Learning for Improving Performance of OOK Modulation Over FSO Turbulent Channels. *IEEE Access* **2020**, *8*, 155275–155284. [CrossRef]
491. Prasad, P.; K, A.M.; Jain, A. An Investigation on the Error Performance of NOMA-aided Hybrid Satellite-Terrestrial FSO System. In Proceedings of the 2023 International Conference on Communication, Circuits, and Systems (IC3S), Bhubaneswar, India, 26–28 May 2023; pp. 1–5. [CrossRef]
492. Lv, L.; Yang, Z.; Fang, Y.; Guizani, M. Adaptive Interleaver and Rate-Compatible PLDPC Code Design for MIMO FSO-RF Systems. *IEEE Trans. Veh. Technol.* **2024**, *73*, 15757–15762. [CrossRef]
493. Chiti, F.; Picchi, R.; Pierucci, L. Metropolitan Quantum-Drone Networking and Computing: A Software-Defined Perspective. *IEEE Access* **2022**, *10*, 126062–126073. [CrossRef]
494. Nguyen, H.; Krishna Yoga Utama, I.B.; Jang, Y.M. Enabling Technologies and New Challenges in IEEE 802.15.7 Optical Camera Communications Standard. *IEEE Commun. Mag.* **2024**, *62*, 90–95. [CrossRef]
495. *IEEE Std 802.15.7-2018 (Revision of IEEE Std 802.15.7-2011)*; IEEE Standard for Local and Metropolitan Area Networks–Part 15.7: Short-Range Optical Wireless Communications. IEEE: New York, NY, USA; 2019; pp. 1–407. [CrossRef]
496. *IEEE 802.15.7-2018*; 802.15.7-2018 IEEE LANMAN—Part 15.7: Short-Range Optical Wireless Communications (PODRL). IEEE: New York, NY, USA; 2018.
497. Thieu, M.D.; Pham, T.L.; Nguyen, T.; Jang, Y.M. Optical-RoI-Signaling for Vehicular Communications. *IEEE Access* **2019**, *7*, 69873–69891. [CrossRef]
498. *Co-Location Longitudinally Compatible Interfaces for Free Space Optical Systems*; Recommendation itu-t-g.640; ITU-T; ITU: Geneva, Switzerland; 2006.
499. *Fixed Service Applications Using Free-Space Optical Links*; Report itu-r f.2106-1; ITU-R; ITU: Geneva, Switzerland; 2010.
500. Kouhini, S.M.; Kottke, C.; Ma, Z.; Freund, R.; Jungnickel, V.; Müller, M.; Behnke, D.; Vazquez, M.M.; Linnartz, J.P.M.G. LiFi Positioning for Industry 4.0. *IEEE J. Sel. Top. Quantum Electron.* **2021**, *27*, 1–15. [CrossRef]
501. Kaufmann, M.; Mas-Machuca, C.; Müller, M.; Behnke, D.; Stobbelaar, P.; Linnartz, J.P.; Riegel, M.; Schulz, D.; Jungnickel, V. Techno-Economics of LiFi in IoT Applications. In Proceedings of the 2022 International Conference on Optical Network Design and Modeling (ONDM), Warsaw, Poland, 16–19 May 2022; pp. 1–6. [CrossRef]
502. Hamza, A.S.; Deogun, J.S.; Alexander, D.R. Classification Framework for Free Space Optical Communication Links and Systems. *IEEE Commun. Surv. Tutor.* **2019**, *21*, 1346–1382. [CrossRef]
503. *Propagation Data Required for the Design of Terrestrial Free-Space Optical Links* ; Recommendation itu-r p.1817-1; ITU-R; ITU: Geneva, Switzerland, 2012.
504. Nebuloni, R.; Verdugo, E. FSO Path Loss Model Based on the Visibility. *IEEE Photonics J.* **2022**, *14*, 1–9. [CrossRef]
505. *Prediction Methods Required for the Design of Terrestrial Free-Space Optical Links*; Recommendation itu-r p.1814; ITU-R; ITU: Geneva, Switzerland, 2007.
506. *IEC 60825-1:2014*; International Standard, Edition 3.0; Safety of Laser Products—Part 1: Equipment Classification and Requirements. International Electrotechnical Commission: Geneva, Switzerland, 2014.
507. Al-Gailani, S.A.; Mohd Salleh, M.F.; Salem, A.A.; Shaddad, R.Q.; Sheikh, U.U.; Algeelani, N.A.; Almohamad, T.A. A Survey of Free Space Optics (FSO) Communication Systems, Links, and Networks. *IEEE Access* **2021**, *9*, 7353–7373. [CrossRef]
508. Chiti, F.; Picchi, R.; Pierucci, L. Advanced Control Architectures for Quantum Satellite Temporal-Networking. *IEEE Access* **2024**, *12*, 43410–43421. [CrossRef]

509. Al-Hraishawi, H.; Rehman, J.U.; Razavi, M.; Chatzinotas, S. Characterizing and Utilizing the Interplay Between Quantum Technologies and Non-Terrestrial Networks. *IEEE Open J. Commun. Soc.* **2024**, *5*, 1937–1957. [CrossRef]
510. Alimi, I.A.; Patel, R.K.; Aboderin, O.; Abdalla, A.M.; Gbadamosi, R.A.; Muga, N.J.; Pinto, A.N.; Teixeira, A.L. Network-on-Chip Topologies: Potentials, Technical Challenges, Recent Advances and Research Direction. In *Network-on-Chip*; Alimi, I.A., Aboderin, O., Muga, N.J., Teixeira, A.L., Eds.; IntechOpen: Rijeka, Croatia, 2021; Chapter 3. [CrossRef]
511. Alimi, I.A.; Abdalla, A.M.; Olapade Mufutau, A.; Pereira Guiomar, F.; Otung, I.; Rodriguez, J.; Pereira Monteiro, P.; Teixeira, A.L., Energy Efficiency in the Cloud Radio Access Network (C-RAN) for 5G Mobile Networks. In *Optical and Wireless Convergence for 5G Networks*; John Wiley & Sons, Ltd.: Hoboken, NJ, USA, 2019; Chapter 11, pp. 225–248. [CrossRef]
512. Alimi, I.A.; Patel, R.K.; Mufutau, A.O.; Muga, N.J.; Pinto, A.N.; Monteiro, P. Towards a Sustainable Green Design for Next-Generation Networks. *Wirel. Pers. Commun.* **2021**, *121*, 1123–1138. [CrossRef]

sensors

MDPI

Review

A Review of Hybrid VLC/RF Networks: Features, Applications, and Future Directions

Lisandra Bravo Alvarez [1], Samuel Montejo-Sánchez [2,*], Lien Rodríguez-López [3], Cesar Azurdia-Meza [4] and Gabriel Saavedra [1,5]

1 Department of Electrical Engineering, Universidad de Concepción, Edmundo Larenas 219, Concepción 4030000, Chile; lisanbravo@udec.cl (L.B.A.); gasaavedra@udec.cl (G.S.)
2 Programa Institucional de Fomento a la Investigación, Desarrollo e Innovación, Universidad Tecnológica Metropolitana, Ignacio Valdivieso 2409, Santiago 8940000, Chile
3 Facultad de Ingeniería, Arquitectura y Diseño, Universidad San Sebastián, Lientur 1457, Concepción 4030000, Chile; lien.rodriguez@uss.cl
4 Department of Electrical Engineering, Universidad de Chile, Santiago 8370451, Chile; cazurdia@ing.uchile.cl
5 Millennium Institute for Research in Optics, Universidad de Concepción, Concepción 160-C, Chile
* Correspondence: smontejo@utem.cl

Abstract: The expectation for communication systems beyond 5G/6G is to provide high reliability, high throughput, low latency, and high energy efficiency services. The integration between systems based on radio frequency (RF) and visible light communication (VLC) promises the design of hybrid systems capable of addressing and largely satisfying these requirements. Hybrid network design enables complementary cooperation without interference between the two technologies, thereby increasing the overall system data rate, improving load balancing, and reducing non-coverage areas. VLC/RF hybrid networks can offer reliable and efficient communication solutions for Internet of Things (IoT) applications such as smart lighting, location-based services, home automation, smart healthcare, and industrial IoT. Therefore, hybrid VLC/RF networks are key technologies for next-generation communication systems. In this paper, a comprehensive state-of-the-art study of hybrid VLC/RF networks is carried out, divided into four areas. First, indoor scenarios are studied considering lighting requirements, hybrid channel models, load balancing, resource allocation, and hybrid network topologies. Second, the characteristics and implementation of these hybrid networks in outdoor scenarios with adverse conditions are analyzed. Third, we address the main applications of hybrid VLC/RF networks in technological, economic, and socio-environmental domains. Finally, we outline the main challenges and future research lines of hybrid VLC/RF networks.

Keywords: hybrid networks; optical camera communication (OCC); radio frequency (RF); visible light communication (VLC); internet of things (IoT)

1. Introduction

In recent years, the limitations of the radio frequency (RF) spectrum for mobile communications have become very evident [1]. This is a great challenge to overcome in light of the traffic load demands associated with fifth-generation (5G) mobile communication, with alarming projections beyond 5G and into the sixth generation (6G). Due to these limitations, the dense deployment of RF access points leads to high competition for available channels [2], which entails degradation of the quality of service (QoS).

As an alternative solution, visible light communication (VLC) has been proposed, an optical wireless communication technology that uses the visible light spectrum with wavelengths between 375 and 780 nm [3]. This technology uses light-emitting diodes (LEDs) that produce incoherent light to illuminate and transmit data simultaneously, achieving high data rates of hundreds of megabits per second [4]. With the goal of secure and efficient communications, the VLC Consortium (VLCC) was established to promote and standardize

VLC technology in 2007 [5]. VLC is characterized by high security, as information cannot be filtered [6]. In addition, it is immune to RF interference, meaning that the system can be used freely in environments sensitive to electromagnetic signals [7]. VLC has experienced rapid development and has attracted much interest from researchers [8]. In 2011, the Visible Light Communications Task Group created the IEEE 802.15.7 standard [9] that establishes three physical (PHY) VLC operation modes, as shown in Table 1. The three PHY modes can coexist with each other, thereby mitigating LED flicker and reducing dimming [10]. In 2019, the ITU-T Telecommunication Standardization Sector finalized recommendation G.9991 that established the first commercially ready light fidelity (Li-Fi) standard [11]. Both IEEE 802.15.7 and G.9991 have now converged into IEEE 802.11bb to refine the medium access control (MAC) and physical layer (PHY) of VLC [12]. One of the essential features of the IEEE 802.11bb standard [13] is that there is a single medium access control (MAC) sublayer common to all physical layers. This feature enables easier interoperability between the different physical layers, making cooperation between RF and visible light communications technologies possible [2].

Table 1. IEEE 802.15.7 (PHY) parameters.

	Data Rate	Light Source	Modulation
PHY I	11.67–266.6 [Kbps]	only one	OOK [1], VPPM [2]
PHY II	1.25–96 [Mbps]	only one	OOK, VPPM
PHY III	12–96 [Mbps]	multiple	CSK [3]

[1] OOK: On–off keying. [2] VPPM: Variable pulse position modulation. [3] CSK: Color Shift Keying.

On the other hand, the implementation of the B5G cellular networks has brought with it an increase in mobile traffic with applications such as virtual reality, augmented reality, and online video games [14]. Ultra-high capacity wireless connectivity is considered a key technology to support end-to-end delivery in the era of beyond 5G and 6G [15,16]. Therefore, there is a need for a network infrastructure capable of supporting multiple new services that demand ultra-high transmission speeds, device connectivity, and high quality of experience (QoE) [17].

In the last few years, researchers have shown great interest in hybrid VLC/RF systems, and this type of network is expected to be the basis for the development of B5G and the next generations of cellular networks [18–20]. Hybrid VLC/RF networks are presented as a complementary technology capable of realizing cooperation between RF and VLC networks without interfering with each other [18]. They are able to significantly increase the system data rate and achieve better load balancing by distributing the data load between the two networks [19]. Additionally, they have a very low implementation cost due to the extensive use of LEDs in lighting infrastructure around the globe that can be reused for mobile communication [20].

In this sense, a complete search was carried out in the Scopus database [21] using the keywords of Visible Light Communication/Ratio Frequency. A total of 727 English-language papers were collected from the years 2012–2022. A bibliometric analysis was performed using the VOSviewer software tool [22] (see Figure 1). The analyzed items were the published documents, keywords, country distribution, and affiliation.

Figure 1 shows a clear increase in the number of research papers published in the last decade in the field of hybrid VLC/RF networks, which demonstrates the growing scientific interest in this area of research. The countries appearing in the search according to the affiliation of the main authors were China (257), the United Kingdom (99), India (92), the USA (54), and Turkey (44). The universities with more than five publications related to the VLC/RF field were Ozyegin University, Istanbul; Southeast University, China; and Northumbria University, United Kingdom. This analysis shows that research related to this area is greater in the northern hemisphere than in the southern hemisphere, Chile being one of the countries with the lowest number of publications in this area (2).

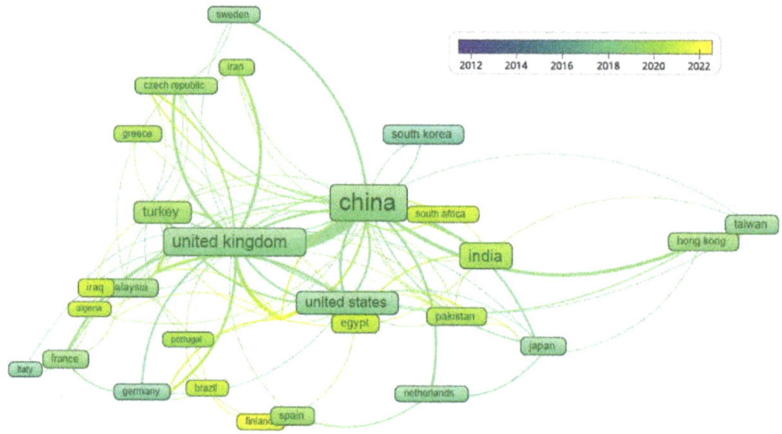

Figure 1. Bibliometric analysis of the VLC/RF networks studies by countries.

The emergent Internet of Things (IoT) has boosted the need for innovative communication solutions that provide reliable and efficient data transmission methods. An intelligent, efficient, and reliable system can be obtained by combining VLC and RF technologies [23]. This integrated system can support various IoT applications, such as smart lighting, home automation, e-healthcare, and industrial IoT. For example, in smart lighting the VLC network can be used to provide high-speed communication through luminaries while the RF network can be used to control luminaries from a remote location [24]. Similarly, in e-healthcare applications the VLC network can provide reliable communications between medical devices, while the RF network can transmit data to healthcare providers or emergency services. Therefore, the implementation of hybrid VLC/RF networks offers a reliable and efficient solution for IoT applications, especially indoors [25].

Due to the great interest shown in hybrid VLC/RF networks, several authors have conducted reviews of the history of these networks [2,17,26–29]. However, the dynamic and changing characteristics depending on the deployment scenario and the extensive evolution of VLC/RF networks in recent times have produced the need to delve deeper into issues that have not been fully analyzed by the existing reviews. Table 2 compares this work with the most prominent reviews in the last ten years relating to hybrid VLC/RF networks. We used the markers ⋆ and ★ to represent partially and deeply analyzed topics, respectively. In our criterion, we mark as partially analyzed those topics that were only commented on in a general way without going deeper into characteristics, problems, and possible solutions. On the other hand, the papers marked as deeply analyzed are those that we consider to address each of the topics mentioned in detail, providing characteristics of the technologies, pending challenges, an analysis of what has been achieved by other authors, and possible solutions that can be carried out to optimize VLC/RF networks.

Therefore, in this work an extensive study of the state of the art of hybrid VLC/RF networks in ideal indoor scenarios in the scientific literature is carried out and the characteristics and implementation of these networks in outdoor scenarios with harsh conditions are analyzed. The main applications of hybrid VLC/RF networks in the technological, economic, and social-environmental fields are considered. In addition, the main challenges faced by hybrid VLC/RF networks and future research directions are discussed.

Table 2. Comparison between surveys of hybrid VLC/RF networks; ⋆ means partially analyzed and ★ means deeply analyzed.

Ref.	Year	Lighting Requirements	Load Balancing	Network Selection	Hybrid Channel	Hybrid Topology	Outdoor	Applications
[27]	2016	⋆						
[28]	2018							⋆
[29]	2018	⋆	⋆					
[30]	2019							⋆
[17]	2020		⋆					★
[2]	2021	★	⋆					⋆
[26]	2021			⋆		⋆		⋆
[31]	2022	⋆	⋆					
This paper		★	★	★	★	★	★	★

The rest of this paper is organized as follows. In Section 2, the state of the art of hybrid VLC/RF networks in indoor scenarios with ideal conditions is reviewed, including analysis of aspects such as lighting requirements, hybrid channel model, load balancing, network selection, and resource allocation, and hybrid network topologies. In Section 3, a review of the state of the art of hybrid VLC/RF networks in outdoor scenarios with harsh conditions is presented. Section 4 analyzes the possible applications of hybrid VLC/RF networks in the technological, economic, and social-environmental fields. Section 5 establishes the main challenges and future directions of these hybrid networks. Finally, the conclusions of this review are presented in Section 6.

2. Hybrid VLC/RF Networks in Indoor Scenarios under Ideal Conditions

In this section, a review of the state-of-the-art of hybrid VLC/RF networks in indoor scenarios with ideal conditions is conducted. A hybrid VLC/RF network in an indoor scenario can be composed of LED arrays located on the ceiling or walls of a room that provides the VLC connection and one or several radio frequency access points (APs) that provide the RF connection. Figure 2 shows an example of a hybrid VLC/RF network in an indoor scenario. Here, users inside the room can simultaneously connect to both networks using the VLC connection through the existing lighting and wireless access to the AP for the RF connection. Being an enclosed room, the illumination is provided by LED lights only, and there is no interference from outside lighting or ambient noise to affect the connection.

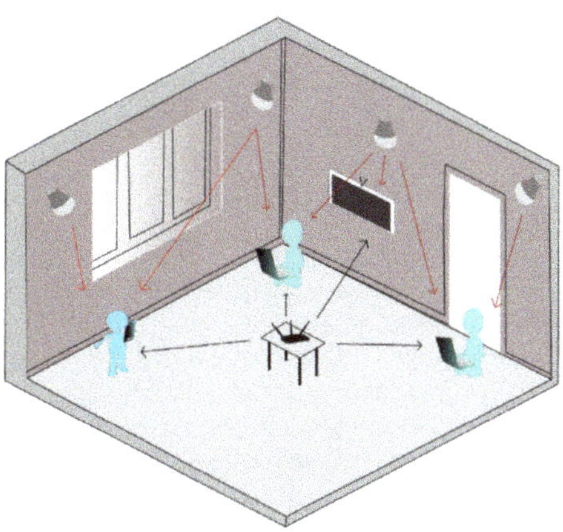

Figure 2. Hybrid VLC/RF network in an indoor scenario.

2.1. Lighting Requirements

Through proper room lighting design, the energy efficiency of the VLC/RF hybrid network can be improved. The goal is to achieve the desired lighting pattern with minimum energy consumption. This requires optimizing the placement and control of LED output power levels. Lighting quality is essential in these hybrid networks because all information and electronic signals are transmitted through light [32]. A bad lighting design can affect the quality of the signal; therefore, it is necessary to take into account different aspects such as the distance between the devices, the intensity of illumination, environmental interference, and power management. The illumination level stipulated by the International Organization for Standardization (ISO) for an indoor environments is 300 to 1500 lx. Generally, the greatest illumination is in the center of the room rather than in the corners, so careful design is required to provide uniform illumination throughout the room. In [2], a lighting scheme with the following requirements is established: $ilum(x,y) = \sum_{i=1}^{n_{LEDs}} \frac{I_0}{d_i^2} cos^l(\theta_i) cos(\phi_i)$, where I_0 is the illumination intensity of the LEDs, and d_i is the Euclidean distance between the ith LED and the arbitrary location (x,y) of the user. The number of LEDs is denoted by n_{LEDs}, and l is the Lambertian emission order. The radiation and angle incidence at the (x,y) location of the user is denoted by θ_i y ϕ_i, respectively. Figure 3 shows an example of lighting inside a $5\,m^2$ room illuminated by 4 LEDs placed uniformly on the ceiling. As the intensity of light increases, so does the data rate. However, the relationship between illumination and throughput is not always linear, as interference between different lighting devices and physical obstacles can affect link quality and reduce throughput. To reduce interference between devices, it is necessary that lighting and data transmission share the same frequency spectrum in the VLC network.

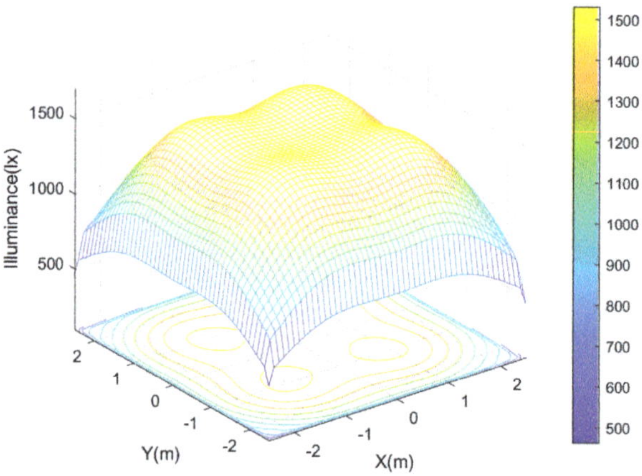

Figure 3. Illuminance level of a $5\,m^2$ room with 4 LED lights.

The versatility of the lighting system lies in its ability to serve as an ideal vehicle for VLC technology, paving the way for the development of various Internet of Things (IoT) applications. By blending the potential of smart sensors with lighting, it becomes possible to create cutting-edge IoT solutions that are both unique and resourceful [33]. Notably, the lighting system can revolutionize several urban applications, from creating smart homes to building energy-efficient offices and even revolutionizing urban infrastructure through smart streetlights. This combination of light and technology opens up endless possibilities in the IoT realm, enabling the development of novel innovative solutions that cater to myriad industrial and societal needs [34].

In [35], an algorithm based on a fast game theory for optimizing energy efficiency under illumination constraints was presented. The authors formulated a generalized Nash equilibrium problem in which each user maximizes its own VLC energy efficiency while subject to the shared illumination constraint. In [36], an electrical and optical power allocation scheme was proposed taking into account the lighting constraints in order to maximize the multi-user sum rate (throughput). It was shown that with increasing luminous flux the maximum summation rate exhibits a downward parabolic trend due to the limited dynamic range of LEDs. Alternatively, a higher data rate can be achieved with a higher correlated color temperature. In [37], a simple approach to the illumination design problem based on the virtual LED concept was presented. The authors formulated an optimization problem with constraints to minimize energy consumption while maintaining near-uniform illumination throughout the room. In addition, they derived the number of LEDs and the luminous intensities of the LEDs required to achieve the desired lighting constraint.

2.2. Hybrid Channel Model

In an indoor scenario, the impulse response of a VLC channel is modeled through a Geometric Based Model, which can be approximated for simplicity to a Lambertian model, which depends on half the beamwidth angle of the transmitting LED [38]. The total channel impulse response is provided by the sum of the line-of-sight (LOS) impulse response and the non-line-of-sight (NLOS) impulse response [39] using recursive methods. When the LEDs are very close to the receiver, a LOS link is achieved, promoting a channel model without temporal dispersion. When there are multiple signal reflections due to walls or obstacles within the room or the distance between the emitter and receiver is very large, NLOS links are created. To calculate the impulse response of these links, it is necessary to take into account the signal tails caused by time dispersion and multiple paths. This is because the information transmitted by the LED may reflect off stage walls or other blocking object before reaching the receiver [40].

Because LEDs produce light with non-coherent characteristics and have a limited dynamic range, the most conducive modulation scheme for VLC is intensity modulation (IM)/direct detection (DD) [41]. When an optical OFDM signal is used in an IM/DD system, intensity modulation of the carrier signal is performed. Therefore, the transmitted signal must be real and of positive value in the time domain. Consequently, it is necessary to use real transformations such as the discrete cosine transform (DCT) or the fast Walsh Hadamard transform (FWHT), among others, to obtain a real signal. If the optical OFDM signal cannot be matched to the linear region of the LED voltage–current relationship, clipping of the signal occurs, causing nonlinear distortions. To overcome the distortions induced by this nonlinearity, peak-to-average power ratio (PAPR) reduction techniques or predistortion techniques must be used [42].

An electromagnetic wave incident on an object can be reflected, refracted, scattered, diffracted, or absorbed by it. This depends on the frequency of the incident wave and the physicochemical properties of the object. This interaction determines the level of opacity or transparency of the object at different wavelengths. When white light falls on a surface, certain wavelengths are absorbed while others are reflected, determining the object's color. Opaque objects such as wood, metal, and plastic restrict the propagation of visible light [43]; therefore, when designing a VLC communication channel the behavior of visible light wavelengths interacting with such materials strongly influences the decision-making process. Conversely, an object can be transparent at certain wavelengths, allowing light to pass through it easily.

On the other hand, the RF communication channel model consists of the path loss model and the spatial channel model, usually composed of the lumped channel impulse responses and angular dispersion statistics [44]. In an indoor scenario, the received signal arrives through multiple simultaneous paths that are vectorially combined to obtain a resulting oscillatory signal. This multi-path can be caused by phenomena such as reflection, refraction, diffraction, or scattering. The most common way to include these phenomena is by means of the channel impulse response model, which is presented in a temporal way, as it can have discrete values or be a stochastic variable [45]. Due to the symmetry properties of the RF channel, the RF channel model can be Hermitian. The ratio between the imaginary part and the real part of the channel impulse response defines the statistical distribution in which the signal fades follow at each coefficient, that is, each delay can be considered a different variable, typically Rayleigh, Rice, or simply Gaussian depending on whether or not there is a predominant component [46]. The possibility of Doppler effects due to mobility between transmitter and receiver must be taken into account. In addition, if an orthogonal frequency division multiplexing with a multiple inputs–multiple outputs (OFDM-MIMO) system is used, spatial diversity must be taken into account, making it necessary to introduce parameters at one end related to the correlation between the different antennas [47].

Table 3 shows a comparison of the communication channel characteristics of VLC and RF networks. Although these two technologies have very different communication channel models, it is possible to design a hybrid model because there is no interference between the two networks. In [48], a robust channel model of a hybrid VLC/RF system was developed for optimal channel selection. Figure 4 shows a hybrid model of a VLC/RF communication channel composed of mobile terminals at each end equipped with hybrid VLC/RF frequency bands. The RF channel has a Hermitian model with WLAN frequencies (2.4 GHz/5 GHz), while the VLC channel approximates a Lambertian model with frequency bands on the order of THz. The wireless communication channels are shared simultaneously between the two technologies to transmit (T) and receive (R) information. The authors of [49] described a hybrid model of an imperfect VLC/RF channel, which they used to analyze the energy efficiency of these hybrid systems. Among the imperfections of the hybrid channel, they took into account the uncertainty caused by noise and quantization errors in the digital–analog process.

Table 3. Comparison of VLC and RF channel models.

Parameters	VLC Channel Model	RF Channel Model
Model	Lambertian	Hermitian
Propagation mode	Visible light	Electromagnetic wave
Modulation technique	IM/DD, NOMA [1]	OFDM, OFDM-MIMO, NOMA
Coverage	Minor	Major
Complexity	Simple	complex
Symmetry	No	Yes
Links	Download	Download/Upload
Doppler effect	No	Yes
Interference inter symbols	Minor	Major

[1] NOMA: Non-Orthogonal-Multiple-Access.

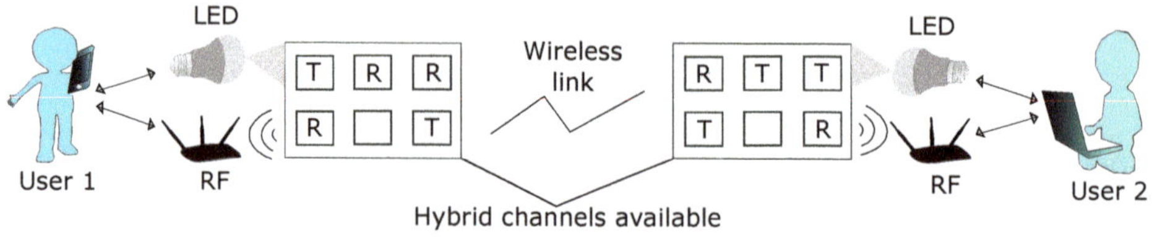

Figure 4. Hybrid model of a VLC/RF communication channel.

Due to the presence of hybrid frequency bands and the dynamic conditions of the hybrid communication channel, optimal channel design is a very important and difficult challenge for VLC/RF systems. Optimizing the transmission channel of the VLC/RF network is essential for the development of IoT systems because smart devices often require a stable connection for proper operation. A reliable and robust connection can help extend the lifetime of IoT devices and their applicability.

2.3. Load Balancing

One of the most important elements in hybrid networks is load balancing, as it manages traffic and resources in the network. Load balancing helps to evenly distribute a load of operations and traffic among each device connected to the hybrid network. This reduces network congestion, prevents overheating of devices, and improves performance in terms of speed, latency, and reliability [50].

Part of the literature on VLC/RF systems focuses on the development of algorithms to optimize network load balancing. The goal is for each user to find the best VLC or RF access point which provides the highest data rate depending on the location and mobility of each user. In [51], a protocol combined with horizontal and vertical handover mechanisms was proposed for a mobile terminal to solve the mobility of users between access points and the OFDMA system. In addition, a VLC network scheme was presented to solve multi-user access problems at each access point in the network. The authors of [52] proposed a load-balancing algorithm to manage hybrid VLC/RF network resources and determine the association of users to each system, and additionally proposed a decentralized optimization method based on Lagrangian multipliers. In [53], a bidirectional selection handover algorithm that takes load balancing into account was developed to improve the overall performance of the hybrid system. Such an algorithm selects the suitable target AP based on the hybrid technique of fuzzy analytic hierarchy process and order preference technique based on the similarity to the ideal solution.

Figure 5 shows a load balancing scheme of a hybrid VLC/RF network in an indoor scenario. Here, users are associated with an access point, either VLC (red triangles) or RF (blue squares), depending on the location and possible movement of each one. The association of users with the AP only takes place within the association zone (yellow shadow). In order to save network resources, access points outside of this zone (far from the users) do not transmit.

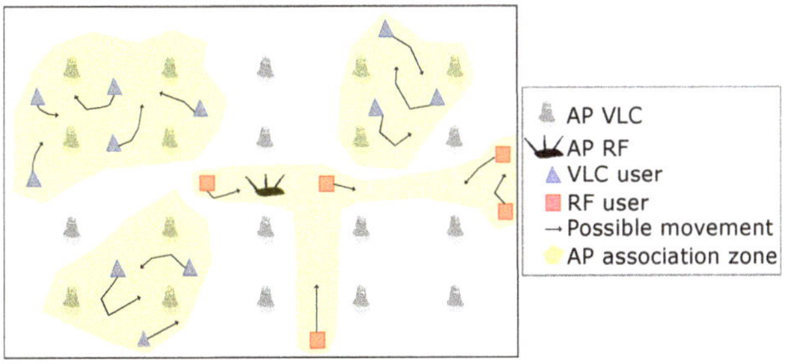

Figure 5. Load balancing scheme of a hybrid VLC/RF network in an indoor scenario.

Another part of the research focuses on the objective of minimizing the handover latency and increasing the overall network throughput using load-balancing techniques such as Handover. Handover in hybrid VLC/RF networks is the process through which a user switches from one network to another without losing connectivity. Thus, if a user connected to the VLC network moves to an area without access to the VLC network, the device automatically connects to the available RF network [54]. Similarly, if the user returns to a VLC access zone, the device automatically reconnects to the VLC network without losing connectivity. During this process, the quality of the received signal and the availability of the networks must be monitored to ensure that the connection transfer from one network to another is optimal. In [55], the authors proposed a dynamic handover scheme based on fuzzy logic with the objective of reducing the handover overhead. This scheme takes into account channel state information, user rate, and desired data rate to determine whether or not to request a handover. If a mobile terminal moves from one access point to another, a handover of the mobile load occurs between the access points. This handover can be classified as horizontal if the change occurs intra-system or vertical if the change occurs inter-system. Vertical handover is a fundamental type of handover to improve the performance of hybrid VLC/RF networks. The authors of [56] formulated a vertical handover scheme as a Markov decision process problem. Moreover, they adopted a dynamic approach to obtain a compromise between the switching cost and the delay requirement. On the other hand, the authors of [57] proposed an algorithm based on decision selection, taking into account dynamic network parameters and actual traffic preferences. The degree of rationality of each decision was calculated by combining the results of the analytical hierarchy process and a two-person cooperative game model. Recently, the authors of [58] proposed an automated vertical handover algorithm for VLC-WiFi networks using the received signal strength values. This algorithm selects the next most likely base station to transmit based on the hidden Markov model.

Horizontal handover is a fundamental technique in hybrid VLC/RF networks. If a user is moving from one location to another connected to one of the two networks of the system (VLC or RF), it is necessary to perform a horizontal handover to move the mobile load to the next access point of the same technology. In [59], an adaptive horizontal and vertical handover mechanism for hybrid VLC/MMW systems was proposed. The mechanism was based on a centralized system architecture and used the signal-to-interference plus noise ratio levels of the APs as a performance metric. The authors of [60] proposed a solution to the horizontal and vertical handover of a VLC/WiFi network to address problems due to user mobility, obstacles, and shadowing.

Table 4 shows a summary of studies carried out in the literature on load balancing in hybrid VLC/RF networks in the last ten years. Although this topic has been extensively studied in recent years, the constant evolution and dynamics of cellular networks make it necessary to constantly update load-balancing algorithms.

2.4. Network Selection and Resource Allocation

The main objective of communication networks is to provide users with personalized demand, high quality of service (QoS), and high overall system performance. Achieving this goal is a constant challenge in hybrid VLC/RF networks because of the need for an elaborate balance between user demand and network resources [61]. Another fundamental challenge in these networks is the problem of associating mobile terminals to each base station for packet transfer due to the density of access points, their heterogeneity, differences in coverage area, and the limitation of available resources [62].

For all these reasons, it is very important to realize an approach to address the problem of network association and resource allocation and to obtain better information about the users and resources in the environment. The authors of [63] proposed a network selection scheme for a hybrid VLC/RF system in an indoor scenario, taking into account the context-aware information of asymmetric downlink and uplink characteristics and network performance. In addition, a context-dependent learning algorithm was proposed based on the information about the location of the users and the type of traffic they require. In [64], the authors proposed a knowledge transfer reinforcement learning algorithm to perform network selection in a hybrid VLC/RF system. This algorithm reveals and exploits contextual information about networks, load distribution, and network traffic characteristics. The authors of [65] showed another approach to network selection by designing a fuzzy logic model with relatively low computational complexity, in which all users to be connected to the RF network are first selected and then all remaining users are assigned to the VLC network. In [66], a network selection process was proposed for a hybrid system composed of VLC and RF, in which the transmitter sends the transmission data over the link that best guarantees the quality of service. The average arrival data rate at the transmitter buffer and the non-asynchronous limits of the buffer delay are used to evaluate the performance of the proposed algorithm.

Table 4. Summary of the studies of load balancing in hybrid VLC/RF networks.

Ref	Year	Hybrid Network	Contributions
[51]	2014	VLC/OFDM	A combined protocol with horizontal and vertical handover mechanisms is proposed for the mobile terminal to solve the mobility of users between different access points and the OFDMA system. A VLC network scheme and its frame format are presented to solve multi-user access problems at each access point. A new metric is defined to evaluate the capacity of the hybrid network based on the assumption of homogeneous Poisson point process distribution of users.
[67]	2015	LIFI/WIFI	Load balancing is studied taking into account user mobility and handover signaling overhead. A dynamic load balancing scheme is proposed considering system throughput and fairness. The service areas of different APs are studied and the performance of each AP is analyzed using the proposed load-balancing scheme.
[56]	2015	VLC/WIFI	An efficient vertical handover scheme is investigated by formulating it as a Markov decision process problem. A dynamic approach is adapted to obtain a compromise between the switching cost and the delay requirement. The proposed scheme determines whether to perform vertical handover given the queue length and the wireless optical channel condition.
[68]	2015	RF/VLC	A cooperative load balancing that achieves proportional fairness using centralized and distributed resource allocation algorithms is studied. Performance is analyzed in terms of both throughput and fairness under various cell formation scenarios.
[55]	2016	LIFI/RF	A dynamic handover scheme based on fuzzy logic is proposed, which uses channel state information, user rate and desired data rate to determine if the handover is necessary. The objective is to reduce the handover overhead by properly assigning users to the RF or LiFi access point.
[57]	2017	VLC/Femto	An approach that takes into account multiple attributes, including dynamic network parameters and actual traffic preferences, is performed using the analytical hierarchy process. The degree of rationality of each decision is calculated by combining the results and the decision with the highest degree of rationality is selected.
[69]	2017	LIFI/RF	A dynamic load balancing scheme that takes into account the handover overhead is proposed to improve the overall system performance. A joint optimization algorithm and a separate optimization algorithm, which jointly and separately optimize the access point allocation and resource allocation, are proposed.
[50]	2018	VLC/RF	An iterative algorithm is proposed to distribute users into APs and to distribute the powers of the APs to their users. An optimization problem is formulated to allocate the power of each AP to the connected users in order to maximize the total achievable data rate. Two approaches are proposed to perform joint AP and load balancing: a main approach that considers the exact interference information for all users, and a suboptimal approach that aims to decrease the complexity of the first approach by considering only the approximate interference information of the users.
[70]	2018	LIFI/RF	A handover algorithm based on the received signal strength indicator is considered. A theoretical analysis of the handover probability based on a Markov chain model is presented.
[71]	2019	LIFI/WIFI	An overview of the concept of hybrid handover in LiFi/WFi networks is given and related work done in this field is discussed.
[52]	2020	VLC/RF	A load balancing algorithm is proposed to manage the hybrid network resources and determine the association of users to each system. A decentralized optimization method based on Lagrangian multipliers is proposed.
[53]	2021	VLC/RF	A bidirectional selection handover algorithm is developed that takes load balancing into account to improve system performance. The appropriate destination AP is selected based on the hybrid technique of fuzzy analytic hierarchy process and order preference technique by similarity to the ideal solution for each user that needs to handover.
[72]	2021	VLC/WLAN	A vertical handover technique that uses the user's location information to make a handover decision is proposed. The proposed algorithm is contrasted with the immediate vertical handover algorithm and the permanent vertical handover algorithm with two different dwell times.
[26]	2022	VLC/MMW	An event-triggered adaptive handover mechanism for centralized hybrid systems is proposed. The handover variables are self-adjusted using rule-based algorithms with the number of connected users and the data transmission rate as input variables.
[60]	2022	LIFI/WIFI	A higher layer handover solution known as "Li-Wi" is proposed to address handover issues due to mobility, shadows and obstacles. Robust connectivity with shorter handover outage duration and high network throughput is provided.

In hybrid VLC/RF networks, resource allocation is closely related to network selection, as the selection of one of the two networks depends on the allocation of resources to each network, just as the allocation of resources to each network has a significant impact on network selection to optimize system performance [73]. An optimal combination of resource allocation and network selection enables overall system performance in terms of data rate, delay, and energy efficiency. Several types of research on resource allocation in hybrid VLC/RF networks are found in the literature [20,74–76]. A study of resource allocation in two different scenarios of hybrid VLC/RF networks to maximize user throughput was recently performed in [74]. In the first scenario, a hybrid system was modeled in which VLC is powered by light waves to radiate only optical energy and the throughput maximization problem is formulated by optimizing the uplink time allocations. In the second scenario, time-switching is performed to simultaneously transfer information and power over the lightwave of the VLC network and a multi-objective optimization problem is formulated through a set of Pareto-optimal resource allocation policies. On the other hand, in [20] a resource allocation scheme that maximizes the data rate, taking into account user fairness was proposed for a VLC/RF network, with both networks served by a common backhaul and the channel state information able to be either perfect or imperfect. The authors of [75] developed a resource allocation algorithm with the objective of maximizing the energy efficiency (throughput per unit power) of the hybrid system. The results obtained by simulating the algorithm show that the energy efficiency can be increased if all the VLC access points are evenly distributed within the scenario. Lastly, in [76] a centralized algorithm for joint downlink resource allocation in a hybrid VLC/RF network was developed.

2.5. Hybrid Network Topologies

Network topologies are primarily used to describe the physical or logical structure of a network, i.e., the topology defines how and where the various devices in the network are connected. The correct design of the topology of a hybrid network is very important because it allows for optimization of the location and interconnection between the different access points of the network [77]. The main topologies designed for hybrid VLC/RF networks are separated into six major groups: symmetric with interference, symmetric without interference, asymmetric, dual transmission, separate transmission, and hybrid transmission, as shown in Figure 6.

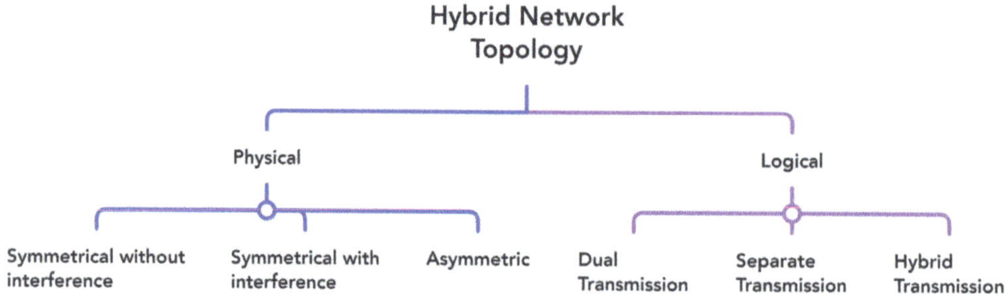

Figure 6. Types of hybrid topologies in the literature.

Symmetrical without interference is a topology in which both networks are independent. Both the VLC network and the RF network can be used for downlink and uplink connection, as shown in Figure 7A. There is no interference between the networks [56,63,65,67,69,75,78–86].

In symmetric with interference topologies, the VLC and RF networks are linked in the uplink; therefore, they are not independent and interference occurs between them. Figure 7B shows a schematic of a symmetrical topology with interference. In the downlink, both networks can be used simultaneously without interference, while in the uplink only the

RF network is used to transmit information from both networks, which causes interference between RF users and VLC users transmitting over the RF network [87–91].

Figure 7C shows an asymmetric topology scheme. The downlink and uplink connections are routed in different directions for the two networks. The VLC network is only used for the downlink connection, while the RF network is only used for the uplink [52,77,92–104].

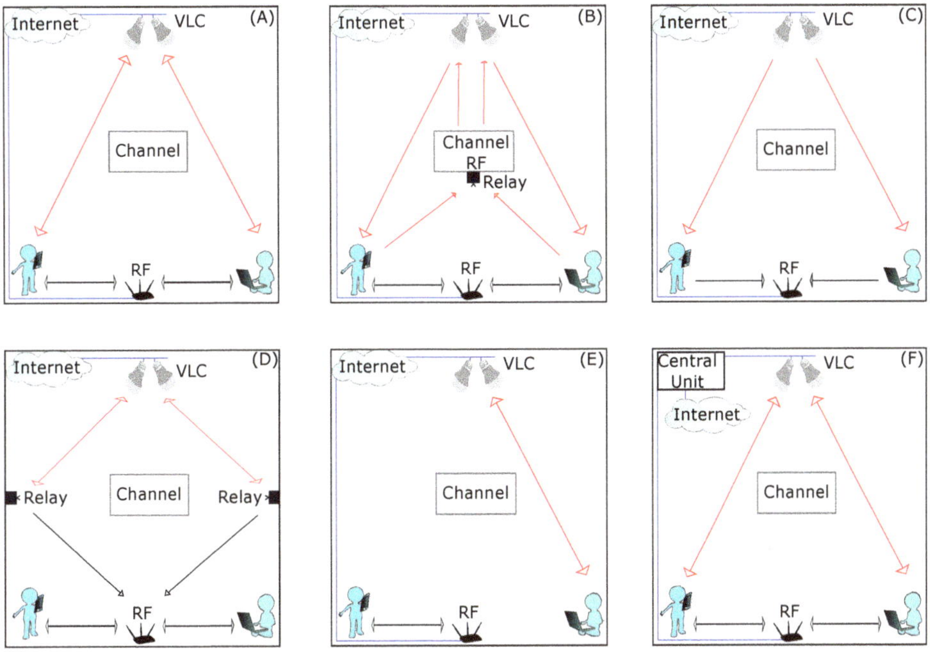

Figure 7. Different hybrid network topologies: (**A**) symmetrical without interference, (**B**) symmetric with interference, (**C**) asymmetric, (**D**) dual transmission, (**E**) separate transmission, (**F**) hybrid transmission.

In dual transmission topologies only one of the two networks can be connected. The VLC network is used for the connection between the VLC access points and the RF access point via a relay, while the RF network is used for the connection between the RF access point and the users. Both networks transmit on the downlink and uplink, as shown in Figure 7D [18,105–110].

In separate transmission, the two networks are separated into independent connections. As shown in Figure 7E, users can choose between one of the two networks depending on the location and movement of each user as well as the SNR of the links. While this topology has great flexibility for users, it is difficult to control and is not optimal [49,111–123].

Figure 7F shows a hybrid transmission topology, in which both networks are interconnected by a control unit. Depending on the location of the users and the network conditions, the control unit allocates network resources to each user, ensuring fairness. The connection can be made by both networks in both the downstream and upstream directions [50,68,78,124–135].

3. Hybrid VLC/RF Networks in Outdoor Scenarios with Harsh Conditions

In an outdoor scenario with harsh conditions (e.g., sunlight, rain, dust, wind) the communication channel of hybrid VLC/RF networks becomes very complex and dynamic. These conditions distort the VLC optical beam, causing amplitude and phase fluctuations along with optical losses. The level of loss of the VLC signal is higher than the losses suffered by the RF network, as the VLC signal is very sensitive to atmospheric conditions such as sun, rain, and fog [136]. There are several factors (scattering, absorption, non-alignment, and turbulence) that can cause signal loss in both the VLC and RF networks, which complicate the design of the hybrid communication channel model [137]. Among the most important phenomena responsible for signal loss in VLC/RF networks are:

- Scattering corresponds to an angular spread of the optical field, which is independent of frequency and wavelength in certain cases. It can be caused by solar radiation, rain, or particles and objects in the environment, which create a divergence in the transmission beam and cause the size of the received transmission beam to be larger than the receiver aperture, resulting in part of the signals being lost.
- Absorption takes place when a collision occurs between the photons of the propagating beam of the VLC network and the particles or molecules existing in the environment. The absorption coefficient depends on the type of molecules in the air and their concentration. There are areas of minimum absorption in the atmosphere called transmission windows. One way to avoid or decrease this phenomenon is to select the wavelengths of the grid to be within the wavelengths of the transmission window.
- The distance between the transmitter and receiver in an outdoor scenario can cause non-alignment of the transmitter and receiver, resulting in a loss of signal power. Non-alignment between the transmitter and receiver can be due to the presence of obstacles, buildings, and trees, as well as changes in the weather. As the distance between the transmitter and receiver increases, these factors must be taken into account.
- Ambient turbulence corresponds to a divergence in the optical signal, which can be caused by random temperature variations, solar radiation, increased wind speed, and atmospheric pressure. Divergence caused by turbulence during transmission results in random variations of signal amplitude and phase, causing fading of the received power and decreased performance of hybrid systems.
- In an outdoor scenario, multiple sources of noise and interference can affect hybrid VLC/RF networks, including shot noise, thermal noise, and environmental interference. Among these, sunlight is considered the most serious co-channel interference source for the VLC network, and thus for the hybrid network, as it can significantly reduce the SINR of the system. One way to reduce the level of sunlight interference in the VLC network is to use lenses and filters on the receivers; however, this is not an optimal solution.

The harsh conditions in outdoor scenarios are important factors to take into account in the design of hybrid VLC/RF networks. A study of the degrading effects of physical phenomena on the quality of VLC communications in an industrial environment, where dust, other particles, and artificial light sources increase signal attenuation and cause optical interference, was carried out in [138]. In such cases, it is necessary to design and implement a robust model capable of considering the existing environmental conditions in order to improve the overall performance of the hybrid system [139]. The use of transmitters and receivers with complementary metal oxide semiconductor-based cameras and image sensors is a robust way to realize VLC network connection in outdoor scenarios, as complementary metal oxide semiconductors are capable of converting the modulated optical signal into an electrical waveform. As in indoor scenarios, a hybrid VLC/RF network can be implemented through different topologies in outdoor scenarios, mainly depending on the type of VLC network link, i.e., line-of-sight (LOS) or non-line-of-sight (NLOS). A very suitable scheme for an outdoor scenario with a line-of-sight link is the Point-to-Point topology. In this topology, the transmitters and receivers are in a fixed position and oriented towards each other. Otherwise, if the link is NLOS, a fuzzy topology can be used in which the transmitter and receiver are not require to be aligned. In this case, a receiver with a panoramic Lambertian pattern that can rotate within the radiation field of the transmitter is used to receive the signal [140].

One of the main challenges in hybrid VLC/RF networks stems from difficult outdoor conditions, mainly due to the high mobility of background noise caused by sunlight or other artificial light sources. A study of the impact of sunlight noise sources on the packet delivery rate in a VLC network was carried out in [141]. The results showed that when the sun is not in the receiver's field of view (FOV) a packet delivery rate of 100% can be achieved for distances of less than 100 meters. On the other hand, Turan performed a time-domain characterization of the VLC channel by conducting real experiments in a static outdoor scenario during different times of the day [142]. The results showed that sunlight both affects the effective useful bandwidth of the VLC channel and reduces the channel gain when direct sunlight is present in the receiver FOV. To counteract the adverse effects of sunlight on the VLC channel, the receiver FOV can be optimized by adjusting the receiving angle. It should be noted that FOV adjustment is usually a compromise between noise reduction and the possible reception angle. To avoid reducing the receiver FOV too much, in [143] the use of a high-pass filter was proposed to filter the sunlight intensity in a one-channel VLC experiment. This filter was composed of a biconvex lens adaptive optical receiver, allowing high reduction of background noise without affecting the FOV angle. In addition, the authors coated the entire area from the optical receiver to the photodiode in order to block unwanted light angles, as shown in Figure 8. Finally, they added a DC blocker after the photodiode to block strong DC components. Experiments showed that data transmission is possible up to a distance of 40 m under perfect lighting conditions; however, the connection distance decreases in strong sunlight.

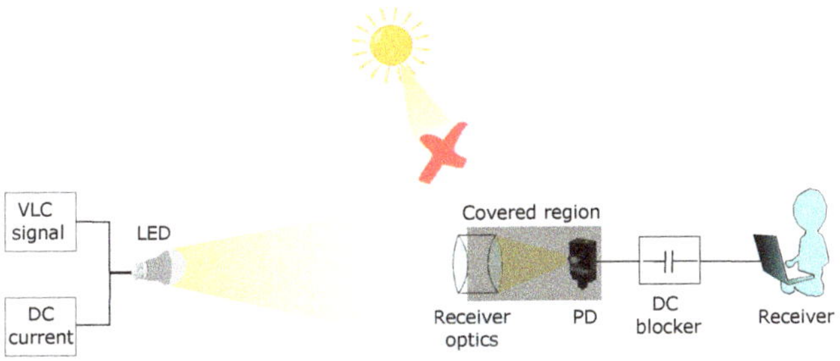

Figure 8. General schematic of a VLC transceiver system with an optical filter.

In visible light communication, due to the propagation and directionality characteristics of light it is important to establish a clear line-of-sight (LOS) link between the transmitter and receiver. However, maintaining this LOS link becomes a challenge when users are in motion. This is especially true in outdoor environments, where light beams can travel long distances and reflected components lose energy, making reliable communication difficult. In [144], the authors suggested that non-line-of-sight (NLOS) communication via ground reflections can be advantageous in VLC. The effectiveness of NLOS communication depends on factors such as weather conditions and ground surface materials, which influence ground reflectivity. Unlike traditional wireless communication, VLC communication does not suffer from multipath fading caused by NLOS reflections. This is due to the inherent spatial diversity resulting from the carrier wavelength of visible light waves, which is significantly smaller compared to the size of typical receivers [43].

Based on the conditions of the outdoor VLC channel and the solutions proposed by [141–143], we propose designing a hybrid outdoor VLC/RF network as shown in Figure 9. This hybrid VLC/RF network can provide service to users through both networks at times when the solar intensity direction does not coincide with the direct FOV angle, thereby taking advantage of the multiple benefits of hybrid networks. At times when it is not possible to connect through the VLC network due to saturation of the photodiode by high background noise caused by sunlight, the service can be provided only by the RF network to ensure that users are not left without connection at any time. At night, the VLC network can be affected by artificial light sources such as lights used by cars and street lights. These light sources have much lower luminance values than sunlight, and over long distances they should not affect the VLC network connection [145]. In case the artificial lights are at a relatively short distance from the user connected to the VLC network, they can be blocked by the filter of the optical receiver as long as they have a different direction from the FOV angle.

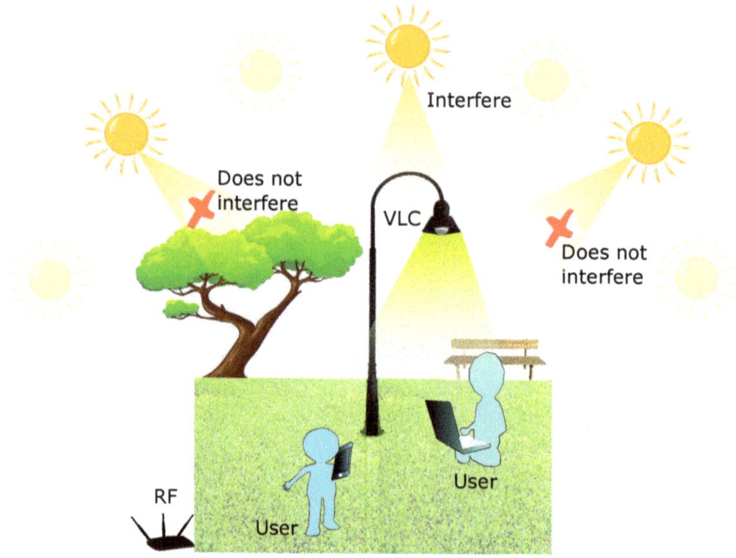

Figure 9. VLC/RF hybrid network in an outdoor scenario for different sun positions.

Due to the attractions and advantages of VLC networks, such as offering greater data security, being less susceptible to interference from external signals, offering greater bandwidth for data transmission, and making efficient use of the frequency spectrum, thereby helping to alleviate congestion in the radio frequency spectrum, multiple studies [140–143,146,147] have been conducted on the study of VLC networks in outdoor scenarios regardless of the presence or absence of sunlight. The deployment of outdoor VLC networks aims to extend the multiple advantages of this technology to all users, regardless of their locations and scenario. A study of the impact of sunlight noise sources on the packet delivery rate in a VLC network was carried out in [141]. In [142], a time-domain characterization of the VLC channel was performed by conducting experiments in a static outdoor scenario during different times of the day. In [143], a high pass filter was proposed to filter the sunlight intensity in a one-channel VLC experiment. Outdoor VLC network access points could be powered by solar panels or other renewable energy sources, meaning that there would be no additional energy costs during the day. In addition, VLC technology could use directional LEDs with a small light cone that transmit only the desired information and would not be occupied by unnecessary lighting. On the other hand, the VLC network saves energy at night, as the lighting LEDs act as data transmitters, eliminating the need for separate devices for lighting and communication. However, outstanding issues related to interference caused by solar illumination on VLC devices must be resolved. Consequently, several investigations on outdoor VLC networks have focused on this issue [140–143].

Other external factors that greatly affect the VLC channel include difficult weather conditions such as fog, rain, and snow. Normally, molecules and particles in the atmosphere interact with the light, deflecting the beams and attenuating the transmitted signal. In adverse weather conditions these effects are much more severe, as fog, rain, and snow are made up of larger particles that have a large impact on the range and reliability of VLC. The study conducted in [148] showed that fog has a greater negative impact on VLC; however, the study conducted in [149] showed that dry snow can be even more detrimental to VLC system performance than fog. For this type of scenario in which weather conditions are adverse, it is possible to establish reliable communication up to 15 m using the highest possible modulation scheme [150].

The hybrid VLC/RF networks and their deployment unlock multiple possibilities for IoT applications in various outdoor scenarios [151]. The high-precision communication capabilities of VLC technology make it an effective solution for interfacing with smart sensors that require accurate real-time data transmission. At the same time, the RF network ensures ubiquitous connectivity and uninterrupted communication. Among the applications with potential for innovative systems are smart outdoor streetlights that autonomously adjust to changing traffic patterns, security cameras that detect movements with high precision, and traffic control systems that can effectively manage traffic congestion in real time. With the integration of VLC/RF networks, exciting new possibilities that expand the realms of IoT become a reality [152].

Several authors have started to investigate this topic, and the first steps in the design of hybrid VLC/RF systems in outdoor scenarios can already be found in the scientific literature [153–156]. In [153], an optimization problem was established to obtain the maximum possible information rate for users through the DC offset and AC peak amplitude of the VLC network access points along with the electrical power supplied to LEDS and relay relays. On the other hand, in [154] the energy efficiency of a Non-Orthogonal-Multiple-Access (NOMA) scheme of hybrid VLC/RF networks with imperfect hybrid channel state information was investigated. It was shown that a non-orthogonal multiple access scheme has higher performance compared to the VLC/RF network in an outdoor scenario where LOS availability is limited. The authors of [155] evaluated a Co-NOMA-based cooperative transmission scheme for hybrid VLC/RF systems in outdoor scenarios where users with weak VLC connections can connect to the network through users with strong VLC connections using the RF network. An increase in the sum rate of the system was demonstrated, as was increased user fairness. Recently, a hybrid FSO-VLC/RF system capable of providing communication services to emergency responders for an outdoor fronthaul network based on a machine learning algorithm was experimentally demonstrated in [156]. A study of outdoor VLC for vehicle-to-vehicle security applications was conducted in [157]. The authors introduce the privacy and security fields in the physical layer of VLC to improve the vehicle-to-vehicle power distribution.

4. VLC/RF Hybrid Network Applications

Thanks to the varied characteristics of VLC and RF networks and their complementarity, there are a wide range of applications in the technological, economic, and social-environmental fields for these types of networks. VLC/RF hybrid networks have demonstrated great improvements in overall system performance in terms of data rate, load balancing, and lower time delay in both indoor and outdoor scenarios [158]. The implementation cost of these networks is relatively low, and they allow for the reuse of the entire existing lighting infrastructure [159]. The energy efficiency of VLC/RF networks and the possibility of distributing multiple access points without interference or environmental waste can facilitate the decrease of energy consumption and environmental pollution, helping to meet the Sustainable Development Goals for the year 2030 established by the UN [160]. Figure 10 shows an example of the applications of VLC/RF networks and the relationship between them.

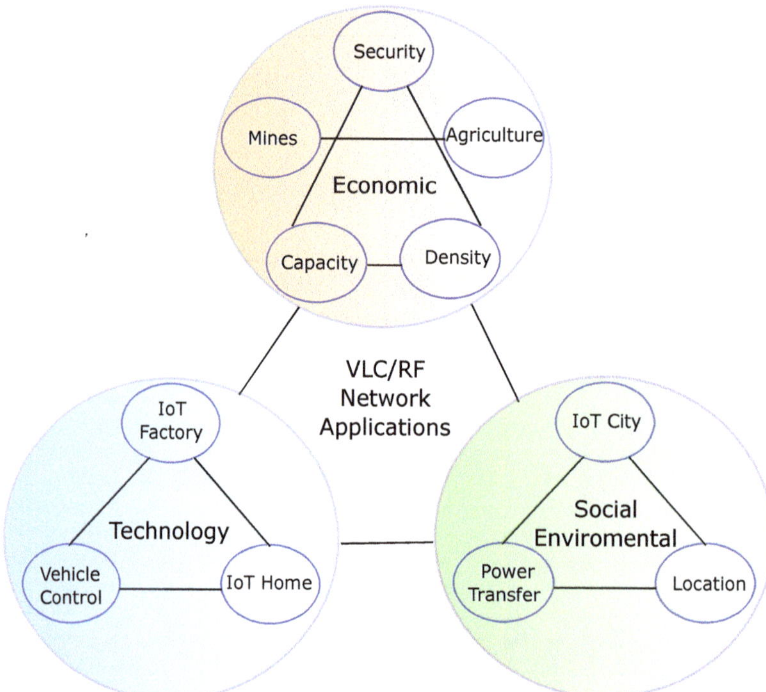

Figure 10. VLC/RF network applications.

4.1. Technological Applications

VLC/RF hybrid technology could be a key tool to achieve smart homes. The data speed and security of VLC networks in conjunction with the ubiquity of RF coverage could be used to interconnect all electronic equipment within smart homes. In this way, any command could be executed on any electronic equipment in the house from a mobile terminal or simply by voice, even if far away from the house. The use of a solar panel on the roof of an IoT smart house represents a way to store energy and transmit information at the same time through the VLC network, which could then be transferred to the interior of the house through the RF network. The authors of [161] investigated the applications of hybrid VLC/RF networks in a three-dimensional indoor IoT system, taking into account temperature sensors, indoor air quality sensors, indoor sensors, and humidity sensors.

IoT applications associated with factories and production have attracted the attention of researchers in recent years [103,161,162]. The goal is to use the performance of VLC/RF networks to interconnect production equipment and automation technologies to enable the generation and sharing of information between all equipment. This can enable big data generation to identify certain production patterns and even predict inefficiencies and future events.

Vehicular and traffic control applications on the road are another promising options for the development of automotive technology. Due to the random nature of car traffic on the road, it is necessary to implement a hybrid technology that allows communication between cars and organizes their passage. Several authors have investigated this topic and proposed different vehicle control applications using hybrid VLC/RF networks [163–165]. In [163], the authors proposed an outdoor cognitive network with electric vehicles using mixed VLC and RF channels to establish interference-free communication between vehicles. On the other hand, in [164] the authors presented a hybrid VLC/RF cooperative system in vehicle-to-vehicle communication networks and analyzed the performance of the hybrid system in terms of outage probability and bit error rate. The authors of [165] investigated resource management in hybrid VLC/RF systems for wireless communication between vehicles and infrastructures.

4.2. Economic Applications

VLC networks offer a great guarantee of security, as the transmitter and receiver must be in the line of sight, the coverage area is very small, and the transmission power is focused only on the light beam. In places where it is necessary to send and receive data that must be protected (e.g., companies, banks, hospitals) the implementation of hybrid VLC/RF networks can provide great security in data transmission [119,127].

In scenarios where there is a high density of users connected to the internet (e.g., stadiums, airports, and parks) the use of multiple RF access points can cause interference and reduce system performance. The use of hybrid VLC/RF networks can solve this problem by reusing the entire existing lighting infrastructure for VLC data transmission. These hybrid networks provide higher data rates, reduce RF interference, and have a very low implementation cost, making them are ideal for high user density scenarios [18,63,89].

Hybrid VLC/RF networks have demonstrated a great improvement in the characteristics and capabilities of wireless communications systems. Research has focused on increasing the data rate [78,84,166] to improve the overall capacity of hybrid system, while other authors have focused on decreasing the system time delay [66,81,167].

Mining communication systems are an application that safeguards the safety of workers in mines and increases the efficiency of their work. In recent years, VLC links have been widely investigated as a complementary technology to RF links for subway mining applications due to their high data rate and freedom from RF interference [168,169]. Subway mines represent a challenging scenario due to their irregular walls and many shadows and dust particles, which produce the scattering phenomenon. For all of these reasons, this industry requires an optimal transverse communication system that can be supplemented by hybrid VLC/RF networks. The use of indoor, motion, and ambient dust particle sensors would be of great help in the implementation of VLC/RF networks in these and similar hostile scenarios.

Technology is a great ally of agriculture, helping to achieve greater efficiency in farming with the objective of supplying the population without depleting the planet's resources. The implementation of hybrid VLC/RF networks in conjunction with sensors dedicated to agricultural activity can be of great help in monitoring and sending information about events concerning crops and their surroundings [170,171].

4.3. Social-Environmental Applications

VLC/RF networks have been presented as a reliable alternative to obtain the location of mobile terminals in deep indoor scenarios (tunnels, mines, and subways) where GPS has no connection. The RF network can be used to obtain the external location of the site, while the VLC network can provide the internal location of the user. Several proposals for positioning applications using VLC/RF networks have been put forward [172–174], and a minimum estimation error of 5.8 cm has been demonstrated.

Several authors have investigated the process of power transfer from a VLC network to the nearest users [118,119]. This power transfer allows the energy transmitted by the LED lights to be stored for later use in other transmissions. This application has demonstrated high energy efficiency for hybrid VLC/RF networks, and could have a large impact on overall energy consumption [175,176].

The growth and development of cities is an aspect that has been taken into account in technological research throughout history. The implementation of hybrid networks, especially VLC/RF networks, represents a key parameter to achieve efficient management of each area of the city, thereby improving the life of its inhabitants. Hybrid VLC/RF technology can allow energy optimization, reduce household consumption, organize the movement of vehicles, and increase the flow of mobile traffic, among many other aspects necessary to achieve the development of sustainable cities [83,131,132].

The main advantages of hybrid VLC/RF networks over conventional RF networks are their ability to increase transmission bandwidth and alleviate congestion in RF networks. In addition, these hybrid networks can be deployed and operated satisfactorily in locations where a pure RF network is insufficient. Table 5 comparatively summarizes the performance of VLC/RF and conventional RF networks for the applications proposed in this section. The ⋆ and ★ marks respectively indicate good and excellent performance considering the criteria of previous works and the corresponding performance metrics.

Table 5. Summary of VLC/RF network performance for different applications. The ⋆ and ★ marks indicate good and excellent performance, respectively.

Applications	VLC/RF Systems	RF Network
IoT Home [161]	★ (Bandwidth, Latency)	⋆
IoT Factory [162]	★ (Bandwidth, Latency)	★ (Coverage)
Vehicle Control [163]	★ (Latency, Coverage)	⋆
Security [119]	★ (Security, Reliability)	⋆
System Capacity [166]	★ (Energy efficiency)	⋆
Mining Systems [168]	★ (Interference, Connectivity)	★ (Coverage)
agriculture [170]	★ (Interference, Connectivity)	⋆
localization [172]	★ (Underground connectivity)	★ (Coverage, Precision)
Power Transfer [118]	★ (Bandwidth, Latency)	⋆
IoT City [131]	★ (Bandwidth, Latency)	⋆

5. Challenges and Future Directions for Hybrid VLC/RF Networks

This section discusses important issues facing hybrid VLC/RF networks as well as their future directions, including the use of Software Defined Networking (SDN) platforms, multiple access techniques, the development of new modulation and coding schemes, the design of efficient resource allocation and handover algorithms, the development of security mechanisms, the hybrid channel model, load balancing, uplink communication in VLC, outdoor scenarios, and Optical Camera Communication (OCC).

The use of SDN platforms has been booming in recent years due to their ability to efficiently manage all access points in the network. Several papers have been presented on SDN [177] in which the advantages of its use in resource allocation and network scalability have been demonstrated. The use of SDN platforms in hybrid VLC/Femtocell networks remains in its infancy, and it remains necessary to study this topic in greater depth. The use of SDN platforms in hybrid VLC/RF networks must overcome challenges such as the coexistence of technologies and hybrid topology management [178]. Hybrid VLC/RF networks must be able to interoperate efficiently with existing technologies, which can pose compatibility and interference management challenges. In addition, the management of hybrid topologies may require a more dynamic and flexible approach to ensure optimal connectivity between the two technologies. Therefore, new standards and algorithms for VLC/RF networks on SDN platforms that facilitate interoperability and coexistence of the technologies are required. This can help minimize compatibility issues and manage interference effectively [179].

Supporting user mobility is vital in hybrid networks. The user may move in the scenario with multiple obstacles, which makes the location of the mobile receiver and the coordination mechanism between the LED transmitter and the RF base station a key challenge for multiple access techniques [51]. The CSMA/CA access technique has been presented separately for single network scenarios instead of hybrid networks [180] making a CSMA/CA access technique especially designed for hybrid VLC/RF networks required in order to support user mobility. On the other hand, non-orthogonal multiple access (NOMA) techniques have been presented in recent works as a promising option for multiple access between users via power multiplexing [97,118].

The communication channel is the medium responsible for transmitting information between the transmitter and receiver. Efficient channel modeling is vital because it reduces multi-project and interference effects [40]. It is necessary to optimally design a hybrid channel model that integrates VLC and RF technologies and takes advantage of the benefits provided by each channel independently. The design of hybrid communication channels for VLC/RF technologies is considered a major challenge due to the marked differences in the communication channel of each network. The channel model, modulation and propagation techniques, data rate, transmission capacity, and coverage are among of the characteristics that differentiate VLC and RF networks [45]. Despite this, because VLC and RF networks do not interfere with each other it is possible to design a common channel for both technologies. The correct selection of the modulation techniques of the networks and the synchronization between the channels available for the transmission of both technologies is fundamental to optimizing the design of a hybrid channel [46]. Although several channel designs have been proposed for hybrid VLC/RF networks, it remains necessary to optimally design a hybrid channel model that integrates VLC and RF technologies and takes advantage of the benefits provided by each channel independently.

A technical concern of hybrid VLC/RF networks is to achieve efficient load balancing. The main question is how to allocate users between VLC access networks and RF femtocells. The first step to finding an answer to this question involves solving the problem of joint association and resource allocation [50]. Although several authors have investigated these problems [53,55–57,59], several parameters remain to be solved before efficient load balancing can be achieved. It is necessary to delve deeper into issues such as the fair distribution of user load; dynamic allocation algorithms could be developed to ensure this balance [51]. In addition, it is necessary to develop load-balancing algorithms that analyze the state of each base station and dynamically assign users to base stations with less load. This can help to avoid congesting the stations and make the best use of the network resources [53]. During a call session a load balancing mechanism must be performed periodically, and users may need to switch to an AP with better service if they are on the move [70].

Due to the particular characteristics of each technology, it is essential to develop new modulation and coding schemes to improve the performance of hybrid VLC/RF networks [181]. While traditional RF networks have limitations such as electromagnetic interference and signal attenuation, VLC networks are not subject to these limitations and can offer higher transmission rates. However, their sensitivity to physical obstacles and variations in light intensity are relevant challenges. It is necessary to develop modulation and coding schemes to overcome these limitations of hybrid RF/VLC networks. These schemes have to mitigate the adverse effects of interference and attenuation on data transmission in order to improve the quality and reliability of optical communication [182]. These new schemes should be able to adapt to environmental conditions and optimize spectral efficiency to take full advantage of the available bandwidth [183].

Resource allocation and handover are essential to transmit large volumes of data quickly and reliably in hybrid VLC/RF networks [53]. These algorithms are crucial to the effective and balanced allocation of transmission resources such as bandwidth, transmitting power, and processing capacity. Moreover, they ensure continuous and uninterrupted transmission by managing the handover of data between VLC and RF technologies. Resource allocation and handover have a direct impact on the quality of service offered to users by maximizing network capacity and avoiding congestion. In addition, proper resource allocation helps to reduce energy consumption and costs associated with data transmission. Although several algorithms have been proposed for hybrid VLC/RF networks [26,60,66], there is a need to design new algorithms that optimize resource usage and decrease handover latency between networks. The use of tools such as artificial intelligence can be of great help in the evolution towards more efficient algorithms.

The growing dependence on communication technologies and the need to protect the integrity and confidentiality of the information being transmitted makes security mechanisms in hybrid VLC/RF networks of vital importance today [184]. Multiple vulnerabilities and security risks can arise from the use of different communication technologies in hybrid VLC/RF networks. Therefore, it is necessary to develop security mechanisms that protect the privacy of the information transmitted, prevent unauthorized interception of data, and prevent attacks and malicious manipulation. These security mechanisms should include data encryption techniques, user and device authentication, and intrusion detection and prevention. In addition, the integrity of transmitted data must be guaranteed, while access and control policies must be established to ensure that only authorized users can access the network [185,186].

To date, VLC systems lack acceptable uplink performance; for this reason, hybrid networks are forced to use different direct and return communication paths [93,111]. For uplinks, these systems require the beam to be in a fixed direction; thus, shifting the MTs can greatly impair system performance and make the link inadequate. One of the main reasons for this is that smart mobile devices have limited power; it has not yet been possible to integrate a light source for communications [115,116]. Therefore, supporting uplink transmission using VLC is a challenge in the implementation of hybrid systems.

Outdoor scenarios are characterized by their very complex and dynamic communication channel. There are several factors that cause fading of the transmission signal, such as scattering, absorption, non-alignment, turbulence, and sunlight. Sunlight is considered to be the most severe co-channel interference source for VLC networks, and coupled with other sources of signal attenuation results in a very low SNR. Another major challenge for the VLC/RF network is the support of mobility in outdoor scenarios where obstacles and time changes can affect the direct link between the transmitter and receiver. To counteract the adverse effects of sunlight on the VLC channel, the receiver FOV can be optimized by adjusting the receiving angle. However, FOV adjustment is usually a compromise between noise reduction and the possible reception angle [155]. To avoid reducing the receiver FOV too much, in [143] the authors proposed using a high-pass filter to filter the sunlight intensity. In addition, it was proposed to coat the entire area from the optical receiver to the photodiode in order to block unwanted light angles [143]. An optimal design of the communication channel of VLC/RF networks in harsh environments capable of adapting to the existing environmental conditions and enhancing overall system performance is required [153,154].

Hybrid VLC/RF networks could be a key point in long-distance (satellite) communication. This could be achieved through the use of solar panels designed in such a way that they can receive satellite information via sunlight or detecting light from lasers [140]. This is a very interesting open research topic, as it would make sunlight a source of illumination, energy, and data transmission.

Optical camera-to-camera communication technology (OCC) is a branch of visible light communication that has emerged in recent years. OCC communication, like VLC, takes place between an optical source and a camera composed of image sensors. OCC can be implemented in any type of camera, smartphone, or any device that has integrated image sensors [187]. No changes are needed to the current infrastructure of imaging sensors that act as signal receivers; therefore, the complexity and installation cost are very low. Several papers have presented OCC technology as a fundamental key for the development of 5G communication systems in both indoor and outdoor scenarios [188–190]. Therefore, it would be very interesting to combine OCC technology with hybrid VLC/RF networks as a way to improve system performance.

Hybrid VLC/RF networks have a great deal of potential for IoT applications. In the future, these networks are expected to be used to improve connectivity in areas with weak RF signals [191]. Additionally, hybrid networks can overcome indoor positioning challenges by combining the high resolution of VLC with the ubiquity of RF networks. Another area in which these networks can be applied is smart agriculture, where VLC can be used to monitor crop growth and RF can be used for data communication. The integration of VLC and RF technologies can lead to the development of more efficient energy-saving solutions. In summary, we anticipate a bright future for hybrid VLC/RF networks in IoT applications in light of their ability to solve many of today's limitations and offer innovative solutions for a wide range of industries [192].

6. Conclusions

Interest in hybrid networks has grown exponentially in recent years due to the limitations of RF systems. In particular, hybrid VLC/RF networks will become relevant for mobile communications beyond 5G and 6G, as such integration meets the required KPIs. Study of the VLC/RF hybrid network framework in an indoor scenario demonstrates the importance of lighting requirements, network load balancing, resource allocation, and network selection for correct deployment of these systems. Optimal design of the hybrid channel model can help to mitigate the effects of interference and multi-path signal loss. Outdoor scenarios are characterized by several factors that cause fading of the transmission signal, such as scattering, absorption, non-alignment, and turbulence. An optimal channel for VLC/RF networks must adapt to harsh environmental conditions and maintain appropriate general performance of the system under such conditions. The study of hybrid VLC/RF networks is very increasing, as demonstrated by the multiple technological, economic, and socio-environmental applications presented for these hybrid systems to date. The combination of VLC and RF has revolutionized the potential of Internet of Things (IoT) systems, unlocking a world of innovative applications, from improving indoor positioning for seamless navigation to enhancing crop growth in precision agriculture. These hybrid networks hold immense potential to create intelligent applications across multiple sectors, paving the way to a more sustainable future. Despite the many advantages of hybrid VLC/RF networks, it is necessary to continue research regarding their incorporation into next-generation mobile networks and their integration with other technologies such as OCC.

Author Contributions: This manuscript was developed by several authors; the individual contributions of each author are shown below. Conceptualization, L.B.A., S.M.-S. and G.S.; methodology, L.B.A., S.M.-S., L.R.-L. and G.S.; software, L.R.-L.; validation, S.M.-S., G.S. and C.A.-M.; formal analysis, L.B.A., S.M.-S. and G.S.; research, L.B.A.; resources, L.B.A., S.M.-S., L.R.-L. and G.S.; data curation, S.M.-S. and L.R.-L.; writing—preparation of the original draft, L.B.A.; writing—revising and editing, S.M.-S., L.R.-L., G.S. and C.A.-M.; visualization, L.B.A. and L.R.-L.; supervision, S.M.-S. and G.S.; project administration, L.B.A., S.M.-S. and G.S.; obtaining funding, S.M.-S., L.R.-L. and G.S. All authors have read and agreed to the published version of the manuscript.

Funding: This research was funded by the following ANID grants: Beca Doctorado Nacional 21220282, Instituto Milenio de Investigación en Óptica ICN17-012, ANID FONDECYT Iniciación No. 11200659, FONDECYT Regular 1211132 and 1231826, and STIC-AmSud AMSUD220026.

Institutional Review Board Statement: Not applicable.

Informed Consent Statement: Not applicable.

Data Availability Statement: Not applicable.

Acknowledgments: L.B.A. thanks ANID, Doctorado Nacional Grant 21220282; G.S. thanks Instituto Milenio de Investigación en Óptica ICN17-012; S.M.-S. thanks ANID FONDECYT Iniciación No. 11200659; C.A.-M. thanks FONDECYT Regular 1211132 and STIC-AmSud 22-STIC-01.

Conflicts of Interest: The authors declare no conflicts of interest.

Abbreviations

The following abbreviations are used in this manuscript:

5G	Fifth Generation
6G	Sixth Generation
AP	Access Point
B5G	Beyond Fifth Generation
CSMA/CA	Carrier Sense Multiple Access with Collision Avoidance
IM/DD	Intensity Modulation/Direct Detection

IoT	Internet of Things
ISO	International Organization for Standardization
ITU-T	Telecommunication Standardization Sector
LED	Light-Emitting Diodes
LiFi	Light Fidelity
LOS	Line-of-Sight
MAC	Medium Access Control
MIMO	Multiple Input–Multiple Output
MMW	Millimeter Waves
NLOS	Non-Line-of-Sight
NOMA	Non-Orthogonal Multiple access
OCC	Optical Camera Communication
OFDM	Orthogonal Frequency Division Multiplexing
PHY	Physical Layer
QoE	Quality of Experience
QoS	Quality of Service
RF	Radio Frequency
SDN	Software Defined Networking
SNR	Signal-to-Noise Ratio
VLC	Visible Light Communication
VLCC	Visible Light Communication Consortium
WiFi	Wireless Fidelity

References

1. Shehab, M.J.; Kassem, I.; Kutty, A.A.; Kucukvar, M.; Onat, N.; Khattab, T. 5G Networks Towards Smart and Sustainable Cities: A Review of Recent Developments, Applications and Future Perspectives. *IEEE Access* **2021**, *10*, 2987–3006. [CrossRef]
2. Wu, X.; Soltani, M.D.; Zhou, L.; Safari, M.; Haas, H. Hybrid LiFi and WiFi networks: A survey. *IEEE Commun. Surv. Tutor.* **2021**, *23*, 1398–1420. [CrossRef]
3. Khan, L.U. Visible light communication: Applications, architecture, standardization and research challenges. *Digit. Commun. Netw.* **2017**, *3*, 78–88. [CrossRef]
4. Obeed, M.; Salhab, A.M.; Alouini, M.S.; Zummo, S.A. On optimizing VLC networks for downlink multi-user transmission: A survey. *IEEE Commun. Surv. Tutor.* **2019**, *21*, 2947–2976. [CrossRef]
5. Figueiredo, M.; Alves, L.N.; Ribeiro, C. Lighting the wireless world: The promise and challenges of visible light communication. *IEEE Consum. Electron. Mag.* **2017**, *6*, 28–37. [CrossRef]
6. Hammouda, M.; Vegni, A.M.; Haas, H.; Peissig, J. Resource allocation and interference management in OFDMA-based VLC networks. *Phys. Commun.* **2018**, *31*, 169–180. [CrossRef]
7. Cevik, T.; Yilmaz, S. An overview of visible light communication systems. *arXiv* **2015**, arXiv:1512.03568.
8. Hu, F.; Chen, S.; Zhang, Y.; Li, G.; Zou, P.; Zhang, J.; Shen, C.; Zhang, X.; Hu, J.; Zhang, J.; et al. High-speed visible light communication systems based on Si-substrate LEDs with multiple superlattice interlayers. *PhotoniX* **2021**, *2*, 1–18. [CrossRef]
9. IEEE 802.15.7 WPAN Task Group 7 (TG7) Visible Light Communication. Available online: http://www.ieee802.org/15/pub/TG7.html (accessed on 28 August 2023).
10. Rajagopal, S.; Roberts, R.D.; Lim, S.K. IEEE 802.15. 7 visible light communication: Modulation schemes and dimming support. *IEEE Commun. Mag.* **2012**, *50*, 72–82. [CrossRef]
11. Sulayman, I.I.A.; He, R.; Manka, M.; Ning, A.; Ouda, A. LiFi/WiFi Authentication and Handover Protocols: Survey, Evaluation, and Recommendation. In Proceedings of the 2021 International Symposium on Networks, Computers and Communications (ISNCC), London, UK, 4–8 April 2021; pp. 1–6.
12. Khorov, E.; Levitsky, I. Current Status and Challenges of Li-Fi: IEEE 802.11 bb. *IEEE Commun. Stand. Mag.* **2022**, *6*, 35–41. [CrossRef]
13. IEEE 802.11bb Task Group on Light Communications. Available online: https://www.ieee802.org/11/Reports/tgbb_update.htm (accessed on 28 August 2023).
14. Dogra, A.; Jha, R.K.; Jain, S. A survey on beyond 5G network with the advent of 6G: Architecture and emerging technologies. *IEEE Access* **2020**, *9*, 67512–67547. [CrossRef]
15. Jiang, W.; Han, B.; Habibi, M.A.; Schotten, H.D. The road towards 6G: A comprehensive survey. *IEEE Open J. Commun. Soc.* **2021**, *2*, 334–366. [CrossRef]
16. Miramirkhani, F.; Karbalayghareh, M.; Zeydan, E.; Mitra, R. Enabling 5G indoor services for residential environment using VLC technology. *Phys. Commun.* **2022**, *53*, 101679. [CrossRef]
17. Chowdhury, M.Z.; Hasan, M.K.; Shahjalal, M.; Hossan, M.T.; Jang, Y.M. Optical wireless hybrid networks: Trends, opportunities, challenges, and research directions. *IEEE Commun. Surv. Tutor.* **2020**, *22*, 930–966. [CrossRef]

18. Xiao, Y.; Diamantoulakis, P.D.; Fang, Z.; Hao, L.; Ma, Z.; Karagiannidis, G.K. Cooperative hybrid VLC/RF systems with SLIPT. *IEEE Trans. Commun.* **2021**, *69*, 2532–2545. [CrossRef]
19. Chen, J.; Wang, Z. Topology control in hybrid VLC/RF vehicular ad-hoc network. *IEEE Trans. Wirel. Commun.* **2019**, *19*, 1965–1976. [CrossRef]
20. Papanikolaou, V.K.; Diamantoulakis, P.D.; Sofotasios, P.C.; Muhaidat, S.; Karagiannidis, G.K. On optimal resource allocation for hybrid VLC/RF networks with common backhaul. *IEEE Trans. Cogn. Commun. Netw.* **2020**, *6*, 352–365. [CrossRef]
21. Available online: https://www.scopus.com/ (accessed on 8 August 2022) .
22. Available online: http://www.vosviewer.com/ (accessed on 8 August 2022).
23. Abdelhady, A.M.; Amin, O.; Alouini, M.S.; Shihada, B. Revolutionizing optical wireless communications via smart optics. *IEEE Open J. Commun. Soc.* **2022**, *3*, 654–669. [CrossRef]
24. Ullah, M.H.; Gelli, G.; Verde, F. Visible light backscattering with applications to the Internet of Things: State-of-the-art, challenges, and opportunities. *Internet Things* **2023**, *1*, 100768. [CrossRef]
25. Song, Y.; Yu, F.R.; Zhou, L.; Yang, X.; He, Z. Applications of the Internet of things (IoT) in smart logistics: A comprehensive survey. *IEEE Internet Things J.* **2020**, *8*, 4250–4274. [CrossRef]
26. Abuella, H.; Elamassie, M.; Uysal, M.; Xu, Z.; Serpedin, E.; Qaraqe, K.A.; Ekin, S. Hybrid RF/VLC systems: A comprehensive survey on network topologies, performance analyses, applications, and future directions. *IEEE Access* **2021**, *9*, 160402–160436. [CrossRef]
27. Ayyash, M.; Elgala, H.; Khreishah, A.; Jungnickel, V.; Little, T.; Shao, S.; Rahaim, M.; Schulz, D.; Hilt, J.; Freund, R. Coexistence of WiFi and LiFi toward 5G: Concepts, opportunities, and challenges. *IEEE Commun. Mag.* **2016**, *54*, 64–71. [CrossRef]
28. Koonen, T. Indoor optical wireless systems: Technology, trends, and applications. *J. Light. Technol.* **2017**, *36*, 1459–1467. [CrossRef]
29. Li, X.; Zhang, R.; Hanzo, L. Optimization of visible-light optical wireless systems: Network-centric versus user-centric designs. *IEEE Commun. Surv. Tutor.* **2018**, *20*, 1878–1904. [CrossRef]
30. Blinowski, G. Security of visible light communication systems—A survey. *Phys. Commun.* **2019**, *34*, 246–260. [CrossRef]
31. Wang, F.; Yang, F.; Song, J.; Han, Z. Access frameworks and application scenarios for hybrid VLC and RF systems: State of the art, challenges, and trends. *IEEE Commun. Mag.* **2022**, *60*, 55–61. [CrossRef]
32. Upadhya, A. Investigation of RIS Enabled Mixed RF/VLC Relaying with Interference. *Phys. Commun.* **2022**, *55*, 101920. [CrossRef]
33. Mir, M.S.; Majlesein, B.; Guzman, B.G.; Rufo, J.; Giustiniano, D. LED-to-LED based VLC Systems: Developments and open problems. In Proceedings of the Workshop on Internet of Lights, Virtual Event, 25 June 2021; pp. 1–6.
34. Wang, F.; Yang, F.; Pan, C.; Song, J.; Han, Z. Joint illumination and communication optimization in indoor VLC for IoT applications. *IEEE Internet Things J.* **2022**, *9*, 20788–20800. [CrossRef]
35. Tang, X.; Ren, P.; Han, Z. Energy Efficiency Competition for Visible Light Communication under Illumination Constraint. In Proceedings of the 2018 Asia-Pacific Signal and Information Processing Association Annual Summit and Conference (APSIPA ASC), London, UK, 3–6 June 2018; pp. 932–936.
36. Jiang, R.; Wang, Z.; Wang, Q.; Dai, L. Multi-user sum-rate optimization for visible light communications with lighting constraints. *J. Light. Technol.* **2016**, *34*, 3943–3952. [CrossRef]
37. Ali, R.E.; Elgala, H. Energy efficient LED layout optimization for near-uniform illumination. In Proceedings of the Fifteenth International Conference on Solid State Lighting and LED-based Illumination Systems, San Diego, CA, USA, 5–8 August 2016; Volume 9954, pp. 10–16.
38. Lee, K.; Park, H.; Barry, J.R. Indoor channel characteristics for visible light communications. *IEEE Commun. Lett.* **2011**, *15*, 217–219. [CrossRef]
39. Jia, K.; Hao, L. Modeling of multipath channel and performance analysis of MIMO-DCO-OFDM system in visible light communications. *Chin. J. Electron.* **2019**, *28*, 630–639. [CrossRef]
40. Fath, T.; Haas, H. Performance comparison of MIMO techniques for optical wireless communications in indoor environments. *IEEE Trans. Commun.* **2012**, *61*, 733–742. [CrossRef]
41. Saengudomlert, P. Transmit beamforming for line-of-sight MIMO VLC with IM/DD under illumination constraints. In Proceedings of the 2015 12th International Conference on Electrical Engineering/Electronics, Computer, Telecommunications and Information Technology (ECTI-CON), Mexico City, Mexico, 26–30 October 2015; pp. 1–4.
42. Vappangi, S.; Mani, V. Performance analysis of dst-based intensity modulated/direct detection (im/dd) systems for vlc. *IEEE Sens. J.* **2018**, *19*, 1320–1337. [CrossRef]
43. Memedi, A.; Dressler, F. Vehicular visible light communications: A survey. *IEEE Commun. Surv. Tutor.* **2020**, *23*, 161–181. [CrossRef]
44. Brown, C.N.; Mattei, E.; Draganov, A. ChaRRNets: Channel robust representation networks for RF fingerprinting. *arXiv* **2021**, arXiv:2105.03568.
45. Neumann, D.; Wiese, T.; Utschick, W. Learning the MMSE channel estimator. *IEEE Trans. Signal Process.* **2018**, *66*, 2905–2917. [CrossRef]
46. García-Morales, J.; Femenias, G.; Riera-Palou, F. Channel-aware scheduling in FFR-aided OFDMA-based heterogeneous cellular networks. In Proceedings of the 2015 IEEE International Conference on Computer and Information Technology, Ubiquitous Computing and Communications, Dependable, Autonomic and Secure Computing, Pervasive Intelligence and Computing, Berlin, Germany, 15–18 May 2015; pp. 44–51.

47. ElSawy, H.; Hossain, E. Two-tier HetNets with cognitive femtocells: Downlink performance modeling and analysis in a multichannel environment. *IEEE Trans. Mob. Comput.* **2013**, *13*, 649–663. [CrossRef]
48. Fouda, M.M.; Hashima, S.; Sakib, S.; Fadlullah, Z.M.; Hatano, K.; Shen, X. Optimal channel selection in hybrid RF/VLC networks: A multi-armed bandit approach. *IEEE Trans. Veh. Technol.* **2022**, *71*, 6853–6858. [CrossRef]
49. Al Hammadi, A.; Sofotasios, P.C.; Muhaidat, S.; Al-Qutayri, M.; Elgala, H. Non-orthogonal multiple access for hybrid VLC-RF networks with imperfect channel state information. *IEEE Trans. Veh. Technol.* **2020**, *70*, 398–411. [CrossRef]
50. Obeed, M.; Salhab, A.M.; Zummo, S.A.; Alouini, M.S. Joint optimization of power allocation and load balancing for hybrid VLC/RF networks. *J. Opt. Commun. Netw.* **2018**, *10*, 553–562. [CrossRef]
51. Bao, X.; Zhu, X.; Song, T.; Ou, Y. Protocol design and capacity analysis in hybrid network of visible light communication and OFDMA systems. *IEEE Trans. Veh. Technol.* **2013**, *63*, 1770–1778. [CrossRef]
52. Adnan-Qidan, A.; Morales-Céspedes, M.; Armada, A.G. Load balancing in hybrid VLC and RF networks based on blind interference alignment. *IEEE Access* **2020**, *8*, 72512–72527. [CrossRef]
53. Huang, S.; Chuai, G.; Gao, W. Two-Way Selection Handover Algorithm for Load Balancing in Hybrid VLC-RF Networks. In Proceedings of the 2021 IEEE/CIC International Conference on Communications in China (ICCC), Beijing, China, 6–8 February 2021; pp. 1065–1070.
54. Vats, A.; Aggarwal, M.; Ahuja, S. Outage and error performance analysis of dual hop hybrid RF-VLC system with wireless energy harvesting. *Phys. Commun.* **2022**, *55*, 101882. [CrossRef]
55. Wang, Y.; Wu, X.; Haas, H. Fuzzy logic based dynamic handover scheme for indoor Li-Fi and RF hybrid network. In Proceedings of the 2016 IEEE International Conference on Communications (ICC), San Francisco, CA, USA, 5–10 April 2016; pp. 1–6.
56. Wang, F.; Wang, Z.; Qian, C.; Dai, L.; Yang, Z. Efficient vertical handover scheme for heterogeneous VLC-RF systems. *J. Opt. Commun. Netw.* **2015**, *7*, 1172–1180. [CrossRef]
57. Liang, S.; Zhang, Y.; Fan, B.; Tian, H. Multi-attribute vertical handover decision-making algorithm in a hybrid VLC-femto system. *IEEE Commun. Lett.* **2017**, *21*, 1521–1524. [CrossRef]
58. Babalola, O.P.; Balyan, V. Vertical handover prediction based on hidden markov model in heterogeneous VLC-WiFi system. *Sensors* **2022**, *22*, 2473. [CrossRef]
59. Narmanlioglu, O.; Uysal, M. Event-Triggered Adaptive Handover for Centralized Hybrid VLC/MMW Networks. *IEEE Trans. Commun.* **2021**, *70*, 455–468. [CrossRef]
60. Jarchlo, E.A.; Eso, E.; Doroud, H.; Siessegger, B.; Ghassemlooy, Z.; Caire, G.; Dressler, F. Li-Wi: An upper layer hybrid VLC-WiFi network handover solution. *Ad Hoc Netw.* **2022**, *124*, 102705. [CrossRef]
61. Du, Z.; Wu, Q.; Yang, P.; Xu, Y.; Wang, J.; Yao, Y.D. Exploiting user demand diversity in heterogeneous wireless networks. *IEEE Trans. Wirel. Commun.* **2015**, *14*, 4142–4155. [CrossRef]
62. Pantisano, F.; Bennis, M.; Saad, W.; Valentin, S.; Debbah, M. Matching with externalities for context-aware user-cell association in small cell networks. In Proceedings of the 2013 IEEE Global Communications Conference (GLOBECOM), Atlanta, GA, USA, 9–13 December 2013; pp. 4483–4488.
63. Du, Z.; Wang, C.; Sun, Y.; Wu, G. Context-aware indoor VLC/RF heterogeneous network selection: Reinforcement learning with knowledge transfer. *IEEE Access* **2018**, *6*, 33275–33284. [CrossRef]
64. Wang, C.; Wu, G.; Du, Z. Reinforcement learning based network selection for hybrid VLC and RF systems. In Proceedings of the MATEC Web of Conferences EDP Sciences, Kuala Lumpur, Malaysia, 13–14 August 2018; Volume 173, p. 03014.
65. Wu, X.; Basnayaka, D.; Safari, M.; Haas, H. Two-stage access point selection for hybrid VLC and RF networks. In Proceedings of the 2016 IEEE 27th Annual International Symposium on Personal, Indoor, and Mobile Radio Communications (PIMRC), Madrid, Spain, 27–30 June 2016; pp. 1–6.
66. Hammouda, M.; Akın, S.; Vegni, A.M.; Haas, H.; Peissig, J. Link selection in hybrid RF/VLC systems under statistical queueing constraints. *IEEE Trans. Wirel. Commun.* **2018**, *17*, 2738–2754. [CrossRef]
67. Wang, Y.; Haas, H. Dynamic load balancing with handover in hybrid Li-Fi and Wi-Fi networks. *J. Light. Technol.* **2015**, *33*, 4671–4682. [CrossRef]
68. Li, X.; Zhang, R.; Hanzo, L. Cooperative load balancing in hybrid visible light communications and WiFi. *IEEE Trans. Commun.* **2015**, *63*, 1319–1329. [CrossRef]
69. Wang, Y.; Basnayaka, D.A.; Wu, X.; Haas, H. Optimization of load balancing in hybrid LiFi/RF networks. *IEEE Trans. Commun.* **2017**, *65*, 1708–1720. [CrossRef]
70. Purwita, A.A.; Soltani, M.D.; Safari, M.; Haas, H. Handover probability of hybrid LiFi/RF-based networks with randomly-oriented devices. In Proceedings of the 2018 IEEE 87th Vehicular Technology Conference (VTC Spring), Florence, Italy, 20–23 June 2018; pp. 1–5.
71. Sanusi, J.; Idris, S.; Aibinu, A.M.; Adeshina, S.; Obadiah, A.N. Handover in hybrid lifi and wifi networks. In Proceedings of the 2019 15th International Conference on Electronics, Computer and Computation (ICECCO), Abuja, Nigeria, 3–7 June 2019; pp. 1–6.
72. Zeshan, A.; Baykas, T. Location Aware Vertical Handover in a VLC/WLAN Hybrid Network. *IEEE Access* **2021**, *9*, 129810–129819. [CrossRef]
73. Alvarez, L.B.; Montejo-Sánchez, S.; Jara, N.; Rodríguez-López, L.; Saavedra, G. CSMA/CA protocol design in hybrid network of Visible Light Communication and RF Femtocell systems. *Opt. Commun.* **2023**, *537*, 129434. [CrossRef]

74. Zargari, S.; Kolivand, M.; Nezamalhosseini, S.A.; Abolhassani, B.; Chen, L.R.; Kahaei, M.H. Resource allocation of hybrid VLC/RF Systems with light energy harvesting. *IEEE Trans. Green Commun. Netw.* **2021**, *6*, 600–612. [CrossRef]
75. Kafafy, M.; Fahmy, Y.; Abdallah, M.; Khairy, M. Power efficient downlink resource allocation for hybrid RF/VLC wireless networks. In Proceedings of the 2017 IEEE Wireless Communications and Networking Conference (WCNC), Malaga, Spain, 5–9 June 2017; pp. 1–6.
76. Zhang, Y.; Li, Y.; Chen, L.; Wang, N.; Fan, B. Evolution computation based resource allocation for hybrid visible-light and rf femtocell. In Proceedings of the Springer International Conference on Communications and Networking in China, Hangzhou, China, 20–22 April 2020; pp. 43–52.
77. Chen, J.; Wang, Z. Coordination Game Theory-Based Adaptive Topology Control for Hybrid VLC/RF VANET. *IEEE Trans. Commun.* **2021**, *69*, 5312–5324. [CrossRef]
78. Wu, X.; Haas, H. Access point assignment in hybrid LiFi and WiFi networks in consideration of LiFi channel blockage. In Proceedings of the 2017 IEEE 18th International Workshop on Signal Processing Advances in Wireless Communications (SPAWC), Oulu, Finland, 4–6 July 2017; pp. 1–5.
79. Kashef, M.; Ismail, M.; Abdallah, M.; Qaraqe, K.A.; Serpedin, E. Energy efficient resource allocation for mixed RF/VLC heterogeneous wireless networks. *IEEE J. Sel. Areas Commun.* **2016**, *34*, 883–893. [CrossRef]
80. Al-Khori, J.; Nauryzbayev, G.; Abdallah, M.M.; Hamdi, M. Joint beamforming design and power minimization for friendly jamming relaying hybrid RF/VLC systems. *IEEE Photonics J.* **2019**, *11*, 1–18. [CrossRef]
81. Jin, F.; Zhang, R.; Hanzo, L. Resource allocation under delay-guarantee constraints for heterogeneous visible-light and RF femtocell. *IEEE Trans. Wirel. Commun.* **2014**, *14*, 1020–1034. [CrossRef]
82. Tran, H.V.; Kaddoum, G.; Diamantoulakis, P.D.; Abou-Rjeily, C.; Karagiannidis, G.K. Ultra-small cell networks with collaborative RF and lightwave power transfer. *IEEE Trans. Commun.* **2019**, *67*, 6243–6255. [CrossRef]
83. Becvar, Z.; Najla, M.; Mach, P. Selection between radio frequency and visible light communication bands for D2D. In Proceedings of the 2018 IEEE 87th Vehicular Technology Conference (VTC Spring), Porto, Portugal, 3–6 June 2018; pp. 1–7.
84. Pratama, Y.S.M.; Choi, K.W. Bandwidth aggregation protocol and throughput-optimal scheduler for hybrid RF and visible light communication systems. *IEEE Access* **2018**, *6*, 32173–32187. [CrossRef]
85. Saud, M.S.; Katz, M. Implementation of a hybrid optical-RF wireless network with fast network handover. In Proceedings of the European Wireless 2017 23th European Wireless Conference VDE, Madrid, Spain, 25–29 June 2017; pp. 1–6.
86. Amjad, M.; Qureshi, H.K.; Hassan, S.A.; Ahmad, A.; Jangsher, S. Optimization of MAC frame slots and power in hybrid VLC/RF networks. *IEEE Access* **2020**, *8*, 21653–21664. [CrossRef]
87. Zhang, C.; Ye, J.; Shi, J.; Pan, G.; Li, Z.; Ding, Z. Cooperative hybrid VLC-RF systems for WSNs. In Proceedings of the 2018 Asia-Pacific Signal and Information Processing Association Annual Summit and Conference (APSIPA ASC), Honolulu, HI, USA, 12–15 November 2018; pp. 436–441.
88. Little, T.D.; Rahaim, M. Network topologies for mixed RF-VLC HetNets. In Proceedings of the 2015 IEEE Summer Topicals Meeting Series (SUM), Nassau, Bahamas, 13–15 July 2015; pp. 163–164.
89. Zhang, C.; Ye, J.; Pan, G.; Ding, Z. Cooperative hybrid VLC-RF systems with spatially random terminals. *IEEE Trans. Commun.* **2018**, *66*, 6396–6408. [CrossRef]
90. Chowdhury, M.Z.; Hossan, M.; Hasan, M.; Jang, Y.M. Integrated RF/optical wireless networks for improving QoS in indoor and transportation applications. *Wirel. Pers. Commun.* **2019**, *107*, 1401–1430. [CrossRef]
91. Petkovic, M.I.; Narandzic, M.; Vukobratovic, D.; Cvetkovic, A. Mixed RF-VLC relaying system with radio-access diversity. In Proceedings of the 2019 28th Wireless and Optical Communications Conference (WOCC), London, UK, 4–9 July 2019; pp. 1–5.
92. Khreishah, A.; Shao, S.; Gharaibeh, A.; Ayyash, M.; Elgala, H.; Ansari, N. A hybrid RF-VLC system for energy efficient wireless access. *IEEE Trans. Green Commun. Netw.* **2018**, *2*, 932–944. [CrossRef]
93. Wu, W.; Zhou, F.; Yang, Q. Adaptive network resource optimization for heterogeneous VLC/RF wireless networks. *IEEE Trans. Commun.* **2018**, *66*, 5568–5581. [CrossRef]
94. Bao, X.; Dai, J.; Zhu, X. Visible light communications heterogeneous network (VLC-HetNet): New model and protocols for mobile scenario. *Wirel. Netw.* **2017**, *23*, 299–309. [CrossRef]
95. Rahaim, M.B.; Vegni, A.M.; Little, T.D. A hybrid radio frequency and broadcast visible light communication system. In Proceedings of the 2011 IEEE GLOBECOM Workshops (GC Wkshps), Houston, TX, USA, 5–9 December 2011; pp. 792–796.
96. Tabassum, H.; Hossain, E. Coverage and rate analysis for co-existing RF/VLC downlink cellular networks. *IEEE Trans. Wirel. Commun.* **2018**, *17*, 2588–2601. [CrossRef]
97. Han, Y.; Zhou, X.; Yang, L.; Li, S. A bipartite matching based user pairing scheme for hybrid VLC-RF NOMA systems. In Proceedings of the 2018 International Conference on Computing, Networking and Communications (ICNC), Maui, HI, USA, 5–8 March 2018; pp. 480–485.
98. Hsiao, Y.C.; Chen, C.M.; Lin, C. Energy efficiency maximization in multi-user miso mixed RF/VLC heterogeneous cellular networks. In Proceedings of the 2018 15th Annual IEEE International Conference on Sensing, Communication, and Networking (SECON), Hong Kong, China, 11–13 June 2018; pp. 1–9.
99. Hammouda, M.; Akln, S.; Vegni, A.M.; Haas, H.; Peissig, J. Hybrid RF/LC systems under QoS constraints. In Proceedings of the 2018 25th International Conference on Telecommunications (ICT), Beijing, China, 4–8 August 2018; pp. 312–318.

100. Pan, G.; Ye, J.; Ding, Z. Secrecy outage analysis of hybrid VLC-RF systems with light energy harvesting. In Proceedings of the 2017 IEEE 18th International Workshop on Signal Processing Advances in Wireless Communications (SPAWC), Sapporo, Japan, 3–6 July 2017; pp. 1–5.
101. Vats, A.; Aggarwal, M.; Ahuja, S.; Vashisth, S. Hybrid VLC-RF system for real time health care applications. In *Advances in Optical Science and Engineering*; Springer: Berlin/Heidelberg, Germany, 2017; pp. 347–353.
102. Rahaim, M.B.; Little, T.D. Toward practical integration of dual-use VLC within 5G networks. *IEEE Wirel. Commun.* **2015**, *22*, 97–103. [CrossRef]
103. Wu, Z.Y.; Ismail, M.; Serpedin, E.; Wang, J. Data-driven link assignment with QoS guarantee in mobile RF-optical HetNet of things. *IEEE Internet Things J.* **2020**, *7*, 5088–5102. [CrossRef]
104. Arshad, R.; Lampe, L. Stochastic geometry analysis of user mobility in RF/VLC hybrid networks. *IEEE Trans. Wirel. Commun.* **2021**, *20*, 7404–7419. [CrossRef]
105. Zenaidi, M.R.; Rezki, Z.; Abdallah, M.; Qaraqe, K.A.; Alouini, M.S. Achievable rate-region of VLC/RF communications with an energy harvesting relay. In Proceedings of the GLOBECOM 2017 IEEE Global Communications Conference, Singapore, 4–8 December 2017; pp. 1–7.
106. Liao, Z.; Yang, L.; Chen, J.; Yang, H.C.; Alouini, M.S. Physical layer security for dual-hop VLC/RF communication systems. *IEEE Commun. Lett.* **2018**, *22*, 2603–2606. [CrossRef]
107. Namdar, M.; Basgumus, A.; Tsiftsis, T.; Altuncu, A. Outage and BER performances of indoor relay-assisted hybrid RF/VLC systems. *IET Commun.* **2018**, *12*, 2104–2109. [CrossRef]
108. Vats, A.; Aggarwal, M.; Ahuja, S. Outage analysis of AF relayed hybrid VLC-RF communication system for E-health applications. In Proceedings of the 2017 International Conference on Computing, Communication and Automation (ICCCA), Noida, India, 5–6 May 2017; pp. 1401–1405.
109. Al-Khori, J.; Nauryzbayev, G.; Abdallah, M.; Hamdi, M. Physical layer security for hybrid RF/VLC DF relaying systems. In Proceedings of the 2018 IEEE 88th Vehicular Technology Conference (VTC-Fall), Boston, MA, USA, 27–30 August 2018; pp. 1–6.
110. Peng, H.; Li, Q.; Pandharipande, A.; Ge, X.; Zhang, J. End-to-end performance optimization of a dual-hop hybrid VLC/RF IoT system based on SLIPT. *IEEE Internet Things J.* **2021**, *8*, 17356–17371. [CrossRef]
111. Chowdhury, H.; Katz, M. Cooperative data download on the move in indoor hybrid (radio-optical) WLAN-VLC hotspot coverage. *Trans. Emerg. Telecommun. Technol.* **2014**, *25*, 666–677. [CrossRef]
112. Li, Z.; Shao, S.; Khreishah, A.; Ayyash, M.; Abdalla, I.; Elgala, H.; Rahaim, M.; Little, T. Design and implementation of a hybrid rf-vlc system with bandwidth aggregation. In Proceedings of the 2018 14th International Wireless Communications & Mobile Computing Conference (IWCMC) 2018, Limassol, Cyprus, 25–29 June 2018; pp. 194–200.
113. Zhang, W.; Chen, L.; Chen, X.; Yu, Z.; Li, Z.; Wang, W. Design and realization of indoor VLC-Wi-Fi hybrid network. *J. Commun. Inf. Netw.* **2017**, *2*, 75–87. [CrossRef]
114. Duvnjak, F.; Ožegović, J.; Kristić, A. Heterogeneous Wi-Fi and VLC (RF-optical) wireless access architecture. In Proceedings of the 2015 23rd International Conference on Software, Telecommunications and Computer Networks (SoftCOM), Split, Croatia, 5–9 June 2015; pp. 310–314.
115. Kashef, M.; Abdallah, M.; Al-Dhahir, N. Transmit power optimization for a hybrid PLC/VLC/RF communication system. *IEEE Trans. Green Commun. Netw.* **2017**, *2*, 234–245. [CrossRef]
116. Kashef, M.; Abdallah, M.; Al-Dhahir, N.; Qaraqe, K. On the impact of PLC backhauling in multi-user hybrid VLC/RF communication systems. In Proceedings of the 2016 IEEE Global Communications Conference (GLOBECOM), Washington, DC, USA, 4–8 December 2016; pp. 1–6.
117. Kashef, M.; Torky, A.; Abdallah, M.; Al-Dhahir, N.; Qaraqe, K. On the achievable rate of a hybrid PLC/VLC/RF communication system. In Proceedings of the 2015 IEEE Global Communications Conference (GLOBECOM), Washington, DC, USA, 13–16 November 2015; pp. 1–6.
118. Zhou, X.; Li, S.; Zhang, H.; Wen, Y.; Han, Y.; Yuan, D. Cooperative NOMA based VLC/RF system with simultaneous wireless information and power transfer. In Proceedings of the 2018 IEEE/CIC International Conference on Communications in China (ICCC), Beijing, China, 10–15 April 2018; pp. 100–105.
119. Pan, G.; Ye, J.; Ding, Z. Secure hybrid VLC-RF systems with light energy harvesting. *IEEE Trans. Commun.* **2017**, *65*, 4348–4359. [CrossRef]
120. Vats, A.; Aggarwal, M.; Ahuja, S. Modeling and outage analysis of multiple relayed hybrid VLC-RF system. In Proceedings of the 2017 International Conference on Computer, Communications and Electronics (Comptelix), Jaipur, India, 6–9 September 2017; pp. 254–259.
121. Hussain, S.I.; Abdallah, M.M.; Qaraqe, K.A. Hybrid radio-visible light downlink performance in RF sensitive indoor environments. In Proceedings of the 2014 6th International Symposium on Communications, Control and Signal Processing (ISCCSP), Athens, Greece, 21–23 May 2014; pp. 81–84.
122. Alenezi, A.M.; Hamdi, K.A. Reinforcement learning approach for hybrid WiFi-VLC networks. In Proceedings of the 2020 IEEE 91st Vehicular Technology Conference (VTC2020-Spring), Virtual, 25 May–31 July 2020; pp. 1–5.
123. Aboagye, S.; Ngatched, T.M.; Dobre, O.A.; Ibrahim, A. Joint access point assignment and power allocation in multi-tier hybrid RF/VLC HetNets. *IEEE Trans. Wirel. Commun.* **2021**, *20*, 6329–6342. [CrossRef]

124. Wang, J.; Jiang, C.; Zhang, H.; Zhang, X.; Leung, V.C.; Hanzo, L. Learning-aided network association for hybrid indoor LiFi-WiFi systems. *IEEE Trans. Veh. Technol.* **2017**, *67*, 3561–3574. [CrossRef]
125. Yan, C.; Xu, Y.; Shen, J.; Chen, J. A combination of VLC and WiFi based indoor wireless access network and its handover strategy. In Proceedings of the 2016 IEEE International Conference on Ubiquitous Wireless Broadband (ICUWB), Singapore, 7–10 July 2016; pp. 1–4.
126. Zhang, H.; Liu, N.; Long, K.; Cheng, J.; Leung, V.C.; Hanzo, L. Energy efficient subchannel and power allocation for software-defined heterogeneous VLC and RF networks. *IEEE J. Sel. Areas Commun.* **2018**, *36*, 658–670. [CrossRef]
127. Marzban, M.F.; Kashef, M.; Abdallah, M.; Khairy, M. Beamforming and power allocation for physical-layer security in hybrid RF/VLC wireless networks. In Proceedings of the 2017 13th International Wireless Communications and Mobile Computing Conference (IWCMC), Rome, Italy, 4–9 October 2017; pp. 258–263.
128. Al-Khori, J.; Nauryzbayev, G.; Abdallah, M.M.; Hamdi, M. Secrecy performance of decode-and-forward based hybrid RF/VLC relaying systems. *IEEE Access* **2019**, *7*, 10844–10856. [CrossRef]
129. Papanikolaou, V.K.; Diamantoulakis, P.D.; Ding, Z.; Muhaidat, S.; Karagiannidis, G.K. Hybrid VLC/RF networks with non-orthogonal multiple access. In Proceedings of the 2018 IEEE Global Communications Conference (GLOBECOM), Abu Dhabi, United Arab Emirates, 9–13 December 2018; pp. 1–6.
130. Papanikolaou, V.K.; Diamantoulakis, P.D.; Karagiannidis, G.K. User grouping for hybrid VLC/RF networks with NOMA: A coalitional game approach. *IEEE Access* **2019**, *7*, 103299–103309. [CrossRef]
131. Yang, H.; Alphones, A.; Zhong, W.D.; Chen, C.; Xie, X. Learning-based energy-efficient resource management by heterogeneous RF/VLC for ultra-reliable low-latency industrial IoT networks. *IEEE Trans. Ind. Inform.* **2019**, *16*, 5565–5576. [CrossRef]
132. Mach, P.; Becvar, Z.; Najla, M.; Zvanovec, S. Combination of visible light and radio frequency bands for device-to-device communication. In Proceedings of the 2017 IEEE 28th Annual International Symposium on Personal, Indoor, and Mobile Radio Communications (PIMRC), London, UK, 7–15 May 2017; pp. 1–7.
133. Saud, M.S.; Chowdhury, H.; Katz, M. Heterogeneous software-defined networks: Implementation of a hybrid radio-optical wireless network. In Proceedings of the 2017 IEEE Wireless Communications and Networking Conference (WCNC), San Francisco, CA, USA, 19–22 March 2017; pp. 1–6.
134. Li, L.; Zhang, Y.; Fan, B.; Tian, H. Mobility-aware load balancing scheme in hybrid VLC-LTE networks. *IEEE Commun. Lett.* **2016**, *20*, 2276–2279. [CrossRef]
135. Liang, S.; Tian, H.; Fan, B.; Bai, R. A novel vertical handover algorithm in a hybrid visible light communication and LTE system. In Proceedings of the 2015 IEEE 82nd vehicular technology conference (VTC2015-Fall), Boston, MA, USA, 6–9 September 2015; pp. 1–5.
136. Huang, W.; Yang, Y.; Chen, M.; Liu, C.; Feng, C.; Poor, H.V. Wireless Network Optimization for Federated Learning with Model Compression in Hybrid VLC/RF Systems. *Entropy* **2021**, *23*, 1413. [CrossRef]
137. Játiva, P.P.; Azurdia-Meza, C.A.; Zabala-Blanco, D.; Gutiérrez, C.A.; Sánchez, I.; Castillo-Soria, F.R.; Seguel, F. Bit error probability of VLC systems in underground mining channels with imperfect CSI. *AEU-Int. J. Electron. Commun.* **2022**, *145*, 154101. [CrossRef]
138. Schneider, D.; Shrotri, A.; Flatt, H.; Stübbe, O.; Wolf, A.; Lachmayer, R.; Bunge, C.A. Impact of industrial environments on visible light communication. *Opt. Express* **2021**, *29*, 16087–16104. [CrossRef]
139. Liu, C.; Guo, C.; Yang, Y.; Chen, M.; Poor, H.V.; Cui, S. Optimization of user selection and bandwidth allocation for federated learning in VLC/RF systems. In Proceedings of the 2021 IEEE Wireless Communications and Networking Conference (WCNC), Nanjing, China, 29 March–1 April 2021; pp. 1–6.
140. Ndjiongue, A.R.; Ferreira, H.C. An overview of outdoor visible light communications. *Trans. Emerg. Telecommun. Technol.* **2018**, *29*, e3448. [CrossRef]
141. Liu, C.B.; Sadeghi, B.; Knightly, E.W. Enabling vehicular visible light communication (V2LC) networks. In Proceedings of the Eighth ACM International Workshop on Vehicular Inter-Networking, Las Vegas, NV, USA, 6–9 September 2011; pp. 41–50.
142. Turan, B.; Gurbilek, G.; Uyrus, A.; Ergen, S.C. Vehicular VLC frequency domain channel sounding and characterization. In Proceedings of the 2018 IEEE Vehicular Networking Conference (VNC), Berlin, Germany, 5–7 December 2018; pp. 1–8.
143. Amjad, M.S.; Tebruegge, C.; Memedi, A.; Kruse, S.; Kress, C.; Scheytt, J.C.; Dressler, F. Towards an IEEE 802.11 compliant system for outdoor vehicular visible light communications. *IEEE Trans. Veh. Technol.* **2021**, *70*, 5749–5761. [CrossRef]
144. Tebruegge, C.; Memedi, A.; Dressler, F. Empirical characterization of the NLOS component for vehicular visible light communication. In Proceedings of the 2019 IEEE Vehicular Networking Conference (VNC), Los Angeles, CA, USA, 4–6 December 2019; pp. 1–4.
145. Chung, Y.H.; Oh, S.b. Efficient optical filtering for outdoor visible light communications in the presence of sunlight or articifical light. In Proceedings of the 2013 International Symposium on Intelligent Signal Processing and Communication Systems, Okinawa, Japan, 12–15 November 2013; pp. 749–752.
146. Lin, S.H.; Wang, J.Y.; Bao, X.; Li, Y. Outage performance analysis for outdoor vehicular visible light communications. In Proceedings of the 2017 9th International Conference on Wireless Communications and Signal Processing (WCSP), London, UK, 7–9 September 2017; pp. 1–5.
147. Georlette, V.; Bette, S.; Brohez, S.; Pérez-Jiménez, R.; Point, N.; Moeyaert, V. Outdoor visible light communication channel modeling under smoke conditions and analogy with fog conditions. *Optics* **2020**, *1*, 259–281. [CrossRef]
148. Martinek, R.; Danys, L.; Jaros, R. Visible light communication system based on software defined radio: Performance study of intelligent transportation and indoor applications. *Electronics* **2019**, *8*, 433. [CrossRef]

149. Singh, G.; Srivastava, A.; Bohara, V.A. Impact of Weather Conditions and Interference on the Performance of VLC based V2V Communication. In Proceedings of the 2019 21st International Conference on Transparent Optical Networks (ICTON), Angers, France, 9–13 July 2019; pp. 1–4.
150. Danys, L.; Martinek, R.; Slanina, Z.; Kolarik, J.; Jaros, R. The impact of rain on performance of visible light communication. In Proceedings of the Photonics Applications in Astronomy, Communications, Industry, and High-Energy Physics Experiments 2019 SPIE, Los Angeles, CA, USA, 26 May–2 June 2019; Volume 11176, pp. 198–205.
151. Liu, X.; Wei, X.; Guo, L. DIMLOC: Enabling high-precision visible light localization under dimmable LEDs in smart buildings. *IEEE Internet Things J.* **2019**, *6*, 3912–3924. [CrossRef]
152. Liu, X.; Wei, X.; Guo, L.; Song, S. DarkVLP: "Lights-Off" Visible-Light Positioning. *IEEE Internet Things J.* **2021**, *9*, 11071–11084. [CrossRef]
153. Guo, Y.; Xiong, K.; Lu, Y.; Wang, D.; Fan, P.; Letaief, K.B. Achievable information rate in hybrid VLC-RF networks with lighting energy harvesting. *IEEE Trans. Commun.* **2021**, *69*, 6852–6864. [CrossRef]
154. Al Hammadi, A.; Muhaidat, S.; Sofotasios, P.C.; Al Qutayri, M. A robust and energy efficient NOMA-enabled hybrid VLC/RF wireless network. In Proceedings of the 2019 IEEE Wireless Communications and Networking Conference (WCNC), Rome, Italy, 15–19 April 2019; pp. 1–6.
155. Obeed, M.; Dahrouj, H.; Salhab, A.M.; Zummo, S.A.; Alouini, M.S. User pairing, link selection, and power allocation for cooperative NOMA hybrid VLC/RF systems. *IEEE Trans. Wirel. Commun.* **2020**, *20*, 1785–1800. [CrossRef]
156. Song, S.; Liu, X.; Liu, Y.; Wu, J.; Wu, T.; Guo, L. Implementation of Machine Learning-based Emergency Communication Using RoFSO-VLC/RF Convergence link. In Proceedings of the Optical Fiber Communication Conference, San Diego, CA, USA, 6–10 March 2022; Optica Publishing Group: San Diego, CA, USA, 2022; p. W3I-2.
157. Shaaban, R.; Faruque, S. Cyber security vulnerabilities for outdoor vehicular visible light communication in secure platoon network: Review, power distribution, and signal to noise ratio analysis. *Phys. Commun.* **2020**, *40*, 101094. [CrossRef]
158. Abuella, H.; Miramirkhani, F.; Ekin, S.; Uysal, M.; Ahmed, S. ViLDAR—Visible light sensing-based speed estimation using vehicle headlamps. *IEEE Trans. Veh. Technol.* **2019**, *68*, 10406–10417. [CrossRef]
159. Fuada, S.; Adiono, T.; Putra, A.P.; Aska, Y. LED driver design for indoor lighting and low-rate data transmission purpose. *Optik* **2018**, *156*, 847–856. [CrossRef]
160. Sanahuja, J.A. *La Agenda 2030 y los ODS: Sociedades Pacíficas, Justas e Inclusivas Como Pilar de la Seguridad*; CIMAPRESS: Madrid, España, 2019 .
161. Pan, G.; Lei, H.; Ding, Z.; Ni, Q. 3-D hybrid VLC-RF indoor IoT systems with light energy harvesting. *IEEE Trans. Green Commun. Netw.* **2019**, *3*, 853–865. [CrossRef]
162. Wang, Y.; Wu, X.; Haas, H. Distributed load balancing for Internet of Things by using Li-Fi and RF hybrid network. In Proceedings of the 2015 IEEE 26th Annual International Symposium on Personal, Indoor, and Mobile Radio Communications (PIMRC), Riga, Latvia, 3–7 June 2015; pp. 1289–1294.
163. Nauryzbayev, G.; Abdallah, M.; Ansari, I.S.; Al-Dhahir, N.; Qaraqe, K. Outage of cognitive electric vehicle networks over mixed RF/VLC channels with signal-dependent noise and imperfect CSI. *IEEE Trans. Veh. Technol.* **2020**, *69*, 6828–6832. [CrossRef]
164. Abouzohri, E.M.H.; Abdallah, M.M. Performance of hybrid cognitive RF/VLC systems in vehicle-to-vehicle communications. In Proceedings of the 2020 IEEE International Conference on Informatics, IoT, and Enabling Technologies (ICIoT), Tallinn, Estonia, 4–8 August 2020; pp. 429–434.
165. Chen, J.; Wang, Z.; Mao, T. Resource management for hybrid RF/VLC V2I wireless communication system. *IEEE Commun. Lett.* **2020**, *24*, 868–871. [CrossRef]
166. Ma, W.; Zhang, L.; Wu, Z. Location information-aided load balancing design for hybrid LiFi and WiFi networks. In Proceedings of the 2019 International Conference on Computing, Networking and Communications (ICNC), Budapest, Hungary, 12–17 October 2019; pp. 413–417.
167. Rakia, T.; Yang, H.C.; Gebali, F.; Alouini, M.S. Dual-hop VLC/RF transmission system with energy harvesting relay under delay constraint. In Proceedings of the 2016 IEEE Globecom Workshops (GC Wkshps), Berlin, Germany, 12–18 April 2016; pp. 1–6.
168. Palacios Játiva, P.; Azurdia-Meza, C.A.; Sánchez, I.; Zabala-Blanco, D.; Dehghan Firoozabadi, A.; Soto, I.; Seguel, F. An Enhanced VLC Channel Model for Underground Mining Environments Considering a 3D Dust Particle Distribution Model. *Sensors* **2022**, *22*, 2483. [CrossRef] [PubMed]
169. Játiva, P.P.; Azurdia-Meza, C.A.; Sánchez, I.; Seguel, F.; Zabala-Blanco, D.; Firoozabadi, A.D.; Gutiérrez, C.A.; Soto, I. A VLC channel model for underground mining environments with scattering and shadowing. *IEEE Access* **2020**, *8*, 185445–185464. [CrossRef]
170. Kadam, K.; Chavan, G.; Chavan, U.; Shah, R.; Kumar, P. Smart and precision polyhouse farming using visible light communication and internet of things. In Proceedings of the 2nd International Conference Intelligent Computing and Information and Communication ICICC, Rome, Italy, 15–18 October 2017; Springer: Berlin/Heidelberg, Germany, 2018; pp. 247–256.
171. Ramberg, L. *Farmer Knowledge Sharing and Social Networks in Agricultural Extension*; Swedish University of Agricultural Sciences: Upsala, Sweden, 2020.
172. Lee, Y.U.; Kavehrad, M. Two hybrid positioning system design techniques with lighting LEDs and ad-hoc wireless network. *IEEE Trans. Consum. Electron.* **2012**, *58*, 1176–1184. [CrossRef]

173. Konings, D.; Parr, B.; Waddell, C.; Alam, F.; Arif, K.M.; Lai, E.M. HVLP: Hybrid visible light positioning of a mobile robot. In Proceedings of the 2017 24th International Conference on Mechatronics and Machine Vision in Practice (M2VIP), Auckland, New Zealand, 21–23 November 2017; pp. 1–6.
174. Ziyan, J. A visible light communication based hybrid positioning method for wireless sensor networks. In Proceedings of the 2012 Second International Conference on Intelligent System Design and Engineering Application, Sanya, China, 6–7 January 2012; pp. 1367–1370.
175. Kong, J.; Ismail, M.; Serpedin, E.; Qaraqe, K.A. Energy efficient optimization of base station intensities for hybrid RF/VLC networks. *IEEE Trans. Wirel. Commun.* **2019**, *18*, 4171–4183. [CrossRef]
176. Wu, W.; Zhou, F.; Yang, Q. Dynamic network resource optimization in hybrid VLC and radio frequency networks. In Proceedings of the 2017 International Conference on Selected Topics in Mobile and Wireless Networking (MoWNeT), London, UK, 6–8 June 2017; pp. 1–7.
177. Kreutz, D.; Ramos, F.M.; Verissimo, P.E.; Rothenberg, C.E.; Azodolmolky, S.; Uhlig, S. Software-defined networking: A comprehensive survey. *Proc. IEEE* **2014**, *103*, 14–76. [CrossRef]
178. Kobo, H.I.; Abu-Mahfouz, A.M.; Hancke, G.P. A survey on software-defined wireless sensor networks: Challenges and design requirements. *IEEE Access* **2017**, *5*, 1872–1899. [CrossRef]
179. Cox, J.H.; Chung, J.; Donovan, S.; Ivey, J.; Clark, R.J.; Riley, G.; Owen, H.L. Advancing software-defined networks: A survey. *IEEE Access* **2017**, *5*, 25487–25526. [CrossRef]
180. Mai, V.V.; Thang, T.C.; Pham, A.T. CSMA/CA-based uplink MAC protocol design and analysis for hybrid VLC/Wifi networks. In Proceedings of the 2017 IEEE International Conference on Communications Workshops (ICC Workshops), Rome, Italy, 14–18 June 2017; pp. 457–462.
181. Pan, G.; Ye, J.; Zhang, C.; An, J.; Lei, H.; Ding, Z.; Alouini, M.S. Secure cooperative hybrid VLC-RF systems. *IEEE Trans. Wirel. Commun.* **2020**, *19*, 7097–7107. [CrossRef]
182. Vats, A.; Aggarwal, M.; Ahuja, S. End-to-end performance analysis of hybrid VLC-RF system using decode and forward relay in E-health medical applications. *Optik* **2019**, *187*, 297–310. [CrossRef]
183. Yang, L.; Yan, X.; Da Costa, D.B.; Tsiftsis, T.A.; Yang, H.C.; Alouini, M.S. Indoor mixed dual-hop VLC/RF systems through reconfigurable intelligent surfaces. *IEEE Wirel. Commun. Lett.* **2020**, *9*, 1995–1999. [CrossRef]
184. Arfaoui, M.A.; Soltani, M.D.; Tavakkolnia, I.; Ghrayeb, A.; Safari, M.; Assi, C.M.; Haas, H. Physical layer security for visible light communication systems: A survey. *IEEE Commun. Surv. Tutor.* **2020**, *22*, 1887–1908. [CrossRef]
185. Sun, S.; Yang, F.; Song, J.; Han, Z. Optimization on multiuser physical layer security of intelligent reflecting surface-aided VLC. *IEEE Wirel. Commun. Lett.* **2022**, *11*, 1344–1348. [CrossRef]
186. Obeed, M.; Salhab, A.M.; Alouini, M.S.; Zummo, S.A. Survey on physical layer security in optical wireless communication systems. In Proceedings of the 2018 Seventh International Conference on Communications and Networking (ComNet), Madrid, Spain, 6–10 May 2018; pp. 1–5.
187. Shahjalal, M.; Hasan, M.K.; Chowdhury, M.Z.; Jang, Y.M. Future optical camera communication based applications and opportunities for 5G and beyond. In Proceedings of the 2019 International Conference on Artificial Intelligence in Information and Communication (ICAIIC), Irving, CA, USA, 8–12 January 2019; pp. 492–495.
188. Saeed, N.; Guo, S.; Park, K.H.; Al-Naffouri, T.Y.; Alouini, M.S. Optical camera communications: Survey, use cases, challenges, and future trends. *Phys. Commun.* **2019**, *37*, 100900. [CrossRef]
189. Jurado-Verdu, C.; Guerra, V.; Matus, V.; Rabadan, J.; Perez-Jimenez, R. Convolutional autoencoder for exposure effects equalization and noise mitigation in optical camera communication. *Opt. Express* **2021**, *29*, 22973–22991. [CrossRef]
190. Matus, V.; Guerra, V.; Jurado-Verdu, C.; Zvanovec, S.; Perez-Jimenez, R. Wireless sensor networks using sub-pixel optical camera communications: Advances in experimental channel evaluation. *Sensors* **2021**, *21*, 2739. [CrossRef] [PubMed]
191. Raj, R.; Jindal, K.; Dixit, A. Fairness Enhancement of Non-Orthogonal Multiple Access in VLC-Based IoT Networks for Intravehicular Applications. *IEEE Trans. Veh. Technol.* **2022**, *71*, 7414–7427. [CrossRef]
192. Oyewobi, S.S.; Djouani, K.; Kurien, A.M. Visible light communications for internet of things: Prospects and approaches, challenges, solutions and future directions. *Technologies* **2022**, *10*, 28. [CrossRef]

MDPI AG
Grosspeteranlage 5
4052 Basel
Switzerland
Tel.: +41 61 683 77 34

Sensors Editorial Office
E-mail: sensors@mdpi.com
www.mdpi.com/journal/sensors

www.ingramcontent.com/pod-product-compliance
Lightning Source LLC
Chambersburg PA
CBHW061815210926
51850CB00006B/716
9783725851751